Lecture Notes in Computer Science 15539

Advanced Research in Computing and Software Science

Subline of Lecture Notes in Computer Science

More information about this series at https://link.springer.com/bookseries/558

Rastislav Královič · Věra Kůrková
Editors

SOFSEM 2025: Theory and Practice of Computer Science

50th International Conference on Current Trends in Theory and Practice of Computer Science, SOFSEM 2025
Bratislava, Slovak Republic, January 20–23, 2025
Proceedings, Part II

Springer

Editors
Rastislav Královič
Comenius University in Bratislava
Bratislava, Slovakia

Věra Kůrková
Czech Academy of Sciences
Prague, Czech Republic

ISSN 0302-9743 ISSN 1611-3349 (electronic)
Lecture Notes in Computer Science
ISBN 978-3-031-82696-2 ISBN 978-3-031-82697-9 (eBook)
https://doi.org/10.1007/978-3-031-82697-9

This Springer imprint is published by the registered company Springer Nature Switzerland AG
The registered company address is: Gewerbestrasse 11, 6330 Cham, Switzerland

Preface

The *50th International Conference on Current Trends in Theory and Practice of Computer Science* (SOFSEM 2025), organized by the Slovak Society for Computer Science and the Faculty of Mathematics, Physics, and Informatics of the Comenius University in Bratislava, was held in Bratislava on January 20–23, 2025.

SOFSEM is an annual winter conference devoted to the theory and practice of computer science. It is focused on the latest results and developments in fundamental computer and artificial intelligence research. The series of SOFSEM (originally SOFtware SEMinar) conferences began in 1974 as a winter seminar for computer scientists in the former Czechoslovakia. Gradually, SOFSEM has transformed from a regional conference into an international one. Since 1995, its proceedings have been published in the LNCS series of Springer.

SOFSEM's scope has extended to include original research from all areas of foundations of computer science and artificial intelligence. Currently, it includes AI-based algorithms and techniques, nature-inspired computing, machine learning theory, multiagent algorithms and games, neural network theory, parallel and distributed computing, quantum computing, computability, decidability, classical and non-classical models of computation, computational complexity, computational learning, cryptographic techniques and security, data compression, data and pattern mining methods, discrete combinatorial optimization, automata, languages, machine models, rewriting systems, efficient data structures, graph structures and algorithms, logics of computation, robotics, and other relevant theory topics in computing and AI.

The SOFSEM series of conferences was only interrupted in 2022 due to the COVID pandemic, while in 2021, it was organized as a virtual meeting. SOFSEM 2023 was again held as a physical meeting, its venue was Nový Smokovec in High Tatras in Slovakia. It was followed by SOFSEM 2024 held at the University of Trier in Cochem. SOFSEM maintains its tradition of a venue where researchers in all stages of their careers can share their achievements and insights.

For SOFSEM 2025, 109 full papers were submitted. The Program Committee, chaired by Rastislav Královič and Věra Kůrková, selected 48 papers after a thorough peer-review process. The evaluation of the papers was based on quality, originality, and relevance for the conference. The handling of the reviewing process was supported by the EasyChair system. Most papers received three single-blind reviews. The accepted papers are published in the two volumes of these proceedings. The conference included four invited talks:

- Paola Flocchini (University of Ottawa, Canada): *Distributed Computing by Mobile Robots: Exploring the Computational Landscape*
- Erik Jan van Leeuwen (Utrecht University, The Netherlands): *Open Problems and Recent Developments on a Complexity Framework for Forbidden Subgraphs*

- Paul Spirakis (University of Liverpool, UK): *Temporal Graph Realization Problems*
- Ivan Tyukin (King's College London, UK): *The Challenge of Stability, Accuracy, and Robustness in Data-Driven AI*

We are pleased to thank everyone who contributed to the event's success. In particular, we thank the invited speakers for presenting inspirational talks and sharing their insights in discussions with participants. We are grateful to the members of the Steering Committee for keeping the tradition of the high-level SOFSEM series and to all the members of the program committee and the reviewers for their efforts in the reviewing process. We thank the organizers led by Branislav Rovan, Dana Pardubská, and Jana Kostičová for perfect local arrangements. We also thank Springer for sponsoring the Best Paper Award and for publishing these proceedings in the Advanced Research in Computing and Software Science (ARCoSS) of the Lecture Notes in Computer Science (LNCS) series. Last but not least, we thank all authors who contributed to this volume to share their new ideas and results with the community of researchers in the rapidly developing field of foundations of computer science and artificial intelligence.

We hope that you will enjoy reading and find inspiration for your future work in the papers contained in these two volumes.

December 2024

Rastislav Královič
Věra Kůrková

Organization

Steering Committee

Henning Fernau (Chair)	Trier University, Germany
Leszek A. Gąsieniec	University of Liverpool, UK
Serge Gaspers	UNSW Sydney, Australia
Ralf Klasing	CNRS and University of Bordeaux, France
Tiziana Margaria	University of Limerick, Ireland
Mirosław Kutyłowski	NASK – National Research Institute, Poland
Branislav Rovan	Comenius University in Bratislava, Slovakia
Jan van Leeuwen	Utrecht University, The Netherlands
Július Štuller	Czech Academy of Sciences, Czech Republic

Program Committee

Amihood Amir	Bar-Ilan University, Israel and Georgia Tech, USA
Přemysl Brada	University of West Bohemia, Czech Republic
Tiziana Calamoneri	Sapienza University of Rome, Italy
Jérémie Chalopin	LIS, CNRS, Aix-Marseille Université, Université de Toulon, France
Marek Chrobak	University of California, Riverside, USA
Ivana Černá	Masaryk University, Czech Republic
Gianluca De Marco	University of Salerno, Italy
Stefan Dobrev	Slovak Academy of Sciences, Slovakia
Martin Drozda	Slovak University of Technology, Slovakia
Robert Ganian	Vienna University of Technology, Austria
Leszek Gąsieniec	University of Liverpool, UK
Cyril Gavoille	LaBRI, University of Bordeaux, France
Lucjan Hanzlik	CISPA Helmholtz Center for Information Security, Germany
Markus Holzer	Universität Giessen, Germany
Ling-Ju Hung	National Taipei University of Business, Taiwan
Petr Jančar	Palacky University Olomouc, Czech Republic
Galina Jiraskova	Slovak Academy of Sciences, Slovakia
Tomasz Jurdzinski	University of Wrocław, Poland
Petteri Kaski	Aalto University, Finland
Philipp Kindermann	Universität Trier, Germany

Dennis Komm	ETH Zurich, Switzerland
Rastislav Královič (Chair)	Comenius University in Bratislava, Slovakia
Danny Krizanc	Wesleyan University, USA
Věra Kůrková (Chair)	Czech Academy of Sciences, Czech Republic
Giuseppe Liotta	University of Perugia, Italy
Alexei Lisitsa	University of Liverpool, UK
Hsiang-Hsuan Liu	Utrecht University, The Netherlands
Alessio Mansutti	IMDEA Software Institute, Spain
Marco Mesiti	University of Milan, Italy
Xavier Muñoz Lopez	Universitat Politècnica de Catalunya, Spain
Vangelis Paschos	Université Paris Dauphine-PSL, France
Rajeev Raman	University of Leicester, UK
Peter Rossmanith	RWTH Aachen University, Germany
Pawel Sobocinski	Tallinn University of Technology, Estonia
Ulrike Stege	University of Victoria, Canada
Gerth Stölting Brodal	Aarhus University, Denmark

Additional Reviewers

Angara, Prashanti
Ayaziová, Paulína
Baheri, Betis
Bai, Tian
Bakhshi-Khaniki, Hessam
Banik, Aritra
Bednarczyk, Bartosz
Benes, Nikola
Bentert, Matthias
Bercea, Ioana
Berendsohn, Benjamin Aram
Binucci, Carla
Bouchard, Sébastien
Bournez, Olivier
Brötzner, Anna
Burjons, Elisabet
Butman, Moshe
Böckenhauer, Hans-Joachim
Böhm, Martin
Cavalleri, Emanuele
Chistikov, Dmitry
Corò, Federico
D'Elia, Marco
de Castro Mendes Gomes, Guilherme
Defrain, Oscar
Deligkas, Argyrios
Disser, Yann
Dondi, Riccardo
Dreier, Jan
Dudek, Bartlomiej
Döring, Michelle
Erlebach, Thomas
Fasoulakis, Michail
Felsner, Stefan
Fijalkow, Nathanaël
Filakovský, Marek
Fioravantes, Foivos
Förster, Henry
Galby, Esther
Ganty, Pierre
Gargano, Luisa
Garncarek, Paweł
Gehnen, Matthias
Gigante, Nicola
Glazenburg, Erwin
Grüne, Christoph

Haase, Carolina
Han, Yo-Sub
Hansen, Kristoffer Arnsfelt
Hlineny, Petr
Huang, Shang-En
Hörsch, Florian
Itzhaki, Michael
Jana, Satyabrata
Kawahara, Jun
Klemz, Boris
Kobayashi, Yasuaki
Korhonen, Tuukka
Kralovic, Richard
Krekelberg, Bob
Kryven, Myroslav
Král, Pavel
Labourel, Arnaud
Lagarde, Guillaume
Lecroq, Thierry
Lenc, Ladislav
Lin, Chuang-Chieh
Liskiewicz, Maciej
Liu, Fu-Hong
Martínek, Jiří
Masopust, Tomas
Mavronicolas, Marios
McKenzie, Pierre
Migler, Theresa
Mock, Daniel
Mráz, František
Nanoti, Saraswati
Ng, Timothy
Nõmm, Sven
Obdrzalek, Jan
Olchanyi, Maxim
Ortali, Giacomo
Osička, Petr
Paesani, Giacomo
Pashaeibarough, Ali
Perz, Daniel
Prigioniero, Luca
Prusa, Daniel
Rampersad, Narad
Reinhardt, Klaus
Rescigno, Adele
Rinaldi, Francesco
Roy, Shivesh K.
Rysgaard, Casper
S. Sankar, Govind
Sadhukhan, Arpan
Sahu, Abhishek
Salvo, Ivano
Sampaio, Rudini
Saumell, Maria
Sawa, Zdeněk
Schapire, Robert
Schou, Jens Kristian Refsgaard
Seki, Shinnosuke
Shapira, Dana
Sheth, Kshiteej
Sieper, Marie Diana
Skoviera, Martin
Stachowiak, Grzegorz
Stocker, Moritz
Suchý, Ondřej
Svenning, Rolf
Szykuła, Marek
Tantau, Till
Tao, Terence
Tappini, Alessandra
Tu, Ta-Wei
Udwani, Rajan
Unger, Walter
Valencia, Frank
Van Der Merwe, Brink
Vaszil, György
Walen, Tomasz
Walzer, Stefan
Whittington, Philip
Zaborniak, Tristan
Zhang, Qiankun
Zhu, Zixuan
Zink, Johannes

Short Invited Talks

Temporal Graph Realization Problems

George B. Mertzios[1] and Paul G. Spirakis[2]

[1] Department of Computer Science, Durham University, UK
george.mertzios@durham.ac.uk
[2] Department of Computer Science, University of Liverpool, UK
p.spirakis@liverpool.ac.uk

Abstract. In this talk, we introduce the *temporal graph realization* problem with respect to the fastest path durations among its vertices, while we focus on periodic temporal graphs. In the basic version of the problem, given an $n \times n$ matrix D and a $\Delta \in \mathbb{N}$, the goal is to construct a Δ-periodic temporal graph with n vertices such that the duration of a *fastest path* from v_i to v_j is equal to $D_{i,j}$, or to decide that such a temporal graph does not exist. The variations of the problem on static graphs have been well studied and understood since the 1960s, see e.g. [1, 2]. As it turns out, this basic version of the periodic temporal graph realization problem has a very different computational complexity behavior than its static (ie. non-temporal) counterpart [3].

First, the problem is NP-hard in general, but polynomial-time solvable if the so-called underlying graph is a tree. Building upon those results, we investigate its parameterized computational complexity with respect to structural parameters of the underlying static graph which measure the "tree-likeness". We prove a tight classification between such parameters that allow fixed-parameter tractability (FPT) and those which imply W[1]-hardness. We show that our problem is W[1]-hard when parameterized by the *feedback vertex number* (and therefore also any smaller parameter such as *treewidth*, *degeneracy*, and *cliquewidth*) of the underlying graph, while we show that it is in FPT when parameterized by the *feedback edge number* (and therefore also any larger parameter such as *maximum leaf number*) of the underlying graph.

Then, we focus on the *upper bound* variation of the problem where, given an $n \times n$ matrix D, the question is whether there exists a periodic temporal graph on n vertices such that the duration of the fastest temporal path from a vertex u to a vertex v is *at most* $D_{u,v}$ [4]. This constraint with respect to upper bounds appears naturally in *transportation network design* applications where, for example, a road network is given, and the goal is to appropriately schedule periodic travel routes, while not exceeding some desired upper bounds on the travel times. This approach is in contrast to verification applications of the graph realization problems, where *exact* values for the distances (respectively, fastest travel times) are given, following some kind of precise measurement. In this problem variation, we focus only on underlying *tree topologies*, which are fundamental in many transportation network applications.

As it turns out, the periodic upper-bounded temporal tree realization problem (TTR) has a very different computational complexity behavior than both (i) the classic graph realization problem with respect to shortest path distances in static graphs and (ii) the periodic temporal graph realization problem with *exact* given fastest travel times (which was recently introduced). First, we prove that, surprisingly, TTR is NP-hard, even for a constant period Δ and when the input tree G satisfies one of the following conditions: (a) G has a constant diameter, or (b) G has constant maximum degree. In contrast, when we are given exact values of the fastest travel delays, the problem is known to be solvable in polynomial time. Second, we prove that TTR is fixed-parameter tractable (FPT) with respect to the number of leaves in the input tree G, via a novel combination of techniques for totally unimodular matrices and mixed integer linear programming.

Keywords: Temporal graph · Periodic temporal labeling · Fastest temporal path · Graph realization · Temporal connectivity · Parameterized complexity

References

1. Erdős, P., Gallai, T.: Graphs with prescribed degrees of vertices. Mat. Lapok, **11**, 264–274 (1960)
2. Hakimi, S.L., Yau, S.S.: Distance matrix of a graph and its realizability. Q. Appl. Math. **22**(4), 305–317 (1965)
3. Klobas, N., Mertzios, G.B., Molter, H., Spirakis, P.G.: Temporal graph realization from fastest paths. In: Proceedings of the 3rd Symposium on Algorithmic Foundations of Dynamic Networks (SAND), pp. 16:1–16:18. Best Student Paper Award (2024)
4. Mertzios, G.B., Molter, H., Spirakis, P.G.: Realizing temporal transportation trees. CoRR, abs/2403.18513 (2024)

The Challenge of Stability, Accuracy and Robustness in Data-Driven AI

Ivan Tyukin

Department of Computer Science, King's College London, London, WC2R 2LS, UK
ivan.tyukin@kcl.ac.uk

Abstract. In this lecture, we will delve into the theoretical limitations of determining the guaranteed stability and accuracy of neural networks built from empirical data in classification tasks. We will show that there is a large family of tasks and settings in which computing and verifying stability and accuracy is extremely challenging. We will also discuss an intriguing connection of these results with adversarial data and examples and propose a potential way to remedy the issues by enabling the networks to adapt over time.

Keywords: Stability of AI · Accuracy of AI · Verifiability of AI

The problem of reliability and validity of artificial intelligence (AI) systems has been the focus of attention of the scientific community since the Dartmouth Conference of 1956, considered the starting point and place of birth of AI as a new independent discipline. The problem of reliability and validity of AI at that time was initially reflected in the assumption set out by McCarthy, Minsky, Rochester and Shannon that "... any aspect of learning or any other property of intelligence can in principle be so precisely described that a machine can simulate it" [1].

This vision determined the development of science in the field of AI for many decades. For example, formal languages were developed for the analysis and verification of AI systems built on rules, including Isabelle and Lean [2], which are now successfully used for automatic verification of formal statements and proofs of theorems. In non-deterministic cases, when it was not possible to compile a deterministic system of rules describing AI decisions, or when the problems themselves had a significant random component (for example, under the assumption of statistical data generation), the issue of reliability, provability and robustness was addressed in the context of statistical methods and approaches [3, 4].

The rapid development of technologies in the last two decades, coupled with the ever-growing demand for automation of increasingly complex tasks that were previously entirely within the competence of humans, has led to the identification of practical limitations of classical theories and methods that have traditionally been used to justify the reliability and robustness of AI systems. It turned out that many important problems often cannot be effectively solved by deterministic algorithms (a formal system of rules). Examples include the well-known halting problems (determine whether a program will end/stop) and the Traveling Salesman problem.

On the other hand, problems have also emerged in statistical settings, the reliable and robust solution of which faces serious and often insoluble practical problems. This class of problems includes classification tasks in high dimensions, in which the number of significant yet mutually dependent attributes is large. Such problems are common in medicine, finance, and computer vision. Obtaining complete information about the distribution in the absence of any assumptions for problems with 100 relevant attributes requires samples of the order of $2^{100} > 10^{30}$. This limits the application of standard and well-studied statistical approaches (e.g. Bayesian networks and trees, Bayesian decision theory, etc.).

Alternative approaches based on the worst-case analysis (e.g. using the Vapnik-Chervonenkis theory, Rademacher complexity, and covering numbers) do not require knowledge of distributions. However, their practical application presupposes the availability of data, the volume of which directly depends on the complexity of the model, which also creates barriers and limitations. On the one hand, the more complex and expressive the AI, the more accurate and precise the solution is potentially. On the other hand, the more data is required to guarantee and confidently achieve this accuracy. Moreover, for modern AI systems, which are described by millions and trillions of parameters (Chat GPT 4), estimates of provable reliability in the form of bounds on the risk of errors following from these generally accepted theories may require unacceptably large amounts of data, growing proportionally to the number of parameters (see, for example, [6], Theorems 5.2 and 8.9).

To circumvent these limitations, alternative approaches to the analysis and assessment of reliability have been proposed in the last decade. Among these approaches, it is worth noting "explainable AI" (XAI) and the approach of quantitative assessment of uncertainty, which is applied to already built systems (also known as uncertainty quantification). Unfortunately, to date, explainable AI is still far from solving the problem of trust and reliability [5, 7], and effective assessment of the uncertainty of modern large generative models, including calibration, is still in its infancy and is difficult to apply in practice (see, for example, [8], " ... the field is still in its infancy.").

In our work, we explore the possibility of ensuring both, stability and accuracy, for a large class of AI models –feed-forward neural networks. We show that this problem may bring new challenges. These challenges revolve around the computational complexity of producing rigorous stability guarantees in classical distribution-agnostic settings. Strikingly, the issues emerge for both low- and high-dimensional problems. High dimension, naturally, makes verification of stability more difficult. The validation challenge is inherent in static networks whose architecture does not change over time. Therefore a potential resolution of the issue may arise from enabling AI models to change their architecture dynamically in response to detected undesirable behaviors. An example of such changes could be the addition of error correctors [9–11]. This, however, may require new notions of stability and a new theory for assessing asymptotic temporal properties of such adaptive AI models.

References

1. McCarthy, J., Minsky, M., Rochester, N., Shannon C.E.: A Proposal for the Dartmouth Summer Research Project on Artificial Intelligence (1955). http://raysolomonoff.com/dartmouth/boxa/dart564props.pdf
2. Harrison, J., Urban, J., Wiedijk, F.: History of interactive theorem proving. In: Handbook of the History of Logic, vol. 9, pp. 135–214, North-Holland (2014)
3. Devroye, L., Györfi, L., Lugosi, G.: A Probabilistic Theory of Pattern Recognition, vol. 31. Springer Science and Business Media (1997)
4. Vapnik, V.: Statistical Learning Theory. John Wiley and Sons (1998)
5. de Bruijn, H., Warnier, M., Janssen, M.: The perils and pitfalls of explainable AI: strategies for explaining algorithmic decision-making. Gov. Inf. Q. **39**(2), 101666 (2022)
6. Antony, M., Bartlet, P: Neural Network Learning: Theoretical Foundations. Cambridge University Press, Cambridge (1999)
7. Bove, C., Laugel, T., Lesot, M.J., Tijus, C., Detyniecki, M.: Why do Explanations Fail? A Typology and Discussion on Failures in XAI (2024). arXiv preprint arXiv: 2405.13474
8. He, W., Jiang, Z., Xiao, T., Xu, Z., Li, Y.: A Survey on Uncertainty Quantification Methods for Deep Learning (2023). arXiv preprint arXiv:2302.13425
9. Gorban, A., Grechuk, B., Tyukin, I.: Stochastic separation theorems: how geometry may help to correct AI errors. Not. Am. Math. Soc. (1), 25–33 (2023)
10. Gorban, A.N., Golubkov, A., Grechuk, B., Mirkes, E.M., Tyukin, I.Y.: Correction of AI systems by linear discriminants: probabilistic foundations. Inf. Sci. **466**, 303–322 (2018)
11. Tyukin, I.Y., Gorban, A.N., Sofeykov, K.I., Romanenko, I.: Knowledge transfer between artificial intelligence systems. Front. Neurorobotics **12**, 49 (2018)

Contents – Part II

Contributed Papers

Contents – Part I

Invited Talks

Contributed Papers

Contributed Papers

On Pumping Problems for Unary Regular Languages

Hermann Gruber[1], Markus Holzer[2](✉), and Christian Rauch[2]

[1] Planerio GmbH, Theresienhöhe 11A, 80339 München, Germany
h.gruber@planerio.de
[2] Institut für Informatik, Universität Giessen, Arndtstr. 2, 35392 Giessen, Germany
{holzer,christian.rauch}@informatik.uni-giessen.de

Abstract. Recently, the descriptional and computational complexity of various pumping lemmata for general regular languages have been investigated in the literature. There it turned out that in almost all cases tight bounds on the operational complexity of minimal pumping constants for regular languages have been obtained. From the computational perspective it was shown that in most cases the question whether a certain value can serve as a pumping constant w.r.t. a fixed pumping lemma is computationally intractable. Whether similar results can be obtained for restricted regular languages, such as unary regular languages, was left open—a language is unary if the underlying alphabet is a singleton set. Here we fill this gap by considering in detail questions on various pumping lemmata for unary regular languages. While some of the results obtained are similar to those in the general case, we also find significant differences. The results presented here fit well with the previous results and give a mostly complete picture of the problems in question.

1 Introduction

To pump, or not to pump, that is the question, if one tries to prove that a specific language is *not* regular. Pumping lemmata are a main tool in any formal language and automata course for this task—see, [15] for a comprehensive survey of different variants of pumping lemmata for regular, context-free languages and beyond. One of the most common pumping lemmata, yet not the first one [18] of its kind, can be found in [14, page 70, Theorem 11.1].

Lemma 1. *Let L be a regular language over Σ. Then, there is a constant p (depending on L) such that the following holds: If $w \in L$ and $|w| \geq p$, then there are words $x \in \Sigma^*$, $y \in \Sigma^+$, and $z \in \Sigma^*$ such that $w = xyz$ and $xy^tz \in L$ for all $t \geq 0$—it is then said that y can be* pumped *in w.*

Most pumping lemmata describe a necessary condition for a language to be regular, as the above one, but there are also lemmata that characterize the regular languages, by describing a necessary and sufficient condition for languages to be regular. One of the first ones of this kind is attributed to Jaffe [13] and reads as follows:

R. Královič and V. Kůrková (Eds.): SOFSEM 2025, LNCS 15539, pp. 3–16, 2025.
https://doi.org/10.1007/978-3-031-82697-9_1

Lemma 2. *A language L is regular if and only if there is a constant p (depending on L) such that the following holds: If $w \in \Sigma^*$ and $|w| = p$, then there are words $x \in \Sigma^*$, $y \in \Sigma^+$, and $z \in \Sigma^*$ such that $w = xyz$ and*[1]

$$wv = xyzv \in L \iff xy^t zv \in L$$

for all $t \geq 0$ and each $v \in \Sigma^*$.

Recently, the descriptional and computational complexity of problems related to some specific pumping lemmata were considered in a series of papers [2,4–6,8,9]. Roughly speaking this is the study of the minimal pumping constants that satisfy a particular pumping lemma, such as one of the above stated ones, from the aforementioned complexity perspectives. Let $\mathtt{mpc}(L)$ ($\mathtt{mpe}(L)$, respectively) denote the smallest number p that satisfies the condition of Lemma 1 (Lemma 2, respectively) when considering the regular language L. We state one example result for a descriptional complexity problem from [9] and another one for a computational problem [4]: first consider the complementation operation w.r.t. minimal pumping constants. Starting with a language L over the alphabet Σ satisfying $\mathtt{mpc}(L) = n$, one ends up with

$$\mathtt{mpc}(\Sigma^* \setminus L) \in \begin{cases} \{1\}, & \text{if } n = 0, \\ \mathbb{N}_0 \setminus \{1\}, & \text{if } n = 1, \\ \mathbb{N}, & \text{otherwise.} \end{cases}$$

For $\mathtt{mpe}$ the situation is easier, since $\mathtt{mpe}(L) = \mathtt{mpe}(\Sigma^* \setminus L)$. Second, asking the question whether for a language given by a finite automaton and a particular value of p the properties of a pumping lemma is satisfied results in the PUMPING-PROBLEM. For DFAs this problem is coNP-complete regardless whether Lemma 1 or Lemma 2 is considered. For NFAs the situation changes to coNP-hardness and containment in the second level of the polynomial hierarchy for Lemma 1, while becoming PSPACE-complete for Lemma 2. All these results were proven for languages over an alphabet of at least two letters. Thus, the question arises, which results still hold true for unary regular languages? We partially solve this question in this paper in the affirmative. Observe, that the pumping condition from Lemma 1 simplifies to

> "If $w \in L$ and $|w| \geq p$, then there are words $x \in \Sigma^*$ and $y \in \Sigma^+$ such that $w = xy$ and $xy^t \in L$ for $t \geq 0$,"

for unary regular languages, since concatenation is commutative for unary languages. A similar simplification applies to Lemma 2 when considering unary regular languages. Thus, in both cases the decomposition of the words becomes easier, which may affect some of the known pumping problem results for languages over alphabets that are at least binary—see the summary of results shown in the Tables 1 and 3 on page 5 and 10, respectively. Due to space constraints most proofs are omitted.

[1] Observe that the words $w = xyz$ and $xy^t z$, for all $t \geq 0$, belong to the same Myhill-Nerode equivalence class of the language L. Thus, one can say that the pumping of the word y in w *respects equivalence classes*.

2 Preliminaries

We assume the reader to be familiar with the basics in computational complexity theory [16]. In particular we recall the inclusion chain: $\mathsf{P} \subseteq \mathsf{NP} \subseteq \mathsf{PSPACE}$. Here P (NP, respectively) denotes the class of problems solvable by deterministic (nondeterministic, respectively) Turing machines in polytime, and PSPACE refers to the class of languages accepted by deterministic or nondeterministic Turing machines in polynomial space [19]. As usual, the prefix co refers to the complement class. For instance, coNP is the class of problems that are complements of NP problems. Moreover, recall the complexity class Π_2^{P} from the polynomial hierarchy, which can be described by polynomial time bounded oracle Turing machines. Here $\Pi_2^P = \mathsf{coNP}^{\mathsf{NP}}$, where $\mathsf{coNP}^{\mathsf{A}}$ is the set of decision problems solvable in polynomial time by a universal Turing machine with an oracle for some language in the class A. The class NP^{A} is defined analogously on a universal Turing machine. Moreover, let $\Theta_2^{\mathsf{P}} = \mathsf{P}_{||}^{\mathsf{NP}}$, which is the set of all decision problems that can be solved by a deterministic Turing machine in polynomial time with access to an NP oracle such that a list of all queries is formed before any of them is made (non-adaptive queries). The inclusion chain $\mathsf{coNP} \subseteq \Theta_2^{\mathsf{P}} \subseteq \Pi_2^{\mathsf{P}}$ is known [21]. Completeness and hardness are always meant with respect to deterministic many-one logspace reducibilities ($\leq_m^{\log}$) unless stated otherwise.

Next we fix some definitions on finite automata—cf. [7]. A *nondeterministic finite automaton* (NFA) is a quintuple $A = (Q, \Sigma, \cdot, q_0, F)$, where Q is the finite set of *states*, Σ is the finite set of *input symbols*, $q_0 \in Q$ is the *initial state*, $F \subseteq Q$ is the set of *accepting states*, and the *transition function* $\cdot$ maps $Q \times \Sigma$ to 2^Q. Here 2^Q refers to the powerset of Q. The *language accepted* by the NFA A is defined as $L(A) = \{\, w \in \Sigma^* \mid (q_0 \cdot w) \cap F \neq \emptyset \,\}$, where the transition function is recursively extended to a mapping $Q \times \Sigma^* \to 2^Q$ in the usual way. An NFA A is said to be *partial deterministic* if $|q \cdot a| \leq 1$ and *deterministic* (DFA) if $|q \cdot a| = 1$ for all $q \in Q$ and $a \in \Sigma$. In these cases we simply write $q \cdot a = p$ instead of $q \cdot a = \{p\}$. Note that every partial DFA can be made complete by introducing a non-accepting sink state that collects all non-specified transitions. For a DFA, obviously every letter $a \in \Sigma$ induces a mapping from the state set Q to Q by $q \mapsto q \cdot a$, for every $q \in Q$. Finally, a finite automaton is *unary* if the input alphabet Σ is a singleton set, that is, $\Sigma = \{a\}$, for some input symbol a.

The *deterministic state complexity of a finite automaton* A with state set Q is referred to as $\mathtt{sc}(A) := |Q|$ and the *deterministic state complexity of a regular language* L is defined as

$$\mathtt{sc}(L) = \min\{\, \mathtt{sc}(A) \mid A \text{ is a dfa accepting } L, i.e., L = L(A) \,\}.$$

A similar definition applies for the *nondeterministic state complexity of a regular language* by changing DFA to NFA in the definition, which we refer to as $\mathtt{nsc}(L)$. It is well known that

$$\mathtt{nsc}(L) \leq \mathtt{sc}(L) \leq 2^{\mathtt{nsc}(L)},$$

for every regular language L.

A finite automaton is *minimal* if its number of states is minimal with respect to the accepted language. It is well known that each minimal DFA is isomorphic to the DFA induced by the Myhill-Nerode equivalence relation. The *Myhill-Nerode* equivalence relation $\sim_L$ for a language $L \subseteq \Sigma^*$ is defined as follows: for $u, v \in \Sigma^*$ let $u \sim_L v$ if and only if $uw \in L \iff vw \in L$, for all $w \in \Sigma^*$. The equivalence class of u is referred to as $[u]_L$ or simply $[u]$ if the language is clear from the context and it is the set of all words that are equivalent to u w.r.t. the relation $\sim_L$, i.e., $[u]_L = \{ v \mid u \sim_L v \}$. Therefore we refer to the automaton induced by the Myhill-Nerode equivalence relation $\sim_L$ as the minimal DFA for the language L. On the other hand there may be minimal non-isomorphic NFAs for L.

3 Results on Pumping for Unary Finite Automata

At first we summarize some basic properties on the minimal pumping constants w.r.t. Lemmata 1 and 2 for arbitrary regular languages, not necessarily unary languages. The following relations

$$\texttt{mpc}(L) \leq \texttt{mpe}(L) \leq \texttt{sc}(L) \quad \text{and} \quad \texttt{mpc}(L) \leq \texttt{nsc}(L)$$

for every regular language L are known from [2,4,8]; interestingly the two measures mpe and nsc are incomparable [4]. Further simple facts for a regular language $L \subseteq \Sigma^*$ are the following:

1. $\texttt{mpc}(L) = 0$ if and only if $L = \emptyset$,
2. for every nonempty finite language L we have

$$\texttt{mpc}(L) = 1 + \max\{ |w| \mid w \in L \} \leq \texttt{mpe}(L) \leq 2 + \max\{ |w| \mid w \in L \},$$

3. $\texttt{mpc}(L) = 1$ implies that the empty word λ is in L, and
4. $\texttt{mpe}(L) = 1$ if and only if $L = \emptyset$ or $L = \Sigma^*$.

By definition of the minimal pumping constant w.r.t. Lemma 1 we know that for every regular language L there is a nonempty word $w \in L$ with $|w| = \texttt{mpc}(L) - 1$. A similar result is *not* valid for the minimal pumping constant w.r.t. Lemma 2 since one cannot force $w \in L$ in this case. The following observation is immediate by Jaffe's proof, cf. [13], and was mentioned first in [4]—in contrast, there is *no* obvious relation between mpc and the longest path in the finite state device.

Lemma 3. *Let A be a DFA and $L := L(A)$. Then $\texttt{mpe}(L) \leq \ell_A + 1$, where ℓ_A is the length, i.e., number of transitions, of the longest path of the automaton A. If L is a unary language, then $\texttt{mpe}(L) = \texttt{sc}(L)$.*

Thus, the descriptional complexity measures mpe and sc are equivalent for unary finite languages.

3.1 Descriptional Complexity of the Operation Problem

This section is devoted to the descriptional complexity of the operation problem w.r.t. minimal pumping constants for the pumping lemmata introduced here. In general, the operation problem for a regularity preserving n-ary function $\circ$ on languages and a descriptional complexity measure K is given by $g_\circ^K(k_1, k_2, \ldots, k_n)$ as the set of all numbers k such that there are regular languages $L_1, L_2, \ldots, L_n$ with descriptional complexity $K(L_i) = k_i$, for $1 \leq i \leq n$, and $K(\circ(L_1, L_2, \ldots, L_n)) = k$. If only unary regular languages are taken into account, then we write $g_\circ^{K,u}(k_1, k_2, \ldots, k_n)$ for the operational complexity of the $\circ$-operation. Here we assume that $K \in \{\mathtt{mpc}, \mathtt{mpe}\}$. Results on the descriptional complexity of the measures `mpc`, `mpe`, and others can be found in [2,8], but only the operation problem for `mpc` and some other measures, except for `mpe`, was investigated in more detail in [9]—the operation problem is completely untouched for `mpe` yet. The behaviour of the `mpc` measure differs with respect to finiteness/infinity of ranges depending on the considered operation. Our findings are summarized in Table 1, where the set of all natural numbers not including zero is denoted by $\mathbb{N}$; if zero is included, then we write $\mathbb{N}_0$ instead. Moreover, for $n \geq 2$, let $\mathbb{N}_{\geq n}$ refer to the set $\{n, n+1, \ldots\}$. The gray shaded entries in the table are new results.

Table 1. Descriptional complexity of the operation problem for the measures `mpc` and `mpe`. The `mpe`-results for intersection and union are only valid for $m = n$. An upper bound is proved in the paper for general m and n for both cases. Gray shaded entries indicate new results.

	Minimal pumping constant		
	mpc		mpe
Operation	unary	general	unary
Reversal	$\{n\}$	$\{n\}$	$\{n\}$
Prefix	$\{0\}$, if $n = 0$, $\{1, n\}$, otherwise.	$\{0\}$, if $n = 0$, $\{1, 2, \ldots, n\}$, otherwise.	$\{0\}$, if $n = 0$, $\{1, n\}$, otherwise.
Suffix	$\{0\}$, if $n = 0$, $\{1, n\}$, otherwise.	$\{0\}$, if $n = 0$, $\{1, 2, \ldots, n\}$, otherwise.	$\{0\}$, if $n = 0$, $\{1, n\}$, otherwise.
Complement	$\{1\}$, if $n = 0$, $\mathbb{N}_0 \setminus \{1\}$, if $n = 1$, $\mathbb{N} \setminus \{n\}$, otherwise.	$\{1\}$, if $n = 0$, $\mathbb{N}_0 \setminus \{1\}$, if $n = 1$, $\mathbb{N}$, otherwise.	$\{n\}$
Intersection	$\{0\}$, if $m = 0$ or $n = 0$, $\{1\}$, if $m = n = 1$, $\mathbb{N}_{\geq n}$, if $m = n \geq 2$, $\mathbb{N}_0$, otherwise.	$\{0\}$, if $m = 0$ or $n = 0$, $\mathbb{N}_0 \setminus \{2\}$, if $m = n = 1$, $\mathbb{N}_0$, otherwise.	$[1, (\lfloor n/2 \rfloor - 1) \cdot \lfloor n/4 \rfloor - 1] \subseteq \cdot$ $\cdot \cap [n^2 - n + 2, n^2] = \emptyset$
Union	$\{0\}$, if $m = 0$ or $n = 0$, $\{1\}$, if $m = n = 1$, $\mathbb{N}_{\geq n}$, if $m = n \geq 2$, $\mathbb{N}_0$, otherwise.	$\max\{m, n\}$, if $m = 0$ or $n = 0$, $\{1, 2, \ldots, \max\{m, n\}\}$, otherwise.	$[1, (\lfloor n/2 \rfloor - 1) \cdot \lfloor n/4 \rfloor - 1] \subseteq \cdot$ $\cdot \cap [n^2 - n + 2, n^2] = \emptyset$

Before we start our investigation of the operation problem in detail, we state the following auxiliary lemma, which turns out to be quite useful in the forthcoming arguments, since it is a lower bound argument on the minimal pumping constant w.r.t. Lemma 1. The statement is a reformulation of the fact that a word cannot be pumped. Thus, we omit the straightforward proof.

Lemma 4. *Let L be a unary regular language over $\Sigma = \{a\}$. Then $\mathtt{mpc}(L) > \ell$ if there is a word $w = a^\ell$ in L such that*

1. *there is* no *word $a^k \in L$, for $0 \leq k < \ell$, or*
2. *the property $a^k(a^{\ell-k})^* \not\subseteq L$ holds, for every word $a^k \in L$ with $k < \ell$.*

Let us start with the reversal operation. As usual the reversal of a word w is denoted by w^R and the reversal of the language L by L^R. For unary languages L we have $L = L^R$ which implies the following theorem—compare with the general result for $\mathtt{mpc}$ listed in Table 1.

Theorem 5. *$g_R^{K,u}(n) = \{n\}$, for $K \in \{\mathtt{mpc}, \mathtt{mpe}\}$.*

The next two operations we consider are the closure under prefix and suffix, denoted by Pref(.) and Suff(.), respectively. The descriptional complexity of both measures behaves in the same way for these operations.

Theorem 6. *For $K \in \{\mathtt{mpc}, \mathtt{mpe}\}$ we have*

$$g_{\text{Pref}}^{K,u}(n) = g_{\text{Suff}}^{K,u}(n) = \begin{cases} \{0\}, & \text{if } n = 0, \\ \{1, n\}, & \text{otherwise.} \end{cases}$$

We continue our investigation with the complement operation. Let Σ be an alphabet and L be a language over Σ, then we refer to the complement of L as $C_\Sigma(L) = \Sigma^* \setminus L$. Here we find the following situation for unary languages—see Table 1. The results for the two measures are entirely different. In the next and the forthcoming proofs we use the abbreviation $L^{\leq n}$ ($L^{\geq n}$, respectively), to refer to the set $\bigcup_{i=0}^{n} L^i$ ($\bigcup_{i=n}^{\infty} L^i$, respectively), for any language L.

Theorem 7. *It holds*

$$g_{C_\Sigma}^{\mathtt{mpc},u}(n) = \begin{cases} \{1\}, & \text{if } n = 0, \\ \mathbb{N}_0 \setminus \{1\}, & \text{if } n = 1, \\ \mathbb{N} \setminus \{n\}, & \text{otherwise.} \end{cases} \quad \text{and} \quad g_{C_\Sigma}^{\mathtt{mpe},u}(n) = \{n\}.$$

Proof. We consider first the measure $\mathtt{mpc}$. Like in [2], the high-level strategy is, for given integers n and k, either to find a witness language $L = L_{n,k}$ such that $\mathtt{mpc}(L) = n$ and $\mathtt{mpc}(C_\Sigma(L)) = k$, or to prove that no such witness language exists.

We distinguish the cases $n = 0$, $n = 2 = k + 1$, $n = k \geq 1$, and $n \geq 3$ with $n > k$—all other cases can be derived by using $C_\Sigma(C_\Sigma(L)) = L$:

1. For $n = 0$ we know that $L = \emptyset$ which implies that $C_\Sigma(L) = \Sigma^*$ for all alphabets Σ^* which implies $\mathtt{mpc}(C_\Sigma(L)) = 1$.
2. In the case $n = 2 = k + 1$ we set $L = a^+$ which fulfills $\mathtt{mpc}(L) = 2$ since the word a is the only non-pumpable word in L. On the other hand $C_\Sigma(L) = \{\lambda\}$ is a finite language and therefore satisfies $\mathtt{mpc}(C_\Sigma(L)) = 1$.
3. For $n = k \geq 1$ assume that $\mathtt{mpc}(L_{n,k}) = \mathtt{mpc}(C_\Sigma(L_{n,k})) = k = n$. We observe that due to the Lemma 1 there must be a word $w \in L_{n,k}$ such that $|w| = \mathtt{mpc}(L)-1 = n-1$ which in turn implies that $w = a^{n-1}$. Analogously we obtain that $w = a^{n-1} \in C_\Sigma(L_{n,k})$ and therefore $w = a^{n-1} \in L_{n,k} \cap C_\Sigma(L_{n,k}) = \emptyset$ which is a contradiction.
4. If $n \geq 3$ and $n > k$ we set

$$L_{n,k} = \{a\}^{\leq n-1} \setminus (\{a^{k-1}\}) \quad \text{and thus} \quad C_\Sigma(L_{n,k}) = \{a^{k-1}\} \cup \{a\}^{\geq n}.$$

Since $L_{n,k}$ is a finite language we have $\mathtt{mpc}(L_{n,k}) = n$. On the other hand we observe that each word in $C_\Sigma(L_{n,k})$ is pumpable except the word a^{k-1} which is the shortest word in $C_\Sigma(L_{n,k})$. Therefore we obtain $\mathtt{mpc}(C_\Sigma(L_{n,k})) = k$.

Finally, the statement for $\mathtt{mpe}$ follows directly from Lemma 3. $\square$

Now we come to the intersection of two unary languages.

Theorem 8. *We have*

$$g_\cap^{\mathtt{mpc},u}(m,n) = \begin{cases} \{0\}, & \text{if } m = 0 \text{ or } n = 0, \\ \{1\}, & \text{if } m = n = 1, \\ \mathbb{N}_{\geq n}, & \text{if } m = n \geq 2, \\ \mathbb{N}_0, & \text{otherwise.} \end{cases}$$

For the minimal pumping constants w.r.t. Lemma 2 the situation concerning the intersection operation is much more involved. First recall that for languages L defined over arbitrarily large alphabets we have $\mathtt{mpe}(L) \leq \mathtt{sc}(L)$ and in particular $\mathtt{mpe}(L) = \mathtt{sc}(L)$ for unary languages L. Thus, the results on unary languages presented below can be rewritten using $\mathtt{sc}$ instead of $\mathtt{mpe}$ and they remain still valid.

The cross-product construction can be used to determine the automaton accepting the intersection of two regular languages given by automata. Thus, $g_\cap^{\mathtt{mpe},u}(m,n) \subseteq [1, mn]$ and by a result from [11, Lemma 1] we also have the inclusion relation $[1, m] \subseteq g_\cap^{\mathtt{mpe},u}(m,n)$, if $1 \leq m \leq n$. Before we can make it more explicit which values in the interval $[1, mn]$ can be reached, we need some notation. For two sets S_1 and S_2 of numbers from $\mathbb{N}_0$, let

$$S_1 \oplus S_2 := \{\, x + y \mid x \in S_1 \text{ and } y \in S_2 \,\}$$

denote their Minkowski addition. For $m, n \geq 2$ and $k \geq 1$ we define

$$M_{m,n} = \{\, t_1 t_2 \mid m \bmod t_1 = n \bmod t_2 = 0 \,\}, \quad \text{and} \quad M_{1,k} = M_{k,1} = \{1, k\}.$$

Now we are ready for the next theorem, which is a consequence of a result from [10].

Theorem 9. *We have*

$$g_{\cap}^{mpe,u}(m,n) \subseteq \bigcup_{\ell=1}^{m} \bigcup_{k=1}^{n} (M_{k,\ell} \oplus [0, \max\{n-k, m-\ell\}]).$$

The theorem above gives us a rather implicit description of the values that can be possibly obtained by the intersection of two unary regular languages. In order to get a better impression of this result we list the complements of the upper bound sets described in Theorem 9 for small values of m and n in Table 2—the complementation is done w.r.t. to the sets $[1, mn]$. For instance, for $m = 6$ and $n = 5$ we find that the values in the set $\{23, 27, 28, 29\}$ cannot be achieved as mpe-values by the intersection operation on unary regular languages with mpe-complexity m and n, respectively. A closer look reveals that only numbers from the upper range close to the rightmost border of the interval $[1, mn]$ are listed. Since the analysis of the function $g_{\cap}^{\mathtt{mpe},u}$ in general is quite involved, we continue our investigation with a focus on the case $m = n$.

Table 2. The sets $[1, mn] \setminus \bigcup_{\ell=1}^{m} \bigcup_{k=1}^{n} (M_{k,\ell} \oplus [0, \max\{n-k, m-\ell\}])$ for small m and n.

	n					
m	2	3	4	5	6	7
2	$\emptyset$					
3	$\emptyset$	$\{8\}$				
4	$\emptyset$	$\{11\}$	$\{11, 14, 15\}$			
5	$\emptyset$	$\{14\}$	$\{18, 19\}$	$\{18, 19, 22, 23, 24\}$		
6	$\emptyset$	$\{17\}$	$\{22, 23\}$	$\{23, 27, 28, 29\}$	$\{23, 27, 28, 29, 32, 33, 34, 35\}$	
7	$\emptyset$	$\{20\}$	$\{23, 26, 27\}$	$\{32, 33, 34\}$	$\{33, 34, 38, 39, 40, 41\}$	$\{33, 34, 38, 39, 40, 41, 44, 45, 46, 47, 48\}$

Consequently, we are interested in the function $g_{\cap}^{\mathtt{mpe},u}(n,n) \subseteq [1, n^2]$. First, we show that the upper range $[n^2 - n + 2, n^2]$ cannot be reached by $g_{\cap}^{\mathtt{mpe},u}(n,n)$, for large enough n. From Theorem 9 we can derive that the upper range of possible numbers is not attainable—the stated result is also a direct implication of [17, Theorem 4].

Lemma 10. *For $n \geq 2$ we have $g_{\cap}^{mpe,u}(n,n) \cap [n^2 - n + 2, n^2] = \emptyset$.*

Proof. Observe that if L and L' are cyclic, i.e., they are accepted by a DFA which consists of one cycle with n states and an empty tail, then $L \cap L'$ is accepted by a DFA consisting of one cycle with n states and an empty tail, too. One easily observes that if either L or L' is not cyclic, then within the cross-product construction we obtain an automaton accepting $L \cap L'$ with less than n^2 states. Hence, the number n^2 cannot be in $g_{\cap}^{\mathtt{mpe},u}(n,n)$. On the other hand we have that the greatest number in

$$\bigcup_{\ell=1}^{n} \bigcup_{k=1}^{n} (M_{k,\ell} \oplus [0, \max\{n-k, n-\ell\}]) \setminus \{n^2\}$$

is $n \cdot (n-1) + 1 = n^2 - n + 1$ which proves the stated claim with Theorem 9.□

Regarding the lower end of the interval, the next theorem improves the aforementioned result $[1, n] \subseteq g_{\cap}^{\mathsf{mpe},u}(n, n)$ from [11, Lemma 1] to a much larger range:

Theorem 11. *For $n \geq 2$ we have $[1, (\lfloor n/2 \rfloor - 1) \cdot \lfloor n/4 \rfloor - 1] \subseteq g_{\cap}^{\mathsf{mpe},u}(n, n)$.*

With the aid of Theorems 9 and 11 we can tell for each number whether it can be reached or not *via* intersection, except for the numbers in the range

$$[(\lfloor n/2 \rfloor - 1) \cdot \lfloor n/4 \rfloor, n^2 - n + 1].$$

By computing the numbers $g_{\cap}^{\mathsf{mpe},u}(n, n)$, using exhaustive computer search, up to the value $n = 10$, we found that numbers which are not covered by the above theorems satisfy a specific side condition—we may require that the factors in the products in $M_{k,\ell}$ are coprime. Therefore we conjecture the following.

Conjecture 1 *We have $g_{\cap}^{\mathsf{mpe},u}(n, n) = \bigcup_{\ell=1}^{n} \bigcup_{k=1}^{n} (\widehat{M_{k,\ell}} \oplus [0, \max\{n-k, n-\ell\}])$, where*

$$\widehat{M_{k,\ell}} = \begin{cases} \{1, \max\{k, \ell\}\}, & \text{if } k = 1 \text{ or } \ell = 1, \\ \{\, t_1 t_2 \mid k \bmod t_1 = \ell \bmod t_2 = 0, \gcd(t_1, t_2) = 1 \,\}, & \text{otherwise.} \end{cases}$$

Aside from this, for the union operation on regular languages, we can apply De Morgan's law to derive the following statement from Theorems 10 and 11.

Theorem 12. *For $n \geq 2$ we have*

$$[1, (\lfloor n/2 \rfloor - 1) \cdot \lfloor n/4 \rfloor - 1] \subseteq g_{\cup}^{\mathsf{mpe},u}(n, n)$$

and

$$g_{\cup}^{\mathsf{mpe},u}(n, n) \cap [n^2 - n + 2, n^2] = \emptyset.$$

3.2 Computational Complexity of the Pumping-Problem

We will consider the following decision problem [4] related to the pumping lemmata stated in the introduction:

Language-Pumping-Problem or, for short, Pumping-Problem:
Input: a finite automaton A and a natural number p, i.e., an encoding $\langle A, 1^p \rangle$.
Output: Yes, if and only if the statement from Lemma 1 holds for the language $L(A)$ w.r.t. the value p.

For DFAs and NFAs the following results are known: already for DFAs this problem is intractable, namely coNP-complete [4], regardless whether we check for Kozen's [14] or Jaffe's [13] pumping property. This is quite remarkable, since it is a rare example of a computationally intractable property of a given *deterministic* finite automaton. The latter pumping property turned out to be more complex for NFAs, namely PSPACE-complete, while the former was shown to be coNP-hard and contained in Π_2^{P}, the second co-level of the polynomial hierarchy, for nondeterministic finite state devices [4]. Moreover, for NFAs, and even for DFAs, inapproximability results were recently shown for both pumping properties assuming the Exponential Time Hypothesis (ETH) [4,5]. In all cases the involved finite automata were at least binary. Thus, the question arises, whether similar results on the PUMPING-PROBLEM also hold for unary finite state devices. We answer this question in the affirmative. Our findings are summarized in Table 3.

Table 3. Complexity of the PUMPING-PROBLEM for variants of finite state devices. Gray shaded entries indicate new results.

	PUMPING-PROBLEM w.r.t. ...	
Automaton	Lemma 1	Lemma 2
unary DFA	L-complete	
DFA	coNP-complete	
unary NFA	coNP-hard in Π_2^{P}	coNP-hard in Θ_2^{P}
NFA	coNP-hard in Π_2^{P}	PSPACE-compl.

We first prove an auxiliary result on the relation of minimal pumping constants and the universality of unary languages:

Lemma 13. *Let $L \subseteq \{a\}^*$ be a unary regular language. Then the following statements hold:*

1. *If the empty word λ and the word a are in L, then $\mathit{mpc}(L) = 1$ iff $L = a^*$.*
2. *If λ is in L, then $\mathit{mpe}(L) = 1$ if and only if $L = a^*$.*

Let us first focus on the PUMPING-PROBLEM for unary DFAs. The following lemma is quite useful for the study of this problem. Recall that the condition $xy^tz \in L$ simplifies to $xy^t \in L$ since concatenation is commutative for unary languages.

Lemma 14. *Let $A = (Q, \{a\}, \cdot, q_0, F)$ be a unary DFA and $w = xyz$ a word in $L(A)$ with $x \in a^*$ and $y \in a^+$. Then $xy^* \subseteq L(A)$ if and only if every word $x, xy, xy^2, \ldots, xy^\ell$ belongs to $L(A)$, for $\ell = |Q|$.*

Now we are ready for the complexity of the PUMPING-PROBLEM for unary DFAs w.r.t. Lemma 1.

Theorem 15. *Given a unary DFA A and a natural number p, it is* L*-complete w.r.t. weak reductions to decide whether for the language $L(A)$ the statement from Lemma 1 holds for the value p.*

We turn our attention to the PUMPING-PROBLEM for unary DFA s w.r.t. Lemma 2. Here we observe that the problem remains L-complete. However, we need an auxiliary result checking for the Myhill-Nerode classes that appear during pumping—compare with a similar result for arbitrary DFAs and NFAs recently shown in [4].

Lemma 16. *Given a unary DFA $A = (Q, \{a\}, \cdot, q_0, F)$ and two unary words x and y, deciding whether every word in xy^* describes the same equivalence class w.r.t. the Myhill-Nerode relation $\sim_{L(A)}$, is* L*-complete. If the automaton A is a unary NFA, the problem becomes* coNP*-complete.*

The following theorem gives the answer to the computational complexity of the PUMPING-PROBLEM for unary deterministic finite automata.

Theorem 17. *Given a unary DFA A and a natural number p, it is* L*-complete to decide whether for the language $L(A)$ the statement from Lemma 2 holds for the value p.*

Next we study the PUMPING-PROBLEM for unary NFAs. The non-universality problem for unary NFAs automata was shown to be NP-complete [20]. The classic reduction is from 3SAT, and makes use of the Chinese Remainder Theorem. The construction can be adapted to show our next inapproximability result under the assumption of the so-called *Exponential Time Hypothesis (ETH)* [1,12]: there is no algorithm that solves 3-SAT in time $O^*(2^{o(n+m)})$, where n and m are the number of variables and clauses, respectively.

Theorem 18. *Let A be a unary NFA with s states. Then it is impossible to approximate the pumping constant with respect to Lemma 1 within a factor of $o(\sqrt[4]{s \log s})$ if the running time is in $2^{o\left(\sqrt[4]{\frac{s}{(\log s)^3}}\right)}$, assuming the Exponential Time Hypothesis. For the pumping constant with respect to Lemma 2, the inapproximability factor is even $2^{o\left(\sqrt[4]{s(\log s)^3}\right)}$ for the same running time, again assuming the Exponential Time Hypothesis.*

We note that a more efficient subexponential-time reduction for the NFA universality problem is given in [3], which might yield an improved lower bound on inapproximability. Nevertheless, from the previous proof we can deduce an coNP-lower bound for the PUMPING-PROBLEM for unary NFAs, regardless whether we consider Lemma 1 or 2. For the upper bound we explore an auxiliary statement that was shown in [4] and reads as follows:

Lemma 19. *Given an NFA $A = (Q, \Sigma, \cdot, q_0, F)$ and a word w over Σ, the language inclusion problem for w^* in $L(A)$ is* coNP*-complete.*

Now we are ready for our last theorem:

Theorem 20. *Given a unary NFA A and a natural number p, it is* coNP*-hard and it can be decided in* Π_2^{P} *whether for the language* $L(A)$ *the statement from Lemma 1 holds for the value p. In case Lemma 2 has to be fulfilled, the problem can be decided in* Θ_2^{P}.

Proof. Due to space constraints, we only give proofs for the upper bounds. When considering Lemma 1, the upper bound is seen as follows—and literally is taken from [4]: Let $\langle A, p\rangle$ be an input instance of the problem in question, where Q is the state set of A. We construct a coNP Turing machine M with access to a coNP oracle: first the device M deterministically verifies whether $p \geq |Q|$, and if so, it halts and accepts. Otherwise the computation universally guesses ($\forall$-states) a word w with $p \leq |w| < |Q|$. On that particular branch M checks deterministically if w belongs to $L(A)$. If this is *not* the case the computation halts and accepts. Otherwise, M deterministically cycles through all valid decompositions $w = xyz$ with $|y| \geq 1$. Then it constructs a finite automaton B accepting the language quotient $(x^{-1} \cdot L(A)) \cdot z^{-1}$. Here, if A is deterministic, then so is B. Then M decides whether $y^* \subseteq L(B)$ with the help of the coNP oracle—compare Lemma 19. If $y^* \subseteq L(B)$, then the cycling through the valid decompositions is stopped, and the device M halts and accepts. Notice that the latter is the case iff $xy^*z \subseteq L(A)$. Otherwise, i.e., if $y^* \not\subseteq L(B)$, the Turing machine M continues with the next decomposition in the enumeration cycle. Finally, if the cycle computation finishes, the Turing machine halts and rejects, because no valid decomposition of w was found that allows for pumping. In summary, the Turing machine operates universally, runs in polynomial time, and uses a coNP oracle. Thus, the containment in Π_2^{P} follows.

When considering Lemma 2, we cannot apply this algorithm: it would not result in a polynomial time bound, since $|w| \leq 2^{|Q|}$ has to be satisfied. Thus, we proceed differently, and utilize the relation between the deterministic state complexity of a unary regular language and the minimal pumping constant w.r.t. Jaffe's pumping lemma as mentioned in Lemma 3: if L is a unary regular language, then $\mathtt{mpe}(L) = \mathtt{sc}(L)$. Thus, we construct a P Turing machine M with non-adaptive access to an NP oracle that checks for inequivalence on unary finite state devices—recall that the value p from the input is encoded in unary: first M deterministically lists all words a^m that belong to $L(A)$ with $m \leq q$ by simulating the given NFA A with an incremental powerset construction. The words from this list will be used to determine the accepting states of the DFAs constructed next. Every unary DFA consists of a tail and a loop of states. The Turing machine then lists all unary DFAs by cycling through all combinations of tail and loop states such that the overall number of states does not exceed q. Here the above constructed list of words is used to assign the accepting states to these automata appropriately. By simple combinatorics one observes, that there exists only polynomially many unary DFAs that can be constructed in that way. Then the NP oracle is questioned on this list of unary DFAs. Since we are asking for inequivalence, the Turing machine halts and accepts if at least one query is answered negatively; otherwise the Turing machine halts and rejects. It is easy to see that M decides the PUMPING-PROBLEM w.r.t. Jaffe's pumping lemma. To summarize, the Turing machine operates deterministically, runs in polynomial time, and uses an NP oracle such that a list of all queries is formed before any of them is made. Thus, the containment within Θ_2^{P} follows. □

References

1. Cygan, M., et al.: Parameterized Algorithms, Chap. Lower Bounds Based on the Exponential-Time Hypothesis, pp. 467–521. Springer (2015). https://doi.org/10.1007/978-3-319-21275-3_14
2. Dassow, J., Jecker, I.: Operational complexity and pumping lemmas. Acta Inform. **59**, 337–355 (2022). https://doi.org/10.1007/s00236-022-00431-3
3. Fernau, H., Krebs, A.: Problems on finite automata and the exponential time hypothesis. Algorithms **10**(1), 24 (2017). https://doi.org/10.3390/a10010024
4. Gruber, H., Holzer, M., Rauch, C.: The pumping lemma for regular languages is hard. In: Nagy, B. (ed.) Proceedings of the 27th International Conference on Implementation and Application of Automata, pp. 128–140. No. 14151 in LNCS, Springer, Famagusta, Cyprus (2023). https://doi.org/10.1007/978-3-031-40247-0_9
5. Gruber, H., Holzer, M., Rauch, C.: On pumping preserving homomorphisms and the complexity of the pumping problem (extended abstract). In: Fazekas, S. (ed.) Proceedings of the 28th International Conference on Implementation and Application of Automata, pp. 153–165. No. 15015 in LNCS, Springer, Akita, Japan (2024). https://doi.org/10.1007/978-3-031-71112-1_11
6. Gruber, H., Holzer, M., Rauch, C.: The pumping lemma for context-free languages is undecidable. In: Day, J.D., Manea, F. (eds.) Proceedings of the 28th International Conference on Developments in Language Theory, pp. 141–155. No. 14791 in LNCS, Springer, Göttingen, Germany (2024). https://doi.org/10.1007/978-3-031-66159-4_11
7. Harrison, M.A.: Introduction to Formal Language Theory. Addison-Wesley (1978)
8. Holzer, M., Rauch, C.: On Jaffe's pumping lemma, revisited. In: Bordihn, H., Tran, N., Vaszil, G. (eds.) Proceedings of the 25th International Conference on Descriptional Complexity of Formal Systems, pp. 65–78. No. 13918 in LNCS, Springer, Potsdam, Germany (2023). https://doi.org/10.1007/978-3-031-34326-1_5
9. Holzer, M., Rauch, C.: On minimal pumping constants for regular languages. In: Gazdag, Z., Iván, S., Kovásznai, G. (eds.) Proceedings of the 16th International Conference on Automata and Formal Languages, pp. 127–141. No. 386 in EPTCS, Eger, Hungary (2023). https://doi.org/10.4204/EPTCS.386.11
10. Holzer, M., Rauch, C.: The range of state complexities of languages resulting from the cascade product–the unary case. Internat. J. Found. Comput. Sci. **34**(8), 987–1022 (2023). https://doi.org/10.1142/S0129054123430049
11. Hricko, M., Jirásková, G., Szabari, A.: Union and intersection of regular languages and descriptional complexity. In: Mereghetti, C., Palano, B., Pighizzini, G., Wotschke, D. (eds.) Proceedings of the 7th Workshop on Descriptional Complexity of Formal Systems, pp. 170–181. Universita degli Studi di Milano, Como, Italy (2005)

12. Impagliazzo, R., Paturi, R., Zane, F.: Which problems have strongly exponential complexity? J. Comput. System Sci. **63**(4), 512–530 (2001). https://doi.org/10.1006/jcss.2001.1774
13. Jaffe, J.: A necessary and sufficient pumping lemma for regular languages. SIGACT News **10**(2), 48–49 (Sommer 1978). https://doi.org/10.1145/990524.990528
14. Kozen, D.C.: Automata and Computability. Undergraduate Texts in Computer Science, Springer (1997). https://doi.org/10.1007/978-1-4612-1844-9
15. Nijholt, A.: YABBER—yet another bibliography: pumping lemma's. An annotated bibliography of pumping. Bull. EATCS **17**, 34–53 (1982)
16. Papadimitriou, C.H.: Computational Complexity. Addison-Wesley (1994)
17. Pighizzini, G., Shallit, J.: Unary language operations, state complexity and Jacobsthal's function. Internat. J. Found. Comput. Sci. **13**(1), 145–159 (2002). https://doi.org/10.1142/S012905410200100X
18. Rabin, M.O., Scott, D.: Finite automata and their decision problems. IBM J. Res. Dev. **3**, 114–125 (1959). https://doi.org/10.1147/rd.32.0114
19. Savitch, W.J.: Relationships between nondeterministic and deterministic tape complexities. J. Comput. System Sci. **4**(2), 177–192 (1970). https://doi.org/10.1016/S0022-0000(70)80006-X
20. Stockmeyer, L.J., Meyer, A.R.: Word problems requiring exponential time. In: Proceedings of the 5th Symposium on Theory of Computing, pp. 1–9 (1973)
21. Wagner, K.W.: Bounded query classes. SIAM J. Comput. **19**(5), 833–846 (1990). https://doi.org/10.1137/0219058

The Complexity of Graph Exploration Games

Janosch Fuchs, Christoph Grüne(✉), and Tom Janßen

Department of Computer Science, RWTH Aachen University, Aachen, Germany
{fuchs,gruene,janssen}@algo.rwth-aachen.de

Abstract. Graph Exploration problems ask a searcher to explore an unknown environment. The environment is modeled as a graph, where the searcher needs to visit each vertex beginning at some vertex. Treasure Hunt problems are a variation of Graph Exploration, in which the searcher needs to find a hidden treasure, which is located at a designated vertex.

Usually these problems are modeled as online problems, and any online algorithm performs poorly because it has too little knowledge about the instance to react adequately to the requests of the adversary. Thus, the impact of a priori knowledge is of interest. One form of a priori knowledge is an unlabeled map, which is an isomorphic copy of the graph. We analyze Graph Exploration and Treasure Hunt problems with an unlabeled map that is provided to the searcher. For this, we formulate decision variants of both problems by interpreting the online problems as a game between the online algorithm (the searcher) and the adversary. The map, however, is not controllable by the adversary. The question is whether the searcher is able to explore the graph completely or find the treasure for all possible decisions of the adversary.

We analyze these games in multiple settings, with and without costs on the edges, on directed and undirected graphs and with different constraints (allowing multiple visits to vertices or edges) on the solution. We prove *PSPACE*-completeness for most of these games. Additionally, we analyze the complexity of related problems that have additional constraints on the solution.

Keywords: Online Algorithms · Graph Exploration · Computational Complexity · Online Algorithms Complexity · Two-Player Games · PSPACE-completeness

1 Introduction

Graph Exploration problems model situations in which a searcher, like an autonomous robot, has to explore an environment and solve a task. Among

This work is funded by the Deutsche Forschungsgemeinschaft (DFG, German Research Foundation) – GRK 2236/1, WO 1451/2-1. The full version of this paper can be found on arXiv [13].

R. Královič and V. Kůrková (Eds.): SOFSEM 2025, LNCS 15539, pp. 17–30, 2025.
https://doi.org/10.1007/978-3-031-82697-9_2

those tasks are finding the shortest path to a designated point, which is referred to as the Treasure Hunt problem, or exploring the whole environment with minimal resource consumption, which is referred to as the Online Traveling Salesman problem. Thereby, the searcher does not know the environment at the beginning and only obtains local information during exploration.

Typically, Graph Exploration is modeled as an online problem on a graph, that is fixed by the adversary before the online computation starts. The searcher is positioned at a vertex and then the labels of its neighborhood are revealed together with the corresponding incident edges. Kalyanasundaram and Pruhs defined this model as *fixed graph scenario* [16]. Based on the overall obtained knowledge, the online algorithm has to irrevocably decide along which edge the searcher moves. For a worst-case analysis a malicious adversary is presupposed which controls the revelation process and creates the input. The goal of the adversary is to minimize the performance of the online algorithm.

While the online algorithm moves the searcher, the adversary creates the graph and chooses the vertices that are revealed. Therefore, the adversary is able to tailor the instance in his favor to the decisions of any online algorithm. To overcome this asymmetry, different extensions of the online setting exist, in which the online algorithm is equipped with a priori knowledge. Throughout this paper, we introduce an unlabeled map, which is an isomorphic copy of the input graph. Thus, the input graph is not constructed by the adversary and only the revelation order of the vertices is determined by the adversary.

The connection between the online algorithm and the adversary is analogous to two players in an asymmetric two-player game [5,14,21]. The input graph can be considered as the game board. A *turn* of the game consist of a *move of the adversary* followed by *move of the online algorithm*. Specifically, the adversary reveals the neighborhood of the vertex v, on which the searcher is positioned, by revealing the labels of the neighbors and the edges connecting them to v. The labels are recognizable by the online algorithm later in the game. Thereafter, the online algorithm makes a move by choosing an incident edge of v to move the searcher along. The problem is to decide whether the online algorithm has a *winning strategy*, that is, it can compute a feasible sequence of vertices, for all possible moves of the adversary.

Related Work. Among the preliminary work on online path problems are the Online Traveling Salesman Problem and the Canadian Traveler Problem. Papadimitriou and Yannakakis [21] introduced the problem of finding a shortest path in a graph with edge cost uncertainties, which are revealed when the searcher is positioned at an incident edge, as the Canadian Traveler Problem. Further results on this problem are discussed by Bar-Noy and Schieber [3]. Kalyanasundaram and Pruhs [16] introduce the online version of the Traveling Salesman Problem under the fixed graph scenario, which is later referred as Graph Exploration. They also present an algorithm that yields a 16-approximation for undirected planar graphs. Foerster and Wattenhofer [11] used the same model as in Graph Exploration to analyze the Treasure Hunt Problem. Additionally they provide lower and upper bounds on the competitivity of Graph Exploration on directed graphs. Bounds in the undirected case for Graph Exploration are provided by Megow et al. [19]. They also show that the 16-approximation of Kalyanasundaram and Pruhs [16] extends to graphs of bounded genus, where the competitive ratio increases linearly with the genus. This result was further improved and extended to graphs with excluded minors by Baligács et al. [2]. Furthermore, there are constrained variations of the Graph Exploration problem that limit the ability of the searcher. Duncan et al. [9] analyze Graph Exploration where the searcher is tied to the starting point with a tether of fixed length or has a limited fuel tank. They give upper and lower bounds for these settings.

Another branch of Graph Exploration surveys the influence of additional information on the performance of the searcher. A subset of those variations includes some form of a map. Panaite and Pelc [20] focus on a setting, where the searcher has either a labeled map, a labeled copy of the graph with an additional sense of direction, or an unlabeled map, an isomorphic copy of the graph, and compare these models. Furthermore, Dessmark and Pelc [7] use a similar model where the searcher has an unlabeled map and either knows where it starts on this map (anchored map) or not (unanchored map). Additionally, maps are also of interest for the Treasure Hunting Problem. Bouchard et al. [6] analyze the performance gain of using different forms of maps. Instead of using a model of a map, an abstract and general form of information may be used as well, the so-called advice model. The advice is provided as a binary string, whereby the advice complexity is the number of used bits. Dobrev et al. [8] give a lower bound of $\Omega(|V|\log(|V|))$ on the advice complexity when the algorithm has to compute an optimal solution and present an algorithm using linear advice and achieving a constant competitive ratio of 6. Böckenhauer et al. [4] show that $\mathcal{O}(|E|)$ advice bits are sufficient to optimally explore any graph. Besides, Komm [17] et al. analyze the Treasure Hunt Problem with advice. At last, a new branch uses prediction models as source of information. Eberle et al. [10] consider a learning prediction framework with a bounded error to potentially robustify existing algorithms.

With the work on the Canadian Traveler Problem, Papadimitriou and Yannakakis [21] also introduced online graph games with a map. The task is to find a shortest s-t-path where the edge costs are chosen by the adversary. They show *PSPACE*-for deciding whether there is an r-competitive strategy for traversing the graph, where r is a given ratio. This work is also continued by Bar-Noy and Schieber [3] on different variations, where the k-Canadian Traveler Problem remains *PSPACE*-complete. Additionally, Böhm and Veselý [5] show the Online Chromatic Number problem to be *PSPACE*-complete. In there, an unlabeled map is provided to the online algorithm. In a similar setting, Fuchs et al. [14] build a reduction framework which can be applied to graph problems that search for a subset of vertices such as vertex cover, independent set or dominating set. With that they show that online games based on these problems are *PSPACE*-complete. A complexity analysis on a broader set of *PSPACE*-hard combinatorial games can be found in Fraenkel and Goldschmidt's survey [12].

Contribution. We analyze the complexity properties of Graph Exploration problems by taking up the ideas by Papadimitriou and Yannakakis [21], Böhm and Veselý [5] as well as Fuchs et al. [14]. That is, we introduce online games variants of Graph Exploration problems that include an unlabeled map of the graph that the online algorithm can use. On the one hand, we define and analyze the Online Traveling Salesman Game, which is the online game version of the original Graph Exploration problem defined by Kalyanasundaram and Pruhs [16]. It asks whether an online algorithm is able to find a Hamiltonian cycle of small weight in a given graph for all possible reveal decisions of the adversary while having an unlabeled map. On the other hand, we define the online game version of the Treasure Hunt problem. It asks whether an online algorithm is able to find an s-t-path in a given graph for all possible reveal decisions of the adversary while having an unlabeled map.

Furthermore, we analyze variants of both problems: Besides merely asking for the existence of a path or cycle, an additional number $k \in \mathbb{N}$ is introduced, limiting the length of the solution. Additionally, we consider versions of both problems, in which we relax the path constraint to be a trail or a walk as well as constrained versions of these problems such as the metric version of the online traveling salesman game. We show that nearly all of the above mentioned problems are *PSPACE*-complete. The other problems degenerate to simple offline problems such as ONLINE UNDIRECTED S-T WALK GAME, which is the *LOGSPACE*-complete problem USTCON [22].

Paper Summary. In Sect. 2, we define preliminary terms including complexity theoretic concepts and the online game setting. In Sect. 3, we analyze the ONLINE S-T PATH GAME as well as variants in directed and undirected graphs with and without edge costs. In Sect. 4, we summarize the results on ONLINE HAMILTONIAN PATH GAME and the related variants in directed and undirected graphs with and without edge costs. Then, variations of the classical S-T PATH and HAMILTONIAN PATH are studied in the online game context in Sect. 5. At last in Sect. 6, we conclude the paper and present remaining open problems. For space reasons, proofs marked with ($\star$) are deferred to the full version of the paper.

2 Preliminaries

As usual, we define a walk as a sequence of connected edges. A trail is a walk where all edges are distinct, and a path is a trail such that no vertex occurs more than once. We also refer to s-t walks (resp. trails, paths) to indicate the two endpoints of the walks (resp. trails, paths). With $N(v)$ we refer to the open neighborhood of vertex v.

Search Sequences. A search sequence is a (valid) solution to an instance of a Graph Exploration problem. Intuitively, a search sequence is a walk, which does not contain cycles consisting only of vertices that have occured in the same walk before. That is in every cyclic subwalk, a previously non-visited vertex has to be included in the walk.

Definition 1 (Search Sequence). *For a graph G, a* search sequence *is a sequence of arcs or edges $e_1, \ldots, e_{n-1}$ in G for which there is a sequence $v_1, \ldots, v_n$ of vertices in G, such that $e_i = \{v_i, v_{i+1}\}$. Furthermore, for all subsequences $v_j, \ldots, v_k$ of $v_1, \ldots, v_n$ with $v_j = v_k$ it holds that $\{v_1, \ldots, v_{j-1}\} \subsetneq \{v_1, \ldots, v_k\}$. If $v_1 = v_n$, we call S a* cyclic search sequence. *The cost of a search sequence $S = e_1, \ldots, e_{n-1}$ is defined by $cost(S) = \sum_{e \in S} cost(e)$, if the edges have costs assigned, and $cost(S) = |S|$ otherwise.*

In the online setting, the search sequence is determined by the moves of the adversary as well as the moves of the online algorithm. In each step, the online algorithm is located at some vertex v and chooses one of the vertices from $N(v)$ as the target and moves itself towards it. Then the adversary reveals the neighborhood of the target. That is, the adversary reveals the labels of all neighboring vertices $n \in N(v)$ as well as all edge/arc weights. We refer to this model as the *neighborhood reveal model.* Throughout the paper, we call vertices to which the online algorithm moved before *visited* and vertices which are revealed for the first time *new.* Furthermore, we call non-visited vertices which are revealed a subsequent time *known.*

The online algorithm may be restricted to different variations of search sequences. By definition, a search sequence has to be a walk in the graph. We also consider problems that restrict the search sequences to trails or paths. While trails and paths are polynomially bounded in their length by the size of the graph, this is generally not the case for walks. However, a search sequence may not contain any cycle that does not visit previously unvisited vertices. This does not restrict the online algorithm, since traversing a cycle of only visited vertices does not reveal any new vertices and puts the online algorithm back in the position it was before. Thus the length of a search sequence is always polynomially bounded in the size of the input graph.

Complexity Theory. We define a decision problem to be a subset of $\{0,1\}^*$. For two decision problems A and B, we say that A is polynomially reducible to B, if there is a function $f : \{0,1\}^* \to \{0,1\}^*$ computable in polynomial time such that $x \in A$ if and only if $f(x) \in B$. The class *PSPACE* is given by all decision problems that can be decided by a deterministic Turing machine with polynomial space. As for *NP*, a problem is called *PSPACE*-hard, if any other problem in *PSPACE* can be reduced to it by a polynomial reduction. A problem that is both contained in *PSPACE* and is *PSPACE*-hard, is also called *PSPACE*-complete. The canonical *PSPACE*-complete problem is TRUE QUANTIFIED BOOLEAN FORMULA [23] or TQBF for short. For this paper, the game version of TQBF – TQBF GAME – is of most interest.

This game is played by two players: the ∃-player and the ∀-player. The ∃-player controls all ∃-quantified variables and the ∀-player controls all ∀-quantified variables in the order of quantification. That is, a turn consists of a move of the ∃-player followed by a move of the ∀-player, in which they decide the assignment of their variable(s). The ∃-player wins if and only if $\varphi(X_1, \ldots, X_n)$ is satisfied with the assignments of both players.

Definition 2 (TQBF Game).
Given: *A fully quantified Boolean formula* $Q_1X_1 \dots Q_nX_n\varphi(X_1, \dots, X_n)$ *with* $Q_i \in \{\exists, \forall\}$ *for* $i \in \{1, \dots, n\}$.
Question: *Does the* $\exists$*-player have a winning strategy?*

Deciding whether the $\exists$-player has a winning strategy is *PSPACE*-complete by a simple reduction from TQBF. W.l.o.g. we assume φ to be in CNF. Furthermore, we assume clauses to only contain three literals for simplicity, but our constructions also extend to any number of literals per clause.

Online Search Sequence Games. A search sequence problem P^{SSP} has a graph G as input and the feasible solutions are a subset of all search sequences in G. Examples for such problems are S-T PATH and HAMILTONIAN PATH. For any problem P^{SSP}, we define an online game version.

Definition 3 (Online Search Sequence Game).
Given: *A graph G and possibly start and/or end vertices.*
Question: *Does an online algorithm exist, that finds a valid search sequence in G (as defined by* P^{SSP}*) for all strategies of the adversary in the neighborhood reveal model, while the online algorithm knows an unlabeled map of G?*

We also refer to this problem as P_O^{SSP}. Since our definition of search sequences implies them having a length polynomial in the size of the input graph (as argued above), we obtain the following theorem.

Theorem 1 ($\star$). *If* $P^{SSP} \in NP$*, then* $P_O^{SSP} \in$ *PSPACE.*

3 Path Problems

The first class of problems that we analyze are *s-t* path problems. The online versions of these problems can be interpreted as a Treasure Hunt problem. We start our complexity analysis with the ONLINE UNDIRECTED S-T PATH GAME.

Definition 4 (Online Undirected s-t Path Game).
Given: *An undirected graph* $G = (V, E)$*, and two vertices* $s, t \in V$.
Question: *Does an online algorithm exist, that finds a s-t path in G for all strategies of the adversary in the neighborhood reveal model, while the online algorithm knows an unlabeled map of G?*

We show that ONLINE UNDIRECTED S-T PATH GAME is *PSPACE*-complete and derive further results on variations which include *s-t* path, *s-t* trail and *s-t* walk on directed and undirected graphs. Additionally, we survey the online versions of constrained path problems.

Theorem 2. *ONLINE UNDIRECTED S-T PATH GAME is PSPACE-complete.*

Reduction Overview. We show the *PSPACE*-hardness for ONLINE UNDIRECTED S-T PATH GAME by a reduction from TQBF GAME. However, when considering games based on online problems, the online algorithm always chooses the next vertex. We can still model choices of the adversary though, by letting the online algorithm choose between two vertices it cannot distinguish. This way, the online algorithm is able to decide the truth assignment of $\exists$-variables, and the adversary is able to decide the truth assignment of $\forall$-variables.

The following reduction is loosely adapted from the reduction of Li et al. [18]. The variable gadget of our reduction essentially consists of two paths, which end in the same vertex. One path corresponds to assigning the variable the value true, the other the value false. The clause gadget consists of two disconnected vertices, connected by one path for each literal they contain, where the first vertex of each path has an edge to the respective variable gadget. If for a clause at least one variable gadget is set to a value that satisfies the clause, then the online algorithm can identify one of the paths to traverse the clause gadget. The s-t path, the online algorithm needs to find, starts in s, then traverses all variable gadgets (using only one of the two paths), then traverses all clause gadgets, and finally reaches t.

Two important parts of our reduction that use the online nature of the game are additional edges to reveal vertices, making them recognizable for later decisions, and *sinks*. A sink being attached to a vertex v means that there is a vertex connected to v by an edge but to no other vertices. The purpose of a sink is to prevent the online algorithm from choosing a new neighbor of v, as that allows the adversary to trap it in the sink. Edges that reveal vertices are usually used together with sinks, to prevent the online algorithm from traversing them, but still allow it to recognize a vertex later.

Variable Gadget Overview. The variable gadget for variable x_i roughly consists of four parts, using the same numbering as in Fig. 1:

1. There are two paths, one corresponding to assigning the variable the value true, and the other corresponding to the value false. In Fig. 1, these paths start at the vertices v_1^i and w_1^i, and meet at u_6^i. The variable is set to true by choosing the path corresponding to true. This reveals vertices in the clauses that are satisfied, helping the algorithm traversing those clause gadgets later.
2. There are vertices simulating the decision for the variable assignment of the online algorithm or adversary, depending on whether x_i is $\exists$-quantified or $\forall$-quantified. In Fig. 1, this is done by the vertices v^i and w^i. The green dot-dashed edge only exists if x_i is $\exists$-quantified, and reveals which vertex corresponds to which path. On the other hand, if x_i is $\forall$-quantified, the online algorithm cannot distinguish v^i and w^i. Only after choosing one of the two, the online algorithm learns the truth assignment from the map, as the map shows whether the path corresponding to true (resp. false) has clauses attached first. This is necessary, so the online algorithm can make its choices for $\exists$-variables dependent on the choices for $\forall$-variables of the adversary.
3. There is a path of vertices whose purpose is to reveal other parts of the gadget, so the online algorithm can recognize them later. As described above, this

includes revealing the truth assignments that correspond to the two paths, but also further vertices on the paths, indicated by the cyan dot-dot-dashed edges in Fig. 1. This is used together with sinks to prevent the online algorithm from unwanted behavior.

4. There is a small gadget that ensures the online algorithm correctly finds its way through the path that reveals the later parts of the gadget. Furthermore this gadget enforces that the algorithm traverses the entire path of Part 3, to prevent it from producing a mixed variable assignment by using it to switch between the two paths of Part 1.

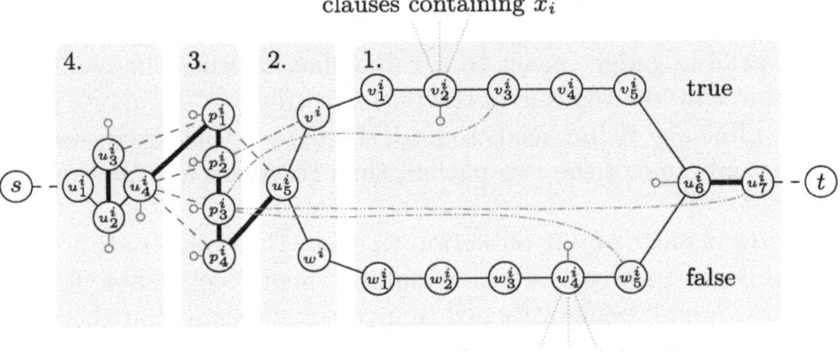

Fig. 1. Variable gadget for the reduction from TQBF GAME to ONLINE S-T PATH GAME for a variable x_i. The thick solid black edges have to be used by the algorithm, while the thin solid black edges are optional. Any non-solid colored edge cannot be used, and only exists for later recognition of the vertex it connects to. These properties are ensured by the sinks (red) that allow the adversary to trap the algorithm whenever it uses a non-solid non-black edge.

Lemma 1 (⋆). *The online algorithm has to traverse the variable gadget for x_i from u^i_1 to u^i_7 and uses either every vertex v^i_j or every vertex w^i_j, and none of the other, respectively. Furthermore, it cannot leave a variable gadget by entering one of the clauses early.*

With the previous lemma, we proved that the online algorithm always has to traverse a variable gadget in a way that assigns it either true or false. It remains to show that the quantifiers are correctly simulated.

Lemma 2 (⋆). *When simulating the TQBF game with the reduction, the following holds. If a variable x_i is ∃-quantified, the online algorithm is able to choose its truth assignment. On the other hand, if x_i is ∀-quantified, the adversary is able to choose its truth assignment, and the online algorithm learns that truth assignment before choosing the next ∃-variable.*

Clause Gadget. The clause gadget for clause C_i consists of two disconnected vertices, connected by one path for every variable they contain. Then, the clause gadget, also shown in Fig. 2, is defined as follows:

- There are two vertices c_1^i and c_2^i, which are not connected and each have a sink attached.
- For each literal $\ell \in C_i$, there are two vertices: $c_\ell^i, c_{\ell'}^i$.
- For each literal $\ell \in C_i$, the vertices $c_1^i, c_\ell^i, c_{\ell'}^i$ and c_2^i form a path.
- For each literal $\ell \in C_i$, let $var(\ell)$ be the index of its corresponding variable. If ℓ is non-negated, there is the edge $\{c_\ell^i, v_2^{var(\ell)}\}$, and if ℓ is negated, there is the edge $\{c_\ell^i, w_4^{var(\ell)}\}$.

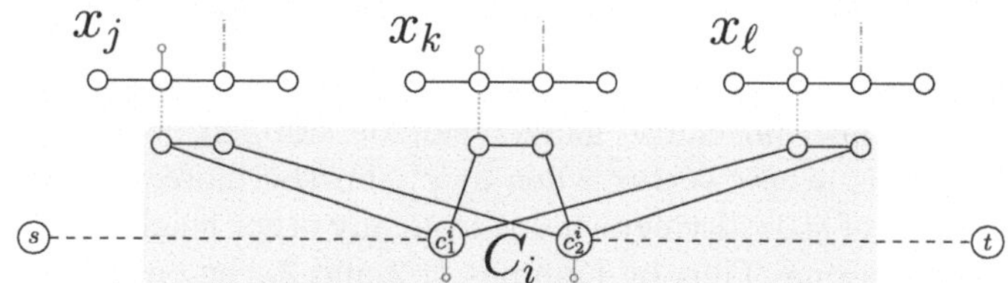

Fig. 2. Clause gadget for the reduction from TQBF GAME to ONLINE S-T PATH GAME. The dashed lines indicate that there might be more gadgets in between. Only parts of the variable gadgets are shown.

Lemma 3 (⋆). *If at least one variable satisfies C_i in the assignment chosen by the online algorithm, it can find a path from c_1^i to c_2^i. Otherwise, the online algorithm loses the game. Furthermore, it cannot use the variable gadgets to enter a not yet visited clause gadget.*

The Complete Reduction. Given a TQBF GAME instance with formula φ containing variables X and clauses C we create an instance of ONLINE UNDIRECTED S-T PATH GAME as follows: For each variable and each clause, a gadget is created as described above, and the clause gadgets are connected to the variable gadgets depending on the variables they contain. Two distinct vertices s and t are created, and the algorithm has to find a path from s to t. The edge $\{s, u_1^1\}$ as well as the edges $\{u_7^i, u_1^{i+1}\}$ for $i \in \{1, \dots, |X|-1\}$ are added. Further, the edges $\{u_7^{|X|}, c_1^1\}$, $\{c_2^i, c_1^{i+1}\}$ for $i \in \{1, \dots, |C|-1\}$ and $\{c_2^{|C|}, t\}$ are introduced. Finally, for $i \in \{1, \dots, |C|-1\}$, the edge $\{c_1^i, c_1^{i+1}\}$ is added (purple and loosely dotted in Fig. 3) as well as the edge $\{c_1^{|C|}, t\}$. This is necessary, since otherwise the next vertex, the algorithm is supposed to choose when at c_2^i, is indistinguishable from the sink attached to it. An example of this construction can be seen in Fig. 3. With this, we can now prove Theorem 2.

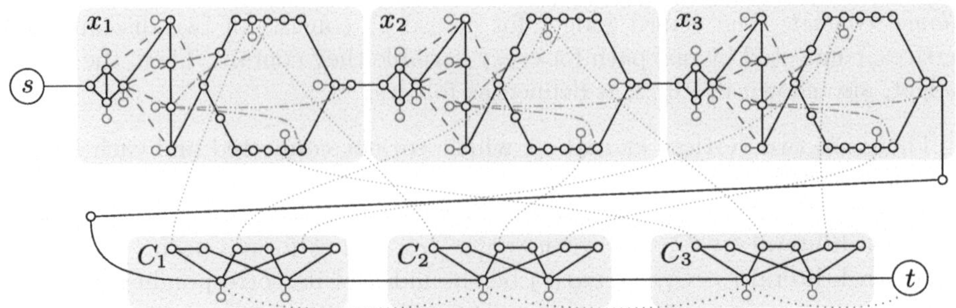

Fig. 3. Sketch of the full construction of the reduction from TQBF GAME to ONLINE UNDIRECTED S-T PATH GAME with the formula $\exists x_1 \forall x_2 \exists x_3 (x_1 \vee x_2 \vee x_3) \wedge (x_1 \vee \overline{x}_2 \vee \overline{x}_3) \wedge (\overline{x}_1 \vee x_2 \vee x_3)$.

Proof (of Theorem 2). Lemma 3 still holds when attaching the additional edges to all clauses and also no clause gadget can be skipped, since c_1^i has a sink attached and c_1^{i+1} is a new vertex when at c_1^i. Furthermore, the neighbor c_1^{i+1} (resp. t for $i = |C|$) of c_2^i is distinguishable from any other neighbor of c_2^i, because it is the only known one. Thus by Lemmas 1, 2 and 3, the online algorithm can find a path from s to t in the graph described above if and only if the TQBF GAME instance has a winning strategy for the $\exists$-player. All our gadgets have constant size. Therefore, our reduction runs in polynomial time, and the claim follows.

3.1 Relaxing the Path Constraint

The goal of this subsection is to analyze all variants of online search sequence games based on s-t path. The variants span over three dimensions: the constraint on the search sequence (path, trail, walk), a cost function on the edges (no costs or unit costs), the underlying graph (directed or undirected).

We begin with the easiest variant, the undirected walk without costs. This problem degenerates to the *LOGSPACE*-complete UNDIRECTED S-T CONNECTIVITY [22] because if the vertices s and t are connected, the online algorithm can explore the complete graph and finally reach t.

All other variants are *PSPACE*-complete. We reuse the presented reduction from TQBF GAME to ONLINE UNDIRECTED S-T PATH GAME and extend it by using different techniques to derive the results for the other variants. For the ONLINE UNDIRECTED S-T TRAIL GAME without costs, we replace every vertex p_j^i in the variable gadget with two vertices p_j^i and q_j^i, connected by an edge $\{p_j^i, q_j^i\}$, with the sinks attached to q_j^i instead of p_j^i. The edges $\{u_3^i, p_1^i\}$, $\{u_4^i, p_j^i\}$ remain unchanged, for $j \in \{1, 2, 3, 4\}$. The edges $\{p_j^i, p_{j+1}^i\}$ are replaced by $\{q_j^i, p_{j+1}^i\}$, for $j \in \{1, 2, 3\}$. All remaining edges starting in p_j^i start in q_j^i instead, for $j \in \{1, 2, 3, 4\}$. This change prevents the algorithm from visiting u_4^i and thus u_5^i multiple times, such that x_i cannot be set to both true and false. Thus, the algorithm needs to travel along the zig-zag path over the p_j^i and q_j^i.

From a slight modification of this reduction, we are also able to deduce the *PSPACE*-completeness of Online Undirected s-t Walk Game with costs. The problem is that the online algorithm is able to leave the sinks once they are visited and is potentially able to go along the colored edges from Fig. 1. Thus, we need to ensure that the algorithm receives large costs, whenever a sink is used. For this, let ℓ be the length of the s-t path in the reduction. We first duplicate all sinks by introducing ℓ additional sinks to all vertices that have a sink. Secondly, we assign every edge costs of 1. Thus, if the algorithm tries to use a colored edge, the adversary can force it to visit all sinks before that.

Table 1. The complexity of online games based on s-t-connectivity problems.

<table>
<tr><th></th><th></th><th>no costs</th><th>unit costs</th></tr>
<tr><td rowspan="2">Path</td><td>undirected</td><td rowspan="2">PSPACE-complete</td><td rowspan="2">PSPACE-complete</td></tr>
<tr><td>directed</td></tr>
<tr><td rowspan="2">Trail</td><td>undirected</td><td rowspan="2">PSPACE-complete</td><td rowspan="2">PSPACE-complete</td></tr>
<tr><td>directed</td></tr>
<tr><td rowspan="2">Walk</td><td>undirected</td><td>LOGSPACE-complete</td><td rowspan="2">PSPACE-complete</td></tr>
<tr><td>directed</td><td>PSPACE-complete</td></tr>
</table>

Directed Graphs. The results for the directed variants follow from the corresponding undirected variant. For this, we modify the reduction such that the edges are directed in the direction of travel and exploration. That is, all edges in Fig. 1 are mapped to arcs that go from left to right. The edge $\{u_2^i, u_3^i\}$ is mapped to two arcs (u_2^i, u_3^i) and (u_3^i, u_2^i) and the edges from $\{p_k^i, p_{k+1}^i\}$ are replaced by arcs (p_k^i, p_{k+1}^i) for $k \in \{1, 2, 3\}$. The clause gadgets are directed in the direction of travel as well. The resulting graph is a directed acyclic graph up to the cycles induced by (u_2^i, u_3^i) and (u_3^i, u_2^i). These cycles, however, are never a subwalk of the resulting walk because it is optimal to use exactly one of the arcs once. Thus, every walk or trail in the graph is also a path because it is impossible to go back.

Overview. All our results for online games based on s-t-connectivity problems are summarized in Table 1. All *PSPACE*-completeness results follow from the completeness of Online Undirected s-t Walk Game with costs, Online Undirected s-t Trail Game without costs and Online Directed s-t Walk Game without costs.

4 Hamiltonian Problems

The second class of problems are Hamiltonian problems, which can be interpreted as decision variants of Graph Exploration problems. These problems ask for a search sequence visiting all vertices of the graph. The corresponding cost variant is the Online Travelling Salesman Game.

For this class of problems, we are also able to construct reductions from TQBF Game. However, we need to use other techniques due to the lack of sinks. Sinks are not possible because the online algorithm has to visit all vertices and if a sink is introduced into the instance, the instance is a trivial NO-instance. In order to circumvent this problem, the adversary needs to force the online algorithm to cut itself off by earlier visited vertices.

For the reduction, we reuse the reduction from Satisfiability to Hamiltonian Path by Arora and Barak [1]. Again, we introduce additional edges to reveal the way through the gadgets such that the online algorithm is able to find a path satisfying the given TQBF formula. With slight modifications, we can also apply these results to other variants of the problem. In Table 2, our results for online games based on Hamiltonian path problems are presented. The fields with ? remain unanswered.

Table 2. The complexity of different Online Hamiltonian Path Games.

<table>
<tr><th></th><th></th><th>no costs</th><th>unit costs</th></tr>
<tr><td rowspan="2">Path</td><td>undirected</td><td rowspan="2">PSPACE-complete</td><td rowspan="2">PSPACE-complete</td></tr>
<tr><td>directed</td></tr>
<tr><td rowspan="2">Trail</td><td>undirected</td><td rowspan="2">?</td><td rowspan="2">PSPACE-complete</td></tr>
<tr><td>directed</td></tr>
<tr><td rowspan="2">Walk</td><td>undirected</td><td>LOGSPACE-complete</td><td rowspan="2">PSPACE-complete</td></tr>
<tr><td>directed</td><td>?</td></tr>
</table>

5 Additional Problems

There are many problems that are closely related to the s-t path problem or the Hamiltonian path problem. Often, they only add simple constraints on the solution. Our previously presented reductions can handle most of these variations by extending them with a small construction or setting a value, representing the constraint, to a specific number.

Thus, we are able to show that the online game versions of the following problems related to s-t-path are also *PSPACE*-complete: Path With Forbidden Pairs, Constrained Shortest Path, Two Disjoint Path, Vertex Disjoint Path as defined in [15]. In general, online games that require to compute an s-t path, optionally with additional constraints, are *PSPACE*-hard as one can drop the additional constraints on the s-t path. For the Hamiltonian path problems, we are able to show that the online game versions of the following problems are *PSPACE*-complete: Stacker-Crane, Rural Postman, Metric TSP, Bottleneck TSP, Longest Path, Longest Cycle as defined in [15].

6 Conclusion

Graph Exploration and Treasure Hunt are interpretable as the online versions of classical s-t path and Hamiltonian path problems. We modeled the Graph Exploration and Treasure Hunt problems with an unlabeled map as online games between the online algorithm and the adversary to obtain decision versions of these problems. Furthermore, we analyzed them from a complexity theoretic perspective and showed that nearly all are *PSPACE*-complete.

It remains open whether the approximation of the discussed problems is *PSPACE*-hard. Another interesting question is the complexity of the existence of undirected online Hamiltonian trails and walks. Additional path problems may be analyzed as well. For example, one may find a path problem that is not directly reducible via the ONLINE S-T PATH GAME because its constraints do not allow a standard s-t path to be a solution.

Besides these open problems concerning graph exploration problems, the online version of other typical combinatorial problems may be analyzed such as PARTITION, SCHEDULING or MATCHING.

References

1. Arora, S., Barak, B.: Computational Complexity - A Modern Approach. Cambridge University Press (2009). http://www.cambridge.org/catalogue/catalogue.asp?isbn=9780521424264
2. Baligács, J., Disser, Y., Heinrich, I., Schweitzer, P.: Exploration of graphs with excluded minors. In: Gørtz, I.L., Farach-Colton, M., Puglisi, S.J., Herman, G. (eds.) 31st Annual European Symposium on Algorithms, ESA 2023, 4–6 September 2023, Amsterdam, The Netherlands. LIPIcs, vol. 274, pp. 11:1–11:15. Schloss Dagstuhl - Leibniz-Zentrum für Informatik (2023). https://doi.org/10.4230/LIPIcs.ESA.2023.11
3. Bar-Noy, A., Schieber, B.: The Canadian traveller problem. In: Aggarwal, A. (ed.) Proceedings of the Second Annual ACM/SIGACT-SIAM Symposium on Discrete Algorithms, pp. 261–270. ACM/SIAM (1991). http://dl.acm.org/citation.cfm?id=127787.127835
4. Böckenhauer, H., Fuchs, J., Unger, W.: Exploring sparse graphs with advice. Inf. Comput. **289**(Part), 104950 (2022). https://doi.org/10.1016/j.ic.2022.104950
5. Böhm, M., Veselý, P.: Online chromatic number is PSPACE-complete. In: Mäkinen, V., Puglisi, S.J., Salmela, L. (eds.) IWOCA 2016. LNCS, vol. 9843, pp. 16–28. Springer, Cham (2016). https://doi.org/10.1007/978-3-319-44543-4_2
6. Bouchard, S., Labourel, A., Pelc, A.: Impact of knowledge on the cost of treasure hunt in trees. Networks **80**(1), 51–62 (2022). https://doi.org/10.1002/net.22075
7. Dessmark, A., Pelc, A.: Optimal graph exploration without good maps. Theor. Comput. Sci. **326**(1–3), 343–362 (2004). https://doi.org/10.1016/j.tcs.2004.07.031
8. Dobrev, S., Královič, R., Markou, E.: Online graph exploration with advice. In: Even, G., Halldórsson, M.M. (eds.) SIROCCO 2012. LNCS, vol. 7355, pp. 267–278. Springer, Heidelberg (2012). https://doi.org/10.1007/978-3-642-31104-8_23
9. Duncan, C.A., Kobourov, S.G., Kumar, V.S.A.: Optimal constrained graph exploration. ACM Trans. Algorithms **2**(3), 380–402 (2006). https://doi.org/10.1145/1159892.1159897

10. Eberle, F., Lindermayr, A., Megow, N., Nölke, L., Schlöter, J.: Robustification of online graph exploration methods. In: Thirty-Sixth Conference on Artificial Intelligence, AAAI, Thirty-Fourth Conference on Innovative Applications of Artificial Intelligence, IAAI, The Twelveth Symposium on Educational Advances in Artificial Intelligence, EAAI, pp. 9732–9740. AAAI Press (2022). https://ojs.aaai.org/index.php/AAAI/article/view/21208
11. Förster, K.-T., Wattenhofer, R.: Directed graph exploration. In: Baldoni, R., Flocchini, P., Binoy, R. (eds.) OPODIS 2012. LNCS, vol. 7702, pp. 151–165. Springer, Heidelberg (2012). https://doi.org/10.1007/978-3-642-35476-2_11
12. Fraenkel, A.S., Goldschmidt, E.: PSPACE-hardness of some combinatorial games. J. Comb. Theory Ser. A **46**(1), 21–38 (1987). https://doi.org/10.1016/0097-3165(87)90074-4
13. Fuchs, J., Grüne, C., Janßen, T.: The complexity of graph exploration games. CoRR abs/2302.08420 (2023). https://doi.org/10.48550/ARXIV.2302.08420
14. Fuchs, J., Grüne, C., Janßen, T.: The complexity of online graph games. In: Fernau, H., Gaspers, S., Klasing, R. (eds.) SOFSEM 2024. LNCS, vol. 14519, pp. 269–282. Springer, Cham (2024).https://doi.org/10.1007/978-3-031-52113-3_19
15. Garey, M.R., Johnson, D.S.: Computers and Intractability: A Guide to the Theory of NP-Completeness. W. H. Freeman (1979)
16. Kalyanasundaram, B., Pruhs, K.: Constructing competitive tours from local information. Theor. Comput. Sci. **130**(1), 125–138 (1994). https://doi.org/10.1016/0304-3975(94)90155-4
17. Komm, D., Královič, R., Královič, R., Smula, J.: Treasure hunt with advice. In: Scheideler, C. (ed.) SIROCCO 2014. LNCS, vol. 9439, pp. 328–341. Springer, Cham (2015). https://doi.org/10.1007/978-3-319-25258-2_23
18. Li, C., McCormick, S.T., Simchi-Levi, D.: The complexity of finding two disjoint paths with min-max objective function. Discret. Appl. Math. **26**(1), 105–115 (1990). https://doi.org/10.1016/0166-218X(90)90024-7
19. Megow, N., Mehlhorn, K., Schweitzer, P.: Online graph exploration: new results on old and new algorithms. Theor. Comput. Sci. **463**, 62–72 (2012). https://doi.org/10.1016/j.tcs.2012.06.034
20. Panaite, P., Pelc, A.: Impact of topographic information on graph exploration efficiency. Networks **36**(2), 96–103 (2000)
21. Papadimitriou, C.H., Yannakakis, M.: Shortest paths without a map. In: Ausiello, G., Dezani-Ciancaglini, M., Della Rocca, S.R. (eds.) ICALP 1989. LNCS, vol. 372, pp. 610–620. Springer, Heidelberg (1989). https://doi.org/10.1007/BFb0035787
22. Reingold, O.: Undirected connectivity in log-space. J. ACM **55**(4), 17:1–17:24 (2008). https://doi.org/10.1145/1391289.1391291
23. Stockmeyer, L.J., Meyer, A.R.: Word problems requiring exponential time: Preliminary report. In: Aho, A.Vet al. (eds.) Proceedings of the 5th Annual ACM Symposium on Theory of Computing, pp. 1–9. ACM (1973). https://doi.org/10.1145/800125.804029

A SUBSET-SUM Characterisation of the A-Hierarchy

Jan Gutleben(✉) and Arne Meier

Institut für Theoretische Informatik, Leibniz Universität Hannover, Appelstrasse 9a, 30167 Hannover, Germany
{gutleben,meier}@thi.uni-hannover.de.de

Abstract. The A-hierarchy is a parametric analogue of the polynomial hierarchy in the context of parameterised complexity theory. We give a new characterisation of the A-hierarchy in terms of a generalisation of the SUBSET-SUM problem.

Keywords: Parameterized Complexity Theory · A-Hierarchy · SUBSET-SUM · Alternation · Model Checking

1 Introduction

Classical worst-case complexity relates the difficulty of a problem in general to its input length. In particular, when dealing with intrinsically hard problems, the input length is a very coarse measure and does not tell much about the "structure" of the problem. That is why parameterised complexity theory was introduced [3]. Here, one considers the complexity of a problem with respect to a *parameter* and aims for a more fine-grained complexity analysis. Such parameterised problems can be formalised as tuples (Q, κ) where Q is a decision problem and κ is a polynomial-time computable function, the *parameterisation*. This function maps inputs to natural numbers that are seen as the parameter of the problem. For instance, the parameter could be the size of a vertex cover in a graph or the number of variables in a formula. Depending on the chosen parameter, the runtime to solve the problem can now be formulated as a function of the parameter *and* the input length. Of course it is important to study parameters that are relevant to practical scenarios; the number of variables in a formula is probably not such a good choice, as it is usually neither constant nor growing slowly with the input length.

A problem is said to be fixed-parameter tractable (FPT) if it can be solved in time $f(\kappa(x)) \cdot n^{O(1)}$ for all inputs x with $|x| = n$ for some computable function f. There exist different hierarchies above FPT, such as the W- and the A-hierarchy. While the hardness of base classes within these hierarchies indicates that a problem is intractable in the parameterized complexity framework, membership in FPT is considered a sign of tractability. The W-hierarchy is defined in terms of weighted circuit satisfiability problems and *weft*, which is the

R. Královič and V. Kůrková (Eds.): SOFSEM 2025, LNCS 15539, pp. 31–44, 2025.
https://doi.org/10.1007/978-3-031-82697-9_3

maximum number of large gates from an input to the output of a circuit [5]. The initial definition of the A-hierarchy was given via a short halting problem of alternating single-tape Turing machines [6]. Yet, further machine characterisations are known [1]. The bridge to predicate logic through the model-checking problem for Σ_ℓ-formulas (predicate formulas in prenex normal form with ℓ alternations) with bounded arity relations was established by Flum and Grohe [6]. The next step to unbounded arity relations is also due to Flum and Grohe [7] (see, in Sect. 2, Lemma 2). From this perspective, the A-hierarchy serves as a direct parametric analogue of the polynomial hierarchy, characterised through model checking problems [8, Prop. 4.28]. That is a reason why this hierarchy is also of independent interest as it bridges the gap between classical complexity theory and parameterised complexity theory.

Contributions. In this study, we present a new characterisation of the A-hierarchy based on generalised SUBSET-SUM problems. First, we will demonstrate that a specific variant of the SUBSET-SUM problem is complete for the third level of the A-hierarchy. Furthermore, we will generalise this variant to identify complete problems for each level of the A-hierarchy. The advantage of this characterization is that these variants of the SUBSET-SUM problem are more intuitive than generic machine problems or model checking problems in predicate logic.

Organisation. We start with a brief introduction of predicate logic and parameterised complexity theory. Afterwards, we define alternating random access machines (ARAMs) and tail-nondeterminism used for characterisations of the A-hierarchy. In the main part, we will give a thorough proof of A[3]-completeness of a certain variant of the SUBSET-SUM problem. Finally, we will generalise this problem to arbitrary alternations and explain how the previous proof can be adapted to show completeness for every level of the A-hierarchy. We finish with a summary and an outlook.

2 Preliminaries

We assume basic familiarity with the concepts of computational complexity theory [10,11], e.g., the polynomial hierarchy and reductions.

Predicate Logic. We will give a brief introduction into first-order logic and assume familiarity with basic concepts [2]. Here, let τ be a *first-order vocabulary* consisting of relation symbols and an equality symbol '='. Denote by VAR a countably infinite set of *first-order variables*. Terms over τ are defined as usual. The set of first-order logic (FO) is defined as

$$\psi ::= t_1 = t_2 \mid R(t_1, \ldots, t_k) \mid \neg R(t_1, \ldots, t_k) \mid \psi \wedge \psi \mid \psi \vee \psi \mid \exists x \psi \mid \forall x \psi,$$

where t_i are terms for $1 \leq i \leq k$, R is a k-ary relation symbol from τ, $k \in \mathbb{N}$, and $x \in \mathrm{VAR}$. Let us denote by $\mathrm{VAR}(\psi)$ the *set of variables* of a formula ψ and by $\mathrm{Fr}(\psi)$ the set of *free variables* of ψ. We call expressions of the form $R(t_1, \ldots, t_k)$ and $t_1 = t_2$ *atoms*. Regarding semantics, we consider FO-formulas in the context of τ*-structures*. An interpretation $\mathcal{A} = (A, \tau^{\mathcal{A}})$ consists of a set A which serves as the *domain* of $\mathcal{A}$ (and, when context allows, we refer to it simply as A instead of $\mathrm{dom}(\mathcal{A})$) and $\tau^{\mathcal{A}}$, which provides an interpretation of the symbols in τ. We say an interpretation $\mathcal{A}$ *models* a formula ψ, $\mathcal{A} \models \psi$ in symbols, if ψ evaluates to true under $\mathcal{A}$. The *model checking* problem in first-order logic is defined as

$$\mathrm{MC}(\Phi) := \{\, \langle \mathcal{A}, \varphi \rangle \mid \mathcal{A} \text{ is a structure}, \varphi \in \Phi \text{ and } \mathcal{A} \models \varphi \,\}.$$

A quantifier-free formula is in Σ_0 or also in Π_0. Let $t, n \in \mathbb{N}_0$. Then, we let for $\varphi \in \Pi_t$ and $\psi \in \Sigma_t$: $\exists x_1 \exists x_2 \cdots \exists x_n \varphi \in \Sigma_{t+1}$, and $\forall x_1 \forall x_2 \cdots \forall x_n \psi \in \Pi_{t+1}$. If Φ is a class of formulas, then $\Phi[r]$ is the subclass of formulas in Φ that contain relations of arity of most r. Furthermore, a Σ_t-formula is *simple* if its quantifier-free part is a conjunction of atoms if t is odd, and a disjunction of atoms if t is even. We then denote by simple-Σ_t the restriction to only simple Σ_t-formulas, and simple-Σ_t^+ to those that are also negation-free.

Parameterised Complexity Theory. For an introduction to the field of parameterised complexity theory, we refer the reader to the textbook by Flum and Grohe [8] or the textbooks by Downey and Fellows [3,4].

A *parameterisation* w.r.t. an alphabet Σ is a function $\kappa\colon \Sigma^* \to \mathbb{N}^+$ which is computable in polynomial time. A *parameterised problem* is a pair (Q, κ), where $Q \subseteq \Sigma^*$ is a decision problem and κ is a parameterisation.

Definition 1. *Let Σ, Γ be alphabets, and (A, κ) and (B, ι) be parameterised problems with $A \subseteq \Sigma^*$ and $B \subseteq \Gamma^*$. Then a function $f\colon \Sigma^* \to \Gamma^*$ is called an* FPT-reduction, *written $(A, \kappa) \leq^{\mathrm{fpt}} (B, \iota)$, if the following is true for all $x \in \Sigma^*$:*

- $x \in A \Leftrightarrow f(x) \in B$
- *There is a computable function g and a polynomial p such that f can be computed in $g(\kappa(x)) \cdot p(x)$ steps.*
- *There exists a computable function $h\colon \mathbb{N}^+ \to \mathbb{N}^+$, such that $\iota(f(x)) \leq h(\kappa(x))$.*

If (Q, κ) is a parameterised problem, then we write $[(Q, \kappa)]^{\mathrm{fpt}}$ for the class of all parameterised problems (Q', κ') such that $(Q', \kappa') \leq^{\mathrm{fpt}} (Q, \kappa)$ and call it the *fpt-closure* of (Q, κ) The parameterised model checking problem then is

$$\text{p-MC}(\Sigma_t) := (\mathrm{MC}(\Sigma_t), \langle \mathcal{A}, \varphi \rangle \mapsto |\varphi|).$$

Finally, we are ready to define the A-hierarchy. For all $t \in \mathbb{N}^+$ let $A[t] := [\text{p-MC}(\Sigma_t)]^{\mathrm{fpt}}$. The union of these classes, $\bigcup_{t \in \mathbb{N}} A[t]$, is called the *A-hierarchy*.

The following lemma states that for p-MC(Σ_t), one can assume the normal form of positive simple Σ_t-formulas.

Lemma 2 ([8, L. 8.10]). *For all* $t \in \mathbb{N}^+$, p-MC(Σ_t) $\leq^{\text{fpt}}$ p-MC(simple-$\Sigma_t^+[2]$).

The lemma above allows both unary and binary relations to appear. We will use a strengthened version of the previous lemma, where only binary relations occur in the formula (and not unary ones). This is achieved via simulating the unary relations by binary relations with a dummy variable.

ARAMs and Tail-Nondeterminism. We now introduce the necessary notions to characterise the class A[t] in terms of ARAMs. ARAMs are a generalisation of RAMs that allow for nondeterministic behaviour. First, we start with classical RAMs and follow the notion of Flum and Grohe [8, pp. 457]. A *Random Access Machine* (RAM) consists of countable infinite many registers $R_0, R_1 \dots$, a finite sequence of instructions $I_1, \dots I_n$, and a program counter that contains a number in $\mathbb{N}^+$. Registers may contain numbers in $\mathbb{N}_0$. The register R_0 is called the *accumulator*. We allow arithmetic instructions: ADD, SUB (cut off at 0), or DIV2 (rounded off) with the usual semantics (notice that arithmetic instructions are executed in $O(1)$ time). The first registers serve for inputs ($x = x_1 \dots x_n$ is in $R_1, \dots, R_n$) and outputs. Acceptance/rejection for decision problems is then via storing 1/0 in R_0. The runtime of algorithms on such machines is then measured with respect to length of input and the number of instructions carried out (no matter how large the involved numbers are).

Definition 3. *An* alternating Random Access Machine *(ARAM) is a RAM with two additional GUESS instructions: EXISTS and FORALL. An ARAM* accepts *an input if for every FORALL instruction the machine accepts on every possible number equal or less than the number in the accumulator; and for every EXISTS instruction there is a number on which the machine accepts.*

We say that an ARAM is 1-*alternating* if there is only one kind of GUESS instructions. We say it is t-*aternating* if there are at most $t-1$ switches between EXISTS and FORALL instructions in every run.

Definition 4. *For a parameterisation* $\kappa\colon \Sigma^* \to \mathbb{N}^+$, *an ARAM is* κ-bounded *if there are computable functions* f, g *and a polynomial* p *such that for the ARAM on all inputs* $x \in \Sigma^*$ *the following is true.*

- *The program needs at most* $f(\kappa(x)) \cdot p(|x|)$ *steps, of which* $g(\kappa(x))$ *are non-deterministic.*
- *The program uses at most the first* $f(\kappa(x)) \cdot p(|x|)$ *registers.*
- *The program only uses numbers less than or equal to* $f(\kappa(x)) \cdot p(|x|)$.

Furthermore, there is a notion of tail-nondeterminism for ARAMs.

Definition 5. *Let* κ *be a parameterisation. A* κ*-bounded ARAM program is called* tail-nondeterministic *if there is a computable function* h *such that the nondeterministic steps of the program are always in the last* $h(\kappa(x))$ *steps.*

Now, we are ready to state a characterisation of the classes of the A-hierarchy in terms of ARAMs.

Theorem 6 (**[8, Thm. 8.8]**). *Let $t \in \mathbb{N}^+$ and (Q, κ) be a parameterised problem. Then $(Q, \kappa) \in A[t]$ if and only if there is a tail-nondeterministic κ-bounded t-alternating ARAM that decides (Q, κ).*

3 Generalised SUBSET-SUM and the A-Hierarchy

In the following, we will consider a variant of the well-known SUBSET-SUM problem. Classically this problem is known to be NP-complete [9]. As a first step, we will give a definition of a variant of the classical problem that is complete for the third level of the A-hierarchy. Afterwards we will generalise this problem to show completeness for every level of the A-hierarchy. We decided to skip the details for the first level as the insights from the construction for the third level better show the interplay between universal and existential quantification. The result for the first level is then a restriction of the presented construction. The problem $\mathrm{ALT_3SUB}$ is defined as follows:

$$\left\{ \langle A_1, A_2, A_3, k, l, m, t\rangle \;\middle|\; \begin{array}{l} A_1, A_2, A_3 \subseteq \mathbb{N}_0,\ k, l, m, t \in \mathbb{N}_0,\ \exists A_1' \subseteq A_1 \\ \text{with } |A_1'| = k,\ \forall A_2' \subseteq A_2 \text{ with } |A_2'| = l, \\ \exists A_3' \subseteq A_3 \text{ with } |A_3'| = m, \text{ such that } \sum_{a \in A_1'} a + \\ \sum_{a \in A_2'} a + \sum_{a \in A_3'} a = t \end{array} \right\},$$

Example 7. Consider $\langle \{0, 3\}, \{1, 2\}, \{2, 3\}, 1, 1, 1, 7\rangle$ as an instance of $\mathrm{ALT_3SUB}$. If this instance is in $\mathrm{ALT_3SUB}$, than there exists a subset of $\{0, 3\}$ with size 1 such that for all subsets of $\{1, 2\}$ with size 1 there exists a subset of $\{2, 3\}$ with size 1 such that all the chosen subsets sum to 7. If $\{0\}$ is selected from $\{0, 3\}$, then every possible sum is smaller as 7. Hence, $\{3\}$ must be chosen. Now every subset with size 1 of $\{1, 2\}$ must be investigated. If $\{1\}$ is chosen, out of the last set $\{3\}$ can be selected. The sum in this case is $3 + 1 + 3 = 7$. The only other subset with a correct size is $\{2\}$. In this case, $\{2\}$ can be chosen from the last set. Since $3 + 2 + 2 = 7$, the chosen sets sum up correctly.

We define $\text{p-ALT}_3\text{SUB} := (\mathrm{ALT_3SUB}, \langle A_1, k, A_2, l, A_3, m, t\rangle \mapsto k + l + m)$ to be the parameterised version of $\mathrm{ALT_3SUB}$, and show that it is A[3]-complete.

Theorem 8. $\text{p-ALT}_3\text{SUB}$ *is* A[3]*-complete.*

We split the proof of the result into the following two lemmas.

Lemma 9. $\text{p-ALT}_3\text{SUB}$ *is in* A[3].

Proof. We give an ARAM program deciding membership. It uses $k+l+m$ registers for the numbers in the sets A_1, A_2, A_3. After these registers, the following ones store the values of $k, l, m, |A_1|, |A_2|, |A_3|$. Then, we need a register `sum` for the respective sum of the subsets. After that register, we use separate registers for the natural numbers in the three sets. The nondeterministic instructions FORALL/EXISTS allow us to guess the required subsets by guessing indices after loading the "offset" $|A_i|$ from the corresponding register into R_0. These numbers are added up one by one into the `sum` register and compared to t.

Regarding the runtime of the ARAM program, after preparing the registers in $O(n)$ steps, guessing of the subsets in $k+l+m$ steps as well as adding up (again $k+l+m$ steps) and comparing with t can be overall done in $O(k+l+m)$ steps. Clearly, the program is 3-alternating and $(k+l+m)$-bounded. Hence, by Theorem 6, we have that p-ALT$_3$SUB $\in$ A[3]. □

The idea of the following lemma is to thoroughly construct numbers over a particular basis such that satisfaction of the model-checking instance formula corresponds to summing up numbers reaching a target sum. In the course of defining these numbers we have to be careful in two ways. Firstly, we need to ensure that no overflow can occur. Secondly, we need to guarantee that, depending on the quantifiers in the formula, there are always choices for the "universal" set. We suggest to consult Fig. 1 first to obtain a vague idea of the numbers used.

Lemma 10. p-ALT$_3$SUB *is* A[3]*-hard.*

Proof. We will show a reduction from p-MC(simple-$\Sigma_3[2]$). For that let $\langle \mathcal{A}, \varphi \rangle$ be an instance of p-MC(simple-$\Sigma_3[2]$). Hence, $\mathcal{A} = (A, \tau)$, where τ only contains binary relations. Furthermore, for $x_{i,j} \in \{x_1, \dots, x_{k+l+m}\}$, we have that

$$\varphi = \exists x_1 \dots \exists x_k \forall x_{k+1} \dots \forall x_{k+l} \exists x_{k+l+1} \dots \exists x_{k+l+m} \bigwedge_{i=1}^{n} \lambda_i(x_{i,1}, x_{i,2}),$$

with λ_i is an atom $\lambda_i = R(x_{i,1}, x_{i,2})$ for some relation $R \in \tau$. W.l.o.g., assume a fixed bijection $I \colon A \to [1, |A|]$ which refers to the index of an element in A.

Note that all numbers defined in the following are in some base D. This base will be later chosen large enough avoiding overflows. Almost all numbers will be of the form $L_1 \dots L_n\ B_1 \dots B_{k+l+m}$ and the rightmost digit is the least significant bit. You can see a summary of the numbers used in the proof in Fig. 1. They will start with n digits, abbreviated by $L_1, \dots, L_n$ (the only exception will be particularly defined numbers at the end, they will start with a larger block of other digits). These digits will yield information on the λ_i's. Succeeding these digits, there will be $k+l+m$ blocks B_j of each $k+l+m+1$ digits. These blocks will contain information on the variables x_j. Altogether, these numbers have $n + (k+l+m+1) \cdot (k+l+m)$ digits.

We will now define the sets A_1, A_2, A_3. We need to encode that a variable is assigned a value from the universe. For that purpose we define for all $a \in A$ and

	L_1	$\cdots L_i \cdots$	L_n	B_1	$\cdots B_{j'} \cdots$	B_j: $B_j^1 \cdots B_j^j \cdots B_j^{j'}$	$\cdots$	$B_j^{k+l+m+1}$	$\cdots$	B_{k+l+m}
					$B_{j'}^j$					
$\mathrm{VAR}(a, x_j) =$	0	$\cdots$	0	0	$\cdots$	$0 \cdots I(a) \cdots 0$	$\cdots$	0	$\cdots$	0
$\mathrm{ATOM}(a, b, x_j, x_{j'}) =$	0	$\cdots 1 \cdots$	0	0	$I(b)$	$0 \cdots 0 \cdots I(a)$	$\cdots$	0	$\cdots$	0
$\mathrm{NORM}(a, x_j) =$	0	$\cdots$	0	0	$\cdots 0$	$(l+1) \cdot \lvert A \rvert$	$\cdots (l+1) \cdot \lvert A \rvert$	1	$0 \cdots$	0
$\mathrm{FIX}(x_j) =$	1	$\cdots$	1	0	$\cdots 0$	$(l+1) \cdot \lvert A \rvert$	$\cdots (l+1) \cdot \lvert A \rvert$	1	$0 \cdots$	0
$\mathrm{FIX}(x_j, d) =$	0	$\cdots$	0	0	$\cdots 0$	$(l+1) \cdot \lvert A \rvert$	$\cdots (l+1) \cdot \lvert A \rvert - d$	1	$0 \cdots$	0

	$1 \cdots r \cdots s$	$L_1 \cdots L_n$	B_1	$\cdots$	B_{k+l+m}
$\mathrm{WAIT}(r) =$	$0 \cdots 1 \cdots 0$	$0 \cdots 0$	0	$\cdots$	0
$\mathrm{NOWAIT} =$	$1 \cdots 1 \cdots 1$	$0 \cdots 0$	0	$\cdots$	0
$t' =$	$1 \cdots 1 \cdots 1$	$1 \cdots 1$	$(l+1) \cdot \lvert A \rvert \cdots (l+1)\lvert A \rvert 1$	$\cdots$	$(l+1) \cdot \lvert A \rvert \cdots (l+1)\lvert A \rvert 1$

Fig. 1. Overview of used numbers in the proof of Theorem 8. Here, $a, b \in A, 1 \leq i \leq n$, $(a, b) \models \lambda_i(x_j, x_{j'})$, $d \in [0, l \cdot |A|]$, $1 \leq j' < j \leq k+l+m$, $1 \leq r \leq s$. By design is $j' \leq j$ for all ATOM-numbers. In case of $j' = j$ the Block B_j has only zeros. Digit B_i^j is the jth digit in block i.

$x_j \in \{x_1, \ldots, x_{k+l+m}\}$ the number $\mathrm{VAR}(a, x_j) := B_1 B_2 \ldots B_{k+l+m}$. Here, the block B_j has the value $I(a)$ at the j-th position, all other positions are 0. These numbers are then distributed into sets VAR_1, VAR_2, and VAR_3 as follows:

- $j \in [1, k] \Rightarrow \mathrm{VAR}(a, x_j) \in \mathrm{VAR}_1$
- $j \in [k+1, k+l] \Rightarrow \mathrm{VAR}(a, x_j) \in \mathrm{VAR}_2$
- $j \in [k+l+1, k+l+m] \Rightarrow \mathrm{VAR}(a, x_j) \in \mathrm{VAR}_3$

We need to ensure that k, l, m elements from the respective VAR-block are chosen. The correctness of Claim 1 follows from the definition of the VAR-numbers.

Claim 1. Let $\mathcal{J}$ be an assignment of $x_1, \ldots, x_{k+l+m}$ and $a_j := \mathcal{J}(x_j)$. Then

$$\sum_{j=1}^{k+l+m} \mathrm{VAR}(a_j, x_j) = B_1 B_2 \ldots B_{k+l+m},$$

where in B_j at position j the value is $I(a_j)$ and otherwise there are only zeros.

Next, we will encode the atomic formulas. For all $a, b \in A$ and $x_j, x_{j'} \in \{x_1, \ldots, x_{k+l+m}\}$ with $j \leq j'$, we define the number

$$\mathrm{ATOM}(a, b, x_j, x_{j'}) := L_1 \ldots L_n B_1 \ldots B_{k+l+m}.$$

For $i \in [1, n]$, L_i is 1 if and only if $x_j, x_{j'}$ are the variables in λ_i and $(a, b) \models \lambda_i$; otherwise $L_i = 0$ for all i. If $j \neq j'$, then all B-blocks are zero except for the j-th and j'-th. Block j has the value $I(a)$ at position j', block j' has the value $I(b)$ at position j and otherwise zeros. If $j = j'$ these values are already encoded in the VAR numbers. In this case all B-Blocks are zero. All these numbers are collected in the set ATOM. The following claim is about the sum corresponding to an assignment that fits to the ATOM- and VAR-numbers.

Claim 2. Let $\mathcal{J}$ be an assignment of $x_1 \dots x_{k+l+m}$. Then $\mathcal{J}$ satisfies the quantifier-free part of φ if and only if

$$\sum_{j=1}^{k+l+m} \left(\mathrm{VAR}(a_j, x_j) + \sum_{j'=j+1}^{k+l+m} \mathrm{ATOM}(a_j, a_{j'}, x_j, x_{j'}) \right) = \underbrace{1 \dots 1}_{n} B_1 \dots B_{k+l+m},$$

where for all $j \in [1, k+l+m]$:

$$B_j := \overbrace{I(\mathcal{J}(x_j)) \dots I(\mathcal{J}(x_j))}^{k+l+m} 0.$$

Proof (of Claim 2). The number is composed as follows. At the beginning it has exactly n ones if $\mathcal{J}$ satisfies the quantifier-free part of φ. The reason for this is that the only summand that can add a 1 at position L_i is $\mathrm{ATOM}(a_j, a_{j'}, x_j, x_{j'})$ if x_j and $x_{j'}$ are the variables in λ_i. There is a 1 added if and only if $(a_j, a_{j'}) \models \lambda_i$. For x_j, at each position $j' \neq j$ of the B_j block, the value $I(\mathcal{J}(x_j))$ was added by adding the values of $\mathrm{ATOM}(a_j, a_{j'}, x_j, x_{j'})$. According to Claim 1, the value $I(\mathcal{J}(x_j))$ is already added by the number $\mathrm{VAR}(a_j, x_j)$. The $(k+l+m+1)$-th position of each block is not increased by any number, so it remains at 0. ■

As it is not clear in advance which variables are assigned to which elements of A, we need to ensure that the sum t can always be reached. For that purpose, we introduce the numbers $\mathrm{NORM}(a, x_j) := B_1 \dots B_{k+l+m}$. For all $j \in [1, k+l+m]$, all B-blocks except B_j consist of zeros. The block B_j has the value $(l+1) \cdot |A| - I(a)$ except for the last position, where a 1 is placed. The set of these numbers is called NORM. Now, we define the target number t as

$$t := \underbrace{1 \dots 1}_{n} B_1 \dots B_{k+l+m}.$$

For all $j \in [1, k+l+m]$ we have that

$$B_j := \underbrace{(l+1) \cdot |A| \dots (l+1) \cdot |A|}_{k+l+m} 1.$$

The first n digits ensure that every atom is satisfied. The remaining $(k+l+m) \cdot (k+l+m+1)$ digits are blocks with the value $(l+1) \cdot |A|$ followed by a 1. The 1 ensures that no more than one NORM-number is chosen for the same variable. The next claim states that the sum t can be reached by choosing numbers from VAR, ATOM, and NORM. It follows by the construction of the numbers.

Claim 3. Let $\mathcal{J}$ be an assignment of $x_1, \dots, x_{k+l+m}$. Then $\mathcal{J}$ satisfies the quantifier-free part of φ if and only if

$$\sum_{j=1}^{k+l+m} \left(\mathrm{VAR}(a_j, x_j) + \mathrm{NORM}(a_j, x_j) + \sum_{j'=j+1}^{k+l+m} \mathrm{ATOM}(a_j, a_{j'}, x_j, x_{j'}) \right) = t.$$

The following claim states that it is impossible to reach the sum t by choosing subsets that are not according to Claim 3.

Claim 4. Let A'_1, A'_2, A'_3 be some chosen subsets. If there exists a variable x_j for which neither $VAR(a, x_j)$ nor $NORM(a, x_j)$ has been chosen then it is impossible to reach the sum t, if only numbers from VAR_1, VAR_2, VAR_3, $ATOM$ or $NORM$ have been chosen.

Proof (of Claim 4). Consider the overall sum. If neither the number $\text{VAR}(a, x_j)$ nor $\text{NORM}(a, x_j)$ has been chosen, then a 0 would be at position j of the B_j-block. This would in turn require two NORM-numbers to be added to reach $(l+1) \cdot |A|$. This is because the largest number that can be added by a NORM-number is $(l+1) \cdot |A| - 1$. This would produce a 2 at the last position of the B_j-block. If no $\text{NORM}(a, x_j)$ has been chosen, at the last position of the B_j-block would be a 0. Using only numbers from ATOM cannot solve this problem either, as they have at position j a 0. ■

As A_1 and A_3 are existentially quantified it is correct that subsets that do not correspond to a valid assignment cannot yield to the sum that is t. However for A_2 all possible subsets must be able to reach the sum t for a valid assignment. If a variable is assigned a value twice in A_2 (which is an invalid assignment but a correct subset), we add numbers to A_3 that allow to still reach the sum t. If one or more variables are not assigned a value in A_2 there must be a variable that has not been assigned as from VAR_2 exactly l numbers have to be chosen. We add numbers that set $L_1 \dots L_n$ to all ones in that case. For all $j \in [k+1, k+l]$, we define

$$\text{FIX}(x_j) := \underbrace{1 \dots 1}_{n} \; B_1 \dots B_{k+l+m}.$$

All B-blocks are zero except for block B_j. This block has the value $(l+1) \cdot |A|$ at every position except a 1 at the last.

Now we need a possibility to set every digit in every B-block to $(l+1) \cdot |A|$. For that purpose, for all $d \in [0, l \cdot |A|]$ and $j \in [1, k+l+m]$ we add the numbers

$$\text{FIX}(x_j, d) := B_1 \dots B_{k+l+m},$$

where all B-blocks consist of zeros except for block B_j, which has the value $(l+1) \cdot |A|$ at every position except a 1 at the last and at j the value $(l+1) \cdot |A| - d$. The set of these numbers is called FIX.

Claim 5. Let $A'_2 \subseteq VAR_2$ and assume that for an x_j no $VAR(a, x_j)$ is in A'_2. For every $j' \neq j$ there exists a $d_{j'}$, such that

$$\sum_{j'=1}^{k} \text{VAR}(a_{j'}, x_{j'}) + \sum_{a \in A'_2} a + \text{FIX}(x_j) + \sum_{j'=1, j \neq j'}^{k+l+m} \text{FIX}(x_{j'}, d_{j'}) = t.$$

Proof (of Claim 5). In this sum, every $L_i = 1$ because of $\text{FIX}(x_{j'})$. Consider the $B_{j'}$-block for $j' \in [1, k+l+m]$. If $j' = j$, by A_2' no number has been added to $B_{j'}$ and the addition of $\text{FIX}(x_{j'})$ ensures that in the block the correct number is written. If $j' \neq j$, between 0 and l numbers of the form $\text{VAR}(a, x_{j'})$ have been added because of A_2'. Thus at position j of the block a number from $[0, l \cdot |A|]$ is written. Choose d_j' for that value. Then the block has the desired form. ■

We let $A_1 = \text{VAR}_1$, $A_2 = \text{VAR}_2$, and $A_3 = \text{VAR}_3 \cup \text{ATOM} \cup \text{NORM} \cup \text{FIX}$. In order to choose subsets of the correct size, we need to select k numbers from VAR_1 and l numbers from VAR_2. For the numbers from A_3 we need to make a case distinction:

1. If a valid choice is made from A_2, then according to Claim 3 the choice for A_3 is as follows: First, m numbers from VAR are chosen. Then for every pair of variables a number from ATOM is chosen. This corresponds to $\frac{(k+l+m)\cdot(k+l+m-1)}{2}$ numbers. Finally, select $k+l+m$ numbers from NORM.
2. If an invalid choice is made from A_2, then according to Claim 5 the choice for A_3 also includes the FIX numbers. Note that in each case the same number of variables must be chosen to reach t. For that reason let the difference between this cases be

$$s = (m + \frac{(k+l+m)\cdot(k+l+m-1)}{2} + k + l + m) - (k + l + m).$$

Observe that s is greater than 0 for all $k, l, m \geq 0$. In the second case, we need some further auxiliary numbers such that in both cases the size of a chosen subset is the same. For all $i \in [0, s]$

$$\text{WAIT}(i) := 1\ \underbrace{0\ldots0\,0\ldots0\,0\ldots0}_{i+n+(k+l+m)\cdot(k+l+m+1)} \quad \text{NOWAIT} := \overbrace{1\ldots1}^{s+1}\ \underbrace{0\ldots0\,0\ldots0\,0\ldots0}_{n+(k+l+m)\cdot(k+l+m+1)}$$

Observe that the sum of the $\text{WAIT}(i)$ equals NOWAIT. These are the WAIT-numbers and are added to A_3. We need one last adjustment of t to

$$t' := \overbrace{1\ldots1}^{s+1} t.$$

The Size of the Basis. The size base D is intended to prevent overflows. We will now present out a worst-case analysis for a single digit to show how an overflow can be avoided. In the proof, we choose $2\cdot(k+l+m) + \frac{(k+l+m)\cdot(k+l+m-1)}{2} + 1$ numbers. The highest value for one digit is $(l+1)\cdot|A|$. This allows us setting

$$D = \left(2\cdot(k+l+m) + \frac{(k+l+m)\cdot(k+l+m-1)}{2} + 1\right)\cdot(l+1)\cdot|A| + 1.$$

The Reduction Function. We are ready to define the reduction function f. Let $f(\langle \mathcal{A}, \varphi \rangle)$ be defined as

$$\left\langle A_1, A_2, \tilde{A}_3, k, l, m + \frac{(k+l+m) \cdot (k+l+m-1)}{2} + k + l + m + 1, t' \right\rangle,$$

where $\tilde{A}_3 = A_3 \cup$ WAIT. We will now show that f is a correct reduction function. Every digit is computable in $O(|\langle \mathcal{A}, \varphi \rangle|)$ time. There are $O(|\varphi|^2)$ digits in a number so every number is computable in $O(|\langle \mathcal{A}, \varphi \rangle|^3)$. There are $O(|\langle \mathcal{A}, \varphi \rangle|^4)$ numbers computed, so f is computable in $O(|\langle \mathcal{A}, \varphi \rangle|^7)$. Let g be the computable function $g(x) = x^2 + 2 \cdot x + 1$. The new parameter is bounded by g:

$$m + \frac{(k+l+m) \cdot (k+l+m-1)}{2} + k + l + m + 1 \leq |\varphi|^2 + 2 \cdot |\varphi| + 1 = g(|\varphi|).$$

Finally, we turn towards the correctness property.

Claim 6. $\langle \mathcal{A}, \varphi \rangle \in \text{MC(simple-}\Sigma_3[2]) \iff f(\langle \mathcal{A}, \varphi \rangle) \in \text{p-ALT}_3\text{SUB}$.

Proof (of Claim 6). "$\Longrightarrow$": Let $\langle \mathcal{A}, \varphi \rangle$ be a positive instance. Hence, $\mathcal{A} \models \varphi$. According to Claim 3, the choice of the subsets A_1, A_2, A_3 is correct and the sum t is reached. Adding NOWAIT yields t' in the correct choice of numbers. If A_2' does not correspond to a valid assignment, then according to Claim 5 the sum t can be reached with the FIX-numbers and t' can be reached by adding WAIT-numbers. As a result, $f(\langle \mathcal{A}, \varphi \rangle) \in \text{p-ALT}_3\text{SUB}$.

"$\Longleftarrow$": We use contraposition to prove that direction. Let $\langle \mathcal{A}, \varphi \rangle$ be a negative instance. Then for all assignments of $x_1, \ldots, x_k$ there is an assignment of $x_{k+1}, \ldots, x_{k+l}$ such that for all assignments of $x_{k+l+1}, \ldots, x_{k+l+m}$ the quantifier-free part, so at least one atomic formula, is not satisfied. If subsets are chosen that obey Claim 3, the sum t cannot be reached. If one selects subsets from A_1 or A_3 that do not correspond to a valid assignment, then according to Claim 4 the sum t cannot be reached by using only numbers from VAR, ATOM, NORM, and FIX. That is because choosing $\text{FIX}(x_j)$ would create a number in the B_j-block that is larger than $(l+1) \cdot |A|$ and t cannot be reached any more.

The only way to reach 1 at every position of the L-blocks is to choose numbers from ATOM. If a number of the form $\text{FIX}(x_j, d)$ is chosen, then the B_j-block would have a number that is larger than $(l+1) \cdot |A|$. As a result, no number from ATOM can be used for that j. Furthermore, we cannot use overflows to achieve this. The FIX-numbers are also insufficient to reach t. Finally, as t cannot be reached, t' cannot be reached either, as the WAIT-numbers have no influence on these digits of t. Hence $f(\langle \mathcal{A}, \varphi \rangle) \notin \text{p-ALT}_3\text{SUB}$. ■

This results in showing that $\text{p-ALT}_3\text{SUB}$ is A[3]-hard. □

Proof (of Theorem 8). By Lemmas 10 and 9 the result is proven. □

We state a generalisation of the problem ALT_3SUB. Also, we consider the complements of the problems of this generalisation. For odd $\ell \in \mathbb{N}^+$, we define the problem $\text{ALT}_\ell\text{SUB}$ as follows:

$$\left\{ \langle A_1, \ldots, A_l, k_1, \ldots, k_\ell, t \rangle \;\middle|\; \begin{array}{l} A_1, \ldots, A_\ell \subseteq \mathbb{N}_0,\ k_1, \ldots, k_\ell, t \in \mathbb{N}_0,\ \exists A'_1 \subseteq \\ A_1 \text{ s.t. } |A'_1| = k_1,\ \forall A'_2 \subseteq A_2 \text{ s.t. } |A'_2| = \\ k_2, \ldots,\ \exists A'_\ell \subseteq A_\ell \text{ s.t. } |A'_\ell| = k_\ell, \text{ and} \\ \sum_{a \in A'_1} a + \cdots + \sum_{a \in A'_\ell} a = t \end{array} \right\}$$

Its parameterised version is p-ALT$_\ell$SUB and is defined as

$$(\mathrm{ALT}_\ell\mathrm{SUB}, \langle A_1, \ldots A_l, k_1, \ldots, k_\ell, t \rangle \mapsto \sum_{i \in [1,\ell]} k_i)$$

Note that for the complement problem CO-ALT$_\ell$SUB of ALT$_\ell$SUB the quantifications of the sets flip (i.e., $\forall$ becomes $\exists$ and vice versa). Analogously, the parameterised version p-CO-ALT$_\ell$SUB is defined.

Corollary 11. *Let* $\ell \in \mathbb{N}^+$. *If* ℓ *is odd then* p-ALT$_\ell$SUB *is* A[ℓ]*-complete under* $\leq^{\mathrm{fpt}}$*-reductions. If* ℓ *is even then* p-CO-ALT$_\ell$SUB *is* A[ℓ]*-complete under* $\leq^{\mathrm{fpt}}$*-reductions.*

Proof. The membership of both problems can be shown analogously to the proof of Lemma 9. In essence the number of alternations needs to be taken into account. If ℓ is odd, the construction from Lemma 10 can be adapted. The VAR-numbers just need to be distributed over more sets. If ℓ is even, we show that p-MC(simple-$\Sigma_\ell[2]$) $\leq^{\mathrm{fpt}}$ p-CO-ALT$_\ell$SUB. We will use the reduction form Lemma 10, so we have to transform the formula into the correct form first. Let $\langle \mathcal{A}, \varphi \rangle$ be a p-MC(simple-$\Sigma_\ell[2]$) instance with $\mathcal{A} = (A, \tau)$. For every $\lambda \in \tau$ let $\bar{\lambda} = (A \times A) \setminus \lambda$ be the complement of R. Then, for $k = \sum_{i \in [1,k]} k_i$ and $x_{i,j} \in \{x_1, \ldots, x_k\}$:

$$\varphi = \exists x_1 \ldots \forall x_k \bigvee_{i=1}^{n} \lambda_i(x_{i,1}, x_{i,2})$$

$$\iff \varphi = \exists x_1 \ldots \forall x_k \bigvee_{i=1}^{n} \neg\bar{\lambda}_i(x_{i,1}, x_{i,2}) \iff \varphi \quad = \neg\forall x_1 \ldots \exists x_k \bigwedge_{i=1}^{n} \bar{\lambda}_i(x_{i,1}, x_{i,2})$$

Now, as in the construction in the proof of Lemma 10, the last quantifier is existential. The negation symbol in front of the formula is equivalent to being in CO-ALT$_\ell$SUB. Furthermore, the evaluation of $\bar{\lambda}_i(x_{i,1}, x_{i,2})$ in the reduction is necessary for the ATOM-numbers. Every $\bar{\lambda}$ is computable in $O(|\mathcal{A}|^2)$, so f is still computable in $O(|\langle \mathcal{A}, \varphi \rangle|^7)$. Otherwise it is a straightforward adaption of the proof of Lemma 10. □

From the characterisation of the polynomial hierarchy in terms of model checking problems [8, Prop. 4.28] we can deduce the following corollary. This is because the reduction presented above is already a polynomial time many-to-one reduction.

Corollary 12. *Let* $\ell \in \mathbb{N}^+$. *If* ℓ *is odd then* ALT$_\ell$SUB *is* Σ^p_ℓ*-complete under* $\leq^p_m$*-reductions. If* ℓ *is even then* CO-ALT$_\ell$SUB *is* Σ^p_ℓ*-complete under* $\leq^p_m$*-reductions.*

4 Conclusion

In this paper, we gave a new and more natural characterisation of the A-hierarchy based on variants of the well-known SUBSET-SUM problem. Intuitively, the problem is defined through alternations of existential and universal quantifiers that quantify over subsets of natural numbers, which must correctly sum to a specific target value. We demonstrated that the parameterised version of this problem—where the parameter is the sum of the sizes of the subsets to be chosen—is complete for every level of the A[]-hierarchy, depending on the number of alternations. Finally, we utilized a well-established connection between the A-hierarchy and the polynomial hierarchy to provide a characterization of the polynomial hierarchy in terms of SUBSET-SUM problems, as well.

Independent of its use in this paper, the SUBSET-SUM generalisation might be helpful in the context of counting problems.

Acknowledgments. We thank the anonymous reviewers for their valuable feedback. The second author was partially supported by a grant of the German Research Foundation (DFG) under the project number 511769688 (ME 4279/3-1).

Disclosure of Interests. The authors have no competing interests to declare that are relevant to the content of this article.

References

1. Chen, Y., Flum, J., Grohe, M.: Machine-based methods in parameterized complexity theory. Theor. Comput. Sci. **339**(2–3), 167–199 (2005). https://doi.org/10.1016/J.TCS.2005.02.003
2. Chiswell, I., Hodges, W.: Mathematical Logic. Oxford Texts in Logic, vol. 3. Clarendon Press (2007)
3. Downey, R.G., Fellows, M.R.: Parameterized Complexity. Monographs in Computer Science. Springer (1999). https://doi.org/10.1007/978-1-4612-0515-9
4. Downey, R.G., Fellows, M.R.: Fundamentals of Parameterized Complexity. Texts in Computer Science. Springer (2013). https://doi.org/10.1007/978-1-4471-5559-1
5. Downey, R.G., Fellows, M.R., Regan, K.W.: Parameterized circuit complexity and the W hierarchy. Theor. Comput. Sci. **191**(1–2), 97–115 (1998). https://doi.org/10.1016/S0304-3975(96)00317-9
6. Flum, J., Grohe, M.: Fixed-parameter tractability, definability, and model-checking. SIAM J. Comput. **31**(1), 113–145 (2001). https://doi.org/10.1137/S0097539799360768
7. Flum, J., Grohe, M.: Model-checking problems as a basis for parameterized intractability. Log. Methods Comput. Sci. **1**(1) (2005). https://doi.org/10.2168/LMCS-1(1:2)2005
8. Flum, J., Grohe, M.: Parameterized Complexity Theory. Texts in Theoretical Computer Science. An EATCS Series. Springer (2006). https://doi.org/10.1007/3-540-29953-X
9. Garey, M.R., Johnson, D.S.: Computers and Intractability: A Guide to the Theory of NP-Completeness. W. H. Freeman (1979)

10. Homer, S., Selman, A.L.: Computability and Complexity Theory. Texts in Computer Science, 2nd edn. Springer (2011). https://doi.org/10.1007/978-1-4614-0682-2
11. Papadimitriou, C.H.: Computational Complexity. Academic Internet Publication (2007)

On the Periodic Decompositions of Multidimensional Configurations

Pyry Herva(✉) and Jarkko Kari

Department of Mathematics and Statistics, University of Turku, Turku, Finland
{pysahe,jkari}@utu.fi

Abstract. We consider d-dimensional configurations, that is, colorings of the d-dimensional integer grid $\mathbb{Z}^d$ with finitely many colors. Moreover, we interpret the colors as integers so that configurations are functions $\mathbb{Z}^d \to \mathbb{Z}$ of finite range. We say that such function is k-periodic if it is invariant under translations in k linearly independent directions. It is known that if a configuration has a non-trivial annihilator, that is, if some non-trivial linear combination of its translations is the zero function, then it is a sum of finitely many periodic functions. This result is known as the periodic decomposition theorem. We prove two different improvements of it. The first improvement gives a characterization on annihilators of a configuration to guarantee the k-periodicity of the functions in its periodic decomposition—for any k. The periodic decomposition theorem is then a special case of this result with $k = 1$. The second improvement concerns so called sparse configurations for which the number of non-zero values in patterns grows at most linearly with respect to the diameter of the pattern. We prove that a sparse configuration with a non-trivial annihilator is a sum of finitely many periodic fibers where a fiber means a function whose non-zero values lie on a unique line.

Keywords: Periodicity · Periodic decomposition · Annihilator · Periodizer · Multidimensional symbolic dynamics

1 Introduction

We work with d-dimensional configurations $c \in \mathcal{A}^{\mathbb{Z}^d}$, that is, colorings of the d-dimensional integer grid $\mathbb{Z}^d$ with a finite set of colors $\mathcal{A}$—where d is any positive integer. In this paper we assume that $\mathcal{A} \subseteq \mathbb{Z}$. We consider configurations c that have non-trivial annihilators meaning that some non-trivial linear combination of translated copies of c is the zero function. It is known that such configurations can be decomposed to sums of finitely many periodic functions [9]—we will refer to this result as the periodic decomposition theorem. In this paper we show that under some additional assumptions the periodic decomposition theorem can be improved. We prove two different improvements of the periodic decomposition theorem.

In our considerations we apply an algebraic approach to multidimensional symbolic dynamics which was developed in [8] to make progress in the famous

R. Královič and V. Kůrková (Eds.): SOFSEM 2025, LNCS 15539, pp. 45–57, 2025.
https://doi.org/10.1007/978-3-031-82697-9_4

still open Nivat's conjecture [14] which links periodicity and sufficiently small rectangular pattern complexity. It is a two-dimensional generalization of the Morse-Hedlund theorem [12]. In the algebraic approach d-dimensional configurations are represented as formal power series with integer coefficients in d variables. An annihilator of a configuration is then a Laurent polynomial such that its formal product with the power series presenting the configuration is the zero power series. A periodizer of a configuration is a Laurent polynomial such that its formal product with the power series presenting the configuration is strongly periodic, that is, it has d linearly independent period vectors.

In our first improvement of the periodic decomposition theorem we consider arbitrary configurations that have periodizers of particular type. This result is very much inspired by similar results in [11]. More precisely, we assume that for some $k \in \{1, \ldots, d\}$ and for every $(k-1)$-dimensional linear subspace V of $\mathbb{R}^d$ the configuration has a periodizer which has exactly one non-zero value in V. We show that the configuration is then a sum of finitely many functions which all have k linearly independent periods (Theorem 3). If $k = 1$, then the statement says just that any configuration with a non-trivial periodizer is a sum of finitely many periodic functions which is exactly the original decomposition theorem. Also, for $k = d$ the statement is known to be true [6]. As a corollary (Corollary 1) of this result we get a similar result as in [11] concerning translational tilings. This corollary is closely related to the still partially open periodic tiling conjecture [10,15]. The periodic tiling conjecture is true in dimensions $d = 1$ [13] and $d = 2$ [2] but fails for sufficiently large d [4].

In the second improvement of the periodic decomposition theorem we consider configurations whose non-zero values are located very "sparsely" in the grid $\mathbb{Z}^d$. We show that if this kind of configuration has a non-zero annihilator, then it is a sum of finitely many periodic configurations whose non-zero values are aligned on unique lines (Theorem 4 and Corollary 2). This version of the periodic decomposition theorem is used in the study of colorings of 1-dimensional Delone sets with non-trivial annihilators and in proving that these colorings are necessarily periodic [5].

Structure of the Paper

We begin with preliminaries in Sect. 2 by presenting the necessary concepts and definitions and some past results relevant to us. In Sects. 3 and 4 we state and prove our improvements of the periodic decomposition theorem and discuss some related results.

2 Preliminaries

2.1 Notation

As usual, we denote by $\mathbb{Z}$, $\mathbb{Q}$, $\mathbb{R}$, and $\mathbb{C}$ the sets of integers, rational numbers, real numbers, and complex numbers, respectively. By $\mathbb{Z}_+$ we mean the set of positive integers and by $\mathbb{N} = \mathbb{Z}_+ \cup \{0\}$ the set of non-negative integers. For any two sets A and B, we denote by A^B the set of all functions from B to A. For two vectors $\mathbf{v} \in \mathbb{R}^d$ and $\mathbf{u} \in \mathbb{R}^d$, we denote by $\langle \mathbf{v}, \mathbf{u} \rangle = \{a\mathbf{v} + b\mathbf{u} \mid a, b \in \mathbb{R}\}$ the linear subspace of $\mathbb{R}^d$ spanned by them. By $\mathbb{G}_k$ we denote the set of all k-dimensional subspaces of $\mathbb{R}^d$ for a fixed $d \in \mathbb{Z}_+$. Clearly, if $k > d$, then $\mathbb{G}_k = \emptyset$.

2.2 Configurations and Periodicity

In this paper we consider mostly elements of the set $\mathbb{Z}^{\mathbb{Z}^d}$, that is, functions

$$c\colon \mathbb{Z}^d \to \mathbb{Z}$$

where d is a positive integer—the dimension. However, in some occasions we also work with the set $\mathbb{C}^{\mathbb{Z}^d}$ and hence most of the definitions and some of the statements are given more generally for the elements of this set. If $c \in \mathbb{Z}^{\mathbb{Z}^d}$ has only finitely many different values, then c is called a (d-dimensional) *configuration*. In particular, configurations are functions of the set $\mathcal{A}^{\mathbb{Z}^d}$ for some finite $\mathcal{A} \subseteq \mathbb{Z}$. The set $\mathcal{A}^{\mathbb{Z}^d}$ is called the *d-dimensional configuration space over $\mathcal{A}$*.

Any function $c \in \mathbb{C}^{\mathbb{Z}^d}$ is $\mathbf{v}$*-periodic* if $c(\mathbf{u} + \mathbf{v}) = c(\mathbf{u})$ for all $\mathbf{u} \in \mathbb{Z}^d$. In this case $\mathbf{v}$ is called a *period* of c. If c has a non-zero period, then it is *periodic*. Let us denote by $\tau^{\mathbf{v}}(c)$ the *translation* of $c \in \mathbb{C}^{\mathbb{Z}^d}$ by a vector $\mathbf{v} \in \mathbb{Z}^d$ defined such that $\tau^{\mathbf{v}}(c)(\mathbf{u}) = c(\mathbf{u} - \mathbf{v})$ for all $\mathbf{u} \in \mathbb{Z}^d$. Hence, c is $\mathbf{v}$-periodic if and only if $\tau^{\mathbf{v}}(c) = c$.

A function $c \in \mathbb{C}^{\mathbb{Z}^d}$ is *k-periodic* if it has k linearly independent periods. In particular, d-periodic functions are called *strongly periodic*. (Note that if $c \in \mathbb{Z}^{\mathbb{Z}^d}$ is strongly periodic, then c is necessarily a configuration.) We say that $c \in \mathbb{C}^{\mathbb{Z}^d}$ is *periodic in direction* $\mathbf{v} \in \mathbb{Q}^d$ if it is $k\mathbf{v}$-periodic for some $k \in \mathbb{Z}$, and we say that it is *V-periodic* for a vector space $V \subseteq \mathbb{R}^d$ if it is periodic in direction $\mathbf{v}$ for all $\mathbf{v} \in V \cap \mathbb{Q}^d$. Clearly, any k-periodic function is V-periodic for some $V \in \mathbb{G}_k$. Finally, $c \in \mathbb{C}^{\mathbb{Z}^d}$ is $\mathbf{v}$*-periodic in a subset* $S \subseteq \mathbb{Z}^d$ if for all $\mathbf{u} \in S$ such that $\mathbf{u} + \mathbf{v} \in S$ it follows that $c(\mathbf{u}) = c(\mathbf{u} + \mathbf{v})$.

A *(finite) pattern* is a function $p \in \mathbb{C}^D$ for some finite shape $D \subseteq \mathbb{Z}^d$. In this case p is also called a *D-pattern*. We say that a function $c \in \mathbb{C}^{\mathbb{Z}^d}$ contains a D-pattern $p \in \mathbb{C}^D$ if $\tau^{\mathbf{t}}(c) \restriction_D = p$ for some $\mathbf{t} \in \mathbb{Z}^d$. We denote by $\mathcal{L}_D(c) = \{\tau^{\mathbf{t}}(c) \restriction_D \mid \mathbf{t} \in \mathbb{Z}^d\}$ the set of all D-patterns of c and by

$$\mathcal{L}(c) = \bigcup_{\text{finite } D \subseteq \mathbb{Z}^d} \mathcal{L}_D(c)$$

the set of all finite patterns of c.

The *support* of $c \in \mathbb{C}^{\mathbb{Z}^d}$ is the set

$$\text{supp}(c) = \{\mathbf{u} \mid c(\mathbf{u}) \neq 0\}.$$

If supp(c) is finite, then c is called *finitely supported*.

Let us denote by $C_m = \{-m, \ldots, m\}^d$ for $m \in \mathbb{N}$ the discrete d-dimensional hypercube of size $(2m+1)^d$ centered at the origin. Any function $c \in \mathbb{C}^{\mathbb{Z}^d}$ is called *sparse* if there exists a positive integer a such that

$$|\operatorname{supp}(c) \cap (C_m + \mathbf{t})| \leq am$$

for all $m \in \mathbb{Z}_+$ and for all $\mathbf{t} \in \mathbb{Z}^d$. Such a is called a sparseness constant of c.

The configuration space $\mathcal{A}^{\mathbb{Z}^d}$ can be made a compact topological space by endowing $\mathcal{A}$ with the discrete topology and considering the product topology it induces on $\mathcal{A}^{\mathbb{Z}^d}$ – the *prodiscrete topology.* This topology is induced by a metric where two configurations are close if they agree on a large area around the origin. So, by compactness in this topology every sequence of configurations has a converging subsequence.

A *subshift* is a closed and translation-invariant subset of the configuration space $\mathcal{A}^{\mathbb{Z}^d}$. The *orbit* of a configuration $c \in \mathcal{A}^{\mathbb{Z}^d}$ is the set $\mathcal{O}(c) = \{\tau^{\mathbf{t}}(c) \mid \mathbf{t} \in \mathbb{Z}^d\}$ of its every translate. The *orbit closure* $\overline{\mathcal{O}(c)}$ is the topological closure of its orbit under the prodiscrete topology. It is the smallest subshift that contains c. It consists of all configurations c' such that $\mathcal{L}(c') \subseteq \mathcal{L}(c)$.

2.3 Algebraic Concepts

As in [8] we present a function $c \in \mathbb{C}^{\mathbb{Z}^d}$ as a formal power series

$$c(X) = \sum_{\mathbf{u} \in \mathbb{Z}^d} c_{\mathbf{u}} X^{\mathbf{u}}$$

in d variables $X = (x_1, \ldots, x_d)$ where we have used the convenient abbreviations $X^{\mathbf{u}} = x_1^{u_1} \cdots x_d^{u_d}$ and $c_{\mathbf{u}} = c(\mathbf{u})$ for a vector $\mathbf{u} = (u_1, \ldots, u_d) \in \mathbb{Z}^d$.

We consider Laurent polynomials with integer coefficients which we call from now on simply polynomials. By a non-trivial polynomial we mean a non-zero polynomial. A multiplication of a formal power series $c(X) = \sum_{\mathbf{u} \in \mathbb{Z}^d} c_{\mathbf{u}} X^{\mathbf{u}}$ with complex coefficients by a polynomial $f = f(X) = \sum_{i=1}^{n} f_i X^{\mathbf{u}_i}$ is well defined as the formal power series

$$fc = f(X)c(X) = \sum_{\mathbf{u} \in \mathbb{Z}^d} \left(\sum_{i=1}^{n} f_i c_{\mathbf{u}} \right) X^{\mathbf{u}+\mathbf{u}_i} = \sum_{\mathbf{u} \in \mathbb{Z}^d} \left(\sum_{i=1}^{n} f_i c_{\mathbf{u}-\mathbf{u}_i} \right) X^{\mathbf{u}}.$$

Thus, the value of fc at $\mathbf{u} \in \mathbb{Z}^d$ is determined by the values of c at $\mathbf{u} - \mathbf{u}_1, \ldots, \mathbf{u} - \mathbf{u}_n$.

By the product of a function $c \in \mathbb{C}^{\mathbb{Z}^d}$ and a polynomial f we mean the above product and denote it conveniently by fc as usual. Since any polynomial f can be thought as a finitely supported function in $\mathbb{Z}^{\mathbb{Z}^d}$, the above product can also be viewed as the discrete convolution of f and c (which is always defined because f is finitely supported). By the support of a polynomial $f = \sum f_{\mathbf{u}} X^{\mathbf{u}}$ we mean the (finite) support of this function, that is, the set

$$\operatorname{supp}(f) = \{\mathbf{u} \mid f_{\mathbf{u}} \neq 0\}.$$

We say that a polynomial f *annihilates* (or is an *annihilator* of) a function $c \in \mathbb{C}^{\mathbb{Z}^d}$ if $fc = 0$. Since multiplying $c \in \mathbb{C}^{\mathbb{Z}^d}$ by the monomial $X^{\mathbf{v}}$ corresponds to translating it by the vector $\mathbf{v}$, that is, $X^{\mathbf{v}}c = \tau^{\mathbf{v}}(c)$, it follows that c is $\mathbf{v}$-periodic if and only if it is annihilated by the difference polynomial $X^{\mathbf{v}} - 1$. The set of every annihilator of c is quite easily seen to be an ideal of the polynomial ring and hence the question whether c is periodic reduces to the question whether its "annihilator ideal" contains a non-trivial difference polynomial.

We say that a polynomial f *periodizes* (or is a *periodizer* of) $c \in \mathbb{C}^{\mathbb{Z}^d}$ if fc is strongly periodic. Clearly, any annihilator of c is also a periodizer of c. On the other hand, if c has a periodizer f, then fc is periodic and hence annihilated by some non-trivial difference polynomial $X^{\mathbf{v}} - 1$. Thus, c has a non-trivial annihilator if and only it has a non-trivial periodizer.

A *line polynomial* f is a polynomial whose support contains at least two points and the points of the support are aligned on a line, that is, there exist vectors $\mathbf{u}, \mathbf{v} \in \mathbb{Z}^d$ such that $\text{supp}(f) \subseteq \mathbf{u} + \mathbb{Q}\mathbf{v}$. Such $\mathbf{v}$ (which is non-zero) is called a *direction* of f and we may say that f is a *line polynomial in direction* $\mathbf{v}$. As usual, we say that two non-zero vectors $\mathbf{v}, \mathbf{v}'$ are *parallel* if $\mathbf{v}' \in \mathbb{Q}\mathbf{v}$ (and vice versa), *i.e.*, if they are linearly dependent over $\mathbb{Q}$. Otherwise they are *non-parallel* in which case they are linearly independent over $\mathbb{Q}$. Clearly, any line polynomial in direction $\mathbf{v}$ is also a line polynomial in any parallel non-zero direction $\mathbf{v}'$.

A function $c \in \mathbb{C}^{\mathbb{Z}^d}$ is a $\mathbf{v}$-*fiber* for a non-zero vector $\mathbf{v} \in \mathbb{Q}^d$ if its support is contained in a line in direction $\mathbf{v}$, that is, if $\text{supp}(c) \subseteq \mathbf{u} + \mathbb{Q}\mathbf{v}$ for some $\mathbf{u} \in \mathbb{Z}^d$. Any $\mathbf{v}$-fiber is called simply a *fiber*. Note that by interpreting polynomials as functions as discussed earlier, line polynomials are finitely supported fibers. By a $\mathbf{v}$-*fiber of* $c \in \mathbb{C}^{\mathbb{Z}^d}$ we mean a function that agrees with c on $\mathbf{u} + \mathbb{Q}\mathbf{v}$ for some $\mathbf{u} \in \mathbb{Z}^d$ and gets value 0 elsewhere. Clearly, any $\mathbf{v}$-fiber of c is a $\mathbf{v}$-fiber.

Note that if a configuration $c \in \mathcal{A}^{\mathbb{Z}^d}$ is annihilated by a line polynomial, then it is necessarily periodic. Indeed, annihilation by a line polynomial in direction $\mathbf{v}$ defines a recurrence relation on each $\mathbf{v}$-fiber of c which implies periodicity of each $\mathbf{v}$-fiber since $\mathcal{A}$ is finite. Moreover, the smallest periods of the $\mathbf{v}$-fibers are bounded and hence also c is $\mathbf{v}$-periodic. In particular, any one-dimensional configuration with a non-trivial annihilator is periodic since all polynomials in only one variable are line polynomials. Conversely, periodicity implies annihilation by a non-trivial difference polynomial which is a line polynomial. Thus, a configuration c is periodic if and only if it is annihilated by a line polynomial.

It is known that if a configuration $c \in \mathcal{A}^{\mathbb{Z}^d}$ has a non-trivial periodizer, then it has an annihilator which is a product of difference polynomials. More precisely, we have the following result.

Theorem 1 ([8,9]). *Let $c \in \mathcal{A}^{\mathbb{Z}^d}$ be a configuration and assume that it has a non-trivial annihilator f. For all $\mathbf{u} \in$ supp(f), there exist pairwise non-parallel vectors $\mathbf{v}_1, \ldots, \mathbf{v}_m \in \mathbb{Z}^d$ such that each $\mathbf{v}_i$ is parallel to $\mathbf{u}_i - \mathbf{u}$ for some $\mathbf{u}_i \in$ supp(f) $\setminus \{\mathbf{u}\}$, and c is annihilated by the polynomial*

$$(X^{\mathbf{v}_1} - 1) \cdots (X^{\mathbf{v}_m} - 1).$$

The multiplication of c by a difference polynomial can be thought as a "discrete derivation" of c. The above theorem says then that if a configuration $c \in \mathcal{A}^{\mathbb{Z}^d}$ has a non-trivial periodizer, then there is a sequence of derivations which annihilates c. So, by "integrating" step by step we have the following periodic decomposition theorem.

Theorem 2 (Periodic decomposition theorem [9]**).** *Let $c \in \mathcal{A}^{\mathbb{Z}^d}$ be a configuration and assume that it has a non-trivial periodizer. There exist pairwise non-parallel vectors $\mathbf{v}_1, \ldots, \mathbf{v}_m$ and functions $c_1, \ldots, c_m \in \mathbb{Z}^{\mathbb{Z}^d}$ such that*

$$c = c_1 + \ldots + c_m$$

and each c_i is $\mathbf{v}_i$-periodic.

Remark 1. The periodic functions c_i in the periodic decomposition of a configuration c may not be configurations, that is, they may get infinitely many different values [7].

3 Periodic Decomposition of Configurations with Specific Annihilators

In this section we prove our first improvement of the periodic decomposition theorem. For the proof we need some lemmas.

Lemma 1. *Let $V \subseteq \mathbb{R}^d$ be a linear subspace of $\mathbb{R}^d$, and let φ and ψ be line polynomials in non-parallel directions $\mathbf{v}_1$ and $\mathbf{v}_2$, respectively. Also, assume that*

$$\langle \mathbf{v}_1, \mathbf{v}_2 \rangle \cap V = \{\mathbf{0}\}.$$

If $c' \in \mathbb{C}^{\mathbb{Z}^d}$ is a V-periodic function annihilated by ψ, then there exists $c \in \mathbb{C}^{\mathbb{Z}^d}$ such that the following conditions hold:

- $\varphi c = c'$,
- $\psi c = 0$, *and*
- *c is V-periodic.*

Moreover, if $c' \in \mathbb{Z}^{\mathbb{Z}^d}$ and φ is a difference polynomial, then the coefficients of c can be chosen to be integers, that is, $c \in \mathbb{Z}^{\mathbb{Z}^d}$.

Lemma 2. *Let $V \subseteq \mathbb{R}^d$ be a linear subspace of $\mathbb{R}^d$. Assume that $c \in \mathbb{C}^{\mathbb{Z}^d}$ is V-periodic and that there exist $m \geq 1$ and line polynomials $\varphi_1, \ldots, \varphi_m$ in pairwise non-parallel directions $\mathbf{v}_1, \ldots, \mathbf{v}_m$, respectively, such that c is annihilated by the product $\varphi_1 \cdots \varphi_m$. Also, assume that for all $i, j \in \{1, \ldots, m\}, i \neq j$ it holds that*

$$\langle \mathbf{v}_i, \mathbf{v}_j \rangle \cap V = \{\mathbf{0}\}.$$

Then there exist functions $c_1, \ldots, c_m \in \mathbb{C}^{\mathbb{Z}^d}$ such that

$$c = c_1 + \ldots + c_m$$

where each c_i is annihilated by φ_i and each c_i is V-periodic.

Moreover, if the line polynomials $\varphi_1, \ldots, \varphi_m$ are difference polynomials and $c \in \mathbb{Z}^{\mathbb{Z}^d}$, then we may choose $c_1, \ldots, c_m \in \mathbb{Z}^{\mathbb{Z}^d}$.

The following lemma is adapted from the proof of Theorem 3 in [6].

Lemma 3 ([6]). *Let $V \in \mathbb{R}^d$ be a proper linear subspace of $\mathbb{R}^d$ ($\dim(V) \leq d-1$) and let $c \in \mathbb{C}^{\mathbb{Z}^d}$. Assume that c has a periodizer g such that $supp(g) \cap V = \{\mathbf{0}\}$. Then c has also an annihilator f such that $supp(f) \cap V = \{\mathbf{0}\}$.*

Now, we are ready to prove our first improvement of the periodic decomposition theorem.

Theorem 3. *Let c be a d-dimensional configuration and let $k \in \{1, \ldots, d\}$. Assume that for every $(k-1)$-dimensional linear subspace $V \subseteq \mathbb{R}^d$ the configuration c has a periodizer f such that $supp(f) \cap V = \{\mathbf{0}\}$. Then there exist k-periodic functions $c_1, \ldots, c_m \in \mathbb{Z}^{\mathbb{Z}^d}$ such that*

$$c = c_1 + \ldots + c_m.$$

Proof. We prove the claim by induction on k. If $k = 1$, then the assumption implies that the configuration c has a non-trivial periodizer and hence by Theorem 2 it is a sum of periodic functions.

Let $k \in \{1, \ldots, d-1\}$ and assume that the claim holds for k. Let us prove that the claim holds then also for $k+1$. So, assume that for all $V \in \mathbb{G}_k$ the configuration c has a periodizer f such that $\operatorname{supp}(f) \cap V = \{\mathbf{0}\}$. Since every $(k-1)$-dimensional subspace is contained in a k-dimensional subspace, there exists such periodizer, in particular, also for all $V \in \mathbb{G}_{k-1}$. Thus, by the induction hypothesis there exist k-periodic functions $e_1, \ldots, e_l \in \mathbb{Z}^{\mathbb{Z}^d}$ such that

$$c = e_1 + \ldots + e_l.$$

For each $i \in \{1, \ldots, l\}$, let $V_i \in \mathbb{G}_k$ be such that e_i is V_i-periodic. We may assume that for all $i \neq j$ we have $V_i \neq V_j$ since the sum of two V-periodic functions is also V-periodic. Indeed, if for some $i \neq j$ we have $V_i = V_j$, then we replace the functions e_i and e_j by the function $e_i + e_j$ which is also V_i-periodic.

Let us fix an arbitrary $i \in \{1, \ldots, l\}$. By the assumption the configuration c has a periodizer f such that

$$\operatorname{supp}(f) \cap V_i = \{\mathbf{0}\}.$$

By Lemma 3 we may assume that f is an annihilator of c and hence by Theorem 1 we conclude that c has an annihilator

$$(X^{\mathbf{u}_1} - 1) \cdots (X^{\mathbf{u}_r} - 1)$$

such that $\mathbf{u}_j \notin V_i$ for all $j \in \{1, \ldots, r\}$ and the vectors $\mathbf{u}_1, \ldots, \mathbf{u}_r$ are pairwise non-parallel.

For each $j \in \{1, \ldots, l\} \setminus \{i\}$, let $\mathbf{w}_j \in V_j$ be such that $\mathbf{w}_j \notin V_i$ and e_j is annihilated by $X^{\mathbf{w}_j} - 1$. Since $c = e_1 + \ldots + e_l$ is annihilated by the product

$(X^{\mathbf{u}_1} - 1) \cdots (X^{\mathbf{u}_r} - 1)$ and each e_j is annihilated by $X^{\mathbf{w}_j} - 1$, it follows that e_i is annihilated by the polynomial

$$(X^{\mathbf{u}_1} - 1) \cdots (X^{\mathbf{u}_r} - 1) \prod_{j \in \{1,\ldots,l\} \setminus \{i\}} (X^{\mathbf{w}_j} - 1).$$

Thus, we have seen that e_i is annihilated by a polynomial of the form

$$(X^{\mathbf{v}_1} - 1) \cdots (X^{\mathbf{v}_s} - 1)$$

where $\mathbf{v}_j \notin V_i$ for each $j \in \{1, \ldots, s\}$. Let us assume that s is minimal in the sense that if a product $(X^{\mathbf{v}'_1} - 1) \cdots (X^{\mathbf{v}'_t} - 1)$ annihilates e_i such that $\mathbf{v}'_j \notin V_i$ for each $j \in \{1, \ldots, t\}$, then $t \geq s$. We claim that the vectors $\mathbf{v}_1, \ldots, \mathbf{v}_s$ are pairwise non-parallel and $\langle \mathbf{v}_j, \mathbf{v}_{j'} \rangle \cap V_i = \{\mathbf{0}\}$ for all $j, j' \in \{1, \ldots, s\}, j \neq j'$.

(Note that in the above annihilator we may replace any $\mathbf{v}_j$ with $p\mathbf{v}_j$ for any integer p and the obtained polynomial is still an annihilator of e_i. This is due to the fact that if a function is $\mathbf{v}$-periodic, then it is also $p\mathbf{v}$-periodic for any integer p.)

If the vectors $\mathbf{v}_1, \ldots, \mathbf{v}_s$ are not pairwise non-parallel, then for some $j, j' \in \{1, \ldots, s\}, j \neq j'$ the product $\varphi = (X^{\mathbf{v}_j} - 1)(X^{\mathbf{v}_{j'}} - 1)$ is a line polynomial in direction $\mathbf{v}_j$. Since φ annihilates the function

$$\frac{(X^{\mathbf{v}_1} - 1) \cdots (X^{\mathbf{v}_s} - 1)}{(X^{\mathbf{v}_j} - 1)(X^{\mathbf{v}_{j'}} - 1)} e_i,$$

it is periodic in direction $\mathbf{v}_j$, that is, annihilated by $X^{p\mathbf{v}_j} - 1$ for some non-zero $p \in \mathbb{Z}$. Thus, e_i is annihilated by

$$(X^{p\mathbf{v}_j} - 1) \cdot \frac{(X^{\mathbf{v}_1} - 1) \cdots (X^{\mathbf{v}_s} - 1)}{(X^{\mathbf{v}_j} - 1)(X^{\mathbf{v}_{j'}} - 1)}$$

which is a product of $s-1$ non-trivial difference polynomials. This is a contradiction with the minimality of s and hence the vectors $\mathbf{v}_1, \ldots, \mathbf{v}_s$ must be pairwise non-parallel.

Let us then show that $\langle \mathbf{v}_j, \mathbf{v}_{j'} \rangle \cap V_i = \{\mathbf{0}\}$ for all $j, j \in \{1, \ldots, s\}, j \neq j'$. Assume on the contrary that there exist $j, j' \in \{1, \ldots, s\}, j \neq j'$ such that $\langle \mathbf{v}_j, \mathbf{v}_{j'} \rangle \cap V_i \neq \{\mathbf{0}\}$. Then there exist integers p, p' such that $p'\mathbf{v}_{j'} = p\mathbf{v}_j + \mathbf{v}$ for some $\mathbf{v} \in V_i \setminus \{\mathbf{0}\}$. Since $\mathbf{v}_j, \mathbf{v}_{j'} \notin V_i$ we have $p, p' \neq 0$. Now, we replace the term $X^{\mathbf{v}_{j'}} - 1$ in the annihilator $(X^{\mathbf{v}_1} - 1) \cdots (X^{\mathbf{v}_s} - 1)$ of e_i by $X^{p\mathbf{v}_j + \mathbf{v}} - 1$. Let us denote the obtained annihilator of e_i by g. Since e_i is $\mathbf{v}$-periodic, also the function

$$e = \frac{g}{X^{p\mathbf{v}_j + \mathbf{v}} - 1} e_i$$

is $\mathbf{v}$-periodic, that is, $X^{\mathbf{v}} e = e$. Since it is annihilated by $X^{p\mathbf{v}_j + \mathbf{v}} - 1$, we have

$$0 = (X^{p\mathbf{v}_j + \mathbf{v}} - 1)e = X^{p\mathbf{v}_j + \mathbf{v}} e - e = X^{p\mathbf{v}_j} e - e = (X^{p\mathbf{v}_j} - 1)e$$

and hence it is also annihilated by $X^{p\mathbf{v}_j} - 1$. Consequently, we replace the term $X^{p\mathbf{v}_j + \mathbf{v}} - 1$ in g by $X^{p\mathbf{v}_j} - 1$. Finally, the term $(X^{p\mathbf{v}_j} - 1)(X^{\mathbf{v}_j} - 1)$ is replaced by

$X^{q\mathbf{v}_j} - 1$ for suitable $q \in \mathbb{Z}$. Again, we get a contradiction with the minimality of s by obtaining an annihilator of e_i which is a product of $s-1$ non-trivial difference polynomials.

Thus, by Lemma 2 there exist V_i-periodic functions $c_1, \ldots, c_s$ such that for each $j \in \{1, \ldots, s\}$ the function c_j is annihilated by $X^{\mathbf{v}_j} - 1$, that is, $\mathbf{v}_j$-periodic and

$$e_i = c_1 + \ldots + c_s.$$

So, each c_j is $\langle \{\mathbf{v}_j\} \cup V_i \rangle$-periodic and hence $(k+1)$-periodic since $\mathbf{v}_j \notin V_i$.

This same reasoning works for any $i \in \{1, \ldots, l\}$. So, we conclude that each e_i is a sum of $(k+1)$-periodic functions and hence $c = e_1 + \ldots + e_l$ is a sum of $(k+1)$-periodic functions. □

Note that also the converse of the above theorem holds. In other words, if $c_1, \ldots, c_m \in \mathbb{Z}^{\mathbb{Z}^d}$ are k-periodic functions, then their sum $c = c_1 + \ldots + c_m$, has for all $V \in \mathbb{G}_{k-1}$ a periodizer f such that $\text{supp}(f) \cap V = \{\mathbf{0}\}$.

Remark 2. A linear subspace $V \subseteq \mathbb{R}^d$ is *expansive* for a subshift $\mathcal{X} \subseteq \mathcal{A}^{\mathbb{Z}^d}$ if there exists $r \in \mathbb{Z}_+$ such that for any $c, e \in \mathcal{X}$ the condition

$$c(\mathbf{u}) = e(\mathbf{u}) \text{ for all } \mathbf{u} \in V^r$$

implies that $c = e$ where $V^r = \{\mathbf{u} \in \mathbb{Z}^d \mid d(\mathbf{u}, \mathbf{v}) \leq r \text{ for some } \mathbf{v} \in V\}$ is the set of integer vectors within distance r from V. A theorem by Boyle and Lind states that if every $(d-1)$-dimensional subspace is expansive for a subshift $\mathcal{X} \subseteq \mathcal{A}^{\mathbb{Z}^d}$, then $\mathcal{X}$ is finite [3]. If in Theorem 3 we have $k = d$, then it is easily seen that every $(d-1)$-dimensional subspace is expansive for the orbit closure $\overline{\mathcal{O}(c)}$ of c and hence $\overline{\mathcal{O}(c)}$ is finite by the theorem by Boyle and Lind. Thus, c is strongly periodic in this setting by using a similar argument as in [1]. This case is also stated as Corollary 1 in [6]. For $k \in \{1, \ldots, d-1\}$, it may be that the functions in Theorem 3 have infinitely many distinct coefficients, that is, they are not configurations as Remark 1 suggests.

Translational Tilings

A *translational tiling* is a binary configuration $c \in \{0,1\}^{\mathbb{Z}^d}$ such that for some non-empty finite set $D \subseteq \mathbb{Z}^d$ and a polynomial $f = \sum_{\mathbf{u} \in -D} X^{\mathbf{u}}$ we have

$$fc = \mathbb{1} = 1^{\mathbb{Z}^d}.$$

In this case D is called a *tile* and c is a *co-tiler of* D or a (translational) *tiling by the tile* D. Visually, the support of a co-tiler C of a tile D gives the positions where copies of D are placed such that every point of the grid $\mathbb{Z}^d$ belongs to exactly one copy of D.

In [11] the authors say that k tiles $D_1, \ldots, D_k \subseteq \mathbb{Z}^d$ satisfying $\mathbf{0} \in D_i$ for each $i \in \{1, \ldots, k\}$ are *independent* if the vectors $\mathbf{v}_1, \ldots, \mathbf{v}_k$ are linearly independent over $\mathbb{Q}$ with any choice of $\mathbf{v}_i \in D_i \setminus \{\mathbf{0}\}$. It is shown that if k independent tiles have a common co-tiler c, then c is a sum of finitely many k-periodic functions from the set $[a,b]^{\mathbb{Z}^d}$ for some reals $a < b$. In the following we prove a similar result as a corollary of Theorem 3.

Let us say that k polynomials $f_1, \ldots, f_k$ satisfying $\mathbf{0} \in \text{supp}(f_i)$ for each $i \in \{1, \ldots, k\}$ are independent if their supports are independent.

Corollary 1. *Let $c \in \mathcal{A}^{\mathbb{Z}^d}$ be a configuration and let $f_1, \ldots, f_k$ be k periodizers of c satisfying $\mathbf{0} \in \text{supp}(f_i)$ for each $i \in \{1, \ldots, k\}$ with $k \in \{1, \ldots, d\}$. If $f_1, \ldots, f_k$ are independent, then*

$$c = c_1 + \ldots + c_m$$

where each $c_i \in \mathbb{Z}^{\mathbb{Z}^d}$ is k-periodic.

Proof. Let us prove that for every $V \in \mathbb{G}_{k-1}$ the configuration c has a periodizer f such that $\text{supp}(f) \cap V = \{\mathbf{0}\}$. Then the claim follows from Theorem 3.

Assume on the contrary that there exists a $V \in \mathbb{G}_{k-1}$ such that $\text{supp}(f) \cap V \neq \{\mathbf{0}\}$ for any periodizer f of c. This implies that for any periodizer f of c we have either $\text{supp}(f) \cap V = \emptyset$ or $|\text{supp}(f) \cap V| \geq 2$. Since $\mathbf{0} \in \text{supp}(f_i)$, we have $|\text{supp}(f_i) \cap V| \geq 2$ for each $i \in \{1, \ldots, k\}$. Thus, there exist non-zero vectors $\mathbf{v}_1, \ldots, \mathbf{v}_k$ such that $\mathbf{v}_i \in \text{supp}(f_i) \cap V$ for each $i \in \{1, \ldots, k\}$ and hence the vectors $\mathbf{v}_1, \ldots, \mathbf{v}_k$ span a linear subspace of dimension at most $k-1$. This means that the sets $\text{supp}(f_1), \ldots, \text{supp}(f_k)$ are not independent and hence the polynomials $f_1, \ldots, f_k$ are not independent. A contradiction. □

For any tile $D \subseteq \mathbb{Z}^d$, the polynomial $f = \sum_{\mathbf{u} \in -D} X^{\mathbf{u}}$ is a periodizer of any co-tiler c of D. Thus, by the above corollary any common co-tiler of k independent tiles is a sum of finitely many k-periodic functions from the set $\mathbb{Z}^{\mathbb{Z}^d}$. So, we have a similar result as the authors of [11] except that the k-periodic functions in the periodic decomposition of the co-tiler are now functions of the set $\mathbb{Z}^{\mathbb{Z}^d}$ instead of the set $[a,b]^{\mathbb{Z}^d}$.

4 Periodic Decomposition of Sparse Configurations

In this section we consider sparse configurations with annihilators and prove that they are sums of finitely many periodic fibers. We prove the following theorem.

Theorem 4. *Let c be a sparse configuration and assume that it is annihilated by a product $\varphi_1 \cdots \varphi_n$ of line polynomials $\varphi_1, \ldots, \varphi_n$ in pairwise non-parallel directions $\mathbf{v}_1, \ldots, \mathbf{v}_n$, respectively. Then*

$$c = c_1 + \ldots + c_n$$

where each c_i is a sum of finitely many periodic $\mathbf{v}_i$-fibers and $\varphi_i c_i = 0$.

Before presenting the proof of the theorem we state some simple observations concerning sparse configurations.

Lemma 4. *Let* $c_1, \ldots, c_n$ *be sparse configurations and let* $k_1, \ldots, k_n \in \mathbb{Z}$. *Then also* $k_1 c_1 + \ldots + k_n c_n$ *is sparse. In particular, for a sparse configuration* c *and a polynomial* g *also* gc *is sparse.*

Lemma 5. *Every element of the orbit closure* $\overline{\mathcal{O}(c)}$ *of a sparse configuration* c *is also sparse.*

The following lemma is Theorem 4 in the case $n = 1$.

Lemma 6. *Let* c *be a sparse configuration and assume that it is periodic in direction* $\mathbf{v}$ *for some non-zero* $\mathbf{v}$. *Then* c *is a sum of finitely many periodic* $\mathbf{v}$-*fibers.*

In fact, also the converse of the above lemma holds, that is, if a configuration is a sum of finitely many periodic $\mathbf{v}$-fibers, then it is sparse and periodic in direction $\mathbf{v}$. Indeed, clearly any $\mathbf{v}$-fiber e is sparse since

$$\text{supp}(e) \cap (C_m + \mathbf{t}) \subseteq (\mathbf{u} + \mathbb{Q}\mathbf{v}) \cap (C_m + \mathbf{t})$$

for some $\mathbf{u}$ and $|(\mathbf{u} + \mathbb{Q}\mathbf{v}) \cap (C_m + \mathbf{t})| \leq 2m + 1$. By Lemma 4 a sum of finitely many sparse configurations is sparse and hence a sum of finitely many periodic $\mathbf{v}$-fibers is sparse and periodic. Thus, a configuration is sparse and periodic in direction $\mathbf{v}$ if and only if it is a sum of finitely many periodic $\mathbf{v}$-fibers.

The following lemma is Theorem 4 in the case $n = 2$.

Lemma 7. *Let* c *be a sparse configuration and assume that it is annihilated by the product* $\varphi\psi$ *of two line polynomials* φ *and* ψ *in non-parallel directions* $\mathbf{v}$ *and* $\mathbf{u}$, *respectively. Then*

$$c = c_1 + c_2$$

where c_1 *is a sum of finitely many periodic* $\mathbf{v}$-*fibers and* $\varphi c_1 = 0$, *and* c_2 *is a sum of finitely many periodic* $\mathbf{u}$-*fibers and* $\psi c_2 = 0$.

Now, we are ready to prove Theorem 4.

Proof of Theorem 4. The proof is by induction on n. First, let us consider the case $n = 1$. So, c is periodic in direction $\mathbf{v}_1$. Then by Lemma 6 it is a sum of finitely many periodic $\mathbf{v}_1$-fibers.

Assume then that $n > 1$ and that the claim holds for $n - 1$. Consider the configuration

$$c' = \varphi_n c.$$

It is sparse and annihilated by $\varphi_1 \cdots \varphi_{n-1}$. By the induction hypothesis

$$c' = c'_1 + \ldots + c'_{n-1}$$

where each c'_i is a sum of finitely many periodic $\mathbf{v}_i$-fibers and annihilated by φ_i.

So, for any $i \in \{1, \ldots, n-1\}$ the configuration c'_i is periodic in direction $\mathbf{v}_i$, that is, $k_i\mathbf{v}_i$-periodic for some $k_i \in \mathbb{Z} \setminus \{0\}$. Consider the sequence

$$c', X^{k_i\mathbf{v}_i}c', X^{2k_i\mathbf{v}_i}c', X^{3k_i\mathbf{v}_i}c', \ldots.$$

It converges to c_i' and hence $c_i' \in \overline{\mathcal{O}(c')}$. Let e be a limit of a converging subsequence of the sequence

$$c, X^{k_i \mathbf{v}_i} c, X^{2k_i \mathbf{v}_i} c, X^{3k_i \mathbf{v}_i} c, \ldots.$$

By compactness of the configuration space such subsequence exists.

We have $\varphi_n e = c_i'$. Since c is sparse and $e \in \overline{\mathcal{O}(c)}$, by Lemma 5 also e is sparse. Moreover, we have $\varphi_i \varphi_n e = 0$ since $\varphi_i c_i' = 0$. By Lemma 7 we have

$$e = e_i + e_n$$

where e_i is a sum of finitely many periodic $\mathbf{v}_i$-fibers and $\varphi_i e_i = 0$, and e_n is a sum of finitely many $\mathbf{v}_n$-fibers and $\varphi_n e_n = 0$. It follows that $\varphi_n e_i = \varphi_n e = c_i'$.

Now, we choose $c_i = e_i$ for $i \in \{1, \ldots, n-1\}$ and
$c_n = c - c_1 - \ldots - c_{n-1}$. Clearly, $c = c_1 + \ldots + c_n$. Moreover, we have

$$\begin{aligned} \varphi_n c_n &= \varphi_n c - \varphi_n c_1 - \ldots - \varphi_n c_{n-1} \\ &= c' - c_1' - \ldots - c_{n-1}' \\ &= 0. \end{aligned}$$

Since for each $i \in \{1, \ldots, n-1\}$ the configuration c_i is sparse as a sum of finitely many periodic $\mathbf{v}_i$-fibers and c is sparse, also $c_n = c - c_1 - \ldots - c_{n-1}$ is sparse by Lemma 4. Thus, by Lemma 6 the configuration c_n is a sum of finitely many periodic $\mathbf{v}_n$-fibers. □

Theorem 4 together with Theorem 1 yields the following corollary.

Corollary 2. *Let c be a sparse configuration and assume that it has a non-trivial annihilator. Then it is a sum of finitely many periodic fibers.*

Proof. By Theorem 1 the configuration c is annihilated by a polynomial

$$(X^{\mathbf{v}_1} - 1) \cdots (X^{\mathbf{v}_m} - 1)$$

where the vectors $\mathbf{v}_1, \ldots, \mathbf{v}_m$ are pairwise non-parallel and hence by Theorem 4 there exist configurations $c_1, \ldots, c_m$ such that each c_i is a sum of finitely many periodic $\mathbf{v}_i$-fibers and $c = c_1 + \ldots + c_m$. Thus, c is a sum of finitely many periodic fibers. □

Acknowledgement. This research was supported by the Academy of Finland grant 354965.

Disclosure of Interests. The authors have no competing interests to declare that are relevant to the content of this article.

References

1. Ballier, A., Durand, B., Jeandal, E.: Structural aspects of tilings. In: Albers, S., Weil, P. (eds.) 25th International Symposium on Theoretical Aspects of Computer Science. Leibniz International Proceedings in Informatics (LIPIcs), vol. 1, pp. 61–72. Schloss Dagstuhl-Leibniz-Zentrum fuer Informatik, Dagstuhl (2008)

2. Bhattacharya, S.: Periodicity and decidability of tilings of $\mathbb{Z}^2$. Am. J. Math. **142** (2016). https://doi.org/10.1353/ajm.2020.0006
3. Boyle, M., Lind, D.: Expansive subdynamics. Trans. Am. Math. Soc. **349**(1), 55–102 (1997). http://www.jstor.org/stable/2155304
4. Greenfeld, R., Tao, T.: A counterexample to the periodic tiling conjecture (2022). https://doi.org/10.48550/ARXIV.2211.15847, https://arxiv.org/abs/2211.15847
5. Herva, P., Kari, J.: Periodicity and local complexity of Delone sets, in preparation
6. Kari, J.: Expansivity and periodicity in algebraic subshifts. Theory Comput. Syst. (2023)
7. Kari, J., Szabados, M.: An algebraic geometric approach to multidimensional words. In: Maletti, A. (ed.) Algebraic Informatics, pp. 29–42. Springer, Cham (2015)
8. Kari, J., Szabados, M.: An algebraic geometric approach to Nivat's conjecture. In: Proceedings of ICALP 2015, Part II. LNCS, vol. 9135, pp. 273–285 (2015)
9. Kari, J., Szabados, M.: An algebraic geometric approach to Nivat's conjecture. Inf. Comput. **271**, 104481 (2020)
10. Lagarias, J.C., Wang, Y.: Tiling the line with translates of one tile. Invent. Math. **124**, 341–365 (1996)
11. Meyerovitch, T., Sanadhya, S., Solomon, Y.: Periodicity of joint co-tiles in $\mathbb{Z}^d$. arXiv:2301.11255 (2023)
12. Morse, M., Hedlund, G.A.: Symbolic dynamics. Am. J. Math. **60**(4), 815–866 (1938). http://www.jstor.org/stable/2371264
13. Newman, D.J.: Tesselation of integers. J. Number Theory **9**, 107–111 (1977)
14. Nivat, M.: Invited Talk at the 24th International Colloquium on Automata, Languages, and Programming (ICALP 1997) (1997)
15. Stein, S.K.: Algebraic tiling. Am. Math. Monthly (1974)

Quantum Algorithm for the Multiple String Matching Problem

Kamil Khadiev(✉) and Danil Serov

Institute of Computational Mathematics and Information Technologies, Kazan Federal University, Kazan, Russia
kamilhadi@gmail.com

Abstract. Let us consider the Multiple String Matching Problem. In this problem, we consider a long string, denoted by t, of length n. This string is referred to as a text. We also consider a sequence of m strings, denoted by S, which we refer to as a dictionary. The total length of all strings from the dictionary is represented by the variable L. The objective is to identify all instances of strings from the dictionary within the text. The standard classical solution to this problem is Aho-Corasick Algorithm that has $O(n + L)$ query and time complexity. At the same time, the classical lower bound for the problem is the same $\Omega(n + L)$. We propose a quantum algorithm with $O(n + \sqrt{mL \log n} + m \log n)$ query complexity and $O(n + \sqrt{mL \log n} \log b + m \log n) = O^*(n + \sqrt{mL})$ time complexity, where b is the maximal length of strings from the dictionary. This improvement is particularly significant in the case of dictionaries comprising long words. Our algorithm's complexity is equal to the quantum lower bound $O(n+\sqrt{mL})$, up to a log factor. In some sense, our algorithm can be viewed as a quantum analogue of the Aho-Corasick algorithm.

Keywords: Aho-Corasick Algorithm · strings · quantum algorithms · query complexity · search in strings · string matching · multiple string matching

1 Introduction

Let us consider the Multiple String Matching Problem. In this problem, we have a long string t of a length n that we call a text; and a sequence of m strings S that we call a dictionary. The total length of all strings from the dictionary is L, and the maximal length of the dictionary's strings is b. Our goal is to find all positions of strings from the dictionary in the text. The standard classical solution to this problem is Aho-Corasick Algorithm [3] that has $O(n + L)$ time and query complexity.

R. Královič and V. Kůrková (Eds.): SOFSEM 2025, LNCS 15539, pp. 58–69, 2025.
https://doi.org/10.1007/978-3-031-82697-9_5

We are interested in a quantum algorithm for this problem. There are many examples of quantum algorithms [1,6,22,41] that are faster than classical counterparts [19,43]. Particularly, we can find such examples among problems for strings processing [2,4,7,8,20,21,23–31,35,36,40]. One such problem is the String matching problem. That is checking whether a string s is a substring of a string t. It is very similar to the Multiple String Matching Problem but with only one string in the dictionary. Additionally, in the String matching problem, we should find any position of s not all of them. The last difference can be significant for quantum algorithms. For this problem, we can use, for example, Knuth-Morris-Pratt algorithm [15,32] in the classical case with $O(n + k)$ time and query complexity, where n is the length of t and k is the length of s. Other possible algorithms can be Backward Oracle Matching [5], Wu-Manber [44]. In the quantum case, we can use Ramesh-Vinay algorithm [42] with $O\left(\sqrt{n}\log\sqrt{\frac{n}{k}}\log n + \sqrt{k}(\log k)^2\right)$ = query complexity. We can say that query and time complexity of the Ramesh-Vinay algorithm is $O^*(\sqrt{n} + \sqrt{k})$, where O^* notation hides log factors. If we naively apply these algorithms to the Multiple String Matching Problem, then we obtain $O(mn+L)$ complexity in the classical case, and $O\left(mn\log\sqrt{\frac{n}{b}}\log n + m^{1.5}\sqrt{L}(\log b)^2\right) = O^*(mn + m^{1.5}\sqrt{L})$ in the quantum case.

In the classical case, Aho-Corasick Algorithm [3] has better complexity that is $O(n + L)$. Another possible algorithm that can work better in average is the Commentz-Walter algorithm [14]. In the quantum case, we suggest a new algorithm for the Multiple String Matching Problem with $O(n + \sqrt{mL\log n} + m\log n)$ query complexity and $O(n+\sqrt{mL\log n}\log b+m\log n)$ time complexity. So, we can say that both are $O^*(n+\sqrt{mL})$. At the same time, we show that the lower bound for classical complexity is $\Omega(n+l)$, and for quantum complexity, it is $\Omega(n + \sqrt{mL})$. It means, that the Aho-Corasick Algorithm reaches the lower bound, and our algorithm reaches the lower bound up to a log factor. At the same time, if $O(m)$ strings of the dictionary have at least $\omega(\log n)$ length, then our quantum algorithm shows speed-up.

The structure of this paper is as follows: Sect. 2 contains preliminaries, used data structures and algorithms. Section 3 provides a quantum algorithm for the Multiple String Matching Problem. The final Sect. 4 concludes the paper and contains open questions.

2 Preliminaries and Tools

For a string $u = (u_1, \dots, u_M)$, let $|u| = M$ be a length of the string. Let $u[i : j] = (u_i, \dots, u_j)$ be a substring of u from i-th to j-th symbol. Let $u[i] = u_i$ be i-th symbol of u. In the paper, we say that a string u is less than a string v if u precedes v in the lexicographical order.

2.1 The Multiple String Matching Problem

Suppose, we have a string t and a sequence of strings $S = (s^1, \ldots, s^m)$. We call the set S a dictionary, and the string t a text. Let $L = |s^1| + \cdots + |s^m|$ be the total length of the dictionary, and $b = \max\{|s^1|, \ldots, |s^m|\}$ be the maximal length of dictionary's strings. Let an integer $n = |t|$ be the length of t. The problem is to find positions in t for all strings from S. Formally, we want to find a sequence of indexes $\mathcal{I} = (I_1, \ldots, I_m)$, where $I_j = (i_{j,1}, \ldots, i_{j,k_j})$ such that $t[i_{j,x} : i_{j,x} + |s^j| - 1] = s^j$ for each $x \in \{1, \ldots, k_j\}, j \in \{1, \ldots, m\}$. Here k_j is the number of occurrences of s^j in the string t.

2.2 Suffix Array

A suffix array [39] is an array $suf = (suf_1, \ldots, suf_l)$ for a string u where $l = |u|$ is the length of the string. The suffix array is the lexicographical order for all suffixes of u. Formally, $u[suf_i : l] < u[suf_{i+1} : l]$ for any $i \in \{1, \ldots, l-1\}$. Let ConstructSuffixArray(u) be a procedure that constructs the suffix array for the string u. The query and time complexity of the procedure are as follows:

Lemma 1 ([37]). *A suffix array for a string u can be constructed with $O(|u|)$ query and time complexity.*

Let longest common prefix (LCP) of two strings u and s or $LCP(u, s)$ be an index i such that $u_j = s_j$ for each $j \in \{1, \ldots, i\}$, and $u_{i+1} \neq s_{i+1}$.

For a suffix array suf we can define an array of longest common prefixes (LCP array). Let call it $lcp = (lcp_1, \ldots, lcp_{l-1})$, where $lcp_i = LCP(u[suf_i : l], u[suf_{i+1} : l])$ is the longest common prefix of two suffixes $u[suf_i : l]$ and $u[suf_{i+1} : l]$. The array can be constructed with linear complexity:

Lemma 2 ([34]). *An array of longest common prefixes (LCP array) for a string u can be constructed with $O(|u|)$ query and time complexity.*

Using the lcp array, we can compute the LCP of any two suffixes $LCP(u[suf_i : l], u[suf_j : l])$ in constant time due to the algorithm from [11].

Lemma 3 ([11]). *For $i, j \in \{1, \ldots |u|\}$, there is an algorithm that computes the longest common prefix for suffixes $u[suf_i : l]$ and $u[suf_j : l]$ of the string u with $O(1)$ query and time complexity with preprocessing with $O(|u|)$ query and time complexity.*

For $i, j \in \{1, \ldots |u|\}$, let LCPSuf($u, i, j$) be the procedure that returns $LCP(u[suf_i : l], u[suf_j : l])$, and PreprocessingForLCP(u) be the preprocessing procedure for this algorithm and constructing the lcp array.

2.3 Quantum Query Model

One of the most popular computation models for quantum algorithms is the query model. We use the standard form of the quantum query model. Let $f : D \to \{0,1\}, D \subseteq \{0,1\}^M$ be an M variable function. Our goal is to compute it on an input $x \in D$. We are given oracle access to the input x, i.e. it is implemented by a specific unitary transformation usually defined as $|i\rangle|z\rangle|w\rangle \mapsto |i\rangle|z + x_i \pmod 2\rangle|w\rangle$, where the $|i\rangle$ register indicates the index of the variable we are querying, $|z\rangle$ is the output register, and $|w\rangle$ is some auxiliary work-space. An algorithm in the query model consists of alternating applications of arbitrary unitaries which are independent of the input and the query unitary, and a measurement at the end. The smallest number of queries for an algorithm that outputs $f(x)$ with probability $\geq \frac{2}{3}$ on all x is called the quantum query complexity of the function f and is denoted by $Q(f)$. We refer the readers to [1,6,22,41] for more details on quantum computing.

In this paper, we are interested in the query complexity of the quantum algorithms. We use modifications of Grover's search algorithm [13,17] as quantum subroutines. For these subroutines, time complexity can be obtained from query complexity by multiplication to $O(\log D)$, where D is the size of a search space [9,18].

3 Quantum Algorithm for the Multiple String Matching Problem

Let us present a quantum algorithm for the Multiple String Matching Problem. A typical solution for the problem is the Aho-Corasick Algorithm that works with $O(n + L)$ query and time complexity. Here we present an algorithm that works with $O(n + \sqrt{mL \log n} + m \log n)$ query complexity. The algorithm uses ideas from the paper of Manber and Myers [39] and uses quantum strings comparing algorithm [10,21,25].

Firstly, let us construct a suffix array, LCP array, and invoke preprocessing for LCPSuf function for the string t (see Sect. 2 for details). Note that $n = |t|$.

Let us consider a string s^j for $j \in \{1, \dots, m\}$. If s^j is a substring of t starting from an index i, then s^j is a prefix of the suffix $t[suf_i : n]$, i.e. $t[suf_i : suf_i + |s^j| - 1] = s^j$. Since $t[suf_i : n]$ strings are in a lexicographical order according to suf order, all suffixes that have s^j as prefixes are situated sequentially. Our goal is to find two indexes $left_j$ and $right_j$ such that all suffixes $t[suf_i : n]$ has s^j as a prefix for each $i \in \{suf_{left_j}, \dots, suf_{right_j}\}$. So, we can say that $I_j = (suf_{left_j}, \dots, suf_{right_j})$.

There is a quantum algorithm for computing LCP of u and v. Let $\mathrm{QLCP}(u, v)$ be the corresponding function. Its complexity is presented in the next lemma.

Lemma 4 ([20]). *There is a quantum algorithm that implements* $\mathrm{QLCP}(u, v)$ *procedure, and has query complexity* $O(\sqrt{d})$ *and time complexity* $O(\sqrt{d}\log d)$*, where* d *is the minimal index of an unequal symbol of* u *and* v*. The error probability is at most* 0.1.

The algorithm is an application of the First-One Search algorithm [16,20, 33,38] that finds a minimal argument of a Boolean-valued function (predicate) that has the 1-result. The First-One Search algorithm algorithm has an error probability 0.1 that is why we have the same error probability for QLCP procedure. The First-One Search is an algorithm that is based on the Grover Search algorithm [13,17]. So, the difference between query and time complexity is a log factor due to [9,18]. The $\mathrm{QLCP}(u, v)$ procedure is similar to the quantum algorithm that compares two strings u and v in lexicographical order that was independently developed in [10] and in [21,25].

For $i \in \{1, \dots, min(|u|, |v|)\}$, let $\mathrm{QLCP}(u, v, i)$ be a similar function that computes LCP starting from the i-th symbol. Formally, $\mathrm{QLCP}(u, v, i) = \mathrm{QLCP}(u[i : |u|], v[i : |v|])$. Due to Lemma 4, complexity of $\mathrm{QLCP}(u, v, i)$ is following

Corollary 1. *There is a quantum algorithm that implements* $\mathrm{QLCP}(u, v, i)$ *procedure, and has* $O(\sqrt{d-i})$ *query complexity and* $O(\sqrt{d-i}\log(d-i))$ *time complexity, where* d *is the minimal index of an unequal symbol of* u *and* v *such that* $d > i$*. The error probability is at most* 0.1.

Let us present the algorithm that finds $left_j$. Let us call it LEFTBORDER SEARCH(J). The algorithm is based on the Binary search algorithm [15]. Let Le be the left border of the search segment, and Ri be the right border of the search segment. Let $St_i = t[suf_i : n]$ be i-th suffix for $i \in \{1, \dots, n\}$.

Step 1. We assign $Le \leftarrow 1$ and $Ri \leftarrow n$. Let $Llcp \leftarrow \mathrm{QLCP}(St_{Le}, s^j)$ be the LCP of the first suffix and the searching string. Let $Rlcp \leftarrow \mathrm{QLCP}(St_{Ri}, s^j)$ be the LCP of the last suffix and the searching string.
Step 2. If $Llcp < |s^j|$ and $s^j < St_1$, i.e. $Llcp < |s^j|$ and $s^j[Llcp + 1] < St_1[Llcp + 1]$, then we can say that s^j is less than any suffix of t and is not a prefix of any suffix of t. So, we can say that I_j is empty. In that case, we stop the algorithm, otherwise, we continue with Step 3.
Step 3. If $Rlcp < |s^j|$ and $s^j > St_n$, i.e. $Rlcp < |s^j|$ and $s^j[Rlcp + 1] > St_n[Rlcp + 1]$, then we can say that s^j greats any suffix of t and is not a prefix of any suffix of t. So, we can say that I_j is empty also. In that case, we stop the algorithm, otherwise, we continue with Step 4.

Step 4. We repeat the next steps while $Ri - Le > 1$, otherwise we go to Step 9.
Step 5. Let $M \leftarrow \lfloor (Le + Ri)/2 \rfloor$.
Step 6. If $Llcp \geq Rlcp$, then we go to Step 7, and to Step 8 otherwise.
Step 7. Here we compare $LCP(St_L, St_M)$ and $Llcp$. Note that $LCP(St_{Le}, St_M)$ can be computed as LCPSUF(t, Le, M) in $O(1)$ query and time complexity. We have one of three options:

- If $LCP(St_{Le}, St_M) > Llcp$, then all suffixes from St_{Le} to St_M are such that $St_M[Llcp+1] = \cdots = St_{Le}[Llcp+1] \neq s^j[Llcp+1]$, and they cannot have s^j as a prefix. So, we can assign $Le \leftarrow M$ and not change $Llcp$.
- If $LCP(St_{Le}, St_M) = Llcp$, then all suffixes from St_{Le} to St_M has at least $Llcp$ common symbols with s^j. Let us compute $Mlcp = LCP(St_M, s^j)$ using QLCP$(St_M, s^j, Llcp{+}1)$. If $Mlcp = |s^j|$, then we can move the right border to M and update $Rlcp$ because we search the leftmost occurrence of s^j. So, $Ri \leftarrow M$ and $Rlcp \leftarrow Mlcp$. A similar update of R and $Rlcp$ we do if $St_M[Mlcp+1] > s^j[Mlcp+1]$. If $St_M[Mlcp+1] < s^j[Mlcp+1]$, then $Le \leftarrow M$ and $Llcp \leftarrow Mlcp$.
- If $LCP(St_{Le}, St_M) < Llcp$, then all suffixes from St_M to St_R cannot have s^j as a prefix. Therefore, we update $Ri \leftarrow M$, and $Rlcp \leftarrow LCP(St_{Le}, St_M)$.

After that, we go back to Step 4.
Step 8. The step is similar to Step 7, but we compare $LCP(St_M, St_{Ri})$ with $Rlcp$. We have one of three options:

- If $LCP(St_M, St_{Ri}) > Rlcp$, then $Ri \leftarrow M$ and we do not change $Rlcp$.
- If $LCP(St_M, St_{Ri}) = Rlcp$, then we compute $Mlcp = LCP(St_M, s^j)$ using QLCP$(St_M, s^j, Rlcp + 1)$. Then, we update variables exactly by the same rule as in the second case of Step 7.
- If $LCP(St_M, St_{Ri}) < Rlcp$, then we update $L \leftarrow M$, and $Llcp \leftarrow LCP(St_M, St_{Ri})$.

After that, we go back to Step 4.
Step 9. The result of the search is Ri. So, we assign $left_j \leftarrow Ri$.

Similarly, we can compute $right_j$ in a procedure RIGHTBORDERSEARCH(j).

The implementation for LEFTBORDERSEARCH(j) is presented in Algorithm 1, and the complexity is presented in Lemma 5.

Algorithm 1. Quantum algorithm for LeftBorderSearch(j) that searches $left_j$.

```
Llcp ← QLCP(St_1, s^j), Rlcp ← QLCP(St_n, s^j)
if Llcp = |s^j| then
    answer ← 0
else if s^j[Llcp + 1] < St_1[Llcp + 1] then
    answer ← −1
else if s^j[Rlcp + 1] > St_n[Rlcp + 1] then
    answer ← −1
else
    Le ← 1, Ri ← n
    while Ri − Le > 1 do
        M ← ⌊(Le + Ri)/2⌋
        if Llcp ≥ Rlcp then
            if LCPSuf(Le, M) > Llcp then
                Le ← M
            end if
            if LCPSuf(Le, M) = Llcp then
                Mlcp = QLCP(St_M, s^j, Llcp + 1)
                if Mlcp = |s^j| or St_M[Mlcp + 1] > s^j[Mlcp + 1] then
                    Ri ← M, Rlcp ← Mlcp
                else
                    Le ← M, Llcp ← Mlcp
                end if
            end if
            if LCPSuf(Le, M) < Llcp then
                Ri ← M, Rlcp ← LCPSuf(Le, M)
            end if
        else
            if LCPSuf(M, Ri) > Rlcp then
                Ri ← M
            end if
            if LCPSuf(M, Ri) = Rlcp then
                Mlcp = QLCP(St_M, s^j, Rlcp + 1)
                if Mlcp = |s^j| or St_M[Mlcp + 1] > s^j[Mlcp + 1] then
                    Ri ← M, Rlcp ← Mlcp
                else
                    Le ← M, Llcp ← Mlcp
                end if
            end if
            if LCPSuf(M, Ri) < Rlcp then
                Le ← M, Llcp ← LCPSuf(M, Ri)
            end if
        end if
    end while
    answer ← Ri
end if
return answer
```

Lemma 5. *Algorithm 1 implements* LeftBorderSearch(j)*, searches* $left_j$ *and works with* $O(\sqrt{|s^j| \log n} + \log n)$ *query complexity,* $O(\sqrt{|s^j| \log n} \log |s^j| + \log n)$ *time complexity, and at most* 0.1 *error probability.*

Proof. Let us consider *Llcp*. It is always increased and never decreased. Assume that during the algorithm, it has values $Llcp_1, \dots, Llcp_d$. We can say that $Llcp_1 < \dots < Llcp_d$, and $d \leq \log_2 n$ because Binary search do at most $\log_2 n$ steps. Additionally, we can say that $Llcp_d \leq |s^j|$. The function QLCP increases *Llcp*. Query complexity of changing *Llcp* from $Llcp_i$ to $Llcp_{i+1}$ is $O(\sqrt{Llcp_{i+1} - Llcp_i})$ due to Corollary 1. The time complexity is

$$O(\sqrt{Llcp_{i+1} - Llcp_i} \log(Llcp_{i+1} - Llcp_i)) = O(\sqrt{Llcp_{i+1} - Llcp_i} \log(|s^j|)).$$

So, the total query complexity is

$$\sqrt{Llcp_1} + \sum_{i=1}^{d-1} O(\sqrt{Llcp_{i+1} - Llcp_i}) = O\left(\sqrt{d(Llcp_1 + \sum_{i=1}^{d-1}(Llcp_{i+1} - Llcp_i))}\right)$$

due to Cauchy–Bunyakovsky–Schwarz inequality. At the same time, $Llcp_1 + \sum_{i=1}^{d-1}(Llcp_{i+1} - Llcp_i) = Llcp_d \leq |s^i|$. Finally, the complexity of all changing of *Llcp* is $O(\sqrt{d|s^j|}) = O(\sqrt{|s^j| \log n})$. Using the same technique, we can show that the time complexity is $O(\sqrt{|s^j| \log n} \log |s^j|)$ Similarly, we can see that the complexity of all changing of *Rlcp* is $O(\sqrt{d|s^j|}) = O(\sqrt{|s^j| \log n})$. All other parts of the step of Binary search work with $O(1)$ query and time complexity, including LCPSuf due to Lemma 3. They give us additional $O(\log n)$ complexity.

Note that each invocation of QLCP has an error probability 0.1. Therefore, $O(\log n)$ invocations of the procedure have an error probability that is close to 1. At the same time, the whole Algorithm 1 is a sequence of First-One Search procedures such that next invocation of the First-One Search uses results of the previous invocations. Due to [33], such sequence can be converted to an algorithm with the same total complexity and error probability 0.1. □

The whole algorithm is presented in Algorithm 2, and complexity is discussed in Theorem 1

Algorithm 2. Quantum algorithm for the Multiple String Matching Problem.

```
ConstructSuffixArray(t)
PreprocessingForLCP(t)
for j ∈ {1, ..., m} do
    left_j ← LeftBorderSearch(j)
    right_j ← RightBorderSearch(j)
    if left_j = −1 or right_j = −1 then
        I_j ← ()
    else
        I_j ← (suf_{left_j}, ..., suf_{right_j})
    end if
end for
return (I_1, ..., I_m)
```

Theorem 1. *Algorithm 2 solves the Multiple String Matching Problem and works with* $O(n+\sqrt{mL\log n}+m\log n)$ *query complexity,* $O(n+\sqrt{mL\log n}\log b+m\log n)$ *time complexity and at most* 0.1 *error probability.*

Proof. Firstly, complexity of ConstructSuffixArray(t) and PreprocessingForLCP(t) is $O(n)$ due to Lemmas 1, 2, and 3.

Query complexity of LeftBorderSearch(j) and RightBorderSearch(j) is $O(\sqrt{|s^j|\log n}+\log n)$, and time complexity is $O(\sqrt{|s^j|\log n}\log|s^j|+\log n)$. So, the total query complexity of these procedures for all $j\in\{1,\dots,m\}$ is

$$O\left(m\log n+\sum_{j=1}^{m}\sqrt{|s^j|\log n}\right)=O\left(m\log n+\sqrt{\log n}\sum_{j=1}^{m}\sqrt{|s^j|}\right)$$

$$=O\left(m\log n+\sqrt{\log n\cdot m\sum_{j=1}^{m}|s^j|}\right)=O\left(m\log n+\sqrt{mL\log n}\right)$$

We get such a result because of Cauchy–Bunyakovsky–Schwarz inequality. Similarly, time complexity is $O\left(m\log n+\sqrt{mL\log n}\log b\right)$ since $b=max\{|s^1|,\dots,|s^m|\}$.

The total query complexity is $O(n+\sqrt{mL\log n}+m\log n)$, and time complexity is $O(n+\sqrt{mL\log n}\log d+m\log b)$.

Note that, each invocation of LeftBorderSearch(j) and RightBorderSearch(j) has error probability 0.1. Therefore, $O(m)$ invocations of the procedure have an error probability that is close to 1. At the same time, the algorithm is a sequence of First-One Search procedures. Due to [33], such sequence can be converted to an algorithm with the same total complexity and error probability 0.1. □

Let us present the lower bound for the problem.

Theorem 2. *The lower bound for query complexity of the Multiple String Matching Problem is* $\Omega(n+L)$ *in the classical case, and* $\Omega(n+\sqrt{mL})$ *in the quantum case.*

Proof. Let us consider a binary alphabet for simplicity. Let $s^1=$ "1". Then, the complexity of searching positions of s^1 in t is at least as hard as an unstructured search of all positions of 1 among n bits. Due to [12], classical query complexity of this problem is $\Omega(n)$ and quantum query complexity is $\Omega(\sqrt{nt})$, where t is the number of occurrences 1 among n bits. In the worst case, it is also $\Omega(n)$.

Let us consider t as a string of 0s. In that case, if we have any symbol 1 in a dictionary string, then this string does not occur in t. We should find all such strings for forming results for these strings. Number of such strings is at most m. The size of the search space is L. Then, the complexity of searching the answer is at least as hard as an unstructured search of at least m positions of 1 among L bits. Due to [12], classical query complexity of this problem is $\Omega(n)$ and quantum query complexity is $\Omega(\sqrt{mL})$

So, the total lower bound is $\Omega(max(n,L))=\Omega(n+L)$ in the classical case and $\Omega(max(\sqrt{n},\sqrt{mL}))=\Omega(\sqrt{n}+\sqrt{mL})$ in the quantum case. □

4 Conclusion

In the paper, we present a quantum algorithm for the Multiple String Matching Problem that works with $O(n + \sqrt{mL} + m \log n)$ query complexity and error probability 0.1. It is better than the classical counterparts if $O(m)$ strings of the dictionary have at least $\omega(\log n)$ length. In that case, $\sqrt{mL \log n} = o(L)$ and $m \log n = o(L)$. Similar situation for time complexity. We obtain speed up if $m \log n (\log b)^2 = o(L)$, or at lest $O(m)$ strings have length $\omega(\log n \log \log n)$.

In the quantum case the lower bound is $\Omega(n + \sqrt{mL})$ and upper bound is $O^*(n+\sqrt{mL})$. They are equal up to a log factor. The open question is to develop a quantum algorithm with complexity that is equal to the lower bound.

Acknowledgments. We sincerely thank Aliya Khadieva for her help and usefull discussions.

This paper has been supported by the Kazan Federal University Strategic Academic Leadership Program ("PRIORITY-2030").

References

1. Ablayev, F., Ablayev, M., Huang, J.Z., Khadiev, K., Salikhova, N., Wu, D.: On quantum methods for machine learning problems part I: quantum tools. Big Data Min. Anal. **3**(1), 41–55 (2019)
2. Ablayev, F., Ablayev, M., Khadiev, K., Salihova, N., Vasiliev, A.: Quantum algorithms for string processing. In: Badriev, I.B., Banderov, V., Lapin, S.A. (eds.) Mesh Methods for Boundary-Value Problems and Applications. LNCSE, vol. 141, pp. 1–14. Springer, Cham (2022). https://doi.org/10.1007/978-3-030-87809-2_1
3. Aho, A.V., Corasick, M.J.: Efficient string matching: an aid to bibliographic search. Commun. ACM **18**(6), 333–340 (1975). https://doi.org/10.1145/360825.360855
4. Akmal, S., Jin, C.: Near-optimal quantum algorithms for string problems. In: Proceedings of the 2022 Annual ACM-SIAM Symposium on Discrete Algorithms (SODA), pp. 2791–2832. SIAM (2022)
5. Allauzen, C., Crochemore, M., Raffinot, M.: Factor oracle: a new structure for pattern matching. In: Pavelka, J., Tel, G., Bartošek, M. (eds.) SOFSEM 1999. LNCS, vol. 1725, pp. 295–310. Springer, Heidelberg (1999). https://doi.org/10.1007/3-540-47849-3_18
6. Ambainis, A.: Understanding quantum algorithms via query complexity. In: Proceedings of the International Congress of Mathematicians 2018, vol. 4, pp. 3283–3304 (2018)
7. Ambainis, A., et al.: Quantum lower and upper bounds for 2D-grid and Dyck language. In: 45th International Symposium on Mathematical Foundations of Computer Science (MFCS 2020). Leibniz International Proceedings in Informatics (LIPIcs), vol. 170, pp. 8:1–8:14 (2020)
8. Ambainis, A., et al.: Quantum bounds for 2D-grid and Dyck language. Quantum Inf. Process. **22**(5), 194 (2023)
9. Arunachalam, S., de Wolf, R.: Optimizing the number of gates in quantum search. Quantum Inf. Comput. **17**(3&4), 251–261 (2017)

10. Babu, H.M.H., Jamal, L., Dibbo, S.V., Biswas, A.K.: Area and delay efficient design of a quantum bit string comparator. In: 2017 IEEE Computer Society Annual Symposium on VLSI (ISVLSI), pp. 51–56. IEEE (2017)
11. Bender, M.A., Farach-Colton, M.: The LCA problem revisited. In: Gonnet, G.H., Viola, A. (eds.) LATIN 2000. LNCS, vol. 1776, pp. 88–94. Springer, Heidelberg (2000). https://doi.org/10.1007/10719839_9
12. Bennett, C.H., Bernstein, E., Brassard, G., Vazirani, U.: Strengths and weaknesses of quantum computing. SIAM J. Comput. **26**(5), 1510–1523 (1997)
13. Boyer, M., Brassard, G., Høyer, P., Tapp, A.: Tight bounds on quantum searching. Fortschr. Phys. **46**(4–5), 493–505 (1998)
14. Commentz-Walter, B.: A string matching algorithm fast on the average. In: Maurer, H.A. (ed.) ICALP 1979. LNCS, vol. 71, pp. 118–132. Springer, Heidelberg (1979). https://doi.org/10.1007/3-540-09510-1_10
15. Cormen, T.H., Leiserson, C.E., Rivest, R.L., Stein, C.: Introduction to Algorithms. McGraw-Hill (2001)
16. Dürr, C., Heiligman, M., Høyer, P., Mhalla, M.: Quantum query complexity of some graph problems. SIAM J. Comput. **35**(6), 1310–1328 (2006)
17. Grover, L.K.: A fast quantum mechanical algorithm for database search. In: Proceedings of the Twenty-Eighth Annual ACM Symposium on Theory of Computing, pp. 212–219. ACM (1996)
18. Grover, L.K.: Trade-offs in the quantum search algorithm. Phys. Rev. A **66**(5), 052314 (2002)
19. Jordan, S.: Quantum algorithms zoo (2023). http://quantumalgorithmzoo.org/
20. Kapralov, R., Khadiev, K., Mokut, J., Shen, Y., Yagafarov, M.: Fast classical and quantum algorithms for online k-server problem on trees. In: CEUR Workshop Proceedings, vol. 3072, pp. 287–301 (2022)
21. Khadiev, K., Ilikaev, A.: Quantum algorithms for the most frequently string search, intersection of two string sequences and sorting of strings problems. In: International Conference on Theory and Practice of Natural Computing, pp. 234–245 (2019)
22. Khadiev, K.: Lecture notes on quantum algorithms. arXiv preprint arXiv:2212.14205 (2022)
23. Khadiev, K., Bosch-Machado, C.M., Chen, Z., Wu, J.: Quantum algorithms for the shortest common superstring and text assembling problems. Quantum Inf. Comput. **24**(3–4), 267–294 (2024)
24. Khadiev, K., Enikeeva, S.: Quantum version of self-balanced binary search tree with strings as keys and applications. In: International Conference on Micro- and Nano-Electronics 2021, vol. 12157, pp. 587 – 594. International Society for Optics and Photonics, SPIE (2022). https://doi.org/10.1117/12.2624619
25. Khadiev, K., Ilikaev, A., Vihrovs, J.: Quantum algorithms for some strings problems based on quantum string comparator. Mathematics **10**(3), 377 (2022)
26. Khadiev, K., Kravchenko, D.: Quantum algorithm for dyck language with multiple types of brackets. In: Unconventional Computation and Natural Computation, pp. 68–83 (2021)
27. Khadiev, K., Machado, C.M.B.: Quantum algorithm for the shortest superstring problem. In: International Conference on Micro- and Nano-Electronics 2021, vol. 12157, pp. 579 – 586. International Society for Optics and Photonics, SPIE (2022). https://doi.org/10.1117/12.2624618
28. Khadiev, K., Remidovskii, V.: Classical and quantum algorithms for assembling a text from a dictionary. Nonlinear Phenom. Complex Syst. **24**(3), 207–221 (2021)

29. Khadiev, K., Remidovskii, V.: Classical and quantum algorithms for constructing text from dictionary problem. Nat. Comput. **20**(4), 713–724 (2021). https://doi.org/10.1007/s11047-021-09863-1
30. Khadiev, K., Savelyev, N., Ziatdinov, M., Melnikov, D.: Noisy tree data structures and quantum applications. Mathematics **11**(22), 4707 (2023)
31. Khadiev, K., Serov, D.: Quantum property testing algorithm for the concatenation of two palindromes language. In: Cho, D.J., Kim, J. (eds.) UCNC 2024. LNCS, vol. 14776, pp. 134–147. Springer, Cham (2024). https://doi.org/10.1007/978-3-031-63742-1_10
32. Knuth, D.E., Morris, J.H., Jr., Pratt, V.R.: Fast pattern matching in strings. SIAM J. Comput. **6**(2), 323–350 (1977)
33. Kothari, R.: An optimal quantum algorithm for the oracle identification problem. In: 31st International Symposium on Theoretical Aspects of Computer Science, pp. 482–493 (2014)
34. Kasai, T., Lee, G., Arimura, H., Arikawa, S., Park, K.: Linear-time longest-common-prefix computation in suffix arrays and its applications. In: Amir, A. (ed.) CPM 2001. LNCS, vol. 2089, pp. 181–192. Springer, Heidelberg (2001). https://doi.org/10.1007/3-540-48194-X_17
35. Le Gall, F., Seddighin, S.: Quantum meets fine-grained complexity: Sublinear time quantum algorithms for string problems. Algorithmica 1–36 (2022)
36. Le Gall, F., Seddighin, S.: Quantum meets fine-grained complexity: sublinear time quantum algorithms for string problems. In: 13th Innovations in Theoretical Computer Science Conference (ITCS 2022). Schloss Dagstuhl-Leibniz-Zentrum für Informatik (2022)
37. Li, Z., Li, J., Huo, H.: Optimal in-place suffix sorting. In: Gagie, T., Moffat, A., Navarro, G., Cuadros-Vargas, E. (eds.) SPIRE 2018. LNCS, vol. 11147, pp. 268–284. Springer, Cham (2018). https://doi.org/10.1007/978-3-030-00479-8_22
38. Lin, C.Y.Y., Lin, H.H.: Upper bounds on quantum query complexity inspired by the Elitzur-Vaidman bomb tester. Theory Comput. **12**(18), 1–35 (2016)
39. Manber, U., Myers, G.: Suffix arrays: a new method for on-line string searches. In: Proceedings of the First Annual ACM-SIAM Symposium on Discrete Algorithms, SODA 1990, pp. 319–327. Society for Industrial and Applied Mathematics (1990)
40. Montanaro, A.: Quantum pattern matching fast on average. Algorithmica **77**(1), 16–39 (2017)
41. Nielsen, M.A., Chuang, I.L.: Quantum Computation and Quantum Information. Cambridge University Press, Cambridge (2010)
42. Ramesh, H., Vinay, V.: String matching in $O(\sqrt{n}+\sqrt{m})$ quantum time. J. Discrete Algorithms **1**(1), 103–110 (2003)
43. de Wolf, R.: Quantum computing and communication complexity. University of Amsterdam (2001)
44. Wu, S., Manber, U.: A fast algorithm for multi-pattern searching. Technical report TR-94-17, Department of Computer Science, University of Arizona, Tucson, AZ (1994)

Parallel Peeling of Invertible Bloom Lookup Tables in a Constant Number of Rounds

Michael T. Goodrich[1(✉)], Ryuto Kitagawa[1], and Michael Mitzenmacher[2]

[1] University of California, Irvine, USA
goodrich@acm.org
[2] Harvard University, Cambridge, USA

Abstract. Invertible Bloom lookup tables (IBLTs) are a compact way of probabilistically representing a set of n key-value pairs so as to support insertions, deletions, and lookups. If an IBLT is not overloaded (as a function of its size and number of key-value pairs that have been inserted), then reporting all the stored key-value pairs can also be done via a "parallel peeling" process. For the case when the IBLT is represented in a very compact form, this can be implemented to run in $O(\log \log n)$ parallel rounds, with all but inversely polynomial probability, as shown in prior work by Jiang, Mitzenmacher, and Thaler, as well as in Gao's work on parallel peeling algorithms for random hypergraphs. Although $O(\log \log n)$ is practically constant for reasonable values of n, there are nevertheless scenarios (such as in the parallel GPU or MapReduce frameworks) where parallel peeling is desired to run in a constant number of rounds, with failure probabilities that are negligible rather than simply being polynomially small. In this paper, we study simple constant-round parallel peeling algorithms for IBLTs, focusing on negligible failure probabilities based on table size, number of elements stored, and number of hash functions. We prove the surprising result that with $O(n \log n)$ space a one-round parallel peeling process succeeds with high probability while a two-round parallel peeling process succeeds with overwhelming probability. We then provide a time-space trade-off theorem for parallel peeling in a constant k number of rounds while still maintaining overwhelming success probability. We also give several new algorithmic applications of parallel peeling of IBLTs and we experimentally study the effectiveness of our approach in practice.

1 Introduction

An ***invertible Bloom lookup table*** (***IBLT***) [16,23,36] is a probabilistic hash-based data structure that concisely represents a set of key-value pairs to support insertion, deletion, lookup, and (if the IBLT is not overloaded) the listing of all the stored key-value entries. Previous work on IBLTs has focused primarily on applications of IBLTs addressing the distributed computing challenges of set reconciliation, blockchain reconciliation, and network synchronization [16,

R. Královič and V. Kůrková (Eds.): SOFSEM 2025, LNCS 15539, pp. 70–84, 2025.
https://doi.org/10.1007/978-3-031-82697-9_6

17,19,21,30,34]. In these applications, one may have a large distributed data store, and one is interested in synchronizing a set of data files or blocks across multiple servers.

Invertible Bloom Lookup Tables. When an IBLT $\mathcal{B}$ is first created, it initializes k arrays $T_1, \ldots, T_k$ of m cells, so each T_i has m/k cells. Each of the cells in a T_i table stores a constant number of fields, each of which, in turn, corresponds to a single memory word (we define these fields below), which is initially 0. An important feature of an IBLT is that at times the number of key-value pairs in $\mathcal{B}$ can be large (even larger than m), but the space used for $\mathcal{B}$ remains $O(m)$ words. The insert and delete methods never fail, whereas the listEntries method, which lists out all the key-value pairs, only guarantees good probabilistic success when the number of stored key-value pairs, n, is below an appropriate threshold. (For more details about the thresholds, see the prior works by Goodrich and Mitzenmacher [23] or Molloy [31,32].)

An IBLT uses a set of k random hash functions, h_1, h_2, ..., h_k, to determine where key-value pairs are stored. In our case, each key-value pair, (x, y), is placed into cells $T_1[h_1(x)]$, $T_2[h_2(x)]$, ... $T_k[h_k(x)]$, respectively, with fields that support all the IBLT operations. We sometimes refer to this subdivision of the IBLT into k tables as "splitting." Such splitting does not affect the asymptotic behavior in our analysis and can yield other benefits, including ease of parallelization of reads and writes into the hash table, as we show. Each cell contains the following four fields:

- count, which is a (signed) count of the number of key-value pairs that have been mapped to this cell,
- keySum, which is the XOR of every key, x, that has been mapped to this cell,
- valueSum, which is the XOR of every value, y, that has been mapped to this cell,
- hashSum, which is the XOR of a cryptographic hash, $g(x)$, of the key, x, for every key-value pair, (x, y), that has been mapped to this cell. This field is used for error-checking purposes.

Inserting a new key-value pair, (x, y), involves incrementing the count field for each cell, $T_i[h_i(x)]$, and XOR-ing into the other fields the respective parameters. Similarly, deleting a key-value pair, (x, y), involves decrementing the count field for each cell, $T_i[h_i(x)]$, and XOR-ing into the other fields the respective parameters, since every number is its own inverse under the XOR operation. For applications where we are finding symmetric set differences, we also allow for counts to become negative; that is you can delete a key-value pair that has not been inserted (as we described in the introduction). Hence a count of negative 1 can correspond to a cell that contains a single item that has been deleted. However, a count of 1 or negative 1 might correspond to multiple key-value pairs, some of which have been inserted and some of which have been deleted. The hashSum field is meant to provide an additional layer of protection to ensure that a count of 1 or negative 1 truly corresponds to a single key, by comparing the hash of the key to this field. (Note the hashSum field is not necessary if only inserting items, or only deleting items that have previously been inserted.)

The listEntries operation is more interesting. The usual way it is described is in terms of a sequential ***peeling*** process, where we look for a cell with a count field that is 1 or −1, remove and report the key-value pair that is stored there, and repeat. Jiang, Mitzenmacher, and Thaler [27] study a parallel version of this peeling process, but their approach is focused on optimizing the space used, not the number of rounds, and their analysis results in a non-constant number of parallel peeling rounds. Instead, in this paper, we are interested in parameters that result in one or two or some constant, k, number of rounds of parallel peeling. For example, such approaches allow for parallel peeling to be more easily implemented in the MapReduce framework, since we can specify a specific number of peeling rounds (possibly even just one round). We describe pseudocode for a parallel peeling algorithm in Fig. 1.

- listEntries():

```
while there is an i and j such that T_i[j].count = 1 or −1 do
  for each such i and j in parallel do
    if T_i[h_i(x)].hashSum = g(T_i[h_i(x)].keySum) then
      output the pair, (T_i[j].keySum , T_i[j].valueSum)
      if T_i[j].count = 1 then
        call delete(T_i[j].keySum, T_i[j].valueSum), parallelizing the XORs
      else
        call insert(T_i[j].keySum, T_i[j].valueSum), parallelizing the XORs
```

Fig. 1. Listing entries in an IBLT. Note that in the case of a count that is −1 we actually remove its corresponding key-value pair by performing an insert.

One of the unfortunate aspects of the pseudocode for the listEntries algorithm, from a parallel implementation standpoint, is that the outer loop is a while loop that implies a conditional number of iterations, which we call ***peeling rounds***. Implementing parallel peeling can be much easier, however, if we can simply bound the number of parallel peeling rounds to be one or two or some constant, k, and hard code that constant into our implementation, using, say, the MapReduce framework (which also simplifies combining the XORs for colliding cells). This desire, therefore, motivates the analysis that follows, which shows how to set the parameters for an IBLT to guarantee with high probability (or all but negligible probability)[1] that parallel peeling can succeed in one, two, or k rounds.

Set Synchronization. For example, using IBLTs allows one to synchronize two data sets, S_1 and S_2, using storage and communication proportional to the

[1] We say that an event occurs with high probability if the failure probability is $1/n^c$, for some constant, $c > 0$, and with overwhelming probability if the failure probability is negligible, that is, the failure probability is asymptotically less than $1/n^c$ for any constant $c > 0$.

size of symmetric set difference between the two sets, $n = |S_1 \oplus S_2|$, rather than the size of the sets themselves.[2]

At a high level, synchronizing a set S that is intended to be mirrored at two servers, which we call "Alice" and "Bob," works as follows. Alice and Bob hold item sets S_A and S_B, respectively, with $S = S_A \cup S_B$. Alice and Bob each store their items in respective IBLTs, T_A and T_B, each holding m cells, where m is linear in the size of the set difference $S_A \oplus S_B$, and each sends their IBLT to the other. Then, each of Alice and Bob computes an entrywise table difference $T_A - T_B$, which allows both Alice and Bob to list out the elements of $S_A \oplus S_B$ with high probability, so long as the size of that difference is indeed at most a threshold that depends only on and is linear in m. Note that computing the table difference also yields which set (Alice's or Bob's) each element in the symmetric difference belongs to; we therefore say that IBLTs provide *signed* symmetric set differences. While this is an interesting and useful application of IBLTs, in this paper we are interested in applications of IBLTs to another challenge that arises in large-scale distributed computing, namely, for data compression [38] and deduplication [44], which can involve using IBLTs in arguably a more "parallel" way. For instance, in data compression and deduplication applications, one is often interested in computing the symmetric or set differences between all pairs of a collection of sets, not just two. Thus, we desire IBLT difference operations that can be done quickly in parallel, where each difference itself is computed using a parallel algorithm, ideally with a constant number of computation rounds.

Prior Work on Parallel Peeling. Computing the entrywise table difference $T_A - T_B$ between two IBLTs T_A and T_B is easy enough to do in parallel, but listing out the elements of the result in parallel is more challenging. Indeed, the algorithm for performing such a listing is usually described as a sequential ***peeling*** process, where items are removed iteratively one at a time [23,36]. Still, there is prior work on ***parallel peeling*** of various graph structures, e.g., by Cao, Fineman, and Russell [10], Goodrich and Pszona [24], Chang, Pettie, and Zhang [11], Dhulipala, Blelloch, and Shun [14], Ghaffari, Grunau, and Jin [22], Shi, Dhulipala, and Shun [39], and Shi and Shun [40]. More relevant to this paper, Jiang, Mitzenmacher, and Thaler [27] and Gao [20] provide parallel peeling results for random hypergraphs and IBLTs. These results yield a super-constant number of rounds, however, rather than a small constant number of rounds, as would be desired for the applications we explore here. Specifically, to peel n items in the IBLT, their results correspond to $\Theta(\log \log n)$ rounds of parallel peeling.

Our Results. In this paper, we study constant-round parallel peeling both theoretically and experimentally. What makes our study of the IBLT data structure different from prior parallel-peeling results is that here we focus more on optimizing (parallel) time and work and less on space, which was the focus in prior works [20,27]. In particular, we guarantee that with high (and, as we describe later, with overwhelming) probability our parallel peeling process com-

[2] We denote the symmetric set difference of two sets, S_1 and S_2, by $S_1 \oplus S_2$. Recall that $S_1 \oplus S_2 = (S_1 - S_2) \cup (S_2 - S_1)$, which is sometimes alternatively denoted as $S_1 \ominus S_2$ or $S_1 \Delta S_2$.

pletes in a small constant number of rounds—ideally one or two. Past work on parallel peeling for IBLTs [20,27], instead aimed for a constant number of hash functions per item, a linear number of cells in the IBLT table, and an asymptotically good, but super-constant, number of rounds (where additional $O(1)$ terms were not considered consequential) with high, but not overwhelming probability. On the theoretical front, we show that if an IBLT has $O(n \log n)$ cells and uses $O(\log n)$ hash functions (where suitably chosen constant factors are used in the order notation), then, with high probability, we can peel the IBLT in a single parallel round. More surprisingly, we also show that the probability that we would fail to peel such an IBLT in only two rounds is negligible (recall that a function is ***negligible*** if it approaches zero faster than the reciprocal of any polynomial, e.g., see Bellare [4]). Further, we provide a full time-space trade-off between the number of rounds and space needed to achieve constant-round parallel peeling with overwhelming probability. Also, we explicitly describe how to use IBLTs in some interesting applications, including deduplication and symmetric-difference minimum spanning trees.

2 Analysis

In this section, we provide our theoretical analysis, showing that, for a table of $m = O(n \log n)$ cells, parallel peeling succeeds with high probability in one round and with overwhelming probability in two, and we then provide a time-space trade-off while achieving overwhelming probability with a constant number of rounds. We note that in the analysis in this section, we assume that there is no false positive in the peeling process in checking the hashSum field; that is, we assume items have only been inserted, or the fingerprints have been chosen large enough to avoid this issue. Before beginning, we recall that previous work (e.g., [23]) has shown for an IBLT with $m = cn$ cells for some constant c and a constant number of hash functions k, if c is above the threshold where decoding occurs with high probability, the failure probability is $\Theta(n^{-k+2})$.

Analysis of One and Two Rounds. As we are focused on such a small number of rounds, we first explicitly consider the case of a single round and then analyze the probability of parallel peeling succeeding in two rounds. We note that something akin to our analysis for one round appeared previously in work by Eppstein and Goodrich [15], but we nevertheless provide a complete analysis of one round parallel peeling to set the table for our other results, which are novel.

For concreteness, we consider an IBLT with $m = cn \log_2 n$ cells for some constant c. Each of n items hashes to $\log_2 n$ cells (we assume n is a power of two for convenience). As we previously described, we assume the IBLT is ***split***; that is, there are $\log_2 n$ subtables, each with cn cells, and the jth hash of each item is independently and uniformly chosen in the jth subtable. Call a cell ***single*** if it holds exactly one item. We show for sufficiently large constant c each item has at least one hash to a single cell with high probability, implying peelability in a single round.

Theorem 1. *For a table of* $m = cn \log_2 n$ *cells using* $\log_2 n$ *hash functions, where* $c \geq 1$ *is a constant, an IBLT peels in one round with probability at least* $1 - n^{-c'+1}$ *where* $c' = -\log_2(1 - e^{-1/c})$.

Proof: Let $X_{i,j}$ be a random variable such that $X_{i,j} = 1$ if the jth hash of the ith item is not single and 0 otherwise. Since the jth table has cn cells, we have

$$\Pr(X_{i,j} = 0) = (1 - 1/(cn))^{n-1} \geq e^{-1/c}.$$

(The inequality holds as long as $c \geq 1$ for example.) Since subtables are independent, we therefore have the probability that no hash for the ith item is single is at most

$$(1 - e^{-1/c})^{\log_2 n} = n^{\log_2(1-e^{-1/c})} = n^{-c'},$$

for $c' = -\log_2(1 - e^{-1/c})$. By a union bound, all items hash to at least one single cell with probability $n^{-c'+1}$, and one can choose c to obtain any suitably small probability that is inversely polynomial in n. ∎

Thus, in our setup, parallel peeling succeeds in one round with high probability and with reasonable constant factors. For example, one can peel in a single round with probability at least $1 - 1/n$ using slightly less than $3.5n \log_2 n$ cells, and with probability at least $1 - 1/n^2$ with slightly less than $7.5n \log_2 n$ cells. Further, we expect a threshold where the probability that one round suffices jumps toward 1 at slightly over $1.443n \log_2 n$ cells, since $1.443 \approx 1/\ln 2$ and at $c = 1/\ln 2$ we have $c' = 1$.

We note the above proof doesn't really require $c \geq 1$, in that for smaller c we have that $(1-1/(cn))^{n-1} = e^{-1/c}(1-o(1))$ and the proof remains essentially the same. This fact will be helpful in what follows. Also, the proof naturally extends to other scenarios, such as if one uses $\alpha \log n$ hash functions for some constant α (or some larger number of hash functions).

We now consider just the case of two rounds. We use the same setup as before. For an item to be peeled within two rounds, either

- one of its cells is single, or
- one of its cells has that all the other items that hash to that cell are peeled after the first round.

We prove that with just two rounds the failure probability is negligible; specifically, the probability of failing to peel every item is $n^{-\Omega(\log n)}$.

Theorem 2. *For a table of* $m = cn \log_2 n$ *cells using* $\log_2 n$ *hash functions, where* $c > 0$ *is a constant, an IBLT peels in two rounds with probability at least* $1 - n^{-\Omega(\log n)}$.

Proof: Consider a specific item y. Let Y_j be the number of other items that hash to the same cell as y in the jth subtable. We first note that Y_j is at most $(\ln n)^2$ with probability at least $1 - n^{-\Omega(\log n)}$. This follows readily from the Poisson approximation of the number of items that hash to each cell, along with Stirling's approximation. We therefore assume that every cell has at most

$(\ln n)^2$ items that hash to it henceforth, as this conditioning does not affect our arguments further. Let $X_i = 1$ if y is single in the ith table or all other items that share the cell in the ith table all peel after the first round. Otherwise $X_i = 0$. We wish to show that the probability that all $X_i = 0$ is $n^{-\Omega(\log n)}$.

First consider X_1. Let z be some other item in the same cell. Following the same argument as in the proof of Theorem 1, the probability that z is not single in at least one of the other subtables is

$$\left(1 - (1 - 1/(cn))^{n-1}\right)^{\log_2 n - 1},$$

which is at most n^{-c_1}, for some constant c_1. As we have assumed Y_1 is at most $(\ln n)^2$, by a union bound, the probability any of the items is not single in at least one other subtable is at most $(\ln n)^2 n^{-c_1}$, and this shows the probability $X_1 = 0$ is at most $(\ln n)^2 n^{-c_1}$.

If the X_i were all independent, we would now be done. However, the X_i are only "roughly independent"; there are unfortunately some problematic cases to consider. For example, consider $\Pr(X_2 = 1 \mid X_1 = 0)$. That is, what is the probability the second subtable allows us to peel the item y even though the first subtable does not. A difficult subcase showing the dependency is when the same second item hashes to the same location as y in both hash tables. That is, the reason $X_1 = 0$ might be that there is an item z in the second hash table at the same location as y in both the first and second table.

We circumvent the dependency by showing the following two conditions hold:

1. With probability $1 - n^{-\Omega(\log n)}$, at least $(\log_2 n)/2$ subtables have no items in the cell with y that also appear with y in earlier subtables.
2. For each such subtable, the event $X_i = 0$ satisfies $\Pr(X_i = 0 \mid X_1 = 0, X_2 = 0, \ldots, X_{i-1} = 0) = O(n^{-c_2})$, for some constant c_2.

The result would then follow, as (implicitly conditioning on none of the rare $n^{-\Omega(\log n)}$ events we have considered occurring)

$$\Pr(X_1 = 0, X_2 = 0, \ldots, X_{\log_2 n} = 0) =$$

$$\Pr(X_1 = 0) \prod_{i=2}^{\log_2 n} \Pr(X_i = 0 \mid X_1 = 0, X_2 = 0, \ldots, X_{i-1} = 0),$$

and the right hand side would have at least $(\log_2 n)/2$ terms that were $O(n^{-c_2})$.

For the first condition, as we sequentially consider each new subtable, there are at most $O((\log n)^3)$ items that share a cell with y. The probability that in a new subtable any of these elements are in the same cell as y is at most q, where q is $O((\log n)^3)/n$. The probability at least $(\log_2 n)/2$ subtables would fail to avoid such elements is at most

$$\sum_{i=(\log_2 n)/2}^{\log_2 n} \binom{\log_2 n}{i} q^i (1-q)^{\log_2 n - i},$$

which is $n^{-\Omega(\log n)}$.

For the second condition, following previous work, it is useful to think of gathering a history of all the item-cell pairs we have seen as we explore the hash table subtable by subtable. We start with y and its position in all the tables. In the first subtable, we consider all the items that share the cell with y; we then consider the cells for those items in all the other tables (to check if those other items are single in some table, and are hence peeled in the first round), and correspondingly all the items in those cells (which, because we have now seen them, we must consider their effect on the conditioning). Similarly, at the second table, we consider all the items that share the cell with y; we then consider the cells for those items in all the other tables, and correspondingly all the items in those cells. Under our assumption that any cell contains at most $(\ln n)^2$ items, we see that this history can only consist of $O((\log n)^6)$ item-cell pairs throughout the process.

Now consider $\Pr(X_i = 0 \mid X_1 = 0, X_2 = 0, \ldots, X_{i-1} = 0)$ for some subtable i where the items colliding with y are all distinct from other levels. The past history introduces some conditioning in evaluating whether each of these items is single in another subtable, because we know the location of some items in other tables in our history. However, we only know at most $O((\log n)^6)$ item-cell pairs in our history. This has at worst a $1 - o(1)$ effect on the probability $(1 - 1/(cn))^{n-1}$ that an item colliding with y in the ith table is single in another table, as from the calculation in the proof of Theorem 1. In particular, any such new item that collides with y must avoid cells known to contain other items in other subtables. As we have said, there are polylog(n) such cells from our history, so this happens with probability $1 - \text{polylog}(n)/(cn)$ in any given subtable. And it can affect the number of other items that might collide with the new item (instead of $n - 1$ other items may be n − polylog(n), as polylog(n) items might already their position known in a subtable). It remains the case that any new item is single in each subtable with probability $(1 - \text{polylog}(n)/(cn))^{n-\text{polylog}(n)} = e^{-1/c}(1 - o(1))$, and correspondingly each new item is single in some other subtable with probability at least $1 - n^{-c_3}$, for some constant c_3. The probability that all new items in the ith subtable are not single is then $\text{polylog}(n)/n^{c_3} = O(n^{-c_2})$ for some constant c_2, giving the result. ∎

A Time-Space Trade-Off for Constant-Round Peeling. Theorem 2 shows parallel peeling succeeds in two rounds with overwhelming probability. Let us next provide our time-space trade-off.

Theorem 3. *For a table of $m = cn(\log_2 n)^{1/k}$ cells using $(\log_2 n)^{1/k}$ hash functions, where where k is a positive integer constant and $c > 0$ is a constant, an IBLT peels in $k + 1$ rounds with probability at least $1 - n^{-\Omega((\log n)^{1/k})}$.*

Proof: We sketch the proof, since it follows roughly the same conceptual framework as the proof of Theorem 2. Consider a specific item, y. For y not to be peeled in $k + 1$ rounds, in the ith subtable there must be some element z_i that has not been peeled after k rounds in the same cell as y; for each z_i, in each of the other subtables there must be some element that has not been peeled after $k - 1$ rounds in the same cell as z_i, and so on.

Accordingly, the failure of an IBLT to peel corresponds to what is commonly referred to as a witness tree (e.g., [41]), with the root of the tree being an element y not peeled after $k+1$ rounds, connected by labeled edges to children that correspond to elements that share a cell with y in some subtable that are not peeled after k rounds, with the label denoting which subtable y and the other element share a cell. (An element can conceivably be in more than one labeled edge.) These elements have children corresponding to elements that share a cell with them in some other subtable (other than the one in the edge connecting them to the root) that are not peeled after $k-1$ rounds, and so on.

We can again use the fact that there are at most $(\ln n)^2$ elements in any cell with probability at least $1 - n^{-\Omega(\log n)}$ to limit the size of the tree built this way, so that with overwhelming probability the tree has size polylogarithmic in n.

We think of going through this tree in a breadth first manner from the root, and temporarily assume that elements are not repeated as we expand this recursive exploration of the table and likewise ignore dependencies introduced by knowledge of where elements we have seen in the tree are placed. The probability that an element q is not single in at least one of $(\log_2 n)^{1/k} - 1$ subtables is

$$\left(1 - (1 - 1/(cn))^{n-1}\right)^{(\log_2 n)^{1/k} - 1},$$

which is at most $c_1^{(\log_2 n)^{1/k}}$, for some constant $c_1 < 1$ and sufficiently large n. The probability that an element q' in a cell is not peeled after two rounds, implying that in all of the other $(\log_2 n)^{1/k} - 1$ subtables there is some element in the same cell as q' that is not peeled after 1 round, is at most

$$(\ln n)^2 (\log_2 n)^{1/k} \left(c_1^{(\log_2 n)^{1/k}}\right)^{(\log_2 n)^{1/k} - 1} \leq c_2^{(\log_2 n)^{2/k}},$$

for some constant $c_2 < 1$ and sufficiently large n. Continuing in this manner, the probability an element sharing a cell with y in one of its subtables has not been peeled for sufficiently large n is bounded by

$$c_k^{(\log_2 n)^{k/k}} = n^{\log_2 c_k},$$

where $c_k < 1$. The result would correspondingly follow, but for the assumptions that elements in the table are not repeated and that dependencies can be ignored.

As in the proof of Theorem 2, however, because the witness tree is only polylogarithmic in size, this does not affect the asymptotics of the result; the dependencies can be dealt with by using the fact that we only see a polylogarithmic number of elements throughout the tree, and most of the cells will not contain repeated elements from the tree. ∎

3 Applications

In this section, we describe some applications of constant-round parallel peeling algorithms for IBLTs to a simplified deduplication problem.

All-Pairs Signed Symmetric Set Differences. We begin by describing how to use an IBLT for computing the (signed) symmetric set differences between every pair of a collection of N sets, $S_1, S_2, \ldots, S_N$, by adapting an approach of Goodrich and Mitzenmacher [23] to our framework. (Note this is equivalent to finding the set difference between every pair of sets.) Let $n > 1$ be an upper bound for the anticipated size of any difference, $S_i \oplus S_j$. (Note that n can be much smaller than the size of any S_i.) We begin by computing an IBLT, T_{S_i}, for each set, S_i, of $O(n \log n)$ cells, and with $O(\log n)$ hash functions (and subtables), based on the theoretical and/or experimental analysis we provide above. In this case, we store each element, x, from a set, S_i, as a key-value pair, $(x, 1)$, since we are treating the S_i's as sets.

For each pair, (i, j), where $i \neq j$, $i, j = 1, 2, \ldots, N$, we compute a set-difference IBLT, $T_{i,j}$, by computing an indexed difference of the corresponding count fields in T_{S_i} and T_{S_j} and an indexed XOR of the corresponding keySum, valueSum, and hashSum fields. We emphasize that in the results of this section, we assume that cell operations (such as XORing the various fields of a cell) are unit cost, even if the fields are large. (As we have noted, in theory the hashSum field may need to be $\omega(\log n)$ bits for some applications and/or for negligible failure probability; in practice, we expect all fields to fit in one or a small constant number of machine words.) Thus, $T_{i,j}$ is a representation of the signed symmetric set difference,[3] $S_i \oplus S_j$, where each element, x, that is in S_i and not in S_j, adds 1 to its respective count fields, and each element, x, that is in S_j and not in S_i, adds -1 to its respective count fields. We then apply the method described in Sect. 1 to perform a parallel peeling to list of the elements in $T_{i,j}$, optionally noting which elements in the difference are from S_i and which are from S_j.

Theorem 4. *Given N sets, $S_1, S_2, \ldots, S_N$, if the size of the difference between two sets, S_i and S_j, is at most a given size parameter, $n > 1$, then the probability that our algorithm will fail to compute the signed set difference between S_i and S_j in at most two rounds is negligible as a function of n. The total work needed is $O(NM \log n + N^2 n \log n)$.*

Proof: The work bound follows from the work need to insert each element into its set's IBLT and then perform all the set differences. The probability bounds follow by Theorem 2. ■

Deduplication via Difference Encodings. Consider now a simple deduplication [44] problem. Suppose we are given a collection of sets, $\mathcal{S} = \{S_1, S_2, \ldots, S_N\}$. For example, each S_i could represent a file or a database, and its elements could be disk blocks, database rows, or identifiers for disk blocks or database rows derived from a cryptographic hash function, such as SHA-256. In deduplication applications, it is anticipated that there are a lot of common elements among

[3] Recall that we say that $S_i \oplus S_j$ is a ***signed symmetric difference*** if we can separately identify the members of $S_i - S_j$ and $S_j - S_i$, that is, the members of S_i not in S_j and the members of S_j not in S_i.

the sets in $\mathcal{S}$; hence, we would like to represent each S_i concisely, e.g., just in terms of a small number of differences with another set in $\mathcal{S}$.

One such representation is a ***difference encoding*** of $\mathcal{S}$, which is a (re)ordering of the sets in $\mathcal{S}$, as $(S_1, S_2, \ldots, S_N)$, such that, for $i > 1$, each set, S_i, is encoded in terms of the differences between S_i and some S_j where $j < i$. That is, we encode S_i with the signed symmetric difference, $S_i \oplus S_j$. Thus, if there is considerable overlap of elements among the sets in $\mathcal{S}$, then a difference encoding could save a lot of memory space over explicitly representing each set in $\mathcal{S}$. The corresponding optimization problem, therefore, is to find an ordering of the sets and a difference encoding that minimizes the total amount of storage required, over all possible orderings and difference encodings. Finding such a ***minimum difference encoding*** might at first seem like a computationally difficult problem, e.g., since there are $N!$ possible orders and each one can have $(N-1)!$ difference encodings. Nevertheless, as we show below, we can solve this problem quickly in parallel with reasonable work.

Symmetric-Difference MSTs. As has been well-known since at least the 1950 s, e.g., see Restle [37], given a collection of sets, $\mathcal{S} = \{S_1, S_2, \ldots, S_N\}$, and a measure function, m, for sets (such as their size if all the sets are finite), then $d(S_i, S_j) = m(S_i \oplus S_j)$ is a distance metric. In our case, since the sets we are dealing with are finite, we define $m(S)$ to be the size of the set, S, which we denote as $|S|$.

Define a graph, G, which we call the ***symmetric-difference graph***, such that each set, S_i, in $\mathcal{S}$ is associated with a vertex, i, in G, and each edge, (i, j), has weight equal to $|S_i \oplus S_j|$.

Lemma 1. *Finding a minimum difference encoding for a collection of sets, $\mathcal{S} = \{S_1, S_2, \ldots, S_N\}$, can be reduced to finding a minimum spanning tree (MST) in the symmetric-difference graph, G, for $\mathcal{S}$.*

Proof: Let H be a spanning tree of G. We can derive a difference encoding of $\mathcal{S}$ from H by choosing the string associated with a vertex in H (it doesn't matter which one) to be the first string in the order. We then root the tree H at this first vertex and order the remaining vertices according to a preorder traversal of H, and let this be the ordering of the corresponding strings. In this way, we are guaranteed that for each vertex, i, its parent in H appears earlier in the ordering. We then encode every vertex, i, besides the first one in terms of the symmetric difference between S_i and S_j, where j is the parent of i in H. Thus, H corresponds to a difference encoding of $\mathcal{S}$.

Alternatively, let D be a difference encoding of $\mathcal{S}$ and let the corresponding order be $(S_1, S_2, \ldots, S_N)$. For each set, S_i, for $i > 1$, choose the (directed) edge, (i, j), in G such that S_i is encoded as a difference with S_j, and let H be the resulting subgraph of G. Then H has no cycles, because each vertex, $i > 1$, we chose an out-going edge to a vertex for j where $j < i$. In addition, for the same reason, H is connected, with each directed path leading to the first vertex (for S_1). Thus, H is a spanning tree, with $n - 1$ edges.

Therefore, every spanning tree corresponds to a difference encoding and every difference encoding corresponds to a spanning tree; hence, an MST for G will correspond to a minimum difference encoding for $\mathcal{S}$. ■

A major computational bottleneck for solving the above symmetric-difference MST problem is in constructing a representation of the symmetric-difference graph, G. Of course, since we are interested in finding an MST in G, it is sufficient to construct a representation of G such that every edge has weight at most some size threshold, $n > 0$, while keeping G connected. Typically, we desire n to be much smaller than M, the average size of each set in $\mathcal{S}$.

Our method for constructing G is to use IBLTs and our parallel peeling algorithm. Namely, we use our algorithm for the all-pairs set difference problem. We define a reasonable threshold, n, for the maximum expected symmetric difference so that G is connected, and run our all-pairs set difference algorithm of Theorem 4. Note that if our threshold, n, is large enough, then the probability that any of our parallel peeling algorithms fail after two rounds is negligible in n, by Theorem 2; hence, if our peeling algorithm fails for some pair (S_i, S_j), we can safely assume that $|S_i \oplus S_j| > n$. Thus, if after running our all-pairs set difference algorithm, this results in a graph, G, that is not connected, then we double our estimate for n and run it again. Since we double the value of n in each such run and our work bound is at least linear in n, the work needed the runs forms a geometric sequence that is dominated by the work for the last run. This gives us the following:

Theorem 5. *Given a collection, $\mathcal{S} = \{S_1, S_2, \ldots, S_N\}$, of N sets, with average size, M, we can determine a threshold value, n, and we can construct a connected subgraph of the symmetric-difference graph, G, for $\mathcal{S}$, containing every edge with weight at most n in $O(\log n)$ rounds in parallel, with total work $O(NM \log n + N^2 n \log n)$, with a failure probability that is negligible in n.*

Given such a connected subgraph of the symmetric-difference graph, G, we can then compute an MST in G using any known parallel MST algorithm, e.g., see [2,5,13,28]. Note that the above method also applies to a set, $S = \{s_1, s_2, \ldots, s_N\}$, of N character strings of length, $M \geq 1$, each, since we can construct a set, S_i, from each character string, s_i, by defining each element in S_i to be the pair, $(k, s_i[k])$. In this case, the symmetric-difference graph, G, would be defined so that each string s_i corresponds to a vertex and the weight of an edge, (i, j), is the Hamming distance[4] between the strings s_i and s_j. This approach could be used, for example, to concisely encode a set of DNA strings where differences are character swaps or replacements (but not insertions or deletions). In this case, the minimum difference encoding would provide an optimized concise encoding of all the strings in S.

[4] Recall that the Hamming distance between two strings, s and t, is the number of positions, i, where $s[i] \neq t[i]$.

References

1. Adhikari, V.K., Guo, Y., Hao, F., Varvello, M., Hilt, V., Steiner, M., Zhang, Z.L.: Unreeling Netflix: understanding and improving multi-CDN movie delivery. In: IEEE INFOCOM, pp. 1620–1628 (2012). https://doi.org/10.1109/INFCOM.2012.6195531
2. Adler, M., Dittrich, W., Juurlink, B., Kutyłowski, M., Rieping, I.: Communication-optimal parallel minimum spanning tree algorithms. In: 10th ACM Symposium on Parallel Algorithms and Architectures (SPAA), pp. 27–36 (1998)
3. B. Jenkins: A hash function for hash table lookup. https://burtleburtle.net/bob/hash/doobs.html (1997)
4. Bellare, M.: A note on negligible functions. J. Cryptology **15**(4) (2002)
5. Bentley, J.L.: A parallel algorithm for constructing minimum spanning trees. J. Algorithms **1**(1), 51–59 (1980). https://doi.org/10.1016/0196-6774(80)90004-8
6. Bloom, B.: Space/time trade-offs in hash coding with allowable errors. Commun. ACM **13**(7), 422–426 (1970)
7. Böttger, T., Cuadrado, F., Tyson, G., Castro, I., Uhlig, S.: Open Connect everywhere: a glimpse at the internet ecosystem through the lens of the Netflix CDN. SIGCOMM Comput. Commun. Rev. **48**(1), 28–34 (2018). https://doi.org/10.1145/3211852.3211857
8. Broder, A., Mitzenmacher, M.: Network applications of bloom filters: a survey. Internet Math. **1**(4), 485–509 (2004)
9. Brodtkorb, A.R., Hagen, T.R., Schulz, C., Hasle, G.: GPU computing in discrete optimization. Part I: Introduction to the GPU. EURO J. Trans. Logistics **2**(1), 129–157 (2013). https://doi.org/10.1007/s13676-013-0025-1, https://www.sciencedirect.com/science/article/pii/S2192437620600267
10. Cao, N., Fineman, J.T., Russell, K.: Parallel shortest paths with negative edge weights. In: 34th ACM Symposium on Parallelism in Algorithms and Architectures (SPAA), pp. 177–190 (2022)
11. Chang, Y.J., Pettie, S., Zhang, H.: Distributed triangle detection via expander decomposition. In: 13th ACM-SIAM Symposium on Discrete Algorithms (SODA), pp. 821–840. SIAM (2019)
12. Dean, J., Ghemawat, S.: MapReduce: simplified data processing on large clusters. Commun. ACM **51**(1), 107–113 (2008)
13. Dehne, F., Gotz, S.: Practical parallel algorithms for minimum spanning trees. In: 17th IEEE Symposium on Reliable Distributed Systems, pp. 366–371 (1998). https://doi.org/10.1109/RELDIS.1998.740525
14. Dhulipala, L., Blelloch, G., Shun, J.: Julienne: a framework for parallel graph algorithms using work-efficient bucketing. In: 29th ACM Symposium on Parallelism in Algorithms and Architectures (SPAA), pp. 293–304 (2017)
15. Eppstein, D., Goodrich, M.T.: Straggler identification in round-trip data streams via Newton's identities and invertible Bloom filters. IEEE Trans. Knowl. Data Eng. **23**(2), 297–306 (2010) (to appear)
16. Eppstein, D., Goodrich, M.T.: Straggler identification in round-trip data streams via Newton's identities and invertible Bloom filters. IEEE Trans. Knowl. Data Eng. **23**(2), 297–306 (2010)
17. Eppstein, D., Goodrich, M.T., Uyeda, F., Varghese, G.: What's the difference? efficient set reconciliation without prior context. ACM SIGCOMM Comput. Commun. Rev. **41**(4), 218–229 (2011)

18. Fagerjord, A., Kueng, L.: Mapping the core actors and flows in streaming video services: what Netflix can tell us about these new media networks. J. Media Bus. Stud. **16**(3), 166–181 (2019). https://doi.org/10.1080/16522354.2019.1684717
19. Fu, W., Abraham, H.B., Crowley, P.: Synchronizing namespaces with invertible Bloom filters. In: ACM/IEEE Symposium on Architectures for Networking and Communications Systems (ANCS), pp. 123–134 (2015). https://doi.org/10.1109/ANCS.2015.7110126
20. Gao, P.: Analysis of the parallel peeling algorithm: a short proof (2014). arxiv.org/abs/1402.7326
21. Gentili, M.: Set Reconciliation and File Synchronization Using Invertible Bloom Lookup Tables. Ph.D. thesis, Harvard Univ. (2015)
22. Ghaffari, M., Grunau, C., Jin, C.: Improved MPC algorithms for MIS, matching, and coloring on trees and beyond. In: 34th International Symposium on Distributed Computing, pp. 34:1–34:18 (2020)
23. Goodrich, M.T., Mitzenmacher, M.: Invertible bloom lookup tables. In: 49th Annual Allerton Conference on Communication, Control, and Computing (Allerton), pp. 792–799. IEEE (2011). arxiv.org/abs/1101.2245
24. Goodrich, M.T., Pszona, P.: External-memory network analysis algorithms for naturally sparse graphs. In: Demetrescu, C., Halldórsson, M.M. (eds.) ESA 2011. LNCS, vol. 6942, pp. 664–676. Springer, Heidelberg (2011). https://doi.org/10.1007/978-3-642-23719-5_56
25. Hijma, P., Heldens, S., Sclocco, A., van Werkhoven, B., Bal, H.E.: Optimization techniques for GPU programming. ACM Comput. Surv. **55**(11) (2023). https://doi.org/10.1145/3570638
26. Hussain, A., Aleem, M.: GoCJ: Google cloud jobs dataset for distributed and cloud computing infrastructures. Data **3**(4), 38 (2018)
27. Jiang, J., Mitzenmacher, M., Thaler, J.: Parallel peeling algorithms. ACM Trans. Parallel Comput. **3**(1) (2017). https://doi.org/10.1145/2938412
28. Karloff, H., Suri, S., Vassilvitskii, S.: A model of computation for MapReduce. In: ACM-SIAM Symposium on Discrete Algorithms (SODA), pp. 938–948 (2010). https://doi.org/10.1137/1.9781611973075.76, https://epubs.siam.org/doi/abs/10.1137/1.9781611973075.76
29. Karloff, H., Suri, S., Vassilvitskii, S.: A model of computation for MapReduce. In: 21st ACM-SIAM Symposium on Discrete Algorithms (SODA), pp. 938–948 (2010)
30. Mizrahi, A., Bar-Lev, D., Yaakobi, E., Rottenstreich, O.: Invertible Bloom lookup tables with listing guarantees. arXiv:2212.13812 (2022)
31. Molloy, M.: The pure literal rule threshold and cores in random hypergraphs. In: 15th ACM-SIAM Symposium on Discrete Algorithms (SODA), pp. 672–681. Society for Industrial and Applied Mathematics (2004)
32. Molloy, M.: Cores in random hypergraphs and Boolean formulas. Random Struct. Algorithms **27**(1), 124–135 (2005)
33. Odun-Ayo, I., Ajayi, O., Akanle, B., Ahuja, R.: An overview of data storage in cloud computing. In: IEEE International Conference on Next Generation Computing and Information Systems (ICNGCIS), pp. 29–34. IEEE (2017)
34. Ozisik, A.P., Andresen, G., Levine, B.N., Tapp, D., Bissias, G., Katkuri, S.: Graphene: efficient interactive set reconciliation applied to blockchain propagation. In: ACM SIGCOMM, pp. 303–317 (2019). https://doi.org/10.1145/3341302.3342082
35. Palankar, M.R., Iamnitchi, A., Ripeanu, M., Garfinkel, S.: Amazon S3 for science grids: a viable solution? In: ACM Int. Workshop on Data-Aware Distributed Computing, pp. 55–64 (2008). https://doi.org/10.1145/1383519.1383526

36. Pontarelli, S., Reviriego, P., Mitzenmacher, M.: Improving the performance of invertible Bloom lookup tables. Inf. Process. Lett. **114**(4), 185–191 (2014)
37. Restle, F.: A metric and an ordering on sets. Psychometrika **24**(3), 207–220 (1959)
38. Sayood, K.: Introduction to Data Compression. Morgan Kaufmann (2017)
39. Shi, J., Dhulipala, L., Shun, J.: Parallel clique counting and peeling algorithms. In: SIAM Conference on Applied and Computational Discrete Algorithms (ACDA), pp. 135–146 (2021). https://doi.org/10.1137/1.9781611976830.13, https://epubs.siam.org/doi/abs/10.1137/1.9781611976830.13
40. Shi, J., Shun, J.: Parallel algorithms for butterfly computations. In: Symposium on Algorithmic Principles of Computer Systems (APOCS), pp. 16–30 (2020). https://doi.org/10.1137/1.9781611976021.2, https://epubs.siam.org/doi/abs/10.1137/1.9781611976021.2
41. Vöcking, B.: How asymmetry helps load balancing. J. ACM **50**(4), 568–589 (2003). https://doi.org/10.1145/792538.792546
42. Wilder, B.: Cloud Architecture Patterns: Using Microsoft Azure. "O'Reilly Media, Inc." (2012)
43. Wittig, M., Wittig, A.: Amazon Web Services in Action. Simon and Schuster (2018)
44. Xia, W., et al.: A comprehensive study of the past, present, and future of data deduplication. Proc. IEEE **104**(9), 1681–1710 (2016)

The Complexity of Counting Turns in the Line-Based Dial-a-Ride Problem

Antonio Lauerbach(✉), Kendra Reiter, and Marie Schmidt

Department of Computer Science, University of Würzburg, Würzburg, Germany
antonio.lauerbach@stud-mail.uni-wuerzburg.de,
{kendra.reiter,marie.schmidt}@uni-wuerzburg.de

Abstract. Dial-a-Ride problems have been proposed to model the challenge to consolidate passenger transportation requests with a fleet of shared vehicles. The LINE-BASED DIAL-A-RIDE PROBLEM *(*LIDARP*)* is a variant where the passengers are transported along a fixed sequence of stops, with the option of taking shortcuts. In this paper we consider the LIDARP with the objective function to maximize the number of transported requests. We investigate the complexity of two optimization problems: the LIDARP, and the problem to determine the minimum number of turns needed in an optimal LIDARP solution, called the MINTURN *problem.* Based on a number of instance parameters and characteristics, we are able to state the boundary between polynomially solvable and NP-hard instances for both problems. Furthermore, we provide parameterized algorithms that are able to solve both the LIDARP and MINTURN problem.

Keywords: Dial-a-Ride Problem · liDARP · NP-hardness · Parameterized Complexity

1 Introduction

Ridepooling, i.e., to flexibly serve passenger transportation requests with a fleet of shared vehicles, has been recognized as a promising option to replace conventional public transport, in particular in regions with low demand density. In this paper, we study the complexity of the LINE-BASED DIAL-A-RIDE PROBLEM (LIDARP), which combines the spatial aspects of a fixed sequence of stops (utilizing existing infrastructure) with the temporal flexibility of ridepooling. The goal is to reduce mobility-related emissions by efficiently pooling passengers with an improved service quality due to the fixed spatial structure. Here, we consider the objective to maximize the number of transported passengers.

Secondly, we study the complexity of determining the minimum number of turns per vehicle in an optimal LIDARP solution, which we refer to as the MINTURN problem. This problem may be relevant in practice, e.g., when turns of autonomous vehicles need to be supervised by a (remote) operator. It can also be relevant when formulating the LIDARP as a mixed-integer linear program, compare [15].

R. Královič and V. Kůrková (Eds.): SOFSEM 2025, LNCS 15539, pp. 85–98, 2025.
https://doi.org/10.1007/978-3-031-82697-9_7

The remainder of the paper is organized as follows. In Sect. 2, we formally define the LIDARP and the MINTURN problems. We summarize known complexity results and further related research findings from prior work in Sect. 3, before we give an overview of the contribution of this paper in Sect. 4. In the following Sects. 5 and 6, we are able to characterize the boundary between polynomially solvable and NP-hard cases of LIDARP and MINTURN according to instance specifics. In Sect. 7 we provide parameterized algorithms for the LIDARP and MINTURN problem. We conclude by discussing open problems for further research in Sect. 8. The (full) proofs of statements marked with a "$\star$" are in the full version of this paper [13].

2 Problem Definition

We start by defining the LINE-BASED DIAL-A-RIDE PROBLEM (LIDARP) based on the definitions by Reiter et al. [15]: a *line*, given by a sequence H of h *stops*, is operated by k *vehicles*, each with *capacity* c, that transport some of the n *passenger requests* P.

The *(time) distance* between distinct stops $i, j \in H$ is given by $t_{i,j} \in \mathbb{N}$. The vehicles may take *shortcuts* by skipping stops, *wait* at a stop, or *turn* (i.e., change direction with respect to the sequence of stops prescribed by the line) at a stop. To turn, a vehicle needs $t_{\text{turn}} \in \mathbb{N}_0$ time (the *turn time*).

Each passenger submits a *request* $p \in P$ for transportation from an *origin* $o_p \in H$ to a *destination* $d_p \in H$ with $o_p \neq d_p$. If o_p precedes d_p in the sequence of stops, we say that the request p is *ascending*, otherwise it is *descending*. In contrast to [15], we assume that each passenger submits an individual requests, i.e., we do not allow group requests and thus the *passenger load* of each requests is 1. A request may specify a *time window* $[e_p, l_p]$, delimited by an *earliest pick-up time* e_p and a *latest drop-off time* l_p, during which it can be served. We therefore can write a request p as $([o_p, d_p], [e_p, l_p])$. Picking up or dropping off a passenger requires a *service time* of $t_s \in \mathbb{N}_0$ per passenger. Further, a passenger may not leave the vehicle until arriving at their destination. We make the *service promise* that the ride time of a passenger p may not exceed $\alpha \geq 1$ times the *direct time distance* t_{o_p,d_p}, that is, the *maximum ride time* of a passenger is $\alpha \cdot t_{o_p,d_p}$. The *ride time* is measured from the end of pick-up to the beginning of drop-off.

In the LIDARP, we further guarantee that if we pick up a passenger, the passenger is at all times transported towards their destination (regarding the sequence of stops). Consequently, a turn is only allowed for vehicles without passengers on board. This so-called *directionality* property [15] constitutes the main difference between the LIDARP and the 'regular' DARP.

A *tour* r consists of a sequence of *timestamped waypoints*, each waypoint being a pick-up/drop-off of a request with the timestamp corresponding to the start of the pick-up/drop-off. In order for a tour to be *feasible*, the timestamps need to adhere to the time constraints imposed by the time distances, as well as the service and turn times, the service promises, and the time windows which delimit the start of pick-ups and drop-offs. Furthermore, at no point in time may there be more than c passengers in each vehicle and each request may only be served once. If we remove the timestamps from the waypoints, we obtain the *route* underlying a tour. A route is *feasible* if it can be complemented to a feasible tour by adding timestamps.

A given tour r can be decomposed into segments, called *subtours*, with the vehicle turning precisely at the end of each subtour. Note that each subtour is, on its own, a tour, thus inheriting the feasibility definition. We analogously define *subroutes* for routes. Similar to requests, we say that a subtour/subroute is *ascending* if the vehicle drives along the sequence of stops, and *descending* if it drives the sequence in reverse. Thus, due to the directionality property, ascending requests must be served in ascending subtours and descending requests in descending subtours. We denote by $|r|$ the number of turns of a tour r. Note that this corresponds to the number of subtours of r as a vehicle always turns at the end of a subtour.

A collection of tours R is *feasible*, if each tour $r \in R$ is feasible and each request is served at most once. We analogously define *feasible collections of routes*.

A *solution* to the LIDARP is a feasible collection of up to k tours. The LIDARP (as we consider it here) consists of determining a solution that maximizes the number of served requests.

Given a LIDARP instance, the MINTURN problem determines the minimum over the largest number of turns a vehicle has to take in an optimal LIDARP solution for this instance. MINTURN thus determines $\tau := \min_{R \in \mathcal{R}^*} \max_{r \in R} |r|$ where $\mathcal{R}^*$ is the set of optimal solutions for the given LIDARP instance.

Conventions. We assume that time starts at 0 and is integer. We consider all cases as special cases of the LIDARP: to omit the service promise, we set $\alpha = \infty$, the service and turn times can be omitted by setting $t_s = 0$ and $t_{\text{turn}} = 0$, and the time windows can be disabled by setting $e_p = 0$ and $l_p = \infty$ for all $p \in P$. In the case without time windows, a request $p \in P$ is thus specified only by its origin and destination, i.e., $p = ([o_p, d_p])$.

We say that two requests *overlap* if they are in the same direction and the intervals between their respective origin and destination are not interior disjoint.

We assume k and c to be bounded by the number of requests n. Given a positive integer n, we use $[n]$ as shorthand for $\{1, 2, \dots, n\}$.

3 Related Work

The Dial-a-Ride Problem has been extensively studied in the literature, with a focus on modelling approaches using mixed-integer linear programs and solution strategies including both exact strategies and heuristics. The surveys by Ho et al. [8] and Vansteenwegen et al. [17] provide a comprehensive overview.

In [15], Reiter et al. introduce the *Line-Based Dial-a-Ride Problem (liDARP)*, where they aim to find a solution which maximizes a weighted sum of transported requests and saved distance (i.e., the difference between the sum of direct distances of transported passengers and the total distance driven by vehicles). The version of the LIDARP studied here is a special case of that problem, as we consider only one of the objectives. Reiter et al. [15] propose and compare three different mixed-integer linear formulations, including the *subline-based* formulation which explicitly models sequences of turns for each vehicle.

The complexity of DARP on a line with *makespan objective*, minimizing the *completion time* (the time to serve all requests), has been addressed by a number of publications in the literature. We summarize their findings in Table 1, where $o = d$ denotes the setting where all requests' origins are equal to their destinations (equivalent to the TRAVELLING SALESPERSON PROBLEM). Furthermore, some publications ([3,16]) consider *individual* service times t_s per request.

All publications listed in Table 1 fix the vehicles' starting positions, consider a *closed* setting, where the vehicles have to return to their starting position at the end of the day, and require all n requests to be served.

Table 1. Overview of known results for DARP on a line with makespan objective.

#Veh.	Cap.	$o = d$	Time Windows	Complexity	Ref.	Comment
1	1	–	–	polynomial	[14]	
1	1	–	e_p	NP-complete	[2]	Thm 7.6
1	2	–	–	NP-complete	[6]	
1	≥ 2	–	–	NP-complete	[2]	Thm 7.8
1	∞	–	–	polynomial	[14]	
1	≥ 1	✓	e_p	NP-complete	[16]	individual t_s
1	≥ 1	✓	$[e_p, l_p]$	NP-complete	[16]	
2	1	✓	$[e_p, l_p]$	NP-complete	[3]	individual t_s
≥ 1	c	✓	l_p	polynomial	[14]	

We note that de Paepe [14] further showed that the setting with an arbitrary number of vehicles of fixed capacity c, without time windows and where $o = d$, is also polynomially solvable under the objective of minimizing the sum of (weighted) completion times.

Further research has been conducted into examining the complexity of the Dial-a-Ride problem with individual loads per requests, minimizing the sum of driven distances on the half-line and line, as well as on the star, tree, circle, and $\mathbb{R}^d$ with the Euclidean metric [1,14].

Lastly, approximation algorithms for the related Vehicle Scheduling Problem on a line (L-VSP) have also been proposed. Karuno et al. [12] developed a $\frac{3}{2}$-approximation algorithm for the closed L-VSP with a single vehicle and time windows, under minimizing completion times, where the starting vertex is fixed. Allowing for arbitrary starting vertices, Gaur et al. [5] develop a $\frac{5}{3}$-approximation for the L-VSP. Under the objective of minimizing makespan, with an arbitrary starting position, Yu and Liu [18] develop a $\frac{3}{2}$-approximation for the closed, and a $\frac{5}{3}$-approximation algorithm for the open variants.

Considering the Multi-Vehicle Scheduling Problem (MVSP) on a line, Karuno and Nagamochi [11] present a 2-approximation algorithm to minimize the total completion time for a fixed number of vehicles.

4 Our Contribution

In this paper, we study the complexity of the LIDARP and the novel MINTURN problem. Unlike the DARP on a line studied in the literature, we consider an *open* setting, where the vehicles do not have to return to their (arbitrary) starting positions, and maximize the number of served passengers as an objective. Note that most of our results can be transferred to the closed setting.

We consider different instance parameters and characteristics: the number of vehicles k and their capacity c, as well as the presence of time windows, shortcuts, turn times, the service promise, and the service time. Our complexity results for the MINTURN problem are summarized in Table 2 and novel results for the LIDARP are given in Table 3. Note that whether there is a turn time or not appears to be irrelevant for the complexity of the problems: we have no turn time in all hardness results, while all algorithmic results hold for arbitrary turn times. We further show that it is strongly NP-hard to approximate MINTURN with a factor better than 3 (Corollary 1).

Table 2. Overview of novel results for the MINTURN problem presented here.

#Veh.	Cap.	Time Windows	Shortcuts	Service Promise	Service Time	Complexity	Ref.
≥ 1	≥ 1	–	–	✓	–	polynomial	Thm. 2
≥ 1	≥ 1	–	✓	–	✓	polynomial	Thm. 2
≥ 1	1	–	✓	✓	✓	polynomial	Thm. 2
≥ 1	≥ 2	–	–	✓	✓	strongly NP-hard	Thm. 3
≥ 1	≥ 2	–	✓	✓	–	strongly NP-hard	Thm. 4
≥ 1	≥ 1	✓	–	–	–	strongly NP-hard	Thm. 5

Table 3. Overview of novel results for the LIDARP problem presented here.

#Veh.	Cap.	Time Windows	Shortcuts	Service Promise	Service Time	Complexity	Ref.
≥ 1	≥ 1	–	✓	✓	✓	polynomial	Thm. 1
≥ 1	≥ 1	✓	–	–	–	strongly NP-hard	Thm. 5

5 Polynomially Solvable Cases

In this section, we characterize the cases in which the LIDARP and MINTURN problem are polynomially solvable. We begin by showing that it suffices to consider feasible routes, as we can efficiently transform them into feasible tours.

Lemma 1. (⋆). *Given a route, we can check in polynomial time whether it is feasible and, if so, complement it to a feasible tour. If there are no time windows, this can even be done in linear time. If additionally, there is no service promise, the route is feasible as long as it respects capacities.*

Without time windows it even suffices to find feasible subroutes, as any route obtained by *joining* feasible routes, i.e., concatenating the sequences of waypoints of the routes, is feasible.

Lemma 2. (⋆). *Consider a* LIDARP *instance without time windows and a feasible collection R of routes. Joining all routes in R (in arbitrary order) yields a feasible route.*

We now use these lemmas, to show that the LIDARP is polynomially solvable in the absence of time windows, by constructing a feasible tour that serves all requests consecutively.

Theorem 1. (⋆). *If there are no time windows, all requests can be served. A solution for the* LIDARP *serving all requests can be computed in linear time.*

As we see later in Theorem 5, the LIDARP is NP-hard as soon as we have time windows. We therefore now focus on the MINTURN problem. We begin by showing that if we have already determined the (number of) subroutes needed to serve all requests, we can efficiently compute τ for the MINTURN problem.

Lemma 3. *Consider an instance of the* MINTURN *problem without time windows. Let a (b) be the smallest number of feasible ascending (descending) subroutes needed to serve all ascending (descending) requests. Assume w.l.o.g. that $a \geq b$. Then, $\tau = \max\{\lceil \frac{a+b}{k} \rceil, 2 \cdot \lceil \frac{a}{k} \rceil - 1\}$.*

Proof. We observe that we can create a feasible collection of k routes serving all requests by alternatingly joining ascending and descending subroutes into routes of length $2\lceil \frac{a}{k} \rceil$, using each subroute once and adding *artificial subroutes*, that do not serve requests, in case there not enough subroutes. The resulting routes are feasible according to Lemma 2, as all subroutes, including the artificial subroutes, are feasible.

Furthermore, we can prove two lower bounds on the number of turns per vehicle: First, by the pigeonhole principle, there has to be a route consisting of at least $\lceil \frac{a+b}{k} \rceil$ subroutes. Second, as there must be a route containing at least $\lceil \frac{a}{k} \rceil$ ascending subroutes, and ascending and descending subroutes alternate, this route contains at least $2\lceil \frac{a}{k} \rceil - 1$ subroutes. The lower bound is thus $\max\{\lceil \frac{a+b}{k} \rceil, 2\lceil \frac{a}{k} \rceil - 1\}$.

If $\lceil\frac{a+b}{k}\rceil = 2\lceil\frac{a}{k}\rceil$, the upper and lower bound coincide. Otherwise, it holds that $\lceil\frac{a+b}{k}\rceil \leq 2 \cdot \lceil\frac{a}{k}\rceil - 1$. In this case, we add d artificial descending subroutes such that $\frac{a+b+d}{k} = 2\lceil\frac{a}{k}\rceil - 1$. We then combine the subroutes into a feasible collection of k routes serving all requests, each with $2\lceil\frac{a}{k}\rceil - 1$ turns. For this, we first add $\lceil\frac{a}{k}\rceil - 1$ ascending subroutes as well as $\lceil\frac{a}{k}\rceil - 1$ descending subroutes to each route. This leaves exactly k subroutes unassigned, as $a+b+d = (2\lceil\frac{a}{k}\rceil - 1)\cdot k = 2k(\lceil\frac{a}{k}\rceil - 1) + k$. We can thus assign each of these k subroutes to a separate route. Thus, each route is assigned $2\lceil\frac{a}{k}\rceil - 1$ subroutes, with a route either containing $\lceil\frac{a}{k}\rceil$ ascending or $\lceil\frac{a}{k}\rceil$ descending subroutes. By alternating the ascending and descending subroutes in a route, we obtain routes with $2\lceil\frac{a}{k}\rceil - 1$ turns. According to Lemma 2, these routes form a feasible collection. □

To solve the MinTurn problem in the absence of time windows, we thus need to determine the minimum number of feasible ascending and descending subroutes needed to serve all requests. We now show that, if the feasibility of subroutes is determined by the capacity, this can be done in polynomial time.

Lemma 4. *Consider a* MinTurn *instance where subroutes are already feasible if they respect the capacity. Let χ be the maximum number of pairwise overlapping ascending (descending) requests. The minimum number of feasible ascending (descending) subroutes needed to serve all ascending (descending) requests is $\lceil\chi/c\rceil$. Determining χ is possible in polynomial time.*

Proof. The assignment of ascending (descending) requests to seats in subroutes corresponds to coloring the conflict graph of the requests, as no two overlapping requests may occupy the same seat. As the conflict graph is an interval graph, the chromatic number χ corresponds to the maximum number of pairwise overlapping requests [9] and can be determined in polynomial time. Thus, if χ seats are needed, $\lceil\chi/c\rceil$ subroutes are necessary to serve all requests. □

These lemmas imply that a MinTurn instance without time windows is solvable in polynomial time if the feasibility of subroutes is determined solely by the capacity constraints. We use this insight to characterize the cases in which the MinTurn problem is polynomially solvable.

Theorem 2. ($\star$). *Consider an instance of* MinTurn. *Let a (b) be the maximum number of pairwise overlapping ascending (descending) requests and assume w.l.o.g. $a \geq b$. In the following cases of the* MinTurn *problem, we have $\tau = \max\{\lceil\frac{a+b}{k}\rceil, 2\cdot\lceil\frac{a}{k}\rceil - 1\}$, which can be determined in polynomial time:*

1. *without time windows and without service promise*
2. *without time windows, without shortcuts, and without service times*
3. *without time windows and with a capacity of* 1

6 Hardness Results

In this section, we show that all remaining liDARP and MinTurn cases are strongly NP-hard, using reductions from 3-Partition, which is well known to be strongly NP-hard, see [3].

Definition 1. (3-Partition [3]). *Given a finite set S of $n = 3m$ positive integers as well as a bound $T \in \mathbb{N}$ such that $\sum_{s \in S} s = mT$ and $T/4 < s < T/2$ for all $s \in S$, is there a partition of S into m disjoint sets $S_1, \ldots, S_m$ such that $\sum_{s \in S_j} s = T$ for all $j \in [m]$?*

We begin by showing hardness of MinTurn even in the absence of time windows and shortcuts.

Theorem 3. (⋆). *The problem* MinTurn *is strongly NP-hard for all $k \geq 1$ and $c \geq 2$ even without time windows, shortcuts, and turn time.*

Proof. We begin by showing the reduction from 3-Partition for $k = 1$ and $c = 2$. Let (S, m, T) be an instance of 3-Partition and $S = \{s_1, \ldots, s_n\}$.

We create an instance of MinTurn such that the ascending subroutes in an optimal solution of the liDARP with minimum turns per route correspond to a 3-partition of S. An example of such an instance can be seen in Fig. 1.

We begin by having stops $H = \langle 1, \ldots, 4 + 4Tn \rangle$ with unit distance between neighboring stops. We create four types of requests: for each $i \in [n]$ we create s_i *value requests* $P^i_{\mathrm{V}} = \{([4 + T(i-1) + (j-1), 4 + T(i-1) + j]) \mid j \in [s_i]\}$ and m *plug requests* $P^i_{\mathrm{P}} = P^i_{\mathrm{LP}} \cup \{p^i_{\mathrm{P}}\}$. The plug requests consist of $m - 1$ *long plug requests* P^i_{LP}, each being $([4 + T(i-1), 4 + Ti])$, and one *short plug request* $p^i_{\mathrm{P}} = ([4 + T(i-1) + s_i, 4 + Ti])$. We also create m *promise requests* P_{SP}, each being $([1, 4 + 4Tn])$, which are used in combination with the service promise to ensure that the number of value requests in a subroute does not exceed T. Lastly, we create m *filter requests* P_{F}, each being $([2, 3])$, which are used to ensure that each subroute contains exactly one promise request. We set the service time $t_{\mathrm{s}} = 1$ and the service promise $\alpha = 1 + b/a$ with $a = 3 + 4Tn$ and $b = 2(1 + T + n)$. Note that $b < a$ and thus $\alpha < 2$.

We now show that S has a 3-partition if and only if $\tau = 2m - 1$ for the constructed MinTurn instance.

From Theorem 1, we know that in an optimal liDARP solution to the constructed instance all requests are served. As we see in Observation 1, each filter request has to be served by a different subroute, requiring at least m ascending subroutes to serve all requests. We thus need at least $2m - 1$ turns to serve all requests. Assume that we have an optimal solution to the liDARP that uses $2m - 1$ turns. Its route must thus contain m ascending subroutes.

Observation 1: Each ascending subroute serves exactly one filter and one promise request. Indeed, as the service promise α is less than 2 and the service time t_{s} is 1, the maximum ride time of a filter request is less than 2, due to its direct time distance being 1, and would thus be exceeded if another filter request is served by the same subroute. Since the m ascending subroutes serve all requests, each ascending subroute must thus contain exactly one filter request. It follows that each ascending subroute must also contain exactly one promise request, since promise and filter request overlap.

Observation 2: Due to the service promise, the maximum ride time of a promise request is at most b more than the direct time distance. As all other requests lie between the origin and destination of a promise request and the

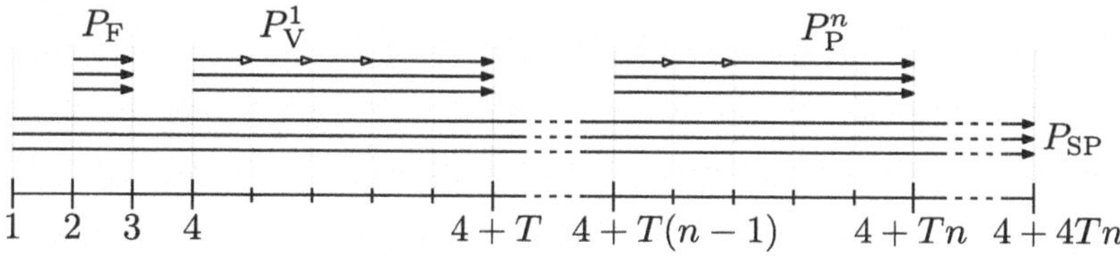

Fig. 1. A MinTurn instance constructed from a 3-Partition instance with $m = 3$ and $T = 5$, as well as $s_1 = 3$ and $s_n = 2$. The arrows represent the requests, with the white tipped arrows being value requests.

service time is 1, at most $1 + T + n$ requests can be transported in a subroute besides a promise request. According to Observation 1, one of these requests must be a filter request.

Observation 3: Each subroute also has to serve exactly one plug request for each $i \in [n]$, as there are m such requests and they overlap each other as well as the promise requests.

Observation 4: Combining Observations 2 and 3, we conclude that each subroute may serve up to T value requests. As the total number of value requests is mT, this means that each subroute transports exactly T such requests.

Observation 5: All requests in P_V^i must be served by the same subroute, the one that serves the short plug request p_P^i, as all the other subroutes contain long plug requests that overlap all requests in P_V^i. That is, for each $i \in [n]$ we have exactly one subroute that serves the s_i value requests P_V^i. In combination with Observation 4 we conclude that, for each subroute r_j, we have an index set I_j, such that all value requests in $\bigcup_{i \in I_j} P_V^i$ are served by the subroute r_j and $\sum_{i \in I_j} s_i = T$. Setting $S_j := \{s_i \mid i \in I_j\}$, we thus obtain a valid 3-partition of S.

Conversely, if there exists a 3-partition $S_1, \ldots, S_m$ of S, we create m ascending subroutes $r_1, \ldots, r_m$ and assign for each $s_i \in S_j$ for $j \in [m]$ the value requests in P_V^i as well as the short plug request in p_P^i to the subroute r_j. To each of the remaining ascending subroutes, we assign one of the long plug requests from P_{LP}^i. Furthermore, we assign one filter and one promise request to each ascending subroute. In this way, all requests are assigned to a subroute. Additionally, the capacities are respected, as no more than 2 requests pairwise overlap in a subroute. By adding $m - 1$ artificial descending subroutes, we connect the ascending subroutes and obtain a route with $2m - 1$ turns. For this route to be feasible it remains to show, according to Lemma 2, that the subroutes are feasible. The only requests which are not transported directly are the promise requests. By construction, each subroute serves, besides the promise request, one filter, n plug, and T value requests. Therefore, the delay in the ride time of a promise request is $2(1 + T + n)$, which is precisely the allowed delay by the service promise, and the subroutes are thus feasible.

For higher capacities and more vehicles we duplicate the promise requests, such that they fill the added seats. Apart from a small adjustment of the service promise and subsequently the direct time distance of promise requests, the construction and correctness are analogous.

Constructing these instances takes pseudo-polynomial time. As 3-PARTITION is strongly NP-hard, it follows that the MINTURN problem is strongly NP-hard. □

The presented reduction can be adapted to show strong NP-hardness for the case without service times but instead with shortcuts, by encoding the values $s \in S$ into detours of the line that can be shortcut by a subroute if it does not serve the corresponding requests.

Theorem 4. (⋆). *The problem* MINTURN *is strongly NP-hard for all $k \geq 1$ and $c \geq 2$ even without time windows, service times and turn times.*

Proof. We begin by proving hardness for $k = 1$ and $c = 2$ before extending it to higher values. Let (S, m, T) be an instance of 3-PARTITION and $S = \{s_1, \ldots, s_n\}$.

As we use shortcuts, we start by describing the layout of the line, which can be seen schematically in Fig. 2.

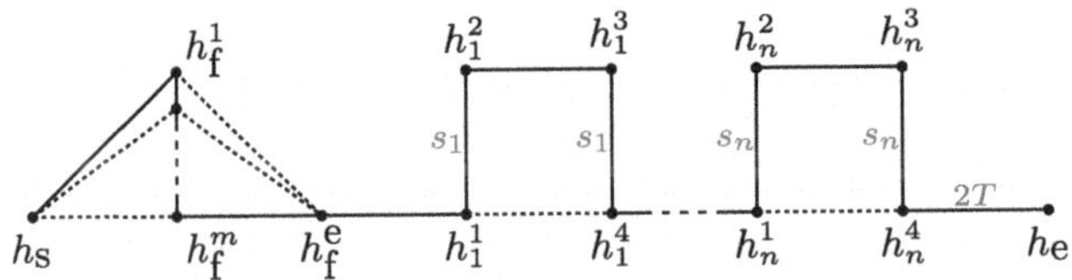

Fig. 2. The layout of the stops constructed for a 3-PARTITION instance. The line is the continuous path, while shortcuts are represented by (dotted) lines. The distances between neighboring stops are given by (blue) labels, with the exception of distances of 1, which are omitted.

The line starts at stop h_s then contains (in the order in which they are listed here) a sequence of stops h_f^j for $j \in [m]$, a stop h_f^e, for each $s_i \in S$ a sequence of stops $h_i^1, h_i^2, h_i^3, h_i^4$, and the final stop h_e. For each $i \in [n]$, the distance between h_i^1 and h_i^2 as well as between h_i^3 and h_i^4 is s_i. The distance between h_n^4 and h_e is $2T$. For all other pairs of subsequent stops, the distances are 1. We have a number of shortcuts: the direct distance from h_s to h_f^j and from h_f^j to h_f^e is 1 for all $j \in [m]$. Furthermore, the direct distance between h_i^1 and h_i^4 is 1 for all $i \in [n]$. The m *promise requests* P_{SP} originate at h_s and end at h_e. Each of the m *filter requests* $p_F^j \in P_F$ for $j \in [m]$ starts at stop h_f^j and goes to h_f^e. For each $i \in [n]$, a *value request* p_i is added that originates at h_i^2 and goes to h_i^3. The service promise is set to $1 + b/a < 2$ with $a = 2 + 2n + 2T$ and $b = 2T$.

It then holds that S has a 3-partition if and only if $\tau = 2m - 1$ for the constructed MINTURN instance. The proof is similar to the proof of Theorem 3.

Differing from the reduction in Theorem 3, we can construct this instance in polynomial time, as our number of stops does not depend on T. It follows that the MINTURN problem is strongly NP-hard. □

By adapting the reductions to use m vehicles instead of one, we can show that the MinTurn problem is strongly NP-hard to approximate with a factor better than 3. This leads to the following corollary.

Corollary 1. (⋆). *The problem* MinTurn *is strongly NP-hard to approximate with a factor better than* 3 *for all* $c \geq 2$ *in the following cases:*

1. *without time windows, shortcuts, and turn times, and*
2. *without time windows, service times, and turn times.*

We now show that as soon as we consider time windows, both the liDARP and MinTurn problem become NP-hard for arbitrary values of k and c.

Theorem 5. (⋆). *Both* liDARP *and* MinTurn *are strongly NP-hard for all* $k \geq 1$ *and* $c \geq 1$ *even without shortcuts, service promise, service times, and turn times.*

Proof. (sketch). Again, we use 3-Partition for the reduction. The main idea of the proof for $k = 1$ and $c = 1$ is to translate the values $s \in S$ into value requests that need time $2s$ to be served. We then use *separator* requests to create m time intervals of length $2T$ during which the value requests must be served if all requests are served. Thus, the assignment of value requests to time intervals corresponds to a 3-partition of S. □

7 Parameterized Algorithms

Seeing as the liDARP and MinTurn problem are NP-hard, we now provide parameterized algorithms for both. Recall that an algorithm is *fixed-parameter tractable* (FPT) w.r.t. a parameter k if its runtime is $f(k) \cdot n^{O(1)}$ for some computable function f, where n is the size of the input. An algorithm is *slice-wise polynomial* (XP) w.r.t. a parameter k if its runtime is in $O(n^{f(k)})$ for some computable function f. When analyzing the runtime, we use the O^* notation, which suppresses polynomial factors in the input size, i.e., a function $g(n)$ is in $O^*(f(n))$ if there is a polynomial p such that $g(n) \in O(f(n) \cdot p(n))$.

As we have shown the liDARP and MinTurn problem to be NP-hard for constant k and c, we have to consider more parameters to obtain parameterized algorithms. We therefore use the number of stops h as well as the maximum time $t := \max_{p \in P} l_p + 1$.

Theorem 6. (⋆). *There exists an FPT-algorithm for* MinTurn *as well as* liDARP *parameterized by* k, c, h *and* t, *with a runtime in* $O^*((h^2 \cdot t^3 \cdot c \cdot k)^{2 \cdot t \cdot c \cdot k})$.

Proof. (sketch). To prove this result, we enumerate all routes that could be part of a liDARP solution. As we can bound the number of requests that a single vehicle can serve in time t by $t \cdot c$, we can enumerate all feasible routes in time $O^*(n^{2 \cdot t \cdot c})$, using the *event-based graph* [4,15]. For all collections of up to k routes, we find the one which serves the most requests and minimizes the maximum turns per route. This results in a runtime of $O^*(n^{2 \cdot t \cdot c \cdot k})$.

To obtain an FPT-runtime, we slightly modify the algorithm such that it reduces inputs to a predetermined size, using the observation that there are at most $h^2 \cdot t^2$ distinct requests which can be served at most $t \cdot c \cdot k$ times each. □

Note that the FPT-algorithm can be adapted easily to work for several other objectives and restrictions, such as maximizing the weighted sum of saved distance and transported requests, as used in [15], or minimizing the makespan while serving all requests, as studied in most complexity papers on DARP on a line, see Table 1.

However, the case without time windows poses some difficulties. In this paper we treat this as a special case of the general case by setting $l_p = \infty$, which implies $t = \infty$. Determining a better bound for l_p is of no use, as it would depend on n. Thus, we now devise an XP-algorithm for the case without time windows whose running time is parameterized by c and h.

Theorem 7. (⋆). *There is an XP-algorithm for the* MinTurn *problem without time windows, parameterized by c and h, with runtime $O^*(n^{h^2} \cdot h^{4 \cdot c \cdot h})$.*

Proof. (sketch). We interpret finding the minimum number of feasible subroutes needed to serve all ascending (descending) requests as a Multiset Multicover problem and apply an algorithm proposed by Hua et al. [10] to solve it. Applying Lemma 3, we obtain the value of τ for MinTurn in time $O^*(n^{h^2} \cdot h^{4 \cdot c \cdot h})$. □

8 Conclusion

We introduce the MinTurn problem and characterize the boundary between polynomial solvability and NP-hardness for the MinTurn and liDARP problem according to instance specifics, including time windows, shortcuts, service promise, and service times. We also show that the MinTurn problem is strongly NP-hard to approximate with factor better than 3. We then provide an FPT-algorithm for the MinTurn and liDARP problem, parameterized by k, c, h and t, which also works for other objectives and restrictions. Finally, we present an XP-algorithm for the MinTurn problem without time windows parameterized by c and h.

An interesting topic for further research would be to analyze the liDARP and MinTurn problem for other objectives and restrictions, such as minimizing the sum of turns (over all vehicles), or allowing an unlimited number of vehicles. It also remains open, whether the parameterized algorithms can be improved, such as if there is an FPT-algorithm parameterized by c and h. For practical applications, such as the subline-based formulation [15], it may be interesting to develop heuristics and approximation algorithms for the MinTurn problem.

Disclosure of Interests. The authors have no competing interests to declare that are relevant to the content of this article.

References

1. Archetti, C., Feillet, D., Gendreau, M., Grazia Speranza, M.: Complexity of the VRP and SDVRP. Trans. Res. Part C: Emerg. Technol. **19**(5), 741–750 (2011). https://doi.org/10.1016/j.trc.2009.12.006
2. Bjelde, A., et al.: Tight bounds for online TSP on the line. ACM Trans. Algorithms **17**(1), 3:1–3:58 (2021). https://doi.org/10.1145/3422362
3. Garey, M.R., Johnson, D.S.: Computers and intractability: a guide to the theory of NP-completeness. W. H. Freeman (1979)
4. Gaul, D., Klamroth, K., Stiglmayr, M.: Event-based MILP models for ridepooling applications. Eur. J. Oper. Res. **301**(3), 1048–1063 (2022). https://doi.org/10.1016/J.EJOR.2021.11.053
5. Gaur, D.R., Gupta, A., Krishnamurti, R.: A 53-approximation algorithm for scheduling vehicles on a path with release and handling times. Inf. Process. Lett. **86**(2), 87–91 (2003). https://doi.org/10.1016/S0020-0190(02)00474-X
6. Guan, D.J.: Routing a vehicle of capacity greater than one. Discret. Appl. Math. **81**(1), 41–57 (1998). https://doi.org/10.1016/S0166-218X(97)00074-7
7. Haugland, D., Ho, S.C.: Feasibility testing for dial-a-ride problems. In: Chen, B. (ed.) Algorithmic Aspects in Information and Management, pp. 170–179. Springer, Berlin, Heidelberg (2010). https://doi.org/10.1007/978-3-642-14355-7_18
8. Ho, S.C., Szeto, W., Kuo, Y.H., Leung, J.M., Petering, M., Tou, T.W.: A survey of dial-a-ride problems: literature review and recent developments. Trans. Res. Part B: Methodological **111**, 395–421 (2018). https://doi.org/10.1016/j.trb.2018.02.001
9. Hougardy, S.: Classes of perfect graphs. Discrete Math. **306**(19), 2529–2571 (2006). https://doi.org/10.1016/j.disc.2006.05.021. creation and Recreation: A Tribute to the Memory of Claude Berge
10. Hua, Q., Wang, Y., Yu, D., Lau, F.C.M.: Dynamic programming based algorithms for set multicover and multiset multicover problems. Theor. Comput. Sci. **411**(26–28), 2467–2474 (2010). https://doi.org/10.1016/J.TCS.2010.02.016
11. Karuno, Y., Nagamochi, H.: A 2-approximation algorithm for the multi-vehicle scheduling problem on a path with release and handling times. In: auf der Heide, F.M. (ed.) Algorithms — ESA 2001, pp. 218–229. Springer Berlin Heidelberg, Berlin, Heidelberg (2001). https://doi.org/10.1016/S0166-218X(02)00596-6
12. Karuno, Y., Nagamochi, H., Ibaraki, T.: Better approximation ratios for the single-vehicle scheduling problems on line-shaped networks. Networks **39**(4), 203–209 (2002). https://doi.org/10.1002/net.10028
13. Lauerbach, A., Reiter, K., Schmidt, M.: The complexity of counting turns in the line-based dial-a-ride problem (2024). https://arxiv.org/abs/2409.15192
14. de Paepe, W., Lenstra, J.K., Sgall, J., Sitters, R.A., Stougie, L.: Computer-aided complexity classification of dial-a-ride problems. INFORMS J. Comput. **16**(2), 120–132 (2004). https://doi.org/10.1287/IJOC.1030.0052
15. Reiter, K., Schmidt, M., Stiglmayr, M.: The line-based dial-a-ride problem. In: Bouman, P.C., Kontogiannis, S.C. (eds.) 24th Symposium on Algorithmic Approaches for Transportation Modelling, Optimization, and Systems (ATMOS 2024), pp. 14:1 – 14:20. Open Access Series in Informatics (OASIcs), Schloss Dagstuhl – Leibniz-Zentrum für Informatik (2024). https://doi.org/10.4230/OASIcs.ATMOS.2024.14
16. Tsitsiklis, J.N.: Special cases of traveling salesman and repairman problems with time windows. Networks **22**(3), 263–282 (1992). https://doi.org/10.1002/net.3230220305

17. Vansteenwegen, P., Melis, L., Aktaş, D., Montenegro, B.D.G., Sartori Vieira, F., Sörensen, K.: A survey on demand-responsive public bus systems. Trans. Res. Part C: Emerg. Technol. **137**, 103573 (2022). https://doi.org/10.1016/j.trc.2022.103573
18. Yu, W., Liu, Z.: Single-vehicle scheduling problems with release and service times on a line. Networks **57**(2), 128–134 (2011). https://doi.org/10.1002/net.20393

Colorful 3-Rainbow Domination

Tetiana Lavynska(✉)

Otto von Guericke University Magdeburg, Magdeburg, Germany
tetiana.lavynska@ovgu.de

Abstract. The k-rainbow domination problem is a well-known variation of the classical domination problem in graph theory. Motivated by applications in facility location problems, we study colorful k-rainbow domination, a variant of rainbow domination, where at least one color must be assigned to each vertex. Given a graph $G = (V, E)$ and a set of k colors $\{1, \ldots, k\}$ we consider a function f that assigns a subset of colors to each vertex, that is $f : V \to 2^{\{1,\ldots,k\}}$. If for each vertex v we have $\bigcup_{x \in N[v]} f(x) = \{1, \ldots, k\}$ and $f(v) \neq \emptyset$, then f is a colorful k-rainbow dominating function of G. The weight of f is $w(f) = \sum_{v \in V} |f(v)|$. The goal of the colorful k-rainbow domination problem is to find a colorful k-rainbow dominating function with the smallest weight. Such weight is called a colorful k-rainbow domination number. For $k = 2$, we give a linear time algorithm that computes an optimal function. For $k \geq 3$, we show that the corresponding decision problem is NP-complete. Moreover, for $k = 3$, it remains NP-complete if the input graph is restricted to be split or bipartite. Furthermore, we compute the colorful 3-rainbow domination number for cliques, bicliques, cycles, and paths. For trees and interval graphs, we give linear time algorithms that compute an optimal function.

Keywords: domination · rainbow domination · facility location problem · NP-complete · split graph · bipartite graph · tree · interval graph

1 Introduction

The domination problem and its variants are classical graph problems with applications in network monitoring and facility location problems. The goal of the classical domination problem is to find the smallest set of vertices that dominates the whole graph G, where a vertex v *dominates* itself and all its neighbors, thus its closed neighborhood $N[v]$. The size of this optimal set is called a domination number and is denoted by $\gamma(G)$. One of the recently introduced variations of the domination problem is rainbow domination. Given a graph $G = (V, E)$ and a set of k colors $\{1, \ldots, k\}$ we consider a function f that assigns a subset of colors to each vertex, that is $f : V \to 2^{\{1,\ldots,k\}}$. If each vertex v that is not assigned to any color has all colors in its neighborhood, then such a color assignment is called k-rainbow dominating function of G. More precisely, if for each

R. Královič and V. Kůrková (Eds.): SOFSEM 2025, LNCS 15539, pp. 99–111, 2025.
https://doi.org/10.1007/978-3-031-82697-9_8

vertex v with $f(v) = \emptyset$ we have $\bigcup_{x \in N[v]} f(x) = \{1, \ldots, k\}$, then f is a k-rainbow dominating function of G. The weight of f is the number of all assigned colors: $w(f) = \sum_{v \in V} |f(v)|$. The minimum weight of the k-rainbow dominating function of G is called a k-rainbow domination number of G and is denoted by $\gamma_{rk}(G)$. The concept of rainbow domination was introduced in 2007 by Bresar and Sumenjak [3]. It was mainly motivated by theoretical questions connected to the domination number of the Cartesian product of graphs and Vizing's Conjecture. Together with its theoretical origin, rainbow domination is also an interesting generalization of domination from a practical point of view. If we want to monitor k different aspects in a network and each aspect needs a different monitoring device, then we can model this situation using k colors. Observe, that in the classical definition of rainbow domination, only vertices without any monitoring devices must be monitored by all k different devices. This creates a situation, in which for some vertices, it is cheaper to place one monitoring device in them than to place all k devices in their neighborhood. In extreme situations, each vertex gets one color so that no vertex must be monitored. To prohibit such behaviour we can change the definition of rainbow domination. If we demand $\bigcup_{x \in N[v]} f(x) = \{1, \ldots, k\}$ for each vertex v instead only for vertices with $f(v) = \emptyset$, then we monitor each vertex by all k monitoring devices. Such change in definition leads to equivalence to classical domination and in this case, the minimum weight of f is equal to $k\gamma(G)$. Since, on the one hand, for each color $c \in \{1, \ldots, k\}$, the subset of vertices with a color set containing this color $\{v | c \in f(v)\}$ is a dominating set and thus cannot be smaller than the size of the minimum dominating set. Hence, $|\{v | c \in f(v)\}| \geq \gamma(G)$ and $w(f) = \sum_{c \in \{1, \ldots, k\}} |\{v | c \in f(v)\}| \geq k\gamma(G)$. On the other hand, we can assign color set $\{1, \ldots, k\}$ to each vertex from a minimum dominating set and empty set to each remaining vertex and get a k-rainbow dominating function.

Let us consider an application in facility location problems. Imagine, that we want to place different kinds of facilities, for example, several supermarkets, pharmacies, and petrol stations, in various villages, such that each village has all the facilities in its neighborhood. To make the allocation fair, each village should get at least one facility. Such a problem can be modeled using the following variation of k-rainbow domination. We demand $\bigcup_{x \in N[v]} f(x) = \{1, \ldots, k\}$ for each vertex v in the graph. Additionally, we demand that for each v in the graph $f(v) \neq \emptyset$. Since each vertex gets at least one color, we call such variation of rainbow domination colorful. The solution to the colorful rainbow domination problem gives us a fair distribution of resources because everyone gets a share in the investment. Despite its practical origin, this variation of rainbow domination has been barely studied [9], and very few results for it exist. We revive this definition, correct some false claims from [9], and give new complexity results and efficient algorithms for the colorful rainbow domination problem.

Related work. Standard k-rainbow domination was introduced by Bresar and Sumenjak 2007 [3]. The authors show that it is NP-complete to decide, whether a given graph has a 2-rainbow dominating function of a given weight. The problem remains NP-complete if we restrict the input graph to be bipartite or chordal.

Furthermore, the authors compute exact values of 2-rainbow domination numbers of several standard graph classes, such as paths, cycles, and suns, and upper and lower bounds for the generalized Petersen graphs. At the same time, Bresar et al. show that 2-rainbow domination is solvable in linear time in trees [2]. Later, Chang et al. consider a general k-rainbow domination for any k and show the same complexity results: the problem is NP-complete in bipartite and in chordal graphs, but solvable in linear time for trees [4]. Hon et al. [7] give polynomial time algorithms for computing 2-rainbow domination number in interval and permutation graphs and k-rainbow domination number in cographs. Yen [13] shows that 2-rainbow domination remains NP-complete in planar graphs.

Different variations of rainbow domination [1,9,10,12] have been recently introduced. Two of them are introduced under the same name: total k-rainbow domination. Ahangar et al. [1] propose the following definition. A k-rainbow dominating function f in a graph with no isolated vertex is called a total k-rainbow dominating function if the subgraph of G induced by the set of vertices that are assigned to no colors has no isolated vertices. For this definition, Jiang and Rao [8] show that the corresponding problem for $k = 2$ is NP-complete in planar bipartite graphs, chordal bipartite graphs, undirected path graphs, and split graphs, and solvable in linear time in trees. Kumbargoudra and Kureethara consider our proposed definition in [9] under the name total k-rainbow domination. The authors compute the total k-rainbow domination number for several graph classes like cliques, bicliques, cycles, and paths. Unfortunately, some of the computed numbers are not correct, as discussed in Sect. 4. Furthermore, no algorithmic or complexity results are given in [9]. Since the term "total rainbow domination" is already used for several different concepts, we propose to use "colorful rainbow domination" for this variant of rainbow domination.

Tepeh [12] proposes another variation of rainbow domination and calls it k-rainbow total domination to distinguish it from the already introduced total k-rainbow domination. A function f is called a k-rainbow total dominating function of G if the following two conditions hold. The first condition is from the classical rainbow domination: for each vertex v with $f(v) = \emptyset$, we have $\bigcup_{x \in N[v]} f(x) = \{1, \ldots, k\}$. The second condition is: for each vertex v with $f(v) = \{i\}$ for some $i \in \{1, \ldots, k\}$ there exists a vertex u in its open neighborhod $N(v)$ such that $i \in f(u)$. Interestingly, this definition also tries to prevent the possible extreme behavior of classical rainbow domination described above, where it is cheaper for some vertices to assign them to one color, such that we do not need to put all colors in their neighborhood. For this definition, if only one color is assigned to a vertex we must also have this color in the color set of at least one of its neighbors.

Our contribution. We consider a variant of k-rainbow domination that we call colorful k-rainbow domination and show that the corresponding problem in general graphs is NP-complete if $k \geq 3$. Furthermore, for $k = 3$, it remains NP-complete if we restrict the input graph to the class of bipartite or split graphs. For $k = 2$, we show how to solve the problem efficiently in general graphs. We compute colorful 3-rainbow domination numbers for paths, cycles, cliques, and bicliques. Moreover, we give linear time algorithms to compute an optimal colorful 3-rainbow domination function if an input graph is a tree or an interval graph.

2 Preliminaries

Definition 1. *Given a graph $G = (V, E)$ and a set of k colors $\{1, \dots, k\}$ we consider a function f that assigns a subset of colors to each vertex, that is $f : V \to 2^{\{1,\dots,k\}}$. If for each vertex v we have $\bigcup_{x \in N[v]} f(x) = \{1, \dots, k\}$ and $f(v) \neq \emptyset$, then f is a colorful k-rainbow dominating function of G. The weight of f is $w(f) = \sum_{v \in V} |f(v)|$. The minimum weight of the colorful k-rainbow dominating function of G is a colorful k-rainbow domination number of G and is denoted by $\gamma_{crk}(G)$.*

Definition 2. *Given a graph G and a number d, the goal of the Colorful k-Rainbow Domination problem is to decide whether $\gamma_{crk}(G) \leq d$. In the optimization version, the goal is to compute an optimal colorful k-rainbow dominating function and its weight $\gamma_{crk}(G)$.*

3 NP-Completeness of Colorful k-Rainbow Domination

A lower bound on the colorful k-rainbow domination number follows from the requirement $\forall v \in V f(v) \neq \emptyset$. Each vertex gets at least one color and thus $|V| \leq \gamma_{crk}(G)$. For $k = 1$, we get trivially $\gamma_{cr1}(G) = |V|$.

Theorem 1. *For a graph $G = (V, E)$ with n being the number of vertices and n_0 being the number of isolated vertices, $\gamma_{cr2}(G) = n + n_0$. We can compute a corresponding optimal colorful 2-rainbow dominating function in linear time.*

Proof. Each isolated vertex must get both colors. Each non-isolated vertex must get at least one color. Thus $\gamma_{cr2}(G) \geq n + n_0$. We show that a colorful 2-rainbow dominating function with weight $n + n_0$ exists. For non-isolated vertices, we consider each connected component separately. In each component with vertices V', we compute a maximal independent set I. We assign color 1 to vertices from I and color 2 to vertices from $V' - I$. Each vertex from I is connected to at least one vertex from $V' - I$ since it is not connected to any vertex from I and V' induces a connected component. Each vertex from $V' - I$ is connected to at least one vertex from I since otherwise it could be added to I which would contradict the maximality of I. Thus, each vertex has both colors in its neighborhood.

Theorem 2. *Colorful 3-Rainbow Domination is NP-complete.*

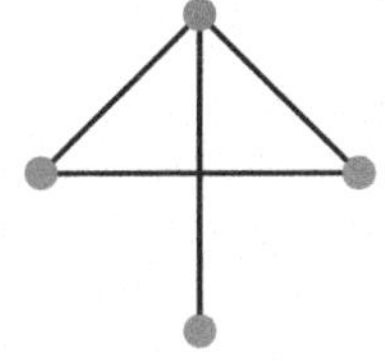

(a) A 3-colorable graph G.

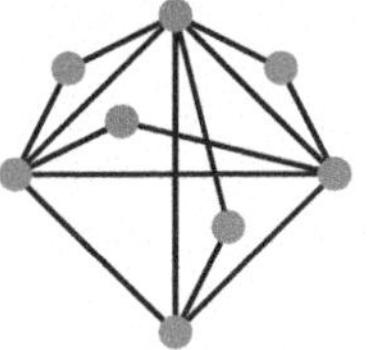

(b) Resulting graph G'.

Fig. 1. An illustration of the reduction from 3-Colorability to Colorful 3-Rainbow Domination.

Proof. We consider the 3-Colorability problem, which is NP-complete [11], and reduce it to the Colorful 3-Rainbow Domination problem. A graph $G = (V, E)$ with $|V| < 3$ is 3-colorable. Otherwise, we construct a new graph G' where we add a new vertex v_{xy} for each edge $\{x, y\} \in E$ and connect it to both old vertices x and y, as illustrated in Fig. 1. Let V_e be the set of these new vertices. We connect pairwise all old vertices from V to form a fully connected graph. Furthermore, we choose d to be equal to the number of vertices in G': $d = |V| + |E|$. Therefore $\gamma_{cr3}(G') \leq d$ only if an optimal colorful 3-rainbow dominating function exists where exactly one color is assigned to each vertex. If G is 3-colorable, we choose a 3-coloring of G such that each color is used at least once and assign the same colors to the old vertices in G'. Each new vertex v_{xy} corresponds to an edge $\{x, y\} \in E$. Therefore, x and y have different colors and we assign the third remaining color to a vertex v_{xy}. Hence, all new vertex have all three colors in their neighborhood. The old vertices form a clique in which all three colors are used, so each old vertex has all three colors in its neighborhood. If for G' a colorful 3-rainbow domination function with weight at most d exists, then each vertex gets exactly one color. This coloring is a proper 3-coloring in the original graph G, such that no two vertices sharing the same edge have the same color. Observe, that vertices from V_e are not connected. Thus, they form an independent set in G'. All old vertices are connected and thus form a clique in G'. Hence, G' is a split graph. This implies the following theorem.

Theorem 3. *Colorful 3-Rainbow Domination in split graphs is NP-complete.*

Theorem 4. *Colorful k-Rainbow Domination for fixed $k \geq 3$ is NP-complete.*

Proof. We consider the k-Colorability problem, which is for any $k \geq 3$ NP-complete [6], and reduce it to the Colorful k-Rainbow Domination problem. A graph $G = (V, E)$ with $|V| < k$ is k-colorable. Otherwise, we construct a new graph G' where we add a new vertex v_{xy} for each edge $\{x, y\} \in E$ and connect it to both old vertices x and y. Furthermore, we add $k - 3$ pendant vertices $v_{xy}^1, \ldots, v_{xy}^{k-3}$ to vertex v_{xy}. Moreover, we add k vertices $u_1, \ldots, u_k$ that all are pairwise connected. We connect each pendant vertex v_{xy}^i to each u_j. Furthermore, we connect pairwise all old vertices from V to form a fully connected graph. We choose d to be equal to the number of vertices in G': $d = |V| + |E|(k-2) + k$.

Therefore $\gamma_{crk}(G') \leq d$ only if an optimal colorful k-rainbow dominating function exists where exactly one color is assigned to each vertex. If G is k-colorable, then we choose a k-coloring of G such that each color is used at least once and assign the same colors to the old vertices in G'. Each new vertex v_{xy} corresponds to an edge $\{x, y\} \in E$. Therefore, x and y have different colors and we distribute the remaining $k - 2$ colors among vertex v_{xy} and all its pendant vertices such that each gets a different color. Each u_i gets color i. Such a color assignment is a colorful k-rainbow dominating function. Let G' have a colorful k-rainbow dominating function that assigns one color to each vertex. Each vertex v_{xy} has $k - 1$ neighbours: $x, y, v_{xy}^1, \dots, v_{xy}^{k-3}$. Hence, each of them must have a different color. Thus, vertices connected by an edge in G have different colors and G is k-colorable.

Theorem 5. *Colorful 3-Rainbow Domination in bipartite graphs is NP-complete.*

Proof. We consider the 3-Colorability problem, which is NP-complete [11], and reduce it to the Colorful 3-Rainbow Domination problem. Given a graph $G = (V, E)$, we construct a new bipartite graph G' where for each edge $\{x, y\} \in E$ we add a new vertex v_{xy} and connect it to both old vertices x and y. Let V_e be the set of such vertices. All old vertices from V form an independent set. Moreover, we add six new vertices $a_1, a_2, a_3, a_4, a_5, a_6$ that form a cycle and connect each old vertex from V to a_2, a_4, a_6, as illustrated in Fig. 2. The constructed graph G' is bipartite with the partition in which one set is $V \cup \{a_1, a_3, a_5\}$ and the other set is $V_e \cup \{a_2, a_4, a_6\}$. We choose d to be equal to the number of vertices in G': $d = |V| + |E| + 6$. Therefore $\gamma_{cr3}(G') \leq d$ only if an optimal colorful 3-rainbow dominating function exists where exactly one color is assigned to each vertex. If G is 3-colorable, we assign the same colors to the old vertices in G'. Each new vertex v_{xy} corresponds to an edge $\{x, y\} \in E$. Therefore, x and y have different colors and we assign the third remaining color to vertex v_{xy}. Thus, each vertex from V_e has all three colors in its neighborhood. Furthermore, vertices a_1 and a_4 get color 1, vertices a_2 and a_5 - color 2, vertices a_3 and a_6 - color 3. Since each vertex from V is connected to vertices a_2, a_4, a_6, there are all three colors in its neighborhood. Moreover, each vertex $a_i, 1 \leq i \leq 6$ has three colors in its neighborhood. If for G' a colorful 3-rainbow domination function with weight at most d exists, then exactly one color is assigned to each vertex. Since each vertex from V_e that corresponds to an edge $\{x, y\} \in E$ has only two neighbors x and y they must have different colors. So the graph G is 3-colorable.

Since, already for $k = 3$, the colorful k-rainbow domination is NP-complete we focus on the Colorful 3-Rainbow Domination problem and analyze for which graph classes it can be solved efficiently.

4 Colorful 3-Rainbow Domination Number

We start with a lower bound on the colorful 3-rainbow domination number.

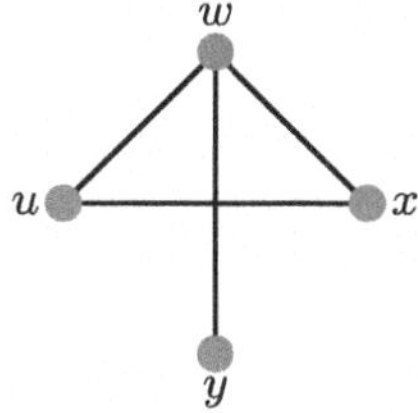

(a) A 3-colorable graph G.

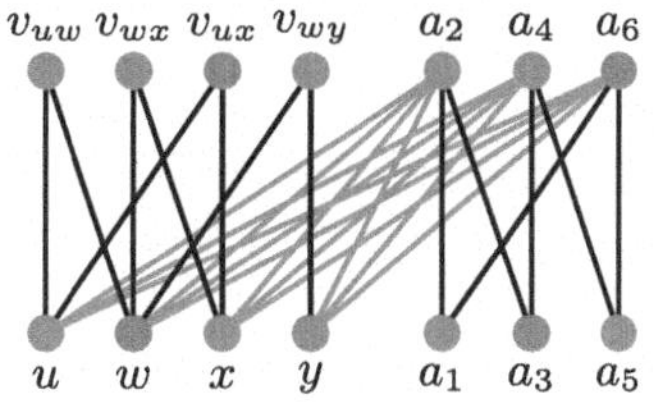

(b) Resulting bipartite graph G'.

Fig. 2. An example illustrating the reduction from 3-Colorability to Colorful 3-Rainbow Domination in bipartite graphs.

Theorem 6. *Let $G = (V, E)$ be a graph with n vertices. Let n_0 be the number of isolated vertices in G. Let n_1 be the number of isolated edges plus the number of vertices that are not adjacent to an isolated edge and have neighbors with degree one. Then $\gamma_{cr3}(G) \geq n + n_1 + 2n_0$.*

Proof. Each isolated vertex must get all three colors. Each non-isolated vertex must get at least one color. Let v be the vertex with degree one and its only neighbor be u. Furthermore, let edge $\{u, v\}$ be not isolated. Both u and v cannot get a single color, otherwise, v would have at most two colors in its neighborhood. Therefore, one of them should get at least two colors. If it is v we can move one of its colors to u. The new function is still a colorful 3-rainbow dominating function and its weight does not increase. Thus, an optimal solution exists where each vertex that is not adjacent to an isolated edge and has a vertex with degree one in its neighborhood gets at least two colors. For each isolated edge, one of its vertices must get one color and the other vertex two colors. Hence, the weight of the optimal colorful 3-rainbow dominating function is at least $n + n_1 + 2n_0$.

In the following, we use function $color(i)$ that is equal to 1, 2, or 3, such that $color(i) \equiv i \mod 3$.

Theorem 7. *Let P_n be a path with n vertices. Then $\gamma_{cr3}(P_n) = n + 2$ if $n \geq 4$ or $n = 1$ and $\gamma_{cr3}(P_n) = n + 1$ if $n = 3$ or $n = 2$.*

Proof. For $n \geq 4$, we have two vertices that have vertices with degree 1 in their neighborhood. Therefore, $\gamma_{cr3}(P) \geq n + 2$. We show, how to assign colors $\{1, 2, 3\}$ such that the weight of the colorful 3-rainbow dominating function is $n + 2$. We enumerate the vertices from 1 to n in the order of their occurrence in the path. Then we assign $color(i)$ to each vertex i for $i \neq 2$ and $i \neq n - 1$, $\{color(2), color(0)\}$ to vertex 2, and $\{color(n-1), color(n+1)\}$ to vertex $n - 1$.

Theorem 8. *Let C_n be a cycle with $n \geq 3$ vertices. Then $\gamma_{cr3}(C_n) = n$ if $n \equiv 0 \mod 3$, $\gamma_{cr3}(C_n) = n+2$ if $n \equiv 1 \mod 3$, and $\gamma_{cr3}(C_n) = n+1$ if $n \equiv 2 \mod 3$.*

Proof. We enumerate the vertices of the cycle from 1 to n in the order of their occurrence in the cycle. To match the lower bound from Theorem 6 a single color must be assigned to each vertex. W.l.o.g. let color 1 be assigned to vertex 1 and color 2 to vertex 2. This implies that color 3 must be assigned to vertex 3, color 1 to vertex 4, and so on. Thus $color(i)$ must be assigned to vertex i, for $1 \leq i \leq n$. For the case $n = 3m$, each vertex i has two neighbors with color sets $\{color(i-1)\}$ and $\{color(i+1)\}$ respectively, and thus has all three colors in its neighborhood. In both other cases, this assignment does not work.

For the case $n = 3m + 2$, vertex 1 has neighbors 2 and n with color sets $\{2\}$ and $\{color(3m+2)\} = \{2\}$. Therefore, no solution with weight n exists. Hence, $\gamma_{cr3}(C_n) \geq n + 1$ and two colors can be assigned to one vertex in the optimal color assignment. We add color 3 to vertex 1 and get an optimal solution.

For the case $n = 3m + 1$, in the original color assignment of $color(i)$ to each vertex i, vertex 1 has neighbors 2 and n with color sets $\{2\}$ and $\{color(3m+1)\} = \{1\}$. So, again no solution with weight n exists and $\gamma_{cr3}(C_n) \geq n + 1$. We show that no solution with weight $n + 1$ exists as well. Such a weight implies, that two colors are assigned to only one vertex. W.l.o.g. let it be vertex 1, with color set $\{1, 2\}$. Thus, color 3 must be assigned to one of the neighbors of vertex 1. W.l.o.g. let it be vertex 2. Color 3 cannot be assigned to vertex 3, otherwise, vertex 3 would not have all three colors in its neighborhood. Thus, w.l.o.g., let color 1 be assigned to vertex 3. Then, color 2 must be assigned to vertex 4, color 3 to vertex 5, and so on till vertex $3m+1$, to which color 2 must be assigned. This vertex then has neighbours $3m$ and 1 with color sets $\{1\}$ and $\{1, 2\}$ respectively. Therefore, vertex $3m + 1$ does not have all three colors in its neighborhood. Hence, no solution with weight $n + 1$ exists. We achieve optimal weight $n + 2$ if we assign $color(i)$ to each vertex $i \neq 1$ and color set $\{1, 2, 3\}$ to vertex 1.

Theorem 9. *Let K_n be a clique with n vertices. Then $\gamma_{cr3}(K_n) = n$ if $n \geq 3$ and $\gamma_{cr3}(K_n) = 3$ if $n = 2$ or $n = 1$.*

Proof. For $n \geq 3$, it is enough to assign color 1 to one vertex, color 2 to one vertex, and color 3 to the rest of the vertices, since all vertices are connected.

Theorem 10. *Let $K_{n,m}$ be a complete bipartite graph (biclique) with n vertices on one side and m vertices on the other. Then (i) $\gamma_{cr3}(K_{n,m}) = n + m$ if $n \geq 3, m \geq 3$, (ii) $\gamma_{cr3}(K_{n,m}) = n + m + 1$ if $n = 2, m \geq 3$, or $m = 2, n \geq 3$, or $n = 1$, or $m = 1$, (iii) $\gamma_{cr3}(K_{n,m}) = n + m + 2$ if $n = 2, m = 2$.*

Proof. Let $U \cup V$ be the partition of the vertex set. For $n \geq 3$ and $m \geq 3$, it is enough to assign, for each set of the partition, color 1 to one vertex, color 2 to one vertex, and color 3 to the rest of the vertices, since all vertices from U are connected to all vertices from V. Now, let $|U| = n = 2$ and $|V| = m \geq 3$. Furthermore, let u_1 and u_2 be the vertices from U and $v_1, \ldots, v_m$ be the vertices from V. Let a colorful 3-rainbow dominating function with weight $n + m$ exist. Exactly one color is assigned to each vertex. Let color 1 be assigned to u_1 and color 2 to u_2. Since each v_i is only connected to u_1 and u_2, color 3 must be assigned to it to have all three colors in its neighborhood. Therefore, vertex u_1 only has colors 1 and 3 in its neighborhood. Hence, no solution with weight $n+m$ exists. We achieve optimal weight $n + m + 1$ if we assign $color(i)$ to each vertex

v_i, color set $\{1, 2\}$ to vertex u_1, and color 3 to vertex u_2. For $n = 2$ and $m = 2$, we have a cycle with $n + m$ vertices. The result follows from Theorem 8. For $m = 1$, Theorem 6 implies that $\gamma_{cr3}(K_{n,m}) \geq n + m + 1$. To get such a weight we assign colors $\{1, 2\}$ to the vertex from U and color 3 to vertices from V.

General values of k. Theorem 9 can be easily generalized for $k \geq 3$. Let K_n be a clique with n vertices. Then $\gamma_{crk}(K_n) = n$ if $n \geq k$ and $\gamma_{crk}(K_n) = k$ if $n < k$. The generalizations of Theorems 7, 8, 10 are more involved, so we omit them here due to space constraints.

Comparison to previous claims. In [9] a notion of total k-rainbow domination is introduced which coincides with colorful k-rainbow domination. The authors compute the total k-rainbow domination number for several graph classes like cliques, bicliques, cycles, and paths. Unfortunately, some of the computed numbers are incorrect. For the sake of concreteness, we consider cliques and paths. In [9] is stated that $\gamma_{crk}(K_n) = k+n-2$, and therefore, for $k = 3$, $\gamma_{cr3}(K_n) = n+1$. As we prove in Theorem 9, for $k = 3, n \geq 3$, $\gamma_{cr3}(K_n) = n$. For paths, we prove in Theorem 7 that for $n \geq 4$, $\gamma_{cr3}(P_n) = n + 2$. In [9] is stated that for $n > 2$

$$\gamma_{crk}(P_n) = \begin{cases} \frac{n(k-1)}{3} + \frac{2n}{3} & \text{if } n \equiv 0 \mod 3, \\ \lfloor \frac{n}{3} \rfloor (k-1) + \frac{2(n-1)}{3} + k - 1 & \text{if } n \equiv 1 \mod 3, \\ \lfloor \frac{n}{3} \rfloor (k-1) + \frac{2(n-2)}{3} + k & \text{if } n \equiv 2 \mod 3. \end{cases}$$

We set $k = 3$ and simplify this formula:

$$\gamma_{cr3}(P_n) = \begin{cases} \frac{4n}{3} & \text{if } n \equiv 0 \mod 3, \\ \frac{4(n-1)}{3} + 2 & \text{if } n \equiv 1 \mod 3, \\ \frac{4(n-2)}{3} + 3 & \text{if } n \equiv 2 \mod 3. \end{cases}$$

These values differ from our values. For example, for $n = 9$, our value is $\gamma_{cr3}(P_4) = n + 2 = 11$ and the value from [9] is $\gamma_{cr3}(P_4) = \frac{4n}{3} = 12$.

5 Efficient Algorithms for Colorful 3-Rainbow Domination

Theorem 11. *Let $G = (V, E)$ be a tree with $n \geq 3$ vertices. Let n_1 be the number of vertices that have neighbors with degree one. Then $\gamma_{cr3}(G) = n + n_1$. We can compute an optimal colorful 3-rainbow dominating function in linear time.*

Proof. Theorem 6 implies that $n + n_1$ is a lower bound on $\gamma_{cr3}(G)$. We give an algorithm, that computes a colorful 3-rainbow dominating function with weight $n + n_1$ in linear time, and thus, show that $\gamma_{cr3}(G) = n + n_1$. We assign two colors to the vertices that have neighbors with degree one and one color to all other vertices. There is at least one vertex in G that in its neighborhood has a vertex with degree 1. Let r be such a vertex. Let $v_1, \ldots, v_j$ be the neighbours of r with degree 1. We assign a number to each vertex, which we call the level of the vertex, as illustrated in Fig. 3. Afterward, we use the level of each vertex to assign colors to it. Vertices $v_1, \ldots, v_j$ get level 0, vertex r gets level 1. To determine the level of remaining vertices we compute a rooted tree from the vertex r using breadth-first search and the following rule. If a parent of v lies on level i then v lies on level $i + 1$. We assign color set $\{1, 2\}$ to root r and color 3 to vertices $v_1, \ldots, v_j$. For all other vertices, the color assignment is as follows. We assign $color(i)$ to each vertex that does not have leaves as children, where i is its level number. Here, function $color()$ is the same as in Sect. 4. Color set $\{color(i), color(i + 2)\}$ is assigned to each vertex that has leaves as children, where i is its level number.

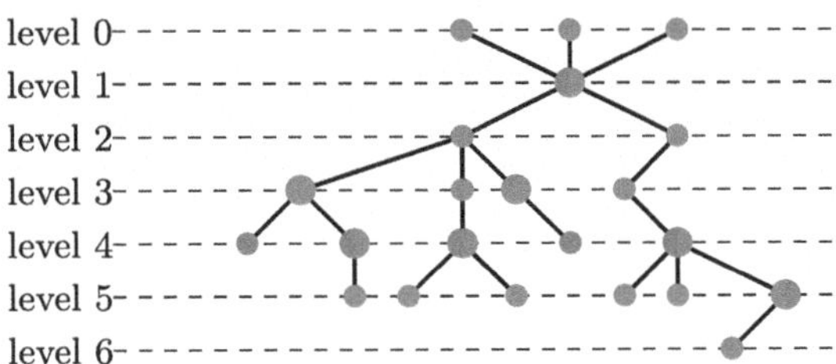

Fig. 3. An example of an optimal colorful 3-rainbow dominating function for a given tree. Here, color 1 is green, color 2 is orange, and color 3 is purple. (Color figure online)

Theorem 12. *Let $G = (V, E)$ be an interval graph with n vertices. Let n_0 be the number of isolated vertices in G. Let n_1 be the number of isolated edges plus the number of vertices that are not adjacent to an isolated edge and have neighbors with degree one. Then $\gamma_{cr3}(G) = n + n_1 + 2n_0$. We can compute an optimal colorful 3-rainbow dominating function in time $O(|V| + |E|)$.*

Proof. We show that for each interval graph, a colorful 3-rainbow dominating function exists with weight equal to the lower bound given in Theorem 6 and give an algorithm that computes an optimal color assignment. We start by eliminating isolated vertices and isolated edges since each of them must get all three colors. Then we solve the problem for each connected component separately.

The proof of Theorem 6 implies that each vertex v that has neighbors with degree one should get two colors. We can simulate this by adding a copy v' for each such vertex v and connecting it to v and all neighbors of v. Now, one of these two colors can be assigned to v and the other one to v'. So we can search for a colorful 3-rainbow dominating function that assigns one color to each vertex in the transformed graph. After such a transformation we still have an interval graph, since in the interval model the corresponding interval to v is simply duplicated. Thus, from now on, we have an interval graph where each vertex has a degree of at least 2.

For each interval graph, a linear ordering of its maximal cliques exists [5], such that each vertex occurs consecutively, and can be computed in time $O(|V|+|E|)$. Let $C_1, C_2, \ldots, C_t$ be such an ordering, see for example Fig. 4.

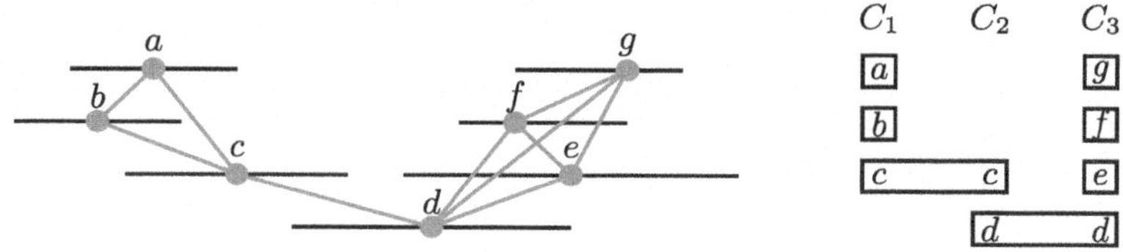

Fig. 4. An interval graph and a required linear ordering of its maximal cliques.

Lemma 1. *For each two consecutive cliques C_i and C_{i+1} a vertex u exists, that is contained in C_i but not in C_{i+1}. Analogously, a vertex v exists, that is contained in C_{i+1} but not in C_i.*

If no u exists, that is contained in C_i but not in C_{i+1}, then each vertex from C_i would be contained in C_{i+1} and C_i would not be maximal.

Each maximal clique contains at least two vertices. We consider two cases. First, we explain how to find a color assignment if each clique contains at least three vertices. Second, we generalize our approach for graphs where some cliques contain only two vertices, using the approach from the first case as a subroutine.
Case 1: Each C_i has at least three vertices. We scan through the ordering of maximal cliques twice. During the first scan, we choose vertices that get color 1. We start with clique C_1, choose some vertex u from it, and assign color 1 to it. Vertex u occurs consecutively in the ordering. Let C_i be the first clique that does not contain u. We choose some vertex v from it and assign again color 1 to it. And so on, until we get to the last clique in the ordering. In such a way color 1 is assigned to exactly one vertex in each clique C_i. Hence, each vertex without color assignment has a neighbor with color 1.

During the second scan through clique ordering, we assign colors 2 and 3. We start with clique C_1 and choose two vertices u and v without color assignment from it and assign color 2 to u and color 3 to v. Since each clique contains at least three vertices and only one vertex per clique gets color 1 such vertices exist. Now let C_k be the last clique in the ordering that contains both u and v. Thus, in the clique C_{k+1} at least one of the colors 2 and 3 is not present. If color 3 is not present but color 2 is, then u is contained in C_{k+1} and v not. Thus C_{k+1} contains one vertex with color 1 and one vertex with color 2. All other vertices do not have colors yet. Since C_{k+1} contains at least three vertices there exists a vertex w that has no color and we can assign color 3 to it. If color 2 is not present but color 3 is, we proceed analogously. If both colors 2 and 3 are not present, then C_{k+1} contains one vertex with color 1 and all other vertices do not have colors. Since C_{k+1} contains at least three vertices there exist two vertices x and y that have no color and we can assign color 2 to x and color 3 to y. Therefore, in each case, C_{k+1} has both colors 2 and 3 and we can proceed to the next clique where at least one of these colors is missing until we get to the last clique. This gives a coloring, in which each clique contains all three colors. The remaining vertices without color assignment can get an arbitrary color. In the end, each vertex has all three colors in its closed neighborhood.

Case 2: Some cliques have only two vertices. Suppose toward a contradiction that the first clique C_1 contains only two vertices. Since one of them is not contained in C_2 and therefore is only contained in C_1, this vertex has only one neighbor. This contradicts the fact, that our interval graph does not contain vertices with degree one. Therefore, C_1 contains at least three vertices.

Let $C_1, \ldots, C_k$ be cliques with at least three vertices, and let C_{k+1} be the first clique with two vertices. Since the graph is connected one of the vertices should also be contained in C_k, let it be x. Due to Lemma 1, the other one should not be contained in C_k, let it be y. Vertex y must also be contained in C_{k+2}. Otherwise, it would have degree one. Due to Lemma 1 some vertex from C_{k+1} is not contained in C_{k+2}. Thus, it only can be x. Let C_{k+1+l} be the first clique in the ordering that again has at least three vertices and cliques $C_{k+1}, \ldots, C_{k+l}$ have two vertices. We continue the above reasoning under usage of Lemma 1 and conclude that the block of maximal cliques with two vertices can only look like a chain: $C_{k+1} = \{x, y\}, C_{k+2} = \{y, z\}, C_{k+3} = \{z, t\} \ldots, C_{k+l-1} = \{u, v\}, C_{k+l} = \{v, w\}$, see Fig. 5. So, our linear ordering of cliques consists of blocks of cliques with at least three vertices and chains of cliques with two vertices in between.

We start with the first block of cliques with at least three vertices $C_1, \ldots, C_k$ and assign colors to vertices using the procedure from the first case. Let x get color $c_1 \in \{1, 2, 3\}$ and let c_2 and c_3 be the other two colors. Then we assign color c_2 to y, color c_3 to z, again color c_1 to t and so on. Let vertex w get color c by this propagation of colors. We arrive at the next block of cliques with at least three vertices that starts at C_{k+l+1}. The only vertex which occurs in this block and has already gotten a color is w. Therefore, for this block, we again use the coloring procedure from the first case. We can ensure that w gets color c by giving color c a role of color 1 in our procedure and choosing w as a vertex with color 1 in the beginning.

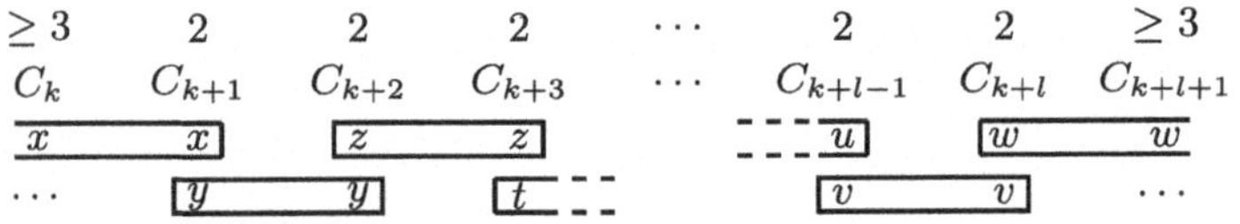

Fig. 5. A chain of maximal cliques with two vertices between the blocks of maximal cliques with at least three vertices.

6 Conclusion

We consider a practical variant of k-rainbow domination with applications in facility location problems. Since, in this variant, each vertex gets at least one color we call it colorful k-rainbow domination. For facility location problems, this requirement leads to a fair distribution of resources because everyone gets a share in the investment. We show that the colorful 3-rainbow domination problem is NP-complete in general graphs and remains NP-complete if we restrict the input graph to the class of bipartite or split graphs. We determine the colorful

3-rainbow domination number for paths, cycles, cliques, and bicliques. Moreover, we give linear time algorithms that compute an optimal colorful 3-rainbow domination function if an input graph is a tree or an interval graph. Furthermore, the colorful k-rainbow domination problem is shown to be NP-complete for fixed $k \geq 3$. For $k = 2$, we show how to solve the problem efficiently. An interesting direction for further research is investigating which other graph classes admit efficient polynomial time algorithms for the colorful 3-rainbow domination problem.

References

1. Ahangar, H.A., Amjadi, J., Rad, N.J., Samodivkin, V.: Total k-rainbow domination numbers in graphs. Commun. Comb. Optim **3**(1), 37–50 (2018). https://doi.org/10.22049/cco.2018.25719.1021
2. Brešar, B., Henning, M.A., Rall, D.F.: Rainbow domination in graphs. Taiwanese J. Math. **12**(1), 213–225 (2008). https://doi.org/10.11650/twjm/1500602498
3. Bresar, B., Sumenjak, T.K.: On the 2-rainbow domination in graphs. Discret. Appl. Math. **155**(17), 2394–2400 (2007). https://doi.org/10.1016/J.DAM.2007.07.018
4. Chang, G.J., Wu, J., Zhu, X.: Rainbow domination on trees. Discret. Appl. Math. **158**(1), 8–12 (2010). https://doi.org/10.1016/J.DAM.2009.08.010
5. Golumbic, M.C.: Chapter 8 - interval graphs. In: Golumbic, M.C. (ed.) Algorithmic Graph Theory and Perfect Graphs, Annals of Discrete Mathematics, vol. 57, pp. 171–202. Elsevier (2004). https://doi.org/10.1016/S0167-5060(04)80056-6
6. Gräf, A., Stumpf, M., Weißenfels, G.: On coloring unit disk graphs. Algorithmica **20**(3), 277–293 (1998). https://doi.org/10.1007/PL00009196
7. Hon, W., Kloks, T., Liu, H., Wang, H.: Rainbow domination and related problems on some classes of perfect graphs. In: Hajiaghayi, M.T., Mousavi, M.R. (eds.) Topics in Theoretical Computer Science - The First IFIP WG 1.8 International Conference, TTCS 2015, Tehran, Iran, August 26-28, 2015, Revised Selected Papers. Lecture Notes in Computer Science, vol. 9541, pp. 121–134. Springer (2015). https://doi.org/10.1007/978-3-319-28678-5_9
8. Jiang, H., Rao, Y.: Total 2-rainbow domination in graphs. Mathematics **10**(12), 2059 (2022). https://doi.org/10.3390/math10122059
9. Kumbargoudra, P., Kureethara, J.: Total k-rainbow domination in graphs. Int. J. Civ. Eng. Technol. **8**(6), 867–875 (2017). https://iaeme.com/MasterAdmin/Journal_uploads/IJCIET/VOLUME_8_ISSUE_6/IJCIET_08_06_094.pdf
10. Shao, Z., Li, Z., Peperko, A., Wan, J., Žerovnik, J.: Independent rainbow domination of graphs. Bull. Malays. Math. Sci. Soc. **42**, 417–435 (2019). https://doi.org/10.1007/s40840-017-0488-6
11. Stockmeyer, L.J.: Planar 3-colorability is polynomial complete. SIGACT News **5**(3), 19–25 (1973). https://doi.org/10.1145/1008293.1008294
12. Tepeh, A.: Total domination in generalized prisms and a new domination invariant. Discuss. Math. Graph Theory **41**(4), 1165–1178 (2021). https://doi.org/10.7151/DMGT.2256
13. Yen, C.K.: 2-rainbow domination and its practical variation on weighted graphs. In: Advances in Intelligent Systems and Applications-Volume 1: Proceedings of the International Computer Symposium ICS 2012 Held at Hualien, Taiwan, December 12–14, 2012, pp. 59–68. Springer (2013). https://doi.org/10.1007/978-3-642-35452-6_8

The Computational Complexity of Equilibria with Strategic Constraints

Bruce M. Kapron(✉) and Koosha Samieefar(✉)

Department of Computer Science, University of Victoria, Victoria, Canada
bmkapron@uvic.ca, koosha021@gmail.com

Abstract. Computational aspects of equilibrium notions for games have been extensively studied, including settings where the goal is to find an equilibrium that possesses some additional properties. Our work extends this direction by considering games in which players are subject to some form of constraint on their strategic choices. We also consider the relationship between Nash equilibria and so-called generalized or social equilibria in this context. Our results demonstrate that the complexity of finding an equilibrium varies significantly between games with slightly different strategic constraints. We also demonstrate that these constraints are useful for modeling problems involving strategic resource allocation and also are of interest from the perspective of behavioral game theory.

Keywords: Game theory · Computational complexity · Nash equilibrium · Hardness results · Constrained equilibrium

1 Introduction

Nash equilibrium [47] is a central concept in game theory and much attention has been given to its computational complexity. While Nash proved the existence of a mixed equilibrium in any strategic game with two or more players, he did not address the complexity of finding such an equilibrium. Starting with [46], where it is shown that the problem is unlikely to be NP-complete, there has been a significant line of work showing the hardness of this problem [15,23,31,49], in particular in relation to the class TFNP of total NP search problems. This culminated in the result of [15], showing that the computational complexity of finding a (mixed) Nash equilibrium for two-player games is as hard as any problem in PPAD, which is a subclass of TFNP capturing the complexity of problems such as Sperner's Lemma and Brouwer's fixed-point theorem.

While there are multiple constant factor approximation algorithms for Nash equilibrium [1,18,19,24,28], the problem of finding an approximate solution is hard in general [15,51,52]. In fact, finding a PTAS for the Nash equilibrium problem has proved challenging [5,25,39,43]. Finally, in a seminal work [52], it was demonstrated that under the assumption of an exponential time hypothesis

R. Královič and V. Kůrková (Eds.): SOFSEM 2025, LNCS 15539, pp. 112–127, 2025.
https://doi.org/10.1007/978-3-031-82697-9_9

for PPAD (PETH), the quasi-polynomial time algorithm for approximate Nash introduced in [43] is optimal.

Beyond the basic Nash equilibrium problem, we can consider equilibria with additional constraints. In this direction, there are two well-known approaches. The goal of the first approach is to find a Nash equilibrium that has certain properties (e.g., achieving a specified social welfare) [17,38,45]. In the second approach, the goal is to find a generalized solution (also called constrained equilibria), i.e., one in which no player would deviate assuming all players only play strategies satisfying the given constraint [2,26,32].

1.1 Equilibrium with Strategic Constraints

Our main focus will be on games where players are trying to avoid "overlap" between their strategies. In the extreme case, players are prohibited from using strategies used by other players, i.e., the intersection of their supports is empty. We call this problem *disjoint Nash* and show that deciding whether a bi-matrix game has such an equilibrium, even under polynomial additive approximation, is NP-complete. We also consider a stronger form of disjoint Nash, that we call *partition Nash*, in which the players' supports must also partition the strategy space. In Sect. 3.2 we apply this notion in the context of strategic resource allocation. We may generalize disjoint Nash by allowing overlap but requiring that mixed strategies be at least 2δ statistically far (i.e., with respect to $L1$ distance) obtaining a problem that we call *δ-far Nash*. It is natural to consider whether there is a threshold approximation error below which the problem δ-far Nash becomes total (has at least one guaranteed solution). While we do not have an answer to this question, we provide a bound in which a solution is guaranteed by the existence of a Nash equilibrium. Viewing the mixed strategies of players as a representative of the strategies of a large number of agents, the statistical difference between the mixed strategies of the players can provide useful information for the analysis of public behavior. In the full version, we apply this perspective to income tax games.

In general, by constraining a player's best response, we can generalize Nash's notion. Debreu extended the notion of equilibrium to allow the possibility that a player's set of available strategies depends on the strategic choices of other players and proved the existence of *generalized* equilibrium (also called *social* or *constrained* equilbrium) for such games under certain assumptions [26]. We will consider disjointness and related constraints in this setting.

1.2 Related Work

The complexity of constrained Nash equilibrium problems has received considerable attention. In contrast to the unconstrained setting, such problems do not necessarily admit a solution. Gilboa and Zemel's work on equilibrium with constraints such as having a guaranteed payoff shows that many of these problems are NP-complete [38]; alternate proofs are provided in [10,16,17,45]. In [17,45], it is shown that the hardness also holds for symmetric games. Furthermore, the hardness can be extended to symmetric and win-lose games [10].

When we have two players, the challenge lies in finding the support of a Nash equilibrium [56] and a solution can be found using linear programming. This approach does not extend beyond two-player games (for the exact case), and in fact, in Nash's original paper, it is shown that there are 3-player games that can have only irrational solutions. Discussion of exact complexity in such cases requires the use of classes such as FIXP [31] and (for Nash equilibrium with additional constraints) $\exists\mathbb{R}$ and ETR [8,9,36].

Beyond questions of exact solutions, limited progress has been made in approximating Nash equilibrium with certain constraints unless they have a direct relationship to the social welfare of the players [4,17,29,40,54]. In fact, to our knowledge, the only positive results for bi-matrix games for finding a Nash equilibrium with constraints are on the payoffs of the players [20,21]. By employing the quasi-polynomial time algorithm of [43], it is possible to approximate *any* Nash equilibria with a constant error by solutions whose support is a logarithmic-size subset of all strategies while maintaining a payoff that is close to the exact solution. This applies to the constrained Nash equilibrium problems considered in this paper which gives a quasi-polynomial upper bound.

Surprisingly, considering the application of generalized equilibrium problems defined by Debreu [26] and Rosen [50] in economics and other areas (see [32,33]), the complexity of these problems has not received much attention until recently [34,35,42,48]. Many researchers from different areas have worked on this notion which in turn explains why it has developed with a number of names in the literature including pseudo-games, social equilibrium, concave games, and common (coupled) constrained equilibrium [2,32,33,55,59]. While much of the work on generalized equilibria focuses on purse strategies, both Charnes [14] and McKinsey [44] provide arguments for the reasonableness of considering constraints on mixed strategies. Furthermore, existing work on finding generalized equilibria focuses on the case of convex constraints, where solutions are guaranteed to exist by well-known mathematical facts [22,26,48,50]. It is easy to observe that most of the problems we consider do not fall into this setting. However, a solution is guaranteed for one of the main problems that we consider.

1.3 Our Contributions

We primarily focus on the computational complexity of two forms of equilibrium with constraints, namely Nash and generalized equilibria with some form of statistical farness constraint. Towards establishing hardness for these problems, we also consider some related problems involving equilibria with constraints. To our knowledge, our work is the first to consider the complexity of equilibria under a set of inherently plausible conditions based on a mutual constraint on players' strategies. Our results demonstrate the richness and subtlety that arise in this setting. Additionally, we consider the applicability of farness constraints to strategic resource allocation, behavioral interpretations of game theory, and population-level games (available in the full version).

Informally stated, we show that approximate versions of disjoint Nash, partition Nash, and δ-far Nash and several related problems are NP-complete. Given the general landscape of Nash equilibria subject to constraints, these results are to be expected, although their proofs require a significant extension of existing techniques.

However, in addition to these NP-completeness results, we provide a variety of computational tractability and hardness results for generalized equilibria with strategic constraints (see Theorem 3). In contrast to the case of Nash equilibrium, the generalized setting is much more subtle, and the problems here vary considerably in their computational complexity. The results we obtain in this setting represent a new direction in computational game theory (see also [22]).

2 Preliminaries

Our focus is on normal-form games with k players, specified by a set $S = [n]$ of *strategies* and a *payoff* (utility) function $U_i : [n]^k \to \mathbb{R}$ for each player $i \in [k]$. A *mixed strategy* is a distribution over S, which may be viewed as an element of the probability simplex Δ_n. A *pure strategy profile* is an element $\mathbf{s} \in n^k$, while a *mixed strategy profile* is an element $\mathbf{x} \in (\Delta_n)^k$. We can extend U_i to mixed strategy profiles by defining $U_i(\mathbf{x})$ as $\mathbb{E}_{\mathbf{s} \leftarrow \mathbf{x}}[U_i(\mathbf{s})]$.

Definition 1. *A strategy profile $\boldsymbol{x}^* = (x_1^*, x_2^*, \ldots, x_k^*)$ is an* ϵ-approximate Nash equilibrium *if for every player i and for all $s \in S$, $U_i(\boldsymbol{x}^*) + \epsilon \geq U_i(s, \boldsymbol{x}^*_{-i})$.*

As discussed in the introduction, we also consider a generalization of Nash equilibrium in which a player's best response may be subject to some constraint.

Definition 2. *Let $\mathcal{G}$ be a game and $\mathcal{R}_i$ a set-valued mapping on $(\Delta_n)^{k-1}$, for each $i \in [k]$ such that $\mathcal{R}_i(\boldsymbol{x}_{-i}) \subseteq \Delta_n$. A mixed strategy profile x^* is a* generalized equilibrium *(constrained equilibrium) if:*

$$\forall x_i \in \mathcal{R}_i\left(\boldsymbol{x}^*_{-i}\right), \qquad U_i\left(\boldsymbol{x}^*\right) \geq U_i\left(x_i, \boldsymbol{x}^*_{-i}\right) \tag{1}$$

Remark 1. A Nash equilibrium which satisfies a given common constraint will also be a generalized equilibrium with respect to that constraint, but in general the converse need not hold.

Remark 2. We only consider the case where all the $\mathcal{R}_i$'s are the same. Equivalently, we might consider that there is one *common constraint* $\mathcal{R}$.

$$\forall x_i \text{ s.t } \mathcal{R}(x_i, \mathbf{x}^*_{-i}) \text{ holds}, \qquad U_i\left(\mathbf{x}^*\right) \geq U_i\left(x_i, \mathbf{x}^*_{-i}\right) \tag{2}$$

Remark 3. The definition of generalized games in [48,50] does not involve games with mixed strategies. Our definition may be viewed as a special case of these definitions.

In this paper, we consider the hardness and tractability of a variety of equilibrium problems in both exact and approximate forms. Approximation will always mean additive approximation (as in Definition 1). When the complexity of approximation depends on the error term ϵ, a distinction arises based on whether ϵ is encoded in binary (so that complexity is a function of $\log(\frac{1}{\epsilon})$) or in unary (so that complexity is a function of $\frac{1}{\epsilon}$). *In the sequel, we adopt the following conventions*: (1) if the approximation is not explicitly mentioned, we are considering the exact version of a problem. (2) unless otherwise noted, the hardness of approximation refers to the (stronger) case of polynomial approximation.

The subclass PPAD of TFNP plays a central role in the complexity of Nash equilibrium.

Definition 3. *The* end-of-the-line (EOTL) *problem is defined: Given a directed graph $\mathcal{G}$, represented by two poly-sized circuits which return the predecessor and the successor of a node (represented in binary), where each vertex has at most one predecessor and successor and a vertex u in $\mathcal{G}$ with no predecessor, find another vertex $v \neq u$ with no predecessor or no successor. A total search problem is in* PPAD *if it is poly-time reducible to EOTL.*

3 Equilibria with Strategic Constraints

We now introduce the definitions and main theorems of problems related to equilibria with strategic constraints. For simplicity, we only investigate the computational aspects of two-player (bi-matrix) games, and we will assume that the utility matrices' entries are in $[0, 1]$ unless otherwise stated. The proofs are easily extendable to normal-form games with constantly many players. We consider the standard notion of *statistical distance* (*total variation distance*) between mixed strategies, viewed as probability distributions on pure strategies.

3.1 Problem Definitions

Definition 4. *Two (mixed) strategies $x \in \Delta_n$ and $y \in \Delta_n$ are δ-far if $||x - y||_1 \geq \delta$, where $||x-y||_1 = \sum_{i=1}^{n} |x_i - y_i|$.* δ-closeness *of strategies can be defined similarly.*

Remark 4. Note that for bi-matrix games, for the mixed strategies x and y for the row and the column player, we simply have $0 \leq ||x - y||_1 \leq 2$.

Definition 5. *A Nash equilibrium $\boldsymbol{x}^* = (x_1^*, \ldots, x_k^*)$ is called a* δ-far Nash equilibrium *if for all $i \neq j$, x_i^* and x_j^* are 2δ-far. The approximate version of this type of equilibrium will be denoted by (δ, ϵ)-far Nash equilibrium*[1].

A specific case of far Nash can also be introduced by prohibiting all players from using the strategies employed by all other players as follows:

[1] We normalized δ to be in the range $[0, 1]$ similarly to ϵ.

Definition 6. *(Disjoint Nash): A strategy profile $\boldsymbol{x} = (x_1, \ldots, x_k)$ is a* disjoint Nash equilibrium *(or simply disjoint Nash) if it is a Nash equilibrium and $\bigcap_{i=1}^{k} P_i = \emptyset$ where $P_i = \text{Supp}(x_i)$*[2]*.*

We say that a mixed strategy profile $(x_1, \ldots, x_k)$ has a disjoint support profile *if for any player i and player j $(i \neq j)$ have disjoint supports.*

Remark 5. When $\delta = 0$, δ-far Nash is just the problem of finding a Nash equilibrium, and when $\delta = 1$ it is exactly disjoint Nash equilibrium. The existence of a disjoint Nash is not necessarily guaranteed, e.g., consider a 2-player, 2-strategy game where $U_1(1,1) = U_2(1,1) = 1$ while for $(i,j) \neq (1,1)$, $U_1(i,j) = U_2(i,j) = 0$.

For the purpose of investigating one of the straightforward applications of equilibria with limited overlapping support, we define the following problem that can be derived by further constraining disjoint Nash. Analogously to Definition 1, we can define the approximate version of these problems (*ϵ-disjoint and ϵ-partition Nash*).

Definition 7. *(Partition Nash): A strategy profile $\boldsymbol{x} = (x_1, \ldots, x_k)$ is a* partition Nash equilibrium *if it is a disjoint Nash equilibrium and $\bigcup_{i=1}^{n} P_i = S$, where $P_i = \text{Supp}(x_i)$.*

We also consider generalized equilibria for this class of constraints. For example, in the generalized version of a disjoint Nash equilibrium (which we call generalized disjoint equilibrium), a player cannot deviate to a mixed strategy whose support has a nonempty intersection with those of the other players.

Definition 8. *For bi-matrix games, a* generalized δ-far equilibrium *is a generalized equilibrium with common constraint:*

$$\mathcal{R}_\delta^n = \{(x,y) \in \Delta_n \times \Delta_n \mid ||x-y||_1 \geq 2\delta\}$$

while a generalized disjoint equilibrium *is a generalized equilibrium with common constraint:*

$$\mathcal{R}_{\text{disjoint}}^n = \{(x,y) \in \Delta_n \times \Delta_n \mid \langle x,y \rangle = 0\}$$

where $\langle x,y \rangle$ denotes the inner product.

Remark 6. Assume that the column player plays y. For the row player (similarly for the column player), we also can consider the relations (see Definition 2):

$$\mathcal{R}_\delta^n(y) = \{x \in \Delta_n \mid ||x-y||_1 \geq 2\delta\}$$

$$\mathcal{R}_{\text{disjoint}}^n(y) = \{x \in \Delta_n \mid \langle x,y \rangle = 0\}$$

[2] $\text{Supp}(x_i)$ denotes the strategies used by player i with non-zero probabilities.

Remark 7. It is easy to observe that all games having exactly two pure strategies have a generalized disjoint equilibrium where both players play one pure strategy. The constrained strategy spaces that we have defined are not convex. As shown by Rosen [50] a solution is guaranteed for convex constraints. Recently, it has been proved that by using a computational version of Kakutani's fixed point theorem, finding an approximate equilibrium (a generalized equilibrium) for a game with convex constraints is PPAD-complete [48]. Our results show that a solution may be guaranteed when constraints are not convex.

In the generalized setting, at least one of the problems we propose, namely approximate generalized disjoint equilibrium, is tractable (Theorem 3 (a)) and has at least one trivial solution. However, we show that one selfish player can degrade the social welfare of all other players. With this drawback in mind, we direct our attention to another type of generalized equilibrium by excluding the possibility of minor probabilities which is defined as follows:

Definition 9. *A* (θ, δ)-restricted far equilibrium problem *is defined as follows: Given a bi-matrix game* $\mathcal{G}$ *find a generalized* δ*-far equilibrium such that all pure strategies played with a positive probability are played with probability greater than* θ*. We can relax this definition by using an additive approximation error* ϵ *and call it* $(\theta, \delta, \epsilon)$-restricted far equilibrium. *For* $\delta = 1$*, we call the problem* (θ, ϵ)-restricted disjoint equilibrium.

Some Related Problems. Dually to notions of farness, we may consider equilibria with closeness constraints. This may also be viewed as a generalization of symmetric equilibrium. Similar to δ-far Nash, the δ-close Nash problem is NP-hard to approximate. The problem of determining if a bi-matrix game has more than one equilibrium is one of the well-known instances of Nash equilibria with constraints [17,38,45]. We observe that given continuity considerations, every bi-matrix game has an infinite number of approximate Nash equilibria. As an approximate analog of this problem, one can consider the problem of discovering two equilibrium conditions with a given statistical distance between them [4,29,30]. In [4] (Theorem 1.4 and 4.1), it is shown that for a sufficiently small approximation error, an instance of this problem is as hard as finding a hidden clique of size $O(\log n)$ in the random graph $G(n, \frac{1}{2})$ with n vertices. Furthermore, under ETH (the exponential time hypothesis), a quasi-polynomial lower bound for a constant approximation error smaller than $\frac{1}{8}$ exists (see problem 3 in [29]). We show that this problem for bi-matrix games is NP-hard to approximate[3]. We also have another variant of Nash equilibrium for which only the exact problem was also investigated in [45]. This variant requires a Nash equilibrium such that all strategies cannot be played with minor probabilities. For brevity, we defer the proof of these results to the full version.

[3] A very obvious difference between this problem and far Nash equilibrium is that far Nash equilibrium focuses on the existence of one equilibrium in which the players' strategies are statistically far apart while the other problem requires at least two Nash equilibria that are statistically far apart.

3.2 Applications

To provide some motivation, we present three potential applications of our constraints, two of which concern the behavioral interpretation of equilibria and are available in the full version.

Strategic Resource Allocation. Problems of resource allocation and fair division have received considerable research attention from a computational perspective, as they are a critical consideration in multi-agent systems. Fair division problems involve a group of agents where each has individual preferences for some collection of resources, typically represented by a utility function. The goal is the allocation of resources in a fair manner. A fundamental underlying question is how to define fairness criteria in the first place. Classical fairness definitions include competitive equilibrium [3] and social Nash welfare [41]. Many fairness notions for indivisible goods have been considered; however, they are typically hard to compute [11–13]. Furthermore, almost all of this research is limited to non-strategic settings.

Strategic fair division introduces participants who may act uncooperatively in order to maximize their own utility. In particular, players may hide their true preferences. In the presence of strategic behavior, the formulation of appropriate fairness criteria is essential. One branch of fair division related to game theory considers equilibrium in games resulting from fair division algorithms [11,13,57]. An *envy-free* allocation is an allocation in which each player receives a share that is, in their eyes, at least as good as the share received by any other agent [37]. There are several different variations of envy-freeness [6,7,58]. We define a variation of envy-freeness in the following strategic allocation problem:

Definition 10. *The game $\mathcal{RV}$ for k players and n items is defined as follows: $\forall i \in [k]$ and $\forall t \in [n]$, $U_i(t, x_{-i}) = \alpha V_i(t) + \beta R_i(x_{-i})$, where α and β are constants. $V_i(t)$ indicates the value of item t for player i and $R_i(x_{-i})$ is a function that is computable in constant time which gives player i's fairness measure for the allocation of items $x_{-i} = (t_1, \ldots, t_{i-1}, t_{i+1}, \ldots, t_n)$ to the other players*[4].

Definition 11. *An allocation is an (δ, ϵ)-*item-wise envy-free *allocation if each player receives the items that are in the support of a ϵ-partition Nash equilibrium x^* of game $\mathcal{RV}$ with the following properties:*

- *Each player $i \in [k]$ has a support of size $\frac{n}{k}$.*
- *For each player $i \in [k]$, we have $||x_i^* - b||_1 \leq \delta$ where b is a given distribution that $\frac{n}{k}$ strategies are assigned with probability $\frac{k}{n}$.*

It is easy to observe that in a $(0, 0)$-*item-wise envy-free* allocation, no player would prefer to exchange any item allocated to another player for any of the items in their own allocation. Note that in an envy-free allocation, each player is only concerned with the value of the items. The following proposition follows from the hardness of partition Nash.

[4] A more complex form of this problem is described in [53].

Proposition 1. *The decision problem of whether there exists* (δ, ϵ)-item-wise envy-free *allocation is NP-complete for some* δ *and* ϵ *poly-bounded in the size of the game.*

Remark 8. Another branch of fair division considers truthful resource allocation mechanisms, in which agents are incentivized to reveal their true valuations for resources. We do not address truthful mechanisms in this paper.

4 Main Results

A number of the problems we consider depend on parameters besides the additive approximation term. These parameters depend on n, which is the size of the strategy space S. We say that such a parameter α is *poly-bounded* if there is some $k > 0$ such that $\alpha(n) = O(\frac{1}{n^k})$. The theorems of this section admit a number of hardness results for other variations of Nash equilibrium including additional information about the parameters. We have not stated them here in the interest of simplifying the presentation. For Nash equilibrium with constraints, we have the following results:

Theorem 1. *For bi-matrix games, the following problems are NP-complete under polynomial approximation:*

(a) Disjoint Nash
(b) Partition Nash
(c) δ*-Far Nash equilibrium, for any poly-bounded* δ

We may consider the question of whether there is a threshold approximation error below which δ-far Nash becomes a total problem. Despite the absence of a definitive answer to this question, we provide a bound within which a δ-far Nash equilibrium is guaranteed to exist in the following theorem:

Theorem 2. *For any game* $\mathcal{G}$*, there exists a* $\delta(n) \geq \frac{1}{n}$ *such that at least one* $(\delta, 4\delta)$*-far Nash equilibrium for* $\mathcal{G}$ *exists. Furthermore, this problem is in PPAD.*

The main complexity results related to generalized equilibrium versions of these problems are the following. These are with respect to bi-matrix games and polynomial approximation except (a).

Theorem 3. *(a) The problem of finding a generalized disjoint equilibrium under exponential approximation is computable in polynomial time.*
(b) There is a poly-bounded δ *for which the problem of finding a* δ*-far generalized equilibrium is PPAD-hard.*
(c) There is a poly-bounded δ *for which the problem of finding a* δ*-far generalized equilibrium for games whose diagonally modified version does not admit a fully mixed Nash equilibrium (see Definition 12) is in PPAD.*
(d) For some poly-bounded θ*, the problem of deciding whether there exists a* θ*-restricted disjoint equilibrium is NP-complete.*

(e) For some poly-bounded θ and any poly-bounded δ, the problem of deciding whether there exists a (θ, δ)-restricted far equilibrium is NP-hard.

Remark 9. The previous assumption (Theorem 3 (c)) can be easily checked by using part of a well-known *support enumeration* algorithm.

5 Technical Overview

5.1 Basic Techniques

Throughout this work, we use two standard techniques (technical lemmas) in multiple situations. For example, to derive hardness results for disjoint Nash through standard NP-hard reductions from well-known NP-hard problems, we need to ensure that the required game solutions have disjoint support. This is done by duplicating some of the strategies so that the players can only play strategies from their *associated* sets. We explain this technique with the following simple example:

Lemma 1. (Simple Duplication). *The problem of finding a Nash equilibrium is poly-time reducible to finding a disjoint Nash equilibrium.*

Proof. (sketch.) Given a game $\mathcal{G}$ with S strategies, we create game $\mathcal{G}'$ with duplicates of S (S_1 and S_2). We prove that a Nash equilibrium in $\mathcal{G}$ will form a disjoint Nash equilibrium in $\mathcal{G}'$ in which the first and second player only select their strategies from S_1 and S_2 respectively.

Remark 10. The preceding technique can only be applied to the exact versions of our problems. We show that, under any approximation scheme that we reviewed, this technique can connect present hardness results on Nash equilibrium to our strategic constraints if applied carefully[5]. Theorem 2 is also a fairly direct corollary of this lemma.

Lemma 2. (Penalty Lemma). *Suppose that (x^*, y^*) is a Nash equilibrium in a bi-matrix game whose utility matrix entries are in the range $[\alpha, \beta]$. For each i, we define x_i^- to be the same as x_i^* except we take $\epsilon_i \leq x_i^*$ from the weight of strategy i and redistribute it to some strategies in $[n]-\{i\}$. Assuming $\sum_{i=1}^{n} \epsilon_i \leq \epsilon$, we have the following[6]:*

1. *(x^-, y^*) and (x^*, y^-) are $2\epsilon(\beta - \alpha)$-Nash equilibria.*
2. *$|U_1(x^*, y^*) - U_1(x^-, y^-)| \leq 2\epsilon(\beta - \alpha)$*
3. *$|U_2(x^*, y^*) - U_2(x^-, y^-)| \leq 2\epsilon(\beta - \alpha)$*

Proof. (sketch.) To prove this lemma, we show that modifying the strategy profile of the players in a Nash equilibrium to another strategy profile can guarantee an approximate Nash equilibrium in which the approximation error is bounded by the statistical distance of the modified strategy and the original Nash equilibrium.

[5] Details are available in the full version.
[6] We assume $\alpha = 0$, $\beta = 1$ unless stated.

To inspect the computational complexity of generalized far equilibrium, we use a technique that penalizes the diagonals of the payoff matrices. Using this approach, we demonstrate that Nash equilibrium with certain properties can be related to generalized equilibrium with farness-related constraints. The following definition identifies a class of games that we will show are guaranteed to have a generalized far equilibrium, under some conditions.

Definition 12. *For a game $\mathcal{G}$, the* diagonally modified *version of $\mathcal{G}$, denoted $\mathcal{D}^M(\mathcal{G})$, is a game where the entries of the payoff matrices $U_i^{\mathcal{D}^M(\mathcal{G})}$ are the same as $\mathcal{G}$ except the diagonal entries as follows:*

$$\forall t \in S : U_i^{\mathcal{D}^M(\mathcal{G})}(t,t) = -M$$

Remark 11. Note that if we find a Nash equilibrium in $\mathcal{D}^{\mathrm{M}}(\mathcal{G})$, it does not necessarily form a Nash equilibrium for the original game $\mathcal{G}$ since we removed all the information that the diagonals of the payoff matrices carry. We might assume that as M becomes sufficiently large, the likelihood of disjoint equilibria increases. While this observation may be useful in most cases, it does not necessarily hold true in general.

5.2 Deriving NP-Hardness

The NP-hardness proofs of the problems stated in this paper are inspired by the generic proof provided by [17] for NP-hardness of deciding the existence of a Nash equilibrium with constraints. This generic reduction (from SAT) improved the NP-completeness results for exact Nash equilibrium with certain constraints given in [38]. However, the hardness of approximation proof given in [17] does not apply to the form of approximation that we consider in this paper (the additive approximation error introduced in [23]). Instead, we start with a reduction given by Schoenebeck and Vadhan [54] (which modifies the proof provided in [17]) showing the approximate Nash equilibrium with a guaranteed payoff for all players is NP-complete. Most NP-hardness reductions for constrained Nash equilibrium problems prove that the existence of a solution for the given SAT instance implies the existence of a desirable equilibrium solution of the constructed game and the absence of an acceptable solution for the first problem implies that *any* existing Nash equilibrium does not satisfy the given constraints. In the absence of a solution for the SAT problem, existing reductions usually force the players to choose a single strategy while our reduction proofs depend on the use of more than one strategy.

Our proofs begin with the hardness of disjoint Nash, obtained by applying a non-trivial combination of the reductions in [17,54]. To clarify the structure of the proof, we break down the proof of [54] into a series of lemmas. We use modest variations of these lemmas to prove the hardness of disjoint Nash by combining the technical lemmas introduced above and the ideas given in both reductions given in [17] (Corollary 6) and [54] (Theorem 8.6). The hardness of partition Nash and the existence of an approximate far Nash in certain cases will be directly derived by using the technical lemmas. Next, by an adaptation of the duplication technique in the constructed game, we force any equilibrium corresponding to a SAT solution to have the farness property along with other properties making it suitable to establish the NP-hardness of far Nash equilibrium. Proving NP-hardness results for the restricted disjoint equilibrium requires a further extension of the games that we construct based on the SAT problem in which we also embed a sub-game that is a generalization of *rock-paper-scissors*. Also, we will require a different analysis of the players' social welfare to establish NP-hardness. For the restricted far equilibrium, we will need to consistently integrate all the techniques and construction elements previously employed. Details are available in the full version.

6 Conclusion and Open Problems

We have presented hardness and feasibility results for games with strategic constraints in normal-form games and argued that these results are useful for modeling a variety of applications. We believe this is a fruitful area for further study, both in theory and in practical applications. Below, we suggest some possible research directions.

By applying the methods outlined in this paper, we may establish a variety of NP-hardness results for alternative forms of Nash equilibrium. Another direction could be whether, under ETH assumption, the lower bounds in [29] could hold for our problem under constant approximation. Furthermore, investigating poly-matrix games that can succinctly represent normal-form games with these constraints could be worthwhile [30]. In addition, exploring the relationship between the close Nash equilibrium notion and imitation games could be of interest. In contrast to generalized disjoint equilibrium, even the inclusion of (δ, ϵ)-generalized far equilibrium in NP cannot be proved easily for any δ. The computational complexity of this problem remains open considering the fact that we showed that an equilibrium in this sense may not exist and approaches such as diagonal modification of the game fail to address the computational complexity of the problem in general. However, the *generalized close equilibrium* notion with the following (convex) common constraint, has a guaranteed solution by [50] which relies on Kakutani's Fixed Point Theorem:

$$\mathcal{R}_\delta^n = \{(x, y) \in \Delta_n \times \Delta_n \mid ||x - y||_1 \leq 2\delta\}$$

Inclusion in PPAD (under approximation) could be proven by using arguments similar to those discussed in [48]. In conclusion, finding communication complexity and inapproximability results for generalized close equilibrium may also be of interest (see also [27,39]). Additionally, it might be worth considering these variations within the context of correlated equilibria. In general, an interesting question is whether there exists a natural variant of Nash equilibrium such that the computational complexity of the problem is in TFNP but not in PPAD.

Acknowledgments. We would like to acknowledge the support of NSERC discovery grant RGPIN-2021-02481.

References

1. Ahmadi, A.A., Zhang, J.: Semidefinite Programming and Nash Equilibria in Bimatrix Games. INFORMS J. Comput. **33**(2), 607–628 (2021) https://doi.org/10.1287/ijoc.2020.0960
2. Altman, E., Solan, E.: Constrained games: the impact of the attitude to adversary's constraints. IEEE Trans. Automat. Contr. **54**(10), 2435–2440 (2009)
3. Arrow, K.J.: An Extension of the Basic Theorems of Classical welfare Economics (1951)
4. Austrin, P., Braverman, M., Chlamtáč, E.: Inapproximability of NP-complete variants of Nash equilibrium. In: Goldberg, L.A., Jansen, K., Ravi, R., Rolim, J.D.P. (eds.) Approximation, Randomization, and Combinatorial Optimization. Algorithms and Techniques, pp. 13–25. Springer Berlin Heidelberg, Berlin, Heidelberg (2011)
5. Babichenko, Y.: Query complexity of approximate Nash equilibria. J. ACM **63**(4), 1–24 (2016)
6. Barbanel, J.B.: Super envy-free cake division and independence of measures. J. Math. Anal. Appl. **197**(1), 54–60 (1996). https://doi.org/10.1006/S0022-247X(96)90006-2. https://www.sciencedirect.com/science/article/pii/S0022247X96900062
7. Barman, S., Ghalme, G., Jain, S., Kulkarni, P., Narang, S.: Fair Division of Indivisible Goods Among Strategic Agents (2019)
8. Berthelsen, M.L.T., Hansen, K.A.: On the computational complexity of decision problems about multi-player Nash equilibria. Theory Comput. Syst. **66**(3), 519–545 (2022). https://doi.org/10.1007/s00224-022-10080-1
9. Bilò, V., Mavronicolas, M.: $\exists\mathbb{R}$-complete Decision Problems about (Symmetric) Nash equilibria in (symmetric) multi-player games. ACM Trans. Econ. Comput. **9**(3), 1–25 (2021). https://doi.org/10.1145/3456758
10. Bilò, V., Mavronicolas, M.: The complexity of computational problems about Nash equilibria in symmetric win-lose games. Algorithmica **83**(2), 447–530 (2021)
11. Brânzei, S., Caragiannis, I., Kurokawa, D., Procaccia, A.: An algorithmic framework for strategic fair division. In: Proceedings of the AAAI Conference on Artificial Intelligence **30**(1) (Feb 2016). https://doi.org/10.1609/aaai.v30i1.10042
12. Brânzei, S., Gkatzelis, V., Mehta, R.: Nash social welfare approximation for strategic agents. Oper. Res. **70**(1), 402–415 (2022). https://doi.org/10.1287/opre.2020.2056
13. Brânzei, S., Miltersen, P.B.: Equilibrium analysis in cake cutting. In: Proceedings of the 2013 International Conference on Autonomous Agents and Multi-Agent Systems, pp. 327–334. AAMAS '13, International Foundation for Autonomous Agents and Multiagent Systems, Richland, SC (2013)
14. Charnes, A.: Constrained games and linear programming. Proc. National Acad. Sci. **39**(7), 639–641 (1953). http://www.jstor.org/stable/88683
15. Chen, X., Deng, X., Teng, S.H.: Settling the complexity of computing two-player Nash equilibria. J. ACM **56**(3), 1–57 (2009)
16. Codenotti, B., Štefankovič, D.: On the computational complexity of Nash equilibria for (0,1) bimatrix games. Inf. Process. Lett. **94**(3), 145–150 (2005). https://doi.org/10.1016/j.ipl.2005.01.010, https://www.sciencedirect.com/science/article/pii/S0020019005000281
17. Conitzer, V., Sandholm, T.: New complexity results about Nash equilibria. Games Econ. Behav. **63**(2), 621–641 (2008)

18. Czumaj, A., Deligkas, A., Fasoulakis, M., Fearnley, J., Jurdziński, M., Savani, R.: Distributed methods for computing approximate equilibria. In: Web and Internet Economics, pp. 15–28. Lecture notes in computer science, Springer Berlin Heidelberg, Berlin, Heidelberg (2016)
19. Czumaj, A., Deligkas, A., Fasoulakis, M., Fearnley, J., Jurdziński, M., Savani, R.: Distributed methods for computing approximate equilibria. Algorithmica **81**(3), 1205–1231 (2019)
20. Czumaj, A., Fasoulakis, M., Jurdzinski, M.: Approximate Nash equilibria with near optimal social welfare. In: Proceedings of the 24th International Conference on Artificial Intelligence, pp. 504–510. IJCAI'15, AAAI Press (2015)
21. Czumaj, A., Fasoulakis, M., Jurdzinski, M.: Approximate plutocratic and egalitarian Nash equilibria: (extended abstract). In: Proceedings of the 2016 International Conference on Autonomous Agents & Multiagent Systems, pp. 1409–1410. AAMAS '16, International Foundation for Autonomous Agents and Multiagent Systems, Richland, SC (2016)
22. Daskalakis, C.: Non-concave Games: A Challenge for Game Theory's Next 100 Years (2022)
23. Daskalakis, C., Goldberg, P.W., Papadimitriou, C.H.: The complexity of computing a Nash equilibrium. SIAM J. Comput. **39**(1), 195–259 (2009)
24. Daskalakis, C., Mehta, A., Papadimitriou, C.: A note on approximate Nash equilibria. Theor. Comput. Sci. **410**(17), 1581–1588 (2009)
25. Daskalakis, C., Papadimitriou, C.H.: On oblivious PTAS's for Nash equilibrium. In: Proceedings of the Forty-First Annual ACM Symposium on Theory of Computing, pp. 75–84. STOC '09, Association for Computing Machinery, New York, NY, USA (2009). https://doi.org/10.1145/1536414.1536427
26. Debreu, G.: A social equilibrium existence theorem. Proc. Natl. Acad. Sci. **38**(10), 886–893 (1952). https://doi.org/10.1073/pnas.38.10.886, https://www.pnas.org/doi/abs/10.1073/pnas.38.10.886
27. Deligkas, A., Fearnley, J., Hollender, A., Melissourgos, T.: Pure-circuit: strong Inapproximability for Ppad. In: 2022 IEEE 63rd Annual Symposium on Foundations of Computer Science (FOCS), pp. 159–170. IEEE Computer Society, Los Alamitos, CA, USA (nov 2022). https://doi.org/10.1109/FOCS54457.2022.00022, https://doi.ieeecomputersociety.org/10.1109/FOCS54457.2022.00022
28. Deligkas, A., Fasoulakis, M., Markakis, E.: A Polynomial-Time Algorithm for 1/2-Well-Supported Nash Equilibria in Bimatrix Games (2022). https://doi.org/10.48550/ARXIV.2207.07007
29. Deligkas, A., Fearnley, J., Savani, R.: Inapproximability results for approximate Nash equilibria. In: Cai, Y., Vetta, A. (eds.) Web and Internet Economics, pp. 29–43. Springer, Berlin Heidelberg, Berlin, Heidelberg (2016)
30. Deligkas, A., Fearnley, J., Savani, R.: Computing constrained approximate equilibria in polymatrix games. In: Algorithmic Game Theory, pp. 93–105. Springer International Publishing (2017). https://doi.org/10.1007/978-3-319-66700-3_8
31. Etessami, K., Yannakakis, M.: On the complexity of Nash equilibria and other fixed points. SIAM J. Comput. **39**(6), 2531–2597 (2010)
32. Facchinei, F., Kanzow, C.: Generalized Nash equilibrium problems. 4OR **5**(3), 173–210 (2007)
33. Facchinei, F., Kanzow, C.: Generalized Nash equilibrium problems. Ann. Oper. Res. **175**(1), 177–211 (2010)
34. Filos-Ratsikas, A., Hansen, K.A., Høgh, K., Hollender, A.: Fixp-membership via convex optimization: games, cakes, and markets. SIAM J. Comput. **0**(0), FOCS21–30–FOCS21–84 (0).https://doi.org/10.1137/22M1472656

35. Filos-Ratsikas, A., Hansen, K.A., Høgh, K., Hollender, A.: PPAD-membership for problems with exact rational solutions: a general approach via convex optimization. In: Proceedings of the 56th Annual ACM Symposium on Theory of Computing. STOC '24, ACM (Jun 2024). https://doi.org/10.1145/3618260.3649645
36. Garg, J., Mehta, R., Vazirani, V.V., Yazdanbod, S.: ETR-completeness for decision versions of multi-player (symmetric) Nash equilibria. In: Automata, Languages, and Programming, pp. 554–566. Springer Berlin Heidelberg (2015). https://doi.org/10.1007/978-3-662-47672-7_45
37. George, G., Stern, M.: Puzzle-math the Big Bang to Black Holes. Viking Press (1958)
38. Gilboa, I., Zemel, E.: Nash and correlated equilibria: some complexity considerations. Games Econom. Behav. **1**(1), 80–93 (1989). https://doi.org/10.1016/0899-8256(89)90006-7, https://www.sciencedirect.com/science/article/pii/0899825689900067
39. Goos, M., Rubinstein, A.: Near-optimal communication lower bounds for approximate Nash equilibria. In: 2018 IEEE 59th Annual Symposium on Foundations of Computer Science (FOCS). IEEE (Oct 2018)
40. Hazan, E., Krauthgamer, R.: How hard is it to approximate the best Nash equilibrium? SIAM J. Comput. **40**(1), 79–91 (2011). https://doi.org/10.1137/090766991
41. Kaneko, M., Nakamura, K.: The Nash social welfare function. Econometrica **47**(2), 423–435 (1979). http://www.jstor.org/stable/1914191
42. Kapron, B.M., Samieefar, K.: On the computational complexity of quasi-variational inequalities and multi-leader-follower games. In: Proceedings of the 23rd International Conference on Autonomous Agents and Multiagent Systems, pp. 2324–2326. AAMAS '24, International Foundation for Autonomous Agents and Multiagent Systems, Richland, SC (2024)
43. Lipton, R.J., Markakis, E., Mehta, A.: Playing large games using simple strategies. In: Proceedings of the 4th ACM Conference on Electronic Commerce - EC '03. ACM Press, New York, New York, USA (2003)
44. McKinsey, J.: Introduction to the Theory of Games. McGraw-Hill Book Co. (1952)
45. McLennan, A., Tourky, R.: Simple complexity from imitation games. Games Econ. Behav. **68**(2), 683–688 (2010)
46. Megiddo, N., Papadimitriou, C.H.: On total functions, existence theorems and computational complexity. Theoret. Comput. Sci. **81**(2), 317–324 (1991). https://doi.org/10.1016/0304-3975(91)90200-l
47. Nash, J.F.: Equilibrium points in n-person games. Proc. Natl. Acad. Sci. **36**(1), 48–49 (1950). https://doi.org/10.1073/pnas.36.1.48
48. Papadimitriou, C., Vlatakis-Gkaragkounis, E.V., Zampetakis, M.: The computational complexity of multi-player concave games and kakutani fixed points. In: Proceedings of the 24th ACM Conference on Economics and Computation, pp. 1045. EC '23, Association for Computing Machinery, New York, NY, USA (2023). https://doi.org/10.1145/3580507.3597812
49. Papadimitriou, C.H.: On the complexity of the parity argument and other inefficient proofs of existence. J. Comput. Syst. Sci. **48**(3), 498–532 (1994)
50. Rosen, J.B.: Existence and uniqueness of equilibrium points for concave N-Person Games. Econometrica **33**(3), 520–534 (1965), http://www.jstor.org/stable/1911749
51. Rubinstein, A.: Inapproximability of Nash equilibrium. In: Proceedings of the Forty-Seventh Annual ACM Symposium on Theory of Computing. ACM, New York, NY, USA (Jun 2015)

52. Rubinstein, A.: Settling the complexity of computing approximate two-player Nash equilibria. SIGecom Exch. **15**(2), 45–49 (2017). https://doi.org/10.1145/3055589.3055596
53. Samieefar, K.: A meta-heuristic approach for strategic fair division problems. In: Proceedings of the 2023 7th International Conference on Intelligent Systems, Metaheuristics & Swarm Intelligence, pp. 87–94. ISMSI '23, Association for Computing Machinery, New York, NY, USA (2023). https://doi.org/10.1145/3596947.3596969
54. Schoenebeck, G.R., Vadhan, S.: The computational complexity of Nash equilibria in concisely represented games. ACM Trans. Comput. Theory **4**(2), 1–50 (2012)
55. Shafer, W., Sonnenschein, H.: Equilibrium in abstract economies without ordered preferences. J. Math. Econ. **2**(3), 345–348 (1975). https://doi.org/10.1016/0304-4068(75)90002-6, https://www.sciencedirect.com/science/article/pii/0304406875900026
56. Stengel, B.v.: Equilibrium Computation for Two-Player Games in Strategic and Extensive Form, pp. 53–78. Cambridge University Press (2007). https://doi.org/10.1017/CBO9780511800481.005
57. Tadenuma, K., Thomson, W.: Games of fair division. Games Econom. Behav. **9**(2), 191–204 (1995). https://doi.org/10.1006/game.1995.1015, https://www.sciencedirect.com/science/article/pii/S0899825685710159
58. Webb, W.A.: An algorithm for super envy-free cake division. J. Math. Anal. Appl. **239**(1), 175–179 (1999). https://doi.org/10.1006/jmaa.1999.6581, https://www.sciencedirect.com/science/article/pii/S0022247X99965812
59. Wieczorek, A.: Constrained and indefinite games and their applications. Instytut Matematyczny Polskiej Akademi Nauk (1983). http://eudml.org/doc/268472

Exact Characterizations of Non-commutative Algebraic Complexity Without Homogeneity

Guillaume Malod(✉)

Université Paris Cité and Sorbonne Université, CNRS, IMJ-PRG, 75013 Paris, France
malod@imj-prg.fr

Abstract. Algebraic complexity theory studies the number of arithmetic operations necessary to compute different polynomials. As in the case of Boolean complexity, there are many open questions: indeed, there are several natural polynomials for which we believe, but are unable to prove, that no efficient computation is possible. When contrasted with the difficulty to obtain lower bounds in general for the main computation model of arithmetic circuits, Nisan's 1991 result is striking: it gives an exact characterization of the complexity of computing any given homogeneous polynomial with a homogeneous algebraic branching program, in the non-commutative setting. As noticed in 2020 by Fijalkow et al., Nisan's theorem can be derived from results by Fliess (1974) in the field of non-commutative formal series. We build on this point of view to extend Nisan's theorem to general (not necessarily homogeneous) polynomials computed by general algebraic branching programs and again obtain an exact characterization. Fijalkow et al. showed how to apply properties of formal tree series to get a similar statement for homogeneous non-associative non-commutative arithmetic circuits and we also extend this characterization to general non-associative non-commutative circuits.

Keywords: algebraic complexity · lower bounds · arithmetic circuits · algebraic branching programs · formal series

1 Introduction

Interest in algebraic complexity theory is partly due to a general strategy, called *geometric complexity theory*, to tackle the main open question of complexity, P versus NP, via algebraic means (see the survey [7] or the website [13]). One of its intermediate goals is to prove that computing the permanent polynomial $\sum_{\sigma \in S_n} \prod_i x_{i,\sigma(i)}$ cannot be efficiently reduced to computing the determinant. This question can naturally be seen as an analogue of P versus NP when using arithmetic circuits as the model of computation, as introduced by Valiant in founding articles [20,21]. Interest in arithmetic circuits was also sparked by recent lower bounds, mostly when considering restricted models of computation, which are detailed in the excellent survey by Saptharishi [18]. Despite the

R. Královič and V. Kůrková (Eds.): SOFSEM 2025, LNCS 15539, pp. 128–141, 2025.
https://doi.org/10.1007/978-3-031-82697-9_10

progress described there, it seems that we are still far from having a sufficient comprehension of algebraic computations, which would let us prove strong general lower bounds.

One possible restriction is to only consider non-commutative computations. In this setting, the characterization given by Nisan [16] is therefore all the more striking: it provides the exact value of the complexity of computing a given polynomial, based on the rank of certain matrices (it gives an exponential complexity for the non-commutative permanent).

Let us describe this more precisely. In the non-commutative ring $\mathbb{F}\langle X\rangle$, variables do not commute so that xy and yx are distinct monomials. Studying non-commutative computations is important in itself, for instance when reasoning about computations over matrices, but also because of applications to *commutative* computations, for example to approximate the commutative permanent using determinants [2,8].

The computational model considered is non-commutative *homogeneous* algebraic branching programs, which are an intermediate model of computation, between the more well-known arithmetic formulas and arithmetic circuits. See Fig. 1 for an example.

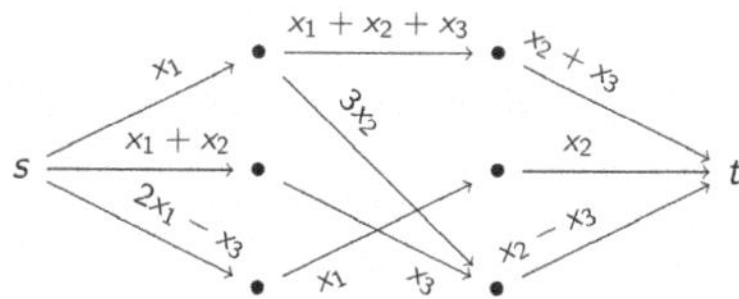

Fig. 1. A homogeneous algebraic branching program

Definition 1. *A non-commutative homogeneous algebraic branching program is a layered directed acyclic graph with a source s and a sink t, where arcs go from a given layer to the next and are weighted with linear forms. The weight of a path is the product of the weights of its arcs, in the order they are encountered. The (necessarily homogeneous) polynomial computed by the program is the sum of the weights of all paths from s to t.*

Consider now a homogeneous polynomial f of degree d. For each $0 \leqslant k \leqslant d$ we define a matrix $M_k(f)$: rows are indexed by non-commutative monomials of degree k, columns by non-commutative monomials of degree $(d-k)$, and the coefficient in row m and column m' is the coefficient of the concatenated monomial mm' in f, as illustrated in Fig. 2.

Theorem 1 (Nisan). *For any homogeneous polynomial f of degree d, the size of a smallest homogeneous algebraic branching program for f is equal to*

$$\sum_{k=0}^{d} \operatorname{rank}(M_k(f)).$$

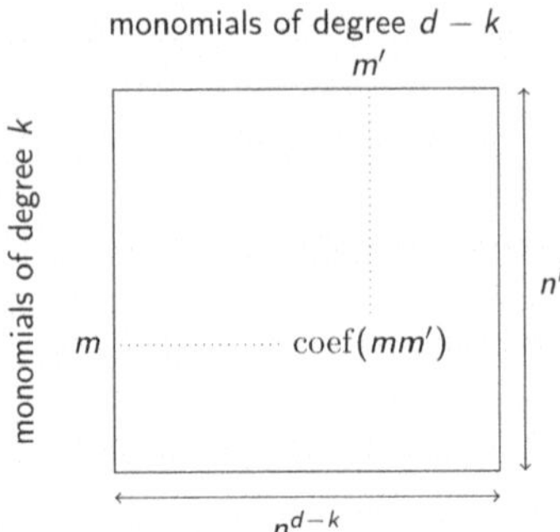

Fig. 2. Definition of Nisan's matrices

More precisely, Nisan shows that the number of vertices in layer k of a minimal homogeneous ABP is exactly the rank of $M_k(f)$. Thus Nisan's result seems intricately linked to the layered structure of the homogeneous ABP.

We will show that a similar result holds for general non-commutative algebraic branching programs, which we define as follows, by omitting the requirement that it be layered and adding the possibility of a constant term in the weight of the arc from source to sink, so that polynomials with a constant term can be computed.

Definition 2. *A* non-commutative algebraic branching program *(ABP) is a directed acyclic graphs with a source s and a sink t, where arcs are weighted with linear forms. The arc from s to t, if it exists, may have a weight which is an affine form. The weight of a path is the product of the weights of its arcs, in the order they are encountered. The polynomial computed by the program is the sum of the weights of all paths from s to t.*

Our goal is to get a characterization for this general definition of algebraic branching programs.

2 Characterization for Algebraic Branching Programs

To prove our characterization, we will need the basic vocabulary and results of non-commutative formal series. We only skim the notions relevant to our proof, but much more could be said about this beautiful theory: the interested reader can refer to the books by Berstel and Reutenauer [4] or Sakarovitch [17].

2.1 Non-commutative Formal Series and Linear Representations

Fix a field K and a finite alphabet A of variables. A non-commutative formal series S is a function $A^* \to K$, where A^* commonly denotes the set of words over the alphabet A. Denote by (S, w) the image of w by S. We can therefore write $S = \sum_{w \in A^*} (S, w)w$. The set of non-commutative formal series is denoted by $K\langle\langle A \rangle\rangle$. The support of a series S is $\{w \in A^* \mid (S, w) \neq 0\}$. The set of series with finite support is also called the set of non-commutative polynomials, $K\langle A \rangle$. Because we are interested in polynomials, we will write 1 for the empty word. The following notion of *recognizability* is common in automata theory and will be very close to our computational model.

Definition 3. *A non-commutative formal series S is recognizable if there exists:*

- *an integer $n \geqslant 1$*
- *a morphism of monoids $\mu : A^* \to K^{n \times n}$*
- *two matrices $\lambda \in K^{1 \times n}$ and $\gamma \in K^{n \times 1}$*

such that, for all words w, $(S, w) = \lambda \mu(w) \gamma$.

Note that it is enough to define $\mu(a)$ for all $a \in A$, and then extend it to words by the morphism property. Such a tuple (λ, μ, γ) is called a *linear representation*, and n is often called the dimension but we will call it the *size*.

Such a linear representation can be seen as an automaton, but we will directly try to see it as branching program. The matrix $\mu(a)$ for a letter a in the alphabet A can be seen as defining a directed graph on the vertices $\{u_1, \dots, u_n\}$, where the arc (u_i, u_j) has weight $\mu(a)_{i,j} a$. We than consider the union of the edges of all these graphs, still on the vertices $\{u_1, \dots, u_n\}$, to which we add a "start" vertex s and an "end" vertex t, with the arcs (s, u_i) of weight $\lambda_{1,i}$ and (u_i, t) of weight $\gamma_{i,1}$. For a walk in this graph, define the weight as the product of the weights of the arcs, in the order encountered. The use of *walk* here means that vertices and arcs can appear several times, if there are cycles, in which case the weights will also be repeated.

Then the equality $(S, w) = \lambda \mu(w) \gamma$ exactly means that the coefficient of a word w in S is the sum of the weights of the walks in this graph going from s to t and producing this word. This also means that the whole series S is the sum of the weights of all walks from s to t in the graph.

We are then very close to the algebraic branching programs we have seen in the introduction. If there are multiple arcs from one vertex to another, we replace them by one arc whose weight is the sum of the weights of the replaced arcs. This means that, apart from the the arcs leaving s or going to t, whose weights are scalars, all other arcs are labeled by linear forms over the variables in A. These "scalar" edges are obviously a difference with the ABPs of our definition, but the main difference is that the graph we have obtained until now may very well contain cycles, even if it is computing a polynomial.

Clearly, for a series whose support is not finite, the graph must contain cycles. But can cycles help to compute polynomials? In other words, is it possible that all minimum-size representations of a given polynomial have cycles? Or is there always at least one minimal representation which is acyclic, and thus yields a branching program?

2.2 Minimal Representation of a Word Series

We will therefore use results on the minimal representations of a series. It is easy to see that $K\langle\langle A\rangle\rangle$ is a vector space over K. For a word u in A^* and a series S in $K\langle\langle A\rangle\rangle$, define the *quotient series* $u^{-1}S$ as $\sum_{w\in A^*}(S, uw)w$. In other words, the coefficient of a word w in the formal series $u^{-1}S$ is the coefficient of uw in the series S: $(u^{-1}S, w) = (S, uw)$. A set $M \subseteq K\langle\langle A\rangle\rangle$ is called *stable* if, for any word u in A^* and for any series S in M, the series $u^{-1}S$ is also in M.

With these definitions we can state the following theorem (see [11] and also [12], both in French; details in English can be found in [4]).

Theorem 2. *[Fliess] A formal series S is recognizable iff there exists a finitely generated and stable linear subspace of $K\langle\langle A\rangle\rangle$ containing S.*

We give the beautiful proof here because we will need the details later.

Proof. Suppose first that S is recognizable via a linear representation (λ, μ, γ). Define S_i by $(S_i, w) = (\mu(w)\gamma)_i$, for $i = 1, \dots, n$. Let M be the subspace generated by the S_i (M is therefore finitely generated). It contains S:

$$(S, w) = \lambda\mu(w)\gamma = \sum_i \lambda_i(\mu(w)\gamma)_i = \sum_i \lambda_i(S_i, w).$$

It is stable. Indeed, for any $x \in A^*$:

$$\begin{aligned}(x^{-1}S_i, w) &= (S_i, xw) = (\mu(xw)\gamma)_i = (\mu(x)\mu(w)\gamma)_i \\ &= \sum_j (\mu(x))_{i,j}(\mu(w)\gamma)_j = \sum_j (\mu(x))_{i,j}(S_j, w),\end{aligned}$$

where the first equality is by definition of the quotient series, the second is by definition of S_i, the third by the morphism property of μ, the fourth by decomposition of the matrix product and the fifth by definition of S_j. Since this equality holds for every w, the series $x^{-1}S_i$ is therefore a linear combination of the S_j's and belongs to M.

Suppose now that we have a stable linear subspace M, containing S and generated by $S_1, \dots, S_n$. Then there are coefficients λ_i such that $S = \sum_i \lambda_i S_i$. For $a \in A, i \in [n]$, by the stability condition we can also express $a^{-1}S_i$ as a linear combination of the S_j's, and we will use the coefficients to define $\mu(a)$: write $a^{-1}S_i$ as $\sum_j \mu(a)_{i,j}S_j$. We also define $\mu(1)$ as the identity matrix. Finally, define $\gamma_j = (S_j, 1)$. Then:

$$\begin{aligned}(S_i, w) &= (w^{-1}S_i, 1) = \sum_j (\mu(w)_{i,j}S_j,\ 1) \\ &= \sum_j \mu(w)_{i,j}\,(S_j, 1) = \sum_j \mu(w)_{i,j}\gamma_j = (\mu(w)\gamma)_i,\end{aligned}$$

where the first equality is by definition of the quotient series, the second is a claim that we prove separately below, the third is by the linear structure of $K\langle\langle A\rangle\rangle$, the fourth is by definition of γ_j and the fifth is the matrix product. Putting everything together, we get: $(S, w) = \sum_i \lambda_i (S_i, w) = (\lambda\mu(w)\gamma)$.

Let us now prove the small claim used for the second equality: for all $i \in [n]$, for all $w \in A^*$, $w^{-1}S_i = \sum_j \mu(w)_{i,j} S_j$. We will proceed by induction on the length of w. The claim is obviously true for words of lengths 0 or 1, by definition of μ. Consider now a word $w = ua$ with $a \in A$. Then:

$$(ua)^{-1}S_i = a^{-1}(u^{-1}S_i) = a^{-1}\left(\sum_j \mu(u)_{i,j}S_j\right) = \sum_j \mu(u)_{i,j}(a^{-1}S_j)$$

$$= \sum_j \mu(u)_{i,j}\left(\sum_k \mu(a)_{j,k}S_k\right) = \sum_{j,k} \mu(u)_{i,j}\mu(a)_{j,k}S_k = \sum_k (\mu(ua))_{i,k}S_k,$$

where the first equality is clear from the definition of quotient series, the second is by induction, the third is an easy to check property of quotient series, the fourth is by definition of $\mu(a)$, and the last is the matrix product.

Considering the proof, we see that the size of a minimal linear representation for S is the smallest dimension of a stable finitely-generated linear subspace containing S. It is therefore also the dimension of the linear space generated by all the $u^{-1}S$ for $u \in A^*$.

The Hankel matrix $H(S)$ of a series is an infinite matrix with rows and columns indexed by words. Define $(H(S))_{u,v}$ as (S, uv): the (u, v)-coefficient of the matrix $H(S)$ is the coefficient of the word uv in S. Another way to see it is that each row of $H(S)$ is indexed by a word u, and the row gives for each v its coefficient in the series $u^{-1}S$. Combining this observation with the preceding comment we get the following result.

Corollary 1. *The size of a minimal linear representation of a series S is the rank of the Hankel matrix.*

The crucial point is that such a minimum-size representation can be built from the Hankel matrix, in particular by considering a basis of its row-space.

In the case of homogeneous polynomials, Fijalkow et al. [9] already noticed that one can recover a homogeneous ABP: their Lemma 8 says that, when considering a homogeneous polynomial of degree d, the automaton obtained is layered and thus acyclic, therefore it is a homogeneous ABP.[1] We will now show that this can also be done in the general case of (not necessarily homogeneous) polynomials.

[1] The homogeneous ABPs considered in [9] have so-called "output values" for the vertices in the last layer. This corresponds to the scalar weights in γ in our presentation and is an artifact which we will be able to eliminate in our generalization.

2.3 From Linear Representations to Algebraic Branching Programs

Consider a degree-d polynomial S whose Hankel matrix $H(S)$ has finite rank n. We start by grouping the rows of $H(s)$ in blocks depending on the length of the word defining them, and we order the blocks so that this length is increasing (we will call the block corresponding to rows defined by words of length k the block of index k). We arrange and order the columns in a similar manner.

The first block, of index 0, only contains one row, indexed by 1, the word of length 0. This row is S itself. The block of index 1 corresponds to all the lines defined by a word of length 1, i.e., a single letter: these rows are the $a^{-1}S$ for a in A. If the degree of S is d, the last block with non-zero rows is the block of index d. For any such row, the only possible non-zero coefficient is the first one, which corresponds to the column indexed by 1.

We will choose a basis for the row space. Suppose that the dimension of this space is n, we will choose lines $u_n, \ldots, u_1$, starting from the block of index d (which is the degree of S) and working on blocks of decreasing index. We must choose exactly one non-zero row in the block of index d: as seen above, all non-zero lines in the block of index d are just scalar multiples of one another.

We then greedily choose additional rows in the block of index $d-1$ such that all together they form a basis of the row space of the blocks of index d and $d-1$.

We keep going, each time expanding a basis of the blocks of index $d, \ldots, p$ to a basis of the blocks of index $d, \ldots, (p-1)$ by choosing appropriate rows in the block of index $p-1$.

At the end, we must choose the only row of the block of index 0, which is S itself, so that $u_1 = 1$. Indeed, this row is the only one which has some non-zero coefficients in the column block of index d and it therefore must appear in any basis of the row space of the matrix.

The resulting basis has the following important property: any (chosen or not) row of the block of index p is a linear combination of chosen rows in blocks of index at least p.

Let us now build the resulting linear representation and see it as a branching program. The vertices of our graph correspond to the rows we have chosen in our basis, let $u_1, \ldots, u_n$ be the words defining these rows, where u_1 is the empty word 1 and u_n defines the only row we have chosen in the block of index d (i.e., we list the defining words in *increasing* order of length): we will also call the vertices of our graph $u_1, \ldots, u_n$.

Recall the construction in the proof of Theorem 2 going from a stable linear subspace containing S to a representation. The S_i we consider now are the $u_i^{-1}S$ for $1 \leqslant i \leqslant k$. To obtain the arcs, we look at how $a^{-1}S_i$ decomposes as a linear combination of the S_i. In our case, we look at $a^{-1}(u_i^{-1}S) = (u_i a)^{-1}S$. This a quotient series obtained with a word of length $p+1$ if the length of u_i is p. As a row, it therefore belongs to the block of index $p+1$ and we know that, by construction of our basis, it will only get non-zero coefficients for rows $u_j^{-1}S$ from blocks of index at least $p+1$. Therefore, in the decomposition $a^{-1}S_i = \sum_j (\mu(a))_{i,j} S_j$, the fact that $\mu(a)_{i,j}$ is non-zero implies that the length of the word u_j defining S_j is greater than the length of u_i, which according to our ordering implies that $j > i$. Thus the graph defined over $u_1, \ldots, u_n$ is acyclic, which is the basic property we need to see it as an algebraic branching program.

The resulting ABP also has the extra vertex s, which sends arcs to $u_1, \ldots, u_n$ with scalar weights in λ, and the vertex t, which receives arcs with weights defined by γ.

Let us start by taking care of γ. In the proof of Theorem 2, we defined γ_i as $(S_i, 1)$, which therefore becomes $(u_i^{-1}S, 1)$ here. More generally, if we have a family $S_1, \ldots, S_n$ which generates a stable subspace containing S, then the coefficients of γ are the constant terms of the S_i's. We will thus try to define a family $T_1, \ldots, T_n$ such that only T_n has a non-zero constant term. To this end, start by defining T_n as the empty word 1. Then, for $1 \leqslant i \leqslant n-1$, let T_i be the series $S_i - (S_i, 1)$ (i.e., we cancel out the constant term).

The set of series $T_1, \ldots, T_n$ can generate the series $S_1, \ldots, S_n$: for $1 \leqslant i \leqslant n-1$, we have $S_i = T_i + (S_i, 1)T_n$; moreover, S_n is just a constant (because we have taken u_n in the block of maximal index d) and thus $S_n = (S_n, 1)T_n$. Therefore the family $T_1, \ldots, T_n$ also generates the row space of $H(S)$ and can be used to build a branching program. We should however check that the resulting program remains acyclic. Again, this boils down to looking at which coefficients will be non-zero when we express $a^{-1}T_i$ as a linear combination of the T_j's:

$$a^{-1}T_i = a^{-1}(S_i - (S_i, 1)) = a^{-1}S_i = a^{-1}(u_i^{-1}S) = (u_i a)^{-1}S,$$

where the second equality just expresses the simple fact that canceling out the constant term in S_i does not change the quotient series $a^{-1}S_i$. Since $u_i a$ is a word of length $p+1$ if the length of u_i is p, the row $(u_i a)^{-1}S$ is a linear combination of S_j's from the blocks of indices at least $p+1$. It is therefore also a linear combination of T_j's from the blocks of indices at least $p+1$, and thus a non-zero arc from T_i to T_j can only exist if $j > i$.

In the acyclic branching program which results from the generating family $T_1, \ldots, T_n$, we will again identify the vertices with the series $T_1, \ldots, T_n$. Now, the associated γ is just $(0, \ldots, 0, 1)$ and the only vertex sending an arc to the output vertex t is T_n, which sends no other arcs as it is of maximal index. Since the arc has weight 1, we can just delete t and the arc.

Let us now look at λ, which corresponds to the coefficients when expressing S as a linear combination of our generating family, which are now the series $T_1, \ldots, T_n$. By construction, $S = S_1$. We have set $T_1 = S_1 - (S_1, 1) = S - (S, 1)$, therefore $S = T_1 + (S, 1) = T_1 + (S, 1)T_n$, because T_n is 1. The only arcs implied by this linear combination are thus from s to T_1 with weight 1 and from s to T_n with weight $(S, 1)$. The first arc is useless and we can delete it. We can then delete s and replace the arc from s to T_n by an arc from T_1 to T_n, also with weight $(S, 1)$. We have thus obtained an algebraic branching program which corresponds exactly to Definition 2. We have therefore proved the following theorem.

Theorem 3. *For a non-commutative polynomial S, the size of a smallest algebraic branching program computing S is exactly the rank of the Hankel matrix of S.*

2.4 Some Simple Consequences

Non-Homogeneous Computations of Homogeneous Polynomials. Consider a homogeneous polynomial S. Then the smallest size of a (general) ABP computing it is given by the rank of the Hankel matrix of S. But this rank is also the smallest size of a *homogeneous* ABP computing the polynomial.

Corollary 2. *For non-commutative homogeneous polynomials, homogeneous and general algebraic branching programs yield the same complexity.*

Note that for *commutative* arithmetic formulas, homogeneity might very well be an important restriction of the computational power. Consider the example of the elementary symmetric polynomials: $S_n^d(x_1, \ldots, x_n)$ is the sum of all the multilinear monomials in $x_1, \ldots, x_n$ of degree exactly d. Ben-Or (refer to [19]) shows that S_n^d has an inhomogeneous formula of depth 3 and size $O(n^2)$ for any d. Homogeneous formulas of polynomial size are not known, though Hrubeš and Yehudayoff [14] have given homogeneous formulas of size $O(d^{\log d} n)$. It is therefore striking that, when computing homogeneous polynomials with ABPs, homogeneous and non-homogeneous complexity should coincide exactly.
Hadamard Product. Arvind et al. [1] define the Hadamard product of non-commutative polynomials. They then show that the complexity (for homogeneous algebraic branching programs) of the Hadamard product of two homogeneous polynomials is upper bounded by the product of their complexity.[2] The notion of Hadamard product already existed for formal series, with the associated result.

Definition 4. *The Hadamard product of two formal series S and T is $(S \odot T, w) = (S, w)(T, w)$.*

Theorem 4 (Schützenberger 1962). *If S and T are recognizable, then so is $S \odot T$.*

Proof. If $S_1, \ldots, S_m$ generate a stable subspace containing S and respectively $T_1, \ldots, T_n$ for T, then the $S_i \odot T_j$ generate a stable subspace for $S \odot T$.

Combining this with our results, we can apply this to the case of (non-homogeneous) polynomials.

Theorem 5. *Let S and T be two non-commutative polynomials, computed by algebraic branching programs of size s and t respectively. Then the polynomial $S \odot T$ can be computed by an algebraic branching program of size at most st.*

---

[2] The proof of their Theorem 4 seems to consider non-homogeneous polynomials but the theorem is stated for polynomials *computed* by homogeneous ABPs, so it is a bit unclear how non-homogeneous polynomials could enter the picture. Decomposing into homogeneous components would also multiply the resulting size by the degree.

Note that Arvind et al. [1] focus more on a constructive version of the result for homogeneous ABPs, as a step towards efficient *polynomial identity testing.*
The Rectangular Permanent. Looking at the Hansel matrix and its rank can serve as a guide to find algorithms. As a brief example, consider the rectangular permanent $\sum_\sigma \prod_i x_{i,\sigma(i)}$, where the summation is over all injections σ from $\{1, 2, \ldots, m\}$ to $\{1, 2, \ldots, n\}$ $(m \leqslant n)$. The rank suggests an algebraic branching program of size $2\binom{n}{\downarrow m/2})$, where we write $\binom{q}{\downarrow r}$ for the sum $\binom{q}{0} + \binom{q}{1} + \cdots + \binom{q}{r}$.

Indeed, there is such a homogeneous ABP, which starts by standard dynamic programming for the first $m/2$ layers, connected to a "reversed" dynamic programming for the last $m/2$ layers. This algebraic branching program yields an algorithm for inputs in an arbitrary semiring with an $O(\binom{n}{m/2})$ space bound, better than the algorithm given by Björklund et al. [5], though the time bound is worse.

3 Tree Series and Non-associative Polynomials

It would be great to have such a characterization for the computation of non-commutative polynomials by arithmetic circuits, because circuits are the general model used in algebraic complexity theory to capture "efficient" computations.

The difficulty of this endeavor can be guessed from the fact that the best known lower bound is just derived from the best bound in the commutative case, i.e., $\Omega(n \log n)$ for an n-variate degree-n polynomial [3].

In fact, an intuition on why we could get such an elegant result for branching programs is that the monomials can be seen as being built sequentially from left to right, and non-commutative computations reflect enough of this structure of computed monomials (the order of variables) to enable the characterization.

Trying to generalize Nisan's theorem, Lagarde et al. [15] identified a particular class of circuits where an exact characterization of complexity could be proved, so-called *unique parse tree circuits.* Then, Fijalkow et al. [9] noticed that these results could be linked to extensions of the previous properties of formal series on words to formal series on trees, in particular via the results of Bozapalidis and Louscou-Bozapalidou [6]. Applying these results, Fijalkow et al. [10] derived an exact characterization for homogeneous non-commutative non-associative computations. In this setting, each monomial comes with parentheses indicating the order in which multiplications are applied and this additional structure attached to monomials is crucial in obtaining an exact characterization. While the authors chose to use exclusively the language of circuits to apply the results of Bozapalidis and Louscou-Bozapalidou, we will once more use the original language of formal series to obtain a general result which does not need homogeneity.

Since the setting and the arguments become more technical, we will only describe the main steps. We would like to emphasize that the main ideas are very similar to those used for ABPs and word series.

3.1 Minimal Representation of a Tree Series

We again fix a field K but consider now a finite ranked alphabet Σ (i.e., a set of symbols with arities). Because we will be interested in circuits, we only need here a set A of variables (symbols of arity 0) and one binary symbol which we will denote by $\vee$ and use in an infix way. We consider trees over the ranked alphabet, in our case binary trees with leaves labeled with variables and internal nodes labeled with $\vee$. Let us write T_A for the set of all such trees over A: in the non-associative non-commutative setting these trees are our monomials. Note that because there is no "empty" tree we do not have a monomial 1 and our formal series and polynomial cannot have a "constant" term. A formal series S is a function $T_A \to K$. Denote by (S,t) the image of a tree t by S. Then $S = \sum_{t \in T_A}(S,t)t$. The set of formal tree series is denoted by $K\{\{\Sigma\}\}$. Again, the support of S is $\{t \in T_\Sigma \mid (S,t) \neq 0\}$ and series with finite support are called polynomials, yielding the set $K\{\Sigma\}$.

Definition 5. *A formal tree series S is recognizable if there exists:*

- *an integer $n \geqslant 1$;*
- *for each $a \in A$, a vector $\mu(a) \in K^n$;*
- *a bilinear map $\mu(\vee) : (K^n)^2 \to K^n$;*
- *a vector $\lambda \in K^n$*

such that, for any tree t, $(S,t) = \lambda \cdot \mu(t)$, where μ is extended to a mapping from T_A to K^n in the obvious way: $\mu(t_1 \vee t_2) = \mu(\vee)(\mu(t_1), \mu(t_2))$.

Then (λ, μ) is called a linear representation and we again call n its size. Note that $K\{\{A\}\}$ is a vector space over K.

Before we can talk about stable sets and characterize recognizable series, we need another notion. A context is a tree over $A \cup \{\square\}$ with exactly one leaf labeled $\square$. More formally, the set of contexts is defined by induction as containing the trivial context $\square$, and, if c is a context and t is a tree, then it also contains the contexts $t \vee c$ and $c \vee t$.

We can therefore extend the definition of μ from trees to contexts, with $\mu(c)$ being a linear map from K^n to K^n. Define $\mu(\square)$ as the identity map. For t a tree and c a context, by induction $\mu(c)$ is a linear map, and so is $\mu(\vee)(\mu(t), \mu(c))$, and we use this to define $\mu(t \vee c)$. Similarly for $\mu(c \vee t)$. From this definition it is clear that for any context c and tree t, $\mu(c[t]) = \mu(c)\mu(t)$.

If c is a context and S a series in $K\{\{A\}\}$, define the quotient series $c^{-1}S$ as $\sum_{t \in T_\Sigma}(S, c[t])t$. In other words, $(c^{-1}S, t) = (S, c[t])$. A set $M \subseteq K\{\{A\}\}$ is called *stable* if, for all contexts c and for all series S in M, the quotient series $c^{-1}S$ is also in M. The following theorem is similar to Theorem 2. Due to lack of space, we omit its proof, which also implies the corollary below.

Theorem 6. *A formal series S is recognizable iff there exists a finitely generated and stable linear subspace of $K\{\{A\}\}$ containing S.*

Corollary 3 (Bozapalidis & Louscou-Bozapalidou). *The size of a minimal linear representation of a series S is the rank of the Hankel matrix.*

3.2 From Linear Representations to Arithmetic Circuits

Once again, most of the work is to show how we can go from a general linear representation of a series to an arithmetic circuit, when the series is a polynomial. Let us first define the notion of arithmetic circuit we will use, which is almost the same as the one in Fijalkow et al. [10].[3]

Definition 6. *A* non-associative non-commutative arithmetic circuit *is a directed acyclic graph, with:*

- *vertices of in-degree* 0*, labeled by a variable, called* input gates*;*
- *vertices of in-degree* 2*, called multiplication gates, where we distinguish the left and right incoming arcs;*
- *vertices of arbitrary in-degree with a weight on each incoming arc, called addition gates;*
- *one addition gate with out-degree* 0*, called the output gate.*

We moreover assume that each arc is either coming from, or going to, an addition gate (but not both). The size of a circuit is the number of addition gates.

Such a circuit computes a non-associative non-commutative polynomial in a way which can be naturally defined by induction on the gates of the circuit.

As in the case of word series, the important direction goes from a Hankel matrix $H(S)$ of finite rank n to a circuit for the polynomial S. Rows in $H(S)$ are now defined by trees instead of words but here too we will choose rows $t_1, \dots, t_n$, after having indexed the blocks, so that a row from a block of index p is a linear combination of rows of index at least p.

To build the circuit, we once again follow [10], with slight modifications:

- The addition gates are identified with the chosen $t_1, \dots, t_n$.
- There is one input gate for each variable in A; input gate a sends an arc to each addition gate t_i with weight $\mu(a)_i$.
- There is one multiplication gate for each (i, j, k) with $1 \leqslant i, j < k \leqslant n$; this gate receives a left arc from the addition gate t_i, a right arc from the addition gate t_j and sends and arc to the addition gate t_k.

With this definition it is clear that we get a circuit with output gate t_n. Indeed, the acyclicity of the graph is enforced by the condition $1 \leqslant i, j < k \leqslant n$ on the indices of addition gates which are linked by a multiplication gate. As the reader can probably guess, this condition stems from our specific choice of rows in the matrix. Finally, as in the case of word series, we can replace our chosen rows to ensure that all λ_i are 0 except for λ_n, whose value can be "pushed down" to multiply the weight of each incoming arc of the output addition gate, so that we get a circuit as in the definition above. We have thus shown the following theorem, with which we will conclude.

Theorem 7. *For a non-associative non-commutative polynomial S, the size of a smallest circuit computing S is exactly the rank of the Hankel matrix of S.*

[3] Once again our constructions spares us from having to consider what they call "output weights" for the addition gates.

References

1. Arvind, V., Joglekar, P.S., Srinivasan, S.: Arithmetic Circuits and the Hadamard Product of Polynomials. In: IARCS Annual Conference on Foundations of Software Technology and heoretical Computer Science, FSTTCS 2009, December 15-17, 2009, IIT Kanpur, India. Ed. by Ravi Kannan and K. Narayan Kumar. Vol. 4. LIPIcs. Schloss Dagstuhl - Leibniz-Zentrum für Informatik, 2009, pp. 25–36. https://doi.org/10.4230/LIPICS.FSTTCS.2009.2304
2. Barvinok, A.: New Permanent Estimators via Non-Commutative Determinants. ArXiv Mathematics e-prints, July (2000).eprint: math/0007153
3. Baur, W., Strassen, V.: The complexity of partial derivatives. Theor. Comput. Sci.**22**(3) 317–330 (1983)
4. Berstel, J., Reutenauer, C.: Noncommutative Rational Series with Applications. Encyclopedia of Mathematics and its Applications. Cambridge University Press (2011). ISBN: 9780521190220. https://books.google.fr/books?id=LL8Nhn72I_8C
5. Björklund, A., et al.: Evaluation of permanents in rings and semirings. In: Inform. Process. Lett. **110**(20) 867–870 (2010). issn: 0020-0190. https://doi.org/10.1016/j.ipl.2010.07.005, https://www.sciencedirect.com/science/article/pii/S0020019010002097
6. Bozapalidis, S., Louscou-Bozapalidou, O.: The rank of a formal tree power series. In: Theor. Comput. Sci. **27**(1), 211–215 (1983). issn: 0304-3975. https://doi.org/10.1016/0304-3975(83)90100-7, https://www.sciencedirect.com/science/article/pii/0304397583901007
7. Bürgisser, P., et al.: An overview of mathematical issues arising in the geometric complexity theory approach to VP $\neq$ VNP. SIAM J. Comput. **40**(4), 1179–1209 (Aug 2011) issn: 0097–5397. https://doi.org/10.1137/090765328
8. Chien, S., Rasmussen, L., Sinclair, A.: Clifford algebras and approximating the permanent. J. Comput. Syst. Sci. **67**(2), 263–290 (2003)
9. Fijalkow, N., Lagarde, G., Ohlmann, P.: Tight bounds using hankel matrix for arithmetic circuits with unique parse trees. In: Electronic Colloquium Computer Complexity. TR18-038 (2018). ECCC: TR18- 038.https://eccc.weizmann.ac.il/report/2018/038
10. Fijalkow, N., Lagarde, G., Ohlmann, P., Serre, O.: Lower bounds for arithmetic circuits via the hankel matrix. Comput. Complex. **30**(2) (2021). https://doi.org/10.1007/s00037-021-00214-1
11. Fliess, M.: Matrices de Hankel. J. de Mathématiques Pures et Appliquées **53**, 197–222 (1974)
12. Fliess, M.: Un codage non commutatif pour certains systèmes échantillonnès non linéaires. Inform. Control **38**(3), 264–287 (1978). https://doi.org/10.1016/S0019-9958(78)90073-6
13. GCT. GCT publications. http://gct.cs.uchicago.edu/
14. Hrubeš, P., Yehudayoff, A.: Homogeneous formulas and symmetric polynomials. Comput. Complex. **20**(3), 559–578 (2011). https://doi.org/10.1007/s00037-011-0007-3
15. Lagarde, G., Malod, G., Perifel, S.: Non-commutative computations: lower bounds and polynomial identity testing. Chic. J. Theor. Comput. Sci. 2019 (2019). http://cjtcs.cs.uchicago.edu/articles/2019/2/contents.html
16. Nisan, N.: Lower bounds for non-commutative computation (Extended Abstract). In: Proceedings of the 23rd ACM Symposium on Theory of Computing, ACM Press, pp. 410–418 (1991)

17. Sakarovitch, J., Thomas, R.: Elements of Automata Theory. Cambridge University Press, isbn: 9781139643795 (2009). https://books.google.fr/books?id=2RX3AQAAQBAJ
18. Saptharishi, R.: A selection of lower bounds in arithmetic circuit complexity. https://github.com/dasarpmar/lowerbounds-survey/releases/latest. Accessed 19 Sept 2024
19. Shpilka, A., Wigderson, A.: Depth-3 arithmetic circuits over fields of characteristic zero. Comput. Complex. **10**(1) 1–27 (2001)
20. Valiant, L.G.: Completeness classes in algebra. In: STOC '79: Proceedings of the eleventh annual ACM symposium on Theory of computing. Atlanta, Georgia, United States: ACM Press, pp. 249–261 (1979)
21. Valiant, L.G.: Reducibility by algebraic projections. In: Logic and Algorithmic: an International Symposium held in honor of Ernst Specker. Vol. 30. Monographies de l'Enseignement Mathémathique, pp. 365–380 (1982)

Roman Hitting Set

Kevin Mann(✉) and Henning Fernau

Universität Trier, Fachbereich 4 – Abteilung Informatikwissenschaften, 54286 Trier, Germany
mann@uni-trier.de, fernau@uni-trier.de

Abstract. Roman domination is one of few examples where the related extension problem is polynomial-time solvable even if the original decision problem is NP-complete. This is interesting as it allows to establish polynomial-delay enumeration algorithms for finding minimal Roman dominating functions, while it is open for more than four decades if all minimal dominating sets can be enumerated efficiently. To find the reason why this is the case, we combine the idea of hitting set with the idea of Roman domination. We hence obtain a new combinatorial problem, called Roman Hitting Set, generalizing Roman Domination in a natural way. This allows us to expand the frontier of polynomial-delay enumerability, as opposed to transversal-hardness. The generalization Roman Hitting Set is insofar interesting as it always allows for polynomial-delay enumeration, and we also give an explicit input-sensitive enumeration algorithm for this problem that is optimal in the sense that we can also give a matching lower bound. Based on Roman Hitting Set, we also discuss consequences for Roman variants of Vertex Cover and Feedback Vertex Set. For minimal Roman vertex covers of size upper-bounded by k, we can also deduce an optimal parameterized enumeration algorithm. Also viewed from Parameterized Complexity, the studies on extension problems are interesting, giving more examples of parameterized problems complete for W[3]. Finally, we also consider implicitly given hypergraphs and prove that, for the example of Roman FVS, our polynomial-delay enumeration algorithm can still be used, avoiding an explicit construction of the hypergraph.

Keywords: Enumeration · Roman Domination · Hitting Set

1 Introduction

The question if all minimal hitting sets can be enumerated with polynomial delay, or even in output-polynomial time, is an open question for more than four decades. This Transversal Hypergraph Problem is equivalent to the question if all minimal dominating sets can be enumerated with polynomial delay, or in output-polynomial time. This question is also equivalent to several enumeration problems in logic and database theory, see [14,20,27,31]. From the point of view of applications, it is quite important to find an affirmative answer:

R. Královič and V. Kůrková (Eds.): SOFSEM 2025, LNCS 15539, pp. 142–156, 2025.
https://doi.org/10.1007/978-3-031-82697-9_11

no user likes to wait "forever" to see the next solution, or to learn that no further solution exist. In order to explore this question, the problem of enumerating minimal dominating sets has been investigated in many special graph classes; several have been identified where enumeration is possible with polynomial delay.

In this paper, we attack this problem from a different side. Namely, in [4] it was shown that all minimal Roman dominating functions can be enumerated with polynomial delay. This was a rather surprising finding as in most other complexity aspects, be it 'classical', parameterized or approximation, Roman Domination and Dominating Set behave quite the same. We only refer to [1,2,8–10,19,32,35,36] for Roman Domination and to [2,7,12,15–17,21,28,34] for Dominating Set. So, we take this as a basis and try to generalize Roman Domination towards Hitting Set. Will the polynomial-delay feature disappear? In passing, we also generalize the enumeration results from [4] considerably.

Let us first explain why Roman domination variations are interesting by themselves. Defense strategies in the Roman Empire have been studied throughout the centuries, mostly from a historical-military perspective, but then also from a more graph-theoretic viewpoint, commencing with an article by Arquilla and Fredricksen [6]. The strategy that we are also discussing in this paper is nowadays known as Roman domination and was introduced in [39] in line with the discussions of Arquilla and Fredricksen, formalizing a military strategic decision going back to Constantine the Great. The concept of Roman domination is a well-studied variation of domination in graphs, certified by its own chapter in the textbook [29], written by Chellali *et al.* Here, armies are placed in different regions. A region is secured if there is at least one army in this region or there are two armies in one neighbored region. A map can be modeled as a graph, where regions correspond to vertices and connections between regions correspond to edges; moreover, the placement of armies is described by a function which maps each vertex to 0, 1 or 2. Such a function is called *Roman dominating* if each vertex with value 0 has a neighbor with value 2. It is clearly desirable to minimize expenses by finding a Roman dominating function that uses the least number of armies; the corresponding decision problem is called ROMAN DOMINATION.

Even though ROMAN DOMINATION and DOMINATING SET behave the same in terms of complexity in a variety of settings this parallelism unexpectedly breaks down for two (mutually related) tasks:

- Can we enumerate all minimal solutions with polynomial delay?
- Can we decide, given a certain part of the solution, if there exists a minimal solution that extends the given pre-solution?

This enumeration question for Roman dominating functions (for two natural notions of minimality) can be solved with polynomial delay, as proven in [4]. This result on Roman domination is based on another result giving a polynomial-time algorithm for the *extension problem(s)* as described in the second item. Namely, the idea is to call an extension test before diving further into branching. This strategy is well-established in the area of enumeration algorithms, dating back to Read and Tarjan [37], but few concrete examples are known; we only refer to the discussion in [33,40]. This makes EXTENSION ROMAN DOMINATION one of few examples where the extension problem is polynomial-time solvable, while the original problem is NP-complete. For more details on extension problems, we refer to the survey [11] that also suggests a general framework to describe such problems and provides further motivations and applications.

The simple scientific question that we want to investigate in this paper is "why": *What causes Roman domination to be feasible with respect to enumeration and extension?* Here, we approach this question by formulating a HITTING SET variation of Roman domination. It is well-known that HITTING SET can be viewed as a generalization of DOMINATING SET by modelling a graph $G = (V, E)$ by the *closed-neighborhood hypergraph* $G_{nb} = (V, (N_G[v])_{v \in V})$. From this perspective, a Roman dominating function $f : V \to \{0, 1, 2\}$ will hit the hyperedge $N_G[v]$, either because $f(w) = 2$ for some $w \in N_G[v]$ or because the index v of the hyperedge $N_G[v]$ obeys $f(v) = 1$. The difference between a vertex and an index will become clearer when we turn our attention to arbitrary hypergraphs.

To wrap up our main results quickly: ROMAN HITTING SET leads to a generalization of ROMAN DOMINATION whose extension version is polynomial-time solvable and hence, minimal Roman hitting sets can be enumerated with polynomial delay. We also show how to define notions like Roman vertex cover and transfer polynomial-delay enumeration results, even in the case of implicitly given (possibly much larger) hypergraphs, for instance, when enumerating all minimal Roman feedback vertex sets. For reasons of space, most proofs can be found in the long version of this paper. We also refer to [23,24].

2 Definitions and Notation

Throughout this paper, we will freely use standard notions from complexity theory without defining them here. This includes notions from parameterized complexity, concerning FPT and the further lower levels of the W-hierarchy up to W[2], as described in textbooks like [18,25].

Let $\mathbb{N}$ denote the set of all nonnegative integers (including 0). For $n \in \mathbb{N}$, we will use the notation $[n] := \{1, \ldots, n\}$. For a finite set A and some $n \in \mathbb{N}$ with $n \leq |A|$, the cardinality of A, $\binom{A}{n}$ denotes the set of all subsets of A of cardinality n, while 2^A denotes the power set of A. For two sets A, B, B^A denotes the set of all mappings $f : A \to B$. If $C \subseteq A$, then $f(C) = \{f(x) \mid x \in C\} \subseteq B$. We denote by $\chi_C \in \{0, 1\}^A$ the characteristic function, where $\chi_C(x) = 1$ holds iff $x \in C$. For two functions $f, g \in \mathbb{N}^A$, we write $f \leq g$ iff $f(a) \leq g(a)$ holds for all $a \in A$. Further, we define the *weight* of f by $\omega(f) = \sum_{a \in A} f(a)$.

We focus on not necessarily simple hypergraphs $H = \left(X, \hat{S} = (s_i)_{i \in I}\right)$ with a finite universe X, also called *vertex* set, and a finite index set I, where for each $i \in I$, $s_i \subseteq X$ is a *hyperedge*. With S we denote the set which includes all hyperedges of the sequence $\hat{S}$, i.e., $S = \{s_i \mid i \in I\}$. Note that the same hyperedge may appear multiple times in the family $\hat{S}$. If this is forbidden, we speak of a *simple hypergraph*. For all $x \in X$, define $\mathbf{S}(x) = \{s_i \in S \mid x \in s_i\}$ as the set of hyperedges that is *hit* by x, defining a function $\mathbf{S} : X \to 2^S$.

A set $D \subseteq X$ is a *hitting set* iff $\mathbf{S}(D) = S$, where $\mathbf{S}(D) = \bigcup_{x \in D} \mathbf{S}(x)$. Define $\mathbf{I} : X \to 2^I$ by $\mathbf{I}(x) = \{i \in I \mid x \in s_i\}$, extending to $A \subseteq X$ by $\mathbf{I}(A) = \bigcup_{x \in A} \mathbf{I}(x)$.

Consider a (simple undirected) graph $G = (V, E)$ as a hypergraph $G = (V, \hat{E})$, where each hyperedge contains exactly two elements. In this case, we call the hyperedges just edges. Then, we consider E as the index set. For each vertex $v \in V$, define its neighborhood as $N(v) = \{u \mid \{v, u\} \in E\}$ and the closed neighborhood as $N[v] = \{v\} \cup N(v)$. For $U \subseteq V$, we use $N[U] = \bigcup_{v \in U} N[v]$ for the closed neighborhood of U. $D \subseteq V$ is a *dominating set* of G if $N[D] = V$.

Let $G = (V, E)$ be a graph. A function $f : V \to \{0, 1, 2\}$ is a *Roman dominating function* (Rdf for short) iff, for each vertex $v \in V$ with $f(v) = 0$, there exists a $u \in N(v)$ with $f(u) = 2$. For $f, g \in \{0, 1, 2\}^V$, let $f \leq_{PO} g$ iff $f(v) = 0$ or $f(v) = g(v)$ for each $v \in V$. A Roman dominating function f is minimal (or PO-minimal) if for each Roman dominating function g with $g \leq f$ (or $g \leq_{PO} f$, respectively), $f = g$ holds. The following combinatorial characterization of minimal Roman dominating functions was the basis of some of our algorithms. To formulate this characterization, we need a further notion. For $D \subseteq V$ and $v \in D$, define the *private neighborhood* of $v \in V$ with respect to D as $P_{G,D}(v) := N[v] \setminus N[D \setminus \{v\}]$. Vertices in $P_{G,D}(v)$ are *private neighbors* of v.

Theorem 1. *[4] Let $G = (V, E)$ be a graph and $f : V \to \{0, 1, 2\}$ be a function. Abbreviate $G' := G\left[f^{-1}(0) \cup f^{-1}(2)\right]$. Then, f is a minimal Rdf iff we find:*

1. $N\left[f^{-1}(2)\right] \cap f^{-1}(1) = \emptyset$,
2. $\forall v \in V_2(f) : P_{G', f^{-1}(2)}(v) \nsubseteq \{v\}$, *also called* privacy condition, *and*
3. $f^{-1}(2)$ *is a minimal dominating set of* G'.

Theorem 2. *[4] Let $G = (V, E)$ be a graph and $f : V \to \{0, 1, 2\}$ be a function. Abbreviate $G' := G\left[f^{-1}(0) \cup f^{-1}(2)\right]$. Then, f is a PO-minimal Rdf iff we find:*

1. $N\left[f^{-1}(2)\right] \cap f^{-1}(1) = \emptyset$,
2. $f^{-1}(2)$ *is a minimal dominating set of* G'.

Consider now the following problems.

Problem name: Roman Domination, or RD for short
Given: A graph $G = (V, E)$ and $k \in \mathbb{N}$
Question: Is there a Rdf f with $\omega(f) \leq k$?

Problem name: (PO-)Extension Roman Domination, (PO-)Ext RD
Given: A graph $G = (V, E)$ and a function $f : V \to \{0, 1, 2\}$
Question: Is there a minimal Rdf g for G with $f \leq_{(PO)} g$?

Somewhat surprisingly, the extension problems in the second box were proven to be polynomial-time solvable in [4]. This implies that (PO-)minimal Rdf can be enumerated with polynomial delay. In order to explore why these results were possible, we generalize these notions and problems for hypergraphs next.

Let $(X, \hat{S} = (s_i)_{i \in I})$ be a hypergraph. We call $R = (R_1, R_2) \in 2^I \times 2^X$ a *Roman hitting set*, or Rhs for short, iff, for all $i \in I$, $s_i \cap R_2 \neq \emptyset$ or $i \in R_1$. If $x \in s_i \cap R_2$ we say x *hits* i or s_i. In the case that $i \in R_1$ we say i *hits itself*. For two tuples $(P_1, P_2), (R_1, R_2) \in 2^I \times 2^X$ we say $(P_1, P_2) \leq (R_1, R_2)$ if $P_1 \subseteq R_1$ and $P_2 \subseteq R_2$. We call a Rhs P *minimal* if $R \leq P$ implies $R = P$ for each Rhs R. Adapting earlier notations, we also write $\omega(R_1, R_2) := |R_1| + 2 \cdot |R_2|$, calling it the *weight* of $(R_1, R_2) \leq (I, X)$. The concept of Rhs gives rise to two problems:

Problem name: ROMAN HITTING SET, or RHS for short
Given: A hypergraph $H = (X, \hat{S} = (s_i)_{i \in I})$ and $k \in \mathbb{N}$
Question: Is there a Rhs (R_1, R_2) of H with $\omega(R_1, R_2) \leq k$?

Problem name: EXTENSION ROMAN HITTING SET, or EXT RHS for short
Given: A hypergraph $H = (X, \hat{S} = (s_i)_{i \in I})$ and some $U = (U_1, U_2) \leq (I, X)$
Question: Is there a minimal Rhs $R = (R_1, R_2)$ of H with $U \leq R$?

For the closed-neighborhood hypergraph G_{nb} of the graph G, $f : V \to \{0, 1, 2\}$ is a Roman domination function of G iff $(f^{-1}(1), f^{-1}(2))$ is a Rhs of G_{nb}. This also proves that RHS is NP-hard by [13,19,36].

3 Combinatorial Properties of Minimal Rhs

In order to pave the ground for our algorithmic results, we are now going to derive a combinatorial characterization of minimal Rhs.

Theorem 3. *Let $H = (X, \hat{S} = (s_i)_{i \in I})$ a hypergraph. Then, a tuple $(R_1, R_2) \leq (I, X)$ is a minimal Rhs iff the following constraints hold:*

1. *$\forall i \in R_1 : s_i \cap R_2 = \emptyset$,*
2. *R_2 is a minimal hitting set on $H' = (X, (s_i)_{i \in I \setminus R_1})$.*

Proof. The "only if"-part, assume there is a minimal Rhs $R = (R_1, R_2)$. Let $i \in R_1$. Assume there is a $x \in s_i \cap R_2$. Then $R' := (R_1 \setminus \{i\}, R_2) \leq (R_1, R_2)$. As $s_i \cap R_2 \neq \emptyset$, i is hit by R'. For $j \in I \setminus \{i\}$, s_j is hit by R' in the same way as by R. Therefore, R is not minimal. Assume R_2 is no minimal hitting set on $H' = (X, (s_i)_{i \in I \setminus R_1})$. If R_2 is not even a hitting set, then R is no Rhs, as there is a hyperedge not hit. Assume there exists $x \in R_2$ such that for each $i \in \mathbf{I}(x)$ there exists a $y_i \in s_i \cap (R_2 \setminus \{x\})$. Define $R' := (R_1, R_2 \setminus \{x\})$. The indices in $I \setminus \mathbf{I}(x)$ are hit by R' in the same way as by R. As for each $i \in \mathbf{I}(x)$, $s_i \cap (R_2 \setminus \{x\})$ is not empty, $\mathbf{I}(x)$ is hit by R'. Therefore, R' is a Rhs and R is not minimal.

Conversely, assume that $R = (R_1, R_2)$ fulfills the constraints. As R_2 is a minimal hitting set on H', $i \in R_1$ or $s_i \cap R_2 \neq \emptyset$ for each $i \in I$. So, R is a Rhs. Assume there is a minimal Rhs $R' = (R_1', R_2')$ with $R' \leq R$ and $x \in R_2 \setminus R_2'$. Then for each $i \in \mathbf{I}(x)$, either $i \in R_1' \subseteq R_1$ ($\lightning$ to Constraint 1) or $s_i \cap (R_2') \subseteq s_i \cap (R_2 \setminus \{x\})$ is not empty ($\lightning$ to Constraint 2). This implies $R_2 = R_2'$. Let $i \in R_1 \setminus R_1'$. Then there is an $x \in s_i \cap R_2'$, $\lightning$ to Constraint 1. Hence, $R' = R$. □

The reader is encouraged to compare this characterization theorem with Theorem 2, using the closed-neighborhood hypergraph.

Algorithm 1 ExtRHS Algorithm

1: **procedure** ExtRHS Solver($X, \hat{S}, R := (R_1, R_2)$)
  Input: set X, $\hat{S} := (s_i)_{i \in I}$, $(R_1, R_2) \leq (I, X)$.
  Output: Is there a minimal Roman hitting set $M := (M_1, M_2)$ with $R \leq M$?
2:   **for** $x \in R_2$ **do**
3:     **for** $i \in I(x)$ **do**
4:       **if** $i \in R_1$ **then return** no
5:     **if** $\mathbf{I}(x) \subseteq \mathbf{I}(R_2 \setminus \{x\})$ **then return** no
6:   $M_1 := R_1$, $M_2 := R_2$
7:   **for** $i \in I \setminus (\mathbf{I}(M_2) \cup M_1)$ **do**
8:     Add i to M_1.
9:   **Return yes** $\{(M_1, M_2)$ is the solution.$\}$

4 Ext Roman HS and PO-Ext Roman Domination

In this section, we prove that Ext RHS is solvable in polynomial time. We do this in two ways: (a) by using the fact that PO-Ext RD is polynomial-time solvable (see [4]) and presenting a polynomial-time reduction, or (b) by giving an explicit polynomial-time algorithm. The reader can find option (a) in the appendix and we make (b) explicit.

Theorem 4. Ext RHS *is poly-time solvable by a reduction from PO-*Ext RD.

Theorem 5. *Algorithm 1 solves* Ext RHS *on* $(X, \hat{S}, (R_1, R_2))$ *in poly-time.*

Proof. Let $H = (X, (s_i)_{i \in I})$ be a hypergraph and let $(R_1, R_2) \leq (I, X)$ be a tuple. Assume that the algorithm returns yes. Then $R_1 \subseteq M_1$ and $R_2 = M_2$, since we define M_1, M_2 as R_1, R_2 in the beginning and never delete any vertices from these sets. Further, we put each index from $I \setminus (\mathbf{I}(M_2) \cup M_1)$ into M_1. Therefore $M = (M_1, M_2)$ is a Rhs at the end of this algorithm. Now we want to check if M is minimal. Assume that there is an $i \in R_1$ such that $s_i \cap M_2 \neq \emptyset$. Since the for-loop in Line 7 does not put such an index into M_1, i has to be in R_1. Because $M_2 = R_2$, Line 4 would have returned no. Assume R_2 is not a minimal hitting set on $(s_i)_{i \in I \setminus M_1}$. Since (M_1, M_2) is a Rhs, M_2 is a hitting set on $(s_i)_{i \in I \setminus M_1}$. If $M_2 = R_2$ is not minimal, then Line 5 will return no.

Assume there is a minimal Rhs (P_1, P_2) on H with $(R_1, R_2) \leq (P_1, P_2)$, but the algorithm returns no. If the algorithm returns no in Line 4, then there exists an $i \in R_1 \subseteq P_1$ with an $x \in s_i \cap R_2 \subseteq s_i \cap P_2$. This contradicts Theorem 3. Assume the algorithm returns no in Line 5. Then there is an $x \in R_2 \subseteq P_2$ such that, for each $i \in \mathbf{I}(x)$, there is a $y \in s_i \cap (R_2 \setminus \{x\})$. Then P_2 is no minimal hitting set on $(s_i)_{i \in I \setminus P_1}$. As these are the only cases in which the algorithm returns no, if the input is a yes-instance, then the algorithm will return yes. □

In section 5, we use this to enumerate all minimal Rhs with polynomial delay.

In fact, the converse of the reduction proof (a) of Theorem 4 is easier to describe: Let $G = (V, E)$ be a graph and $f : V \to \{0, 1, 2\}$ be a mapping. Theorem 6 of [4] implies that, if we solve the EXT RHS on $H = (V, (N[v])_{v \in V})$ with the tuple $(f^{-1}(1), f^{-1}(2))$, then we solve the PO-EXT RD instance (G, f). Therefore, we can view EXT RHS as a generalization of PO-EXT RD. Furthermore, both problems are equivalent under polynomial reductions.

5 Enumerating Minimal Roman Hitting Sets

Now we want to enumerate all minimal Roman hitting sets. We do this by constructing a branching algorithm which is inspired by the $\mathcal{O}\left(\sqrt[3]{3}^n\right)$-time minimal Rdf enumeration algorithm on split graphs [3]. Let $H = (X, \hat{S} = (s_i)_{i \in I})$ be a hypergraph. We construct a set $X' \subseteq X$ which will include all vertices that are already in R_2, i.e., $R_2 \subseteq X'$, or at least can be in R_2. To do this, we need the tuple (R_1, R_2), where $R_1 \subseteq I$ includes all indices and $R_2 \subseteq X'$ includes all elements which are a part of the Rhs that we are going to construct. In the branching process, we will either put an $x \in X'$ into R_2 or *delete* x, which describes the following operation: remove x from X' and then also remove x from all the hyperedges that it belonged to. Therefore, we can consider the hypergraph $H' := (X', (s_i \cap X')_{i \in I})$ as describing the 'current hypergraph' during the computation process. However, we will modify the hyperedges s_i during the computation, so that always $s_i \subseteq X'$ is satisfied, but in terms of the original hyperedge s_i, we can always think of $s_i \cap X'$ as the current counterpart.

We define $I' = I \setminus (R_1 \cup \mathbf{I}(R_2))$ as the set of indices that describe edges that are not yet hit. This also defines a new function $\mathbf{I}'$. If we *add an* $i \in I'$ *to* R_1 (as a further operation), we will also delete all $x \in s_i$ from X' as these vertices cannot be in a minimal Rhs if i is also. Our measure is given by $\mu := \mu(X', \hat{S}', R_1, R_2) := |X' \setminus R_2| + |I'|$. Initially, we have $X' = X$, $I' = I$ and $R_2 = \emptyset$, so that $\mu = |X| + |I|$ at the beginning. As our main result of this section, we are going to show that all minimal Rhs can be enumerated in $\mathcal{O}^*\left(\sqrt[3]{3}^{|X|+|I|}\right)$ time with polynomial delay. We also provide an example hypergraph family showing that there cannot be any significantly better enumeration algorithm as a hypergraph with n vertices and m indices from this family has at least $\sqrt[3]{3}^{n+m}$ many minimal Rhs.

Before each branching step, we check if there is a minimal Rhs (P_1, P_2) on H' with $(R_1, R_2) \leq (P_1, P_2)$ and try to use a reduction rule. Among all applicable branching rules, we will use the rule with the smallest number.

Reduction Rule 1. *If there is an* $x \in X' \setminus R_2$ *with* $\mathbf{I}(x) \subseteq \mathbf{I}(R_2)$, *then delete* x.

Reduction Rule 2. *If there is an* $i \in I'$ *with* $s_i = \emptyset$, *then put* i *into* R_1.

Branching Rule 1. *Let* $i \in I'$ *with* $s_i = \{x\} \subseteq X' \setminus R_2$. *Then we branch as follows: (1) Delete* x. *(2) Add* x *to* R_2.

Branching Rule 2. *Let* $x \in X'$ *with* $|\mathbf{I}'(x)| \geq 3$. *Then we branch as follows: (1) Add* x *to* R_2. *(2) Delete* x.

Branching Rule 3. *Let $x, y \in X'$ with $\mathbf{I}'(x) = \{i\}$, $s_i = \{x, y\}$. Branch like: (1) Add x to R_2, delete y. (2) Add y to R_2, delete x. (3) Add i to R_1, delete x, y.*

Branching Rule 4. *Let $x \in X'$ with $\mathbf{I}'(x) = \{i\}$ with $|s_i| \geq 3$. Then we branch as follows: (1) Add x to R_2 and delete the remaining vertices in s_i. (2) Delete x.*

Branching Rule 5. *Let $x, y \in X'$ with $x \neq y$ and $\mathbf{I}'(x) = \mathbf{I}'(y)$. Then branch as follows: (1) Add x to R_2, delete y. (2) Add y to R_2, delete x. (3) Delete x, y.*

Branching Rule 6. *Let $i \in I'$ with $s_i = \{x, y\}$. Then we branch as follows: (1) Add x to R_2. (2) Add y to R_2, delete x. (3) Add i to R_1, delete x, y.*

Branching Rule 7. *Let $i \in I'$ with $s_i = \{x, y, z\}$. Then we branch as follows: (1) Add x to R_2. (2) Add y to R_2, delete x. (3) Add z to R_2, delete x, y. (4) Add i to R_1, delete x, y and z.*

Branching Rule 8. *Let $x \in X'$ with $\mathbf{I}'(x) = \{i, j\}$, $y \in s_i \setminus \{x\}$ with $\mathbf{I}'(y) = \{i, k\}$. Branch as follows: (1) Delete x. (2) Add x to R_2, delete y. (3) Add x, y to R_2 and delete $s_k \setminus \{y\}$.*

Table 1. Branching vectors and numbers for enumerating minimal Rhs.

Rule	branching vector	branching number
1	$(2, 2)$	$\sqrt{2} \leq 1.4143$
2 & 4	$(4, 1)$	1.3803
3	$(3, 3, 3)$	$\sqrt[3]{3} \leq 1.4423$
5	$(4, 4, 2)$	$\sqrt{2} \leq 1.4143$
6	$(3, 4, 3)$	1.3954
7	$(3, 4, 5, 4)$	1.4253
8	$(1, 4, 8)$	1.4271

Corollary 1. *The branching algorithm is a complete case distinction.*

Theorem 6. *In a hypergraph $H = (X, (s_i)_{i \in I})$ with $|X| = n$, $|I| = m$, all minimal Rhs are enumerable in $\mathcal{O}^*(\sqrt[3]{3}^{n+m})$ time with polynomial delay and space.*

Proof. By Corollary 1, the branching algorithm has a branching rule for each possibility. If we calculate the branching numbers for the branching vectors mentioned in Table 1, we get $\mathcal{O}^*(\sqrt[3]{3}^{n+m})$ as the running time for the algorithm.

This leaves to show that the algorithm runs with polynomial delay and polynomial space. As branching algorithms per se use polynomial space, the only potential problem consists in enumerating solutions multiple times, because this would enforce us to store previous results. Therefore, we show next that there is no minimal Rhs that we count multiple times. The Branching Rules 1, 2 and 4 branch by putting x into R_2 or deleting it. Since we never change such a decision afterwards, these branching rules prevent us from enumerating any minimal Rhs twice. The same happens (with some side effects) in the remaining branchings as these are just asymmetrical branches where we branch again in some cases.

Concerning polynomial delay, the only potential problem is that the algorithm spends exponential time when diving into sub-branches where no minimal Rhs solution exists. We prevent this by evoking our polynomial-time extension check procedure. In order to be able to do so, we only need to show that working on restricted instances, as described by H', suffices.

Claim. Let $H = (X, (s_i)_{i \in I})$ be a hypergraph, $R_1 \subseteq I$ and $R_2 \subseteq X' \subseteq X$. Then there is a solution (P_1, P_2) for Ext RHS with given H, R_1, R_2 such that $P_2 \subseteq X'$ iff there is a solution for Ext RHS with given $H' = (X', (s_i \cap X')_{i \in I}), R_1, R_2$.

Therefore, if we use Algorithm 1 with the parameters $X', (s_i \cap X')_{i \in I}, R = (R_1, R_2)$, we check if there is a minimal Rhs $P = (P_1, P_2)$ with $R \leq P$ and $P_2 \subseteq X'$. Since we do this after each branch and each rule increases R, we check if this branch will enumerate a minimal Rhs. If get a no-answer, we can go back and need not dive into the recursion of the branching algorithm. As we branch at most $|X| + |I|$ many times, this only needs polynomial time on each path of the branching tree, hence ensuring polynomial delay. □

Theorem 7. *There is a hypergraph* $H = (X, (s_i)_{i \in I})$ *with* $\sqrt[3]{3}^{|X|+|I|}$ *minimal Rhs.*

Proof. Define $X = \{x_1, \ldots, x_{2n}\}$ and $I = [n]$ with $s_i = \{x_{2i-1}, x_{2i}\}$ for each $i \in I$. For each $i \in I$, there are three ways to hit i by a minimal Rhs (R_1, R_2): $x_{2i-1} \in R_2$ or $x_{2i} \in R_2$ or $i \in I$. As the sets $s_1, \ldots, s_n$ are pairwise disjoint, for each $i \in I$, each x_{2i-1} and x_{2i} cannot be in R_2 at the same time and the way one hyperedge is hit does not effect the way another hyperedge is hit. Therefore, there are $3^n = \sqrt[3]{3}^{3n} = \sqrt[3]{3}^{|X|+|I|}$ many minimal Rhs. □

6 A Branching Algorithm for Minimum RHS

Now, we present a simple branching algorithm that returns a minimum Rhs. Let $H = (X, (s_i)_{i \in I})$ be a hypergraph. We reuse the idea of the branching algorithm of section 5 (including X', I', R_1, R_2, the measure μ and Reduction Rule 2). We add the idea of finding a minimum Rhs with minimal $|R_2|$.

Reduction Rule 3. *If there is an* $x \in X' \setminus R_2$ *with* $|\mathbf{I}'(x)| \leq 2$*, then delete* x*.*

Reduction Rule 4. *For* $x \in X' \setminus R_2$ *with* $|\{i \in I' \mid s_i = \{x\}\}| \geq 3$*, put* x *in* R_2*.*

Branching Rule 9. *Let* $x \in X'$ *with* $|\mathbf{I}'(x)| = 3$*. Then we branch as follows: (1) Delete* x*. (2) Add* x *to* R_2 *and delete all vertices from* $\bigcup_{i \in \mathbf{I}'(x)} s_i$*.*

Branching Rule 10. *Let $x \in X'$ with $|\mathbf{I}'(x)| \geq 4$. Then we branch as follows: (1) Add x to R_2. (2) Delete x.*

Theorem 8. *For a hypergraph $H = (X, (s_i)_{i \in I})$ with $|X| = n$ and $|I| = m$, a minimum Rhs can be computed in time $\mathcal{O}(1.3248^{n+m})$, using polynomial space.*

Proof. We first check if this branching algorithm is a complete case distinction. By Reduction Rule 2, we know that there is no empty hyperedge in I'. This implies that there has to be at least one element $x \in X'$. $|\mathbf{I}'(x)| \leq 2$ triggers Reduction Rule 3. Reduction Rule 4 and Branching Rule 9 handle the case $|\mathbf{I}'(x)| = 3$. The remaining cases are covered by Branching Rule 10. As Branching Rules 9 and 10 are the only ones and both have a branching vector $(1, 5)$, we get a running time of $\mathcal{O}(1.3248^{n+m})$. Reduction Rule 4 and Branching Rule 9 deal with $x \in X'$ such that $|\mathbf{I}'(x)| = 3$. Branching Rule 10 handles the rest. □

7 Applications

In this section, we consider applications of RHS. First we show how the enumeration algorithm of section 5 can be used to enumerate all minimal Rdfs on special graph classes. Then we show that EXTENSION DOMINATING SET is polynomial-time solvable on split graphs by using the fact that EXT RHS is polynomial time solvable. In the end, we will define and discuss further Roman versions for problems for which HS is a generalization, including cases where the obtained hypergraph can be exponentially larger than the given graph structure.

7.1 Enumeration

The $\mathcal{O}^*(\sqrt[3]{3})$ enumeration algorithm for minimal Rdfs on split graphs of [3] (which cannot be improved beside the polynomial factor) is one example for an application of the enumeration algorithm of section 5. It can be also be observed by using the enumeration algorithm of for minimal Rhs twice. This is interesting as in this way the one algorithm could inspire the other one. To show this, we can use Theorems 1 and 3. The details can be found in the appendix.

Nonetheless, observe that if we use the enumeration algorithm of section 5 "blindly" for enumerating minimal Rdf the running time would be $\mathcal{O}(2.0801^{|V|})$ (and we list too many, also non-minimal Rdfs), clearly worse than the general approach by [4]. This is different from the situation for DOMINATING SET, for which the best enumeration algorithm uses HS (see [26]).

7.2 EXTENSION DOMINATING SET on Split and Co-bipartite Graphs

Theorem 9. *Ext DS is solvable in polynomial time on split graphs.*

This is interesting as it is known by [11] that Ext DS is NP-complete on bipartite graphs. To our knowledge, there is no such result mentioned for co-bipartite graphs, yet. But by [31], it is known that enumeration for minimal dominating sets on co-bipartite graphs is transversal-hard. The construction of this proof can also be used to show the NP-completeness (and W[3]-completeness) of Ext DS on co-bipartite graphs. More precisely, we can state:

Corollary 2. *Ext DS is NP-complete and W[3]-complete when parameterized by the size of the pre-solution even on co-bipartite graphs.*

7.3 Roman Versions for Other Graph Problems

There are problems other than DOMINATING SET for which HS is a natural generalization. Some other examples are VERTEX COVER and EDGE COVER. Let $G = (V, E)$ be a graph. We call $C \subseteq V$ a *vertex cover* if $e \cap C \neq \emptyset$ for each $e \in E$, i.e., $C \subseteq V$ is a vertex cover iff C is a hitting set of the hypergraph $(V, (e)_{e \in E})$. A set $C \subseteq E$ is an *edge cover* if for each $v \in V$, $\mathbf{E}(v) \cap C \neq \emptyset$ with $\mathbf{E}(v) := \{e \in E \mid v \in e\}$, i.e., it is a hitting set on $(E, (\mathbf{E}(v))_{v \in V})$.

Now that we generalized RD for HS, we could use this to define problems like ROMAN VERTEX COVER and ROMAN EDGE COVER. Let $G = (V, E)$ be a graph. We call a tuple $(R_1, R_2) \leq (E, V)$ a *Roman vertex cover* (Rvc) if for each $e \in E$, $e \in R_1$ or $R_2 \cap e$ is not empty. We call a tuple $(R_1, R_2) \leq (V, E)$ a *Roman edge cover* (Rec) if for each $v \in V$, $v \in R_1$ or $R_2 \cap \mathbf{E}(v) \neq \emptyset$.

Problem name: ROMAN VERTEX COVER, or RVC for short
Given: A graph $G = (V, E)$ and $k \in \mathbb{N}$
Question: Is there a Rvc $(R_1, R_2) \subseteq E \times V$ that satisfies $\omega(R_1, R_2) \leq k$?

The related problem ROMAN EDGE COVER is in logspace, cf. the appendix.

Theorem 10. RVC *is NP-complete.*

Theorem 11. *Algorithm 2 solves* RVC *in FPT-time if parameterized by k.*

7.4 Implicitly Given Hypergraphs

Hypergraphs are often not explicitly presented, but are implicitly given. Then, an explicit production of the hypergraph might take exponential time. We will discuss one such example now. More precisely, in the remaining section, we will consider Roman versions of feedback vertex sets. Let $G = (V, E)$ be a graph. $C(G)$ denotes the set of all cycles of graph G. A set $F \subseteq V$ is called a *feedback vertex set* of G if $G[V \setminus F]$ does not contain any cycles. This can also be seen

Algorithm 2 RVC FPT time solver

```
1: procedure RVC SOLVER(G = (V, E), k)
     Input: Graph G, k ∈ ℕ.
     Output: Is there a Roman vertex cover (R1, R2) with ω(R1, R2) ≤ k?
2:   if E = ∅ or |E| = k = 1 then return yes
3:   if k = 0 < |E| or k = 1 < |E| then return no
4:   Let e = {v, u} ∈ E.
5:   if RVC SOLVER((V \ {v}, {e ∈ E | v ∉ e}), k − 2) then return yes
6:   if RVC SOLVER((V \ {u}, {e ∈ E | u ∉ e}), k − 2) then return yes
7:   return RVC SOLVER((V, E \ {e}), k − 1)
```

as a HS variation. Different from VC, the hyperedges are now implicitly given, as these are the cycles. Let us define its Roman version. We call $(R_1, R_2) \leq (C(G), V)$ a *Roman feedback vertex set* (Rfvs for short) if $C(G[V \setminus R_2]) \subseteq R_1$.

Problem name: ROMAN FEEDBACK VERTEX SET, or RFVS for short
Given: A graph $G = (V, E)$ and $k \in \mathbb{N}$
Question: Is there a Rfvs $(R_1, R_2) \subseteq E \times V$ that satisfies $\omega(R_1, R_2) \leq k$?

After we published a TR version of this project [22], this problem was taken up and further studied in [30].

Lemma 1. *Let $G = (V, E)$ be a graph. There is a Rfvs $R = (R_1, R_2)$ with minimum $\omega(R)$ such that each $v \in V \setminus R_2$ is on at most one cycle of $G[V \setminus R_2]$.*

Theorem 12. RFVS *is NP-complete.*

We can also consider the enumeration of minimal Rfvs. Here, we can also use the algorithm of section 5 even if the hyperedges are now given implicitly. We have to be careful with the concept of polynomial delay. For example, $(C(G), \emptyset)$ is a minimal Rfvs for each graph $G = (V, E)$, but $C(G)$ could be exponentially big. To output this, we have to enumerate these. We will avoid this issue by only keeping track of R_2, as we could compute R_1 for a minimal Rfvs by enumerating the cycles on $G[V \setminus R_2]$. If these are exponentially many in terms of the original graph, this is no harm for concepts as polynomial delay. Yet, if one would also try to keep track of R_1 during the run of the algorithm, this could lead to exponential delay. Let us discuss our algorithmic approach for this implicit scenario.

Our extension Algorithm 1, applied to an EXT RFVS instance, would first take a $v \in R_2$ and check if it is contained in any cycle of R_1. This check takes time $|R_1| \cdot |R_2| \cdot |V|$. Then, the algorithm tests for an $x \in V$ if $\mathbf{I}(x) \subseteq \mathbf{I}(R_2 \setminus \{x\})$. In the RFVS setting, we can do this by looking for a cycle (including x) in $G[V \setminus (R_2 \setminus \{x\})]$. If we checked this for each vertex in R_2 and did not return no, we can return yes. Then, the algorithm produces a solution. For our enumeration algorithm, we only need to know if it is a yes-instance and need no solution.

For enumeration, it is important that we can compute in poly-time if a vertex is in no, exactly one, exactly two or more than three cycles. In [5], a concrete algorithm is presented that we can use here (it also outputs the cycles). We employ Reduction Rule 2 only in the end while producing a solution. For using Branching Rule 1, we only examine the graph induced by the vertices which are not in the solution, plus the one vertex which we did not decide so far if it is in R_2. To employ Branching Rule 2, we can use the algorithm of [5]. From now on, we can assume that each vertex (for which we still have to decide if it is in R_2) lies in at most 2 cycles. With the algorithm of [5], we can enumerate these and work on them as they would be edges (there are at most $2|V|$ of them). As in each case we delete at least one vertex, the running time is at most $\mathcal{O}^*(2^n)$ on graphs of order n. This is optimal, as can be seen by looking at $G = K_n$, where all sets of at most $n-2$ vertices can serve as R_2, giving a minimal Rfvs (R_1, R_2). Notice that our poly-delay algorithm for minimal Rfvss is fundamentally different from the one presented for minimal feedback vertex sets in [38].

8 Conclusions

We have generalized the notion of Roman domination towards hypergraphs: ROMAN HITTING SET is a problem that behaves quite like ROMAN DOMINATION, also having a poly-time decidable extension version and hence a poly-delay enumeration algorithm. This allowed to conclude a number of further poly-delay enumerability results, even in the case when the hypergraphs are only implicitly given. It would be interesting to see more examples of implicitly given structures where known output-polynomial enumeration algorithms can be adapted to. From a graph-theoretic perspective, it would be interesting to investigate the graph parameters $\mathsf{vc}_R(G)$ and $\mathsf{fvs}_R(G)$, as it has been previously done for γ_R.

References

1. Aazami, A., Cheriyan, J., Jampani, K.R.: Approximation algorithms and hardness results for packing element-disjoint Steiner trees in planar graphs. Algorithmica **63**(1-2), 425–456 (2012)
2. Abu-Khzam, F.N., Bazgan, C., Chopin, M., Fernau, H.: Data reductions and combinatorial bounds for improved approximation algorithms. Journal of Computer and System Sciences **82**(3), 503–520 (2016)
3. Abu-Khzam, F.N., Fernau, H., Mann, K.: Roman census: Enumerating and counting Roman dominating functions on graph classes. In: Leroux, J., Lombardy, S., Peleg, D. (eds.) 48th International Symposium on Mathematical Foundations of Computer Science, MFCS. Leibniz International Proceedings in Informatics (LIPIcs), vol. 272, pp. 6:1–6:15. Schloss Dagstuhl – Leibniz-Zentrum für Informatik (2023)
4. Abu-Khzam, F.N., Fernau, H., Mann, K.: Minimal Roman dominating functions: Extensions and enumeration. Algorithmica **86**, 1862–1887 (2024)
5. Agarwal, U., Ramachandran, V.: Finding k simple shortest paths and cycles. In: Hong, S. (ed.) 27th International Symposium on Algorithms and Computation, ISAAC. LIPIcs, vol. 64, pp. 8:1–8:12. Schloss Dagstuhl - Leibniz-Zentrum für Informatik (2016)
6. Arquilla, J., Fredricksen, H.: "Graphing" an optimal grand strategy. Military Operations Research pp. 3–17 (Fall 1995), http://hdl.handle.net/10945/38438
7. Athanassopoulos, S., Caragiannis, I., Kaklamanis, C., Kyropoulou, M.: An improved approximation bound for spanning star forest and color saving. In: Královic, R., Niwinski, D. (eds.) Mathematical Foundations of Computer Science 2009, 34th International Symposium, MFCS. LNCS, vol. 5734, pp. 90–101. Springer (2009)
8. Bermudo, S., Fernau, H.: Computing the differential of a graph: hardness, approximability and exact algorithms. Discrete Applied Mathematics **165**, 69–82 (2014)
9. Bermudo, S., Fernau, H.: Combinatorics for smaller kernels: The differential of a graph. Theoretical Computer Science **562**, 330–345 (2015)
10. Bermudo, S., Fernau, H., Sigarreta, J.M.: The differential and the Roman domination number of a graph. Applicable Analysis and Discrete Mathematics **8**, 155–171 (2014)
11. Casel, K., Fernau, H., Ghadikolaei, M.K., Monnot, J., Sikora, F.: On the complexity of solution extension of optimization problems. Theoretical Computer Science **904**, 48–65 (2022)

12. Chen, N., Engelberg, R., Nguyen, C.T., Raghavendra, P., Rudra, A., Singh, G.: Improved approximation algorithms for the spanning star forest problem. Algorithmica **65**(3), 498–516 (2013)
13. Cockayne, E.J., Dreyer Jr., P., Hedetniemi, S.M., Hedetniemi, S.T.: Roman domination in graphs. Discrete Mathematics **278**, 11–22 (2004)
14. Creignou, N., Kröll, M., Pichler, R., Skritek, S., Vollmer, H.: A complexity theory for hard enumeration problems. Discrete Applied Mathematics **268**, 191–209 (2019)
15. Dehne, F., Fellows, M., Fernau, H., Prieto, E., Rosamond, F.: NONBLOCKER: parameterized algorithmics for MINIMUM DOMINATING SET. In: Štuller, J., Wiedermann, J., Tel, G., Pokorný, J., Bielikova, M. (eds.) Software Seminar SOFSEM. LNCS, vol. 3831, pp. 237–245. Springer (2006)
16. Dinur, I., Steurer, D.: Analytical approach to parallel repetition. In: Shmoys, D.B. (ed.) Symposium on Theory of Computing, STOC. pp. 624–633. ACM (2014)
17. Downey, R.G., Fellows, M.R.: Parameterized Complexity. Springer (1999)
18. Downey, R.G., Fellows, M.R.: Fundamentals of Parameterized Complexity. Texts in Computer Science, Springer (2013)
19. Dreyer, P.A.: Applications and Variations of Domination in Graphs. Ph.D. thesis, Rutgers University, New Jersey, USA (2000)
20. Eiter, T., Gottlob, G.: Identifying the minimal transversals of a hypergraph and related problems. SIAM Journal on Computing **24**(6), 1278–1304 (1995)
21. Feige, U.: A threshold of $\ln n$ for approximating set cover. Journal of the ACM **45**, 634–652 (1998)
22. Fernau, H., Mann, K.: Hitting the Romans. Tech. Rep. abs/2302.11417, ArXiv, Cornell University (2023). https://doi.org/10.48550/arXiv.2302.11417
23. Fernau, H., Mann, K.: Perfect Roman domination and unique response Roman domination. Tech. rep., ArXiv, Cornell University (2023)
24. Fernau, H., Mann, K.: Roman hitting functions. In: Rzążewski, P., Bonnet, E. (eds.) 19th International Symposium on Parameterized and Exact Computation, IPEC. LIPIcs, vol. 321, pp. 24:1–24:15. Schloss Dagstuhl - Leibniz-Zentrum für Informatik (2024)
25. Flum, J., Grohe, M.: Parameterized Complexity Theory. Springer (2006)
26. Fomin, F.V., Grandoni, F., Pyatkin, A.V., Stepanov, A.A.: Combinatorial bounds via measure and conquer: Bounding minimal dominating sets and applications. ACM Transactions on Algorithms **5**(1), 1–17 (2008)
27. Gainer-Dewar, A., Vera-Licona, P.: The minimal hitting set generation problem: Algorithms and computation. SIAM Journal of Discrete Mathematics **31**(1), 63–100 (2017)
28. Garey, M.R., Johnson, D.S.: Computers and Intractability. New York: Freeman (1979)
29. Haynes, T.W., Hedetniemi, S., Henning, M.A. (eds.): Topics in Domination in Graphs, Developments in Mathematics, vol. 64. Springer (2020)
30. Jana, S., Modak, S., Saurabh, S., Singanporia, K.: Roman cycle hitting set. In: Kráľ, D., Milanič, M. (eds.) Graph-Theoretic Concepts in Computer Science - 50th International Workshop, WG. LNCS, vol. To appear. Springer (2024)
31. Kanté, M.M., Limouzy, V., Mary, A., Nourine, L.: On the enumeration of minimal dominating sets and related notions. SIAM Journal of Discrete Mathematics **28**(4), 1916–1929 (2014)
32. Li, K., Ran, Y., Zhang, Z., Du, D.: Nearly tight approximation algorithm for (connected) Roman dominating set. Optimization Letters **16**(8), 2261–2276 (2022)

33. Mary, A., Strozecki, Y.: Efficient enumeration of solutions produced by closure operations. Discrete Mathematics & Theoretical Computer Science **21**(3) (2019)
34. Nguyen, C.T., Shen, J., Hou, M., Sheng, L., Miller, W., Zhang, L.: Approximating the spanning star forest problem and its application to genomic sequence alignment. SIAM Journal on Computing **38**(3), 946–962 (2008)
35. Padamutham, C., Palagiri, V.S.R.: Algorithmic aspects of Roman domination in graphs. Journal of Applied Mathematics and Computing **64**, 89–102 (2020)
36. Pagourtzis, A., Penna, P., Schlude, K., Steinhöfel, K., Taylor, D.S., Widmayer, P.: Server placements, Roman domination and other dominating set variants. In: Baeza-Yates, R.A., Montanari, U., Santoro, N. (eds.) Foundations of Information Technology in the Era of Networking and Mobile Computing, IFIP 17th World Computer Congress — TC1 Stream / 2nd IFIP International Conference on Theoretical Computer Science IFIP TCS. pp. 280–291. Kluwer (2002), also available as Technical Report 365, ETH Zürich, Institute of Theoretical Computer Science, 10/2001.
37. Read, R.C., Tarjan, R.E.: Bounds on backtrack algorithms for listing cycles, paths, and spanning trees. Networks **5**, 237–252 (1975)
38. Schwikowski, B., Speckenmeyer, E.: On enumerating all minimal solutions of feedback problems. Discrete Applied Mathematics **117**, 253–265 (2002)
39. Stewart, I.: Defend the Roman Empire. Scientific American pp. 136,137,139 (Dec 1999)
40. Strozecki, Y.: Enumeration complexity. EATCS Bulletin **129** (2019)

Visual Complexity of Point Set Mappings

Wouter Meulemans, Arjen Simons(✉), and Kevin Verbeek

TU Eindhoven, Eindhoven, The Netherlands
{w.meulemans,a.simons1,k.a.b.verbeek}@tue.nl

Abstract. We study the visual complexity of animated transitions between point sets. Although there exist many metrics for point set similarity, these metrics are not adequate for this purpose, as they typically treat each point separately. Instead, we propose to look at translations of entire subsets/groups of points to measure the visual complexity of a transition between two point sets. Specifically, given two labeled point sets A and B in $\mathbb{R}^d$, the goal is to compute the cheapest transformation that maps all points in A to their corresponding point in B, where the translation of a group of points counts as a single operation in terms of complexity. In this paper we identify several problem dimensions involving group translations that may be relevant to various applications, and study the algorithmic complexity of the resulting problems. Specifically, we consider different restrictions on the groups that can be translated, and different optimization functions. For most of the resulting problem variants we are able to provide polynomial time algorithms, or establish that they are NP-hard. For the remaining open problems we either provide an approximation algorithm or establish the NP-hardness of a restricted version of the problem. Furthermore, our problem classification can easily be extended with additional problem dimensions giving rise to new problem variants that can be studied in future work.

Keywords: Group Translations · Visual Complexity · Point Sets

1 Introduction

Visualizations are useful for exploring, analyzing and communicating about complex data [25], leveraging human perceptual and cognitive functions. By visually representing data elements, one may identify patterns, trends and outliers. Typically, each data element is depicted through a visual representation (symbol, glyph or other visual encoding of its properties). As one of the strongest visual variables [25], the position of this representation is often determined by specific properties, such as (geo)spatial location on a map or other attribute encodings. Yet, this position may change, for example, as the data updates in time-varying contexts [4], or when a user selects other attributes to map to the element's position. They may also shift to disambiguate visualizations by applying distortions

W. Meulemans and A. Simons—Partially supported by the Dutch Research Council (NWO) under project number VI.Vidi.223.137.

R. Královič and V. Kůrková (Eds.): SOFSEM 2025, LNCS 15539, pp. 157–171, 2025.
https://doi.org/10.1007/978-3-031-82697-9_12

to remove overlap [22,23]. In such cases, it is important for users to understand the transition between states [4]. They have a mental map [29] of the old state and need to translate this to the new state. Animated transitions may aid this process [15] and are often used to support visual analytics.

We are interested in measuring the *visual complexity* of such transitions, as a proxy to automatically assess the cognitive load they may induce. For simplicity, we assume that each data element is positioned based on a point in Euclidean space (instead of by, e.g., a polygon). The question then, is to measure how complex a transition from one point set to another is. Various approaches already exist to assess point set similarity, including simple forms such as summing (squared) distances and taking the maximum distance [23], or more elaborate measures such as the Hausdorff distance [2] and Earth-mover's distance [9,27].

However, such methods are inadequate for assessing the visual complexity of a transition, as they treat each point "separately": how far does it have to move, given some constraints on the entire set. A simple example illustrates that this is not adequate for capturing transition complexity: imagine two point sets such that each point translates by the exact same vector, from one set to the other. Understanding this transition is fairly straightforward, as the user sees but a single, grouped motion, following the Gestalt principle of Common Fate [31], as shown in Fig. 1.

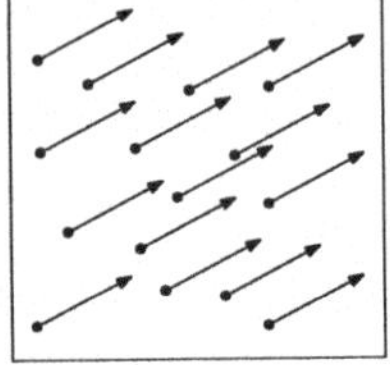

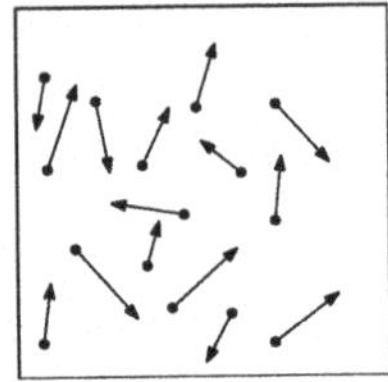

Fig. 1. Grouped vs dispersed motion.

Contributions. We introduce a problem classification for various problem variants that arise in the study of visual complexity of transitions between point sets, based on common motion of points (Sect. 2). The problem variants arise from different constraints on the transitions and methods for measuring the complexity. In the subsequent chapters, we study the algorithmic complexity of cases that arise from the classification; refer to Table 1 for an overview of our results.

Related Work. To improve the understandability of animated transitions between point sets beyond simple linear interpolation, there are roughly two approaches. A first approach is to modify the trajectories along which the points move, to improve the visible structure in the transition by decreasing occlusion and improving grouped motion. Such approaches include trajectory bundling [8], the introduction of waypoints [28] and vector-field based routing [30]. The drawback is that such techniques tend to increase the complexity of object tracking, as nearby objects can be confused [11]. Compared to our work, such techniques take a looser approach to group perception via Common Fate, as they do not build on strict identical movement.

The second approach is to split the animation in several stages – sequential steps. This can be according to a predefined scheme, for example in the case of waypoints [28], or optimized explicitly [24]. Mizuno et al. [24] optimize the stages of changing point locations and set memberships, to minimize gaze shift. That is, the goal is to reduce the amount of movement needed for the eyes to track the moving objects through the stages. This is slightly different from our model, as it focuses more on the tracking aspect of transitions [24], rather than complexity of the mental model that a transition requires.

As mentioned above, our problem relates to point-set similarity. Often, measures are sensitive to applying transformations (such as translations, scaling or rotations) to the objects. When such transformations do not affect a measure, this is called invariance. In this context, our work is conceptually in between measures that are and measures that are not translation invariant: our models aim to capture common translations, while still accounting for such common translations as affecting similarity.

Aligning point sets via a transformation to optimize their similarity, also known as point-set registration, is a well-studied problem, see e.g. [5,16,20] or the survey by Alt and Guibas [3]. For example, optimizing the sum of squared distances under translations only equates to aligning centroids [7] and is useful, for example, in computing grid maps [10]; under different measures such alignment may become more complex [10]. Point-set registrations has applications, for example, also in computer vision to estimate the motion parameters of rigid objects [5]. These techniques can be useful to apply a single transformation followed by individual movement, but cannot leverage common patterns that exist only in subsets of the points. Other methods in this area (e.g. [6,17,26]) primarily focus on finding the best correspondence between point sets without a labeling, and as such different from the problem studied in this work.

Shape-interpolation techniques also aim to ensure coherent movement of points during transformations. Many of these techniques triangulate the input shapes and interpolate between the resulting triangulations [1,14,21]. The goal is often to minimize local distortions, ensuring that nearby points move cohesively. However, the focus lies on the volume being distorted, rather than precise movements of individual points – it aims for similar but not necessarily identical transformations.

2 Problem Classification

In this section we present a problem classification for transitions between point sets based on the common motion of points. The problem variants in the classification arise from different constraints and optimization criteria on the transitions. We discuss potential further extensions/dimensions of the classification in Sect. 6, which then directly introduce new open problems for future work.

Our input consists of two collections of points $A = \{a_1, \dots, a_n\}$ and $B = \{b_1, \dots, b_n\}$ of equal size n in $\mathbb{R}^d$. Our goal is to compute the cheapest transformation from points in A to points in B, where the translation of any subset of points counts as a single operation in terms of complexity, instead of counting all the individual point translations. We refer to such a subset as a *group*[1], and to

[1] Not to be confused with the standard notion of *groups* in abstract algebra.

the corresponding translation as a *group translation*. A single point may be part of multiple groups. We assume that the points in A and B already have a given mapping such that a_i should be translated to b_i for all i. In the following, we use the notation $[n] = \{1, \ldots, n\}$. Formally, a solution to this problem consists of a pair $(\mathcal{F}, \tau)$, where $\mathcal{F} \subseteq \mathcal{P}([n])$ is a family of subsets of $[n]$ indicating the groups via the point indices, and $\tau \colon \mathcal{F} \to \mathbb{R}^d$ describes the translation vectors for the groups defined by $\mathcal{F}$. We also define the subfamilies $\mathcal{F}_i = \{S \in \mathcal{F} \mid i \in S\}$ as the set of groups that contain index i. Since the points in A and B have a given mapping, we can represent the input by a single collection of vectors $\Delta = \{\delta_1, \ldots, \delta_n\}$, where $\delta_i = b_i - a_i$ is the difference in position between a_i and b_i. Now, a solution $(\mathcal{F}, \tau)$ is considered *valid* for an input collection Δ if $\sum_{S \in \mathcal{F}_i} \tau(S) = \delta_i$ for all $i \in [n]$. In the following we will describe two different problem dimensions that can be used to define different variants of the base problem described above.

Family Constraints. We consider several restrictions on the family $\mathcal{F}$ that can be used in a solution. Such restrictions may be imposed or may be natural in a specific context in which this general optimization problem is applied, and each will lead to a different specific optimization problem.

(G)iven: The family of sets $\mathcal{F}$ is specified as part of the input. This is useful when the input data already has an imposed structure that the transformation should follow. To ensure feasibility of the problem, we will assume that $\mathcal{F}$ always contains at least all singleton sets.

(D)isjoint: All sets in $\mathcal{F}$ must be disjoint. This allows for parallel execution of all translations, but strongly restricts the space of valid solutions.

(H)ierarchical: The sets in $\mathcal{F}$ must form a hierarchy, meaning that for any two sets $S, S' \in \mathcal{F}$, either $S \subseteq S'$, $S' \subseteq S$, or $S \cap S' = \emptyset$. This restriction allows for a relatively natural staged transformation, where first the entire collection is translated, and then the translation is broken down into smaller disjoint parts, until all points arrive at their intended destination. Note that, with this restriction, $\mathcal{F}$ can efficiently be represented as a rooted tree, where each leaf corresponds to a single point index, and each internal node represents an element of $\mathcal{F}$ corresponding to the set of all indices in the subtree of that node. We will also use this tree representation in our results.

(F)ree: There is no restriction on $\mathcal{F}$, allowing for the best possible transformation. However, it is not directly clear how to effectively visualize the resulting transition.

Optimization Criteria. We also consider two different complexity measures for the transformations represented by $(\mathcal{F}, \tau)$.

(C)ardinality: The number of subsets in $\mathcal{F}$, or $|\mathcal{F}|$. Note that this complexity measure is not meaningful if $\mathcal{F}$ is specified in the input (the "Given" variant).

(L)ength: The total length translated by all groups, computed as $\sum_{S\in\mathcal{F}} \|\tau(S)\|$, where $\|\cdot\|$ indicates some norm in $\mathbb{R}^d$. Unless specified otherwise, we will assume that $\|\cdot\|$ is the Euclidean norm.

Naming Scheme and Organization. We study the complexity of the variants of our optimization problem described above. We will mostly focus on the 1-dimensional and 2-dimensional cases, as they are most relevant for the applications we have in mind. To refer to a specific problem variants, we use acronyms based on the format: *Minimum* [Complexity measure] [Family constraint] *Transformation*, where the complexity measure and family constraint specify one of the options above, using their first letter (in brackets). For example, the MLHT problem refers to the problem where we want to find a solution $(\mathcal{F}, \tau)$ for which $\mathcal{F}$ represents a hierarchy and where $\sum_{S\in\mathcal{F}} \|\tau(S)\|$ is minimized over all (hierarchical) solutions. We also refer to an optimal solution to this problem for some input Δ simply as an *MLHT* of Δ. Some of our results will require further restrictions on the problem, but these will be stated in the respective results.

In Sect. 3 we consider the variants of the problem where $\mathcal{F}$ is specified in the input (the "Given" variant). Sections 4 and 5 cover the remaining family constraints using the different complexity measures: Sect. 4 focuses on problem variants that use the Length complexity measure, and in Sect. 5 we discuss the variants that use the Cardinality complexity measure. We conclude the paper in Sect. 6, where we also describe several additional problem dimensions of the classification that result in open problems for future work. See Table 1 for an overview of our results. Omitted proofs can be found in the full version on arXiv.

Table 1. Overview of results based on our problem classification.

Constraint	d	Cardinality		Length	
Given	$d = 1$	n/a		$O(n)$ if $\mathcal{F}$ is hierarchical	(Thm. 6)
	$d \geq 1$	n/a		Polynomial	(Thm. 2)
Disjoint	$d \geq 1$	$O(dn \log n)$	(Thm. 1)	$O(dn \log n)$	(Thm. 1)
Hierarchical	$d = 1$	$O(n \log n)$	(Thm. 11)	$O(n \log n)$	(Thm. 7)
	$d \geq 2$	$O(dn \log n)$	(Thm. 11)	NP-Hard	(Thm. 8)
Free	$d = 1$	NP-Hard[2]	(Thm. 12+13)	$O(n \log n)$	(Thm. 7)
	$d \geq 2$	NP-Hard[2]	(Thm. 12)	1.307-approximation	(Thm. 10)

[2]This result is for monotone MCFT (see Sect. 5).

Theorem 1. *We can compute the MCDT and the MLDT of a collection Δ of n vectors in $\mathbb{R}^d$ in $O(dn \log n)$ time.*

As for the Disjoint variants of the problem, we state the results directly here, as they are very straightforward. Note that, if $\mathcal{F}$ is disjoint, then every index i can be in at most one set $S \in \mathcal{F}$, and hence $\mathcal{F}_i = \{S\}$. Thus, for a valid solution, we require that $\tau(S) = \delta_i$ for all $i \in S$. As a result, two points can be part of the same group translation if and only if their translation vectors are exactly the same. To optimize for the Cardinality of $\mathcal{F}$, we can simply sort Δ (say, in lexicographical order) and make groups of equal difference vectors. Note that this also directly fixes τ, and hence it also optimizes for Length.

3 Given Family

In this section we consider the MLGT problem, where we are given both Δ and $\mathcal{F}$ as input. We first state a general result on this problem, before we consider an efficient algorithm for a specific variant in more detail.

Theorem 2. *The MLGT problem in $\mathbb{R}^d$ is a convex optimization problem, and can hence be solved in (weak) polynomial time.*

We now consider a variant of the MLGT problem for which we provide an algorithm that runs in linear time. Specifically, we consider the 1-dimensional version of the problem, and we assume that the given family $\mathcal{F}$ is a hierarchy, including all singleton sets. As already explained in Sect. 2 we can hence represent $\mathcal{F}$ as a rooted tree T (see Fig. 2). Furthermore, Δ now simply consists of a collection of real numbers.

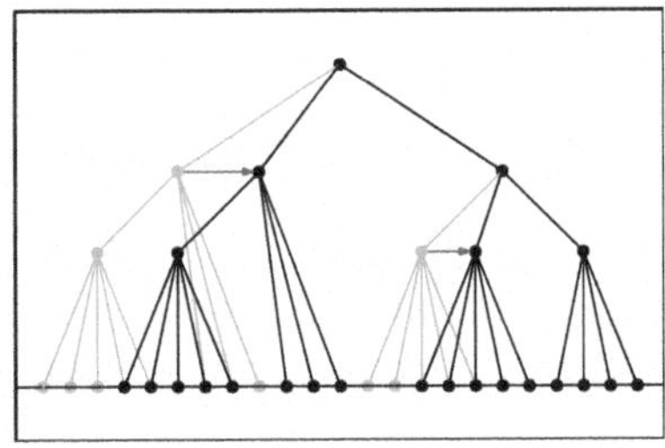

Fig. 2. Translating a group in $\mathcal{F}$ (blue arrow) corresponds to moving an entire subtree. (Color figure online)

This problem variant is closely related to the median-finding problem, which can be seen as follows. Assume that $\mathcal{F}$ consists only of $[n]$ and all singleton sets. If we ignore the cost of the translation of the group $S = [n]$ itself, then the optimal translation for S is exactly a median of all numbers in Δ, because the sum of all individual translations will be minimized. Therefore, this problem can be seen as a generalization of the median-finding problem.

For our algorithm we first define a generalization of a median for a collection of intervals. Let $I = \{[a_1, b_1], \ldots, [a_k, b_k]\}$ be a collection of closed intervals in $\mathbb{R}$. Furthermore, define the distance between $x \in \mathbb{R}$ and an interval $[a, b]$ as:

$$d(x, [a, b]) = \begin{cases} 0, & \text{if } x \in [a, b] \\ x - b, & \text{if } x > b \\ a - x, & \text{if } x < a. \end{cases}$$

Now the *interval median* $M(I)$ of a collection of intervals I is the set of real numbers x for which $\sum_i d(x, [a_i, b_i])$ is minimized.

Lemma 3. *The interval median $M(I)$ of a collection of closed intervals $I = \{[a_1, b_1], \ldots [a_k, b_k]\}$ is itself a closed interval. Furthermore, $\sum_i d(x', [a_i, b_i]) \geq \sum_i d(x, [a_i, b_i]) + d(x', M(I))$ for all $x \in M(I)$ and $x' \in \mathbb{R}$.*

Note that if all intervals consist of a single point, that is, $a_i = b_i$ for all i, then $M(I)$ is simply the interval between the lower and upper median of $\{a_1, \ldots, a_k\}$. We will now extend this concept of an interval median to a hierarchy. Let Δ and $\mathcal{F}$ be our input, where Δ is in $\mathbb{R}$ and $\mathcal{F}$ is a hierarchy including all singleton groups. We represent $\mathcal{F}$ by a rooted tree T where all leaves represent point indices, and all internal nodes represent groups in $\mathcal{F}$ (singleton groups correspond to internal nodes with a single leaf child). For an internal node $v \in T$, let $S(v)$ be the corresponding group in $\mathcal{F}$. For any node $v \in T$, we can now compute $M(v)$ in a bottom-up fashion as follows: If v is a leaf with index i, then $M(v) = [\delta_i, \delta_i]$; otherwise, if v has children $u_1, \ldots, u_k \in T$, then $M(v)$ is the interval median of $\{M(u_1), \ldots, M(u_k)\}$. We can now extend Lemma 3 to trees.

Lemma 4. *Let Δ be a collection of real numbers, and let $\mathcal{F}$ be a given hierarchy represented by a rooted tree T with root r. There exists a nonnegative constant $c \in \mathbb{R}$ such that, for every $x' \in \mathbb{R}$, the minimum value of $\sum_{S \in \mathcal{F}} |\tau(S)|$ over all valid $\tau \colon \mathcal{F} \to \mathbb{R}$ with $\tau(S(r)) = x'$ is at least $d(x', M(r)) + c$.*

Proof. We prove the result by induction on the height of T.
Base Case: T has height 2, and all children $u_1, \ldots, u_k$ have only a single (leaf) child in T. Without loss of generality, we assume that the single child of u_i is the index i, for all $i \in [k]$. For τ to be valid, we require that $\tau(S(u_i)) = \delta_i - x'$, and hence $|\tau(S(u_i))| = d(x', [\delta_i, \delta_i]) = d(x', M(u_i))$. The result now follows directly from Lemma 3 and the construction of $M(r)$, where c constitutes as $\sum_i d(x, [a_i, b_i])$, for some $x \in M(I)$.
Induction Step: Let T be a tree of height $h > 2$, and assume that the statement holds for all trees of height smaller than h. Again, let $u_1, \ldots, u_k$ be the children of r in T. Now consider an optimal valid function τ with $\tau(S(r)) = x'$. If $x' + \tau(S(u_i)) \notin M(u_i)$, then we can adapt $\tau(S(u_i))$ by the minimum amount y such that $x' + \tau(S(u_i)) \in M(u_i)$. By the induction hypothesis on u_i, this will decrease the optimal cost of τ in the subtree rooted at u_i by at least y. Furthermore, the

cost of $\tau(S(u_i))$ can increase by at most y. Thus, we may assume that there exists an optimal valid function τ such that $x' + \tau(S(u_i)) \in M(u_i)$ for all $i \in [k]$. Additionally, we can ensure that $|\tau(S(u_i))| = d(x', M(u_i))$, as the optimal cost in the subtree rooted at u_i is not affected as long as $x' + \tau(S(u_i)) \in M(u_i)$, according to the induction hypothesis. The result now follows directly from Lemma 3 and the construction of $M(r)$.

The result in Lemma 4 can almost directly be used to compute an MLGT for Δ and $\mathcal{F}$. To compute the result efficiently, we use the following lemma.

Lemma 5. *Let $I = \{[a_1, b_1], \ldots, [a_k, b_k]\}$ be a collection of intervals, then the interval median $M(I)$ can be computed in $O(k)$ time.*

The algorithm now simply computes all median intervals in a bottom-up fashion, and then chooses optimal translations inside the median intervals in a top-down fashion. We get the following result.

Theorem 6. *Let Δ be a collection of n real values, and let $\mathcal{F}$ be a given hierarchy. We can compute the MLGT of Δ and $\mathcal{F}$ in $O(n)$ time.*

4 Optimizing Length

In this section we consider the (remaining) variants of the problem that optimize the Length of the transformation, specifically MLHT and MLFT.

We first give a sketch of the optimal solution in the 1D version of the problems, where we are given a collection Δ of real numbers. We define the *span* of a collection $X = \{x_1, \ldots, x_n\}$ of real numbers as $\text{span}(X) = \max_i x_i - \min_i x_i$. The optimal solution for both problems has a length of $\text{span}(\Delta \cup \{0\})$. We can achieve this bound by splitting Δ into a collection Δ^+ containing all values > 0 in Δ and a collection Δ^- containing all values < 0 in Δ. We then sort Δ^+ and iteratively translate all points (using the group $S = \{i \mid \delta_i \in \Delta^+\}$) by the difference between the smallest value in $\{\delta_i \mid i \in S\}$ and the sum of previous translations. After each translation, we remove the index with the smallest value from S, and proceed with the next iteration. We do the same for Δ^-, but then in the other direction. The resulting solution is hierarchical by construction.

Theorem 7. *We can compute an MLFT and an MLHT of a collection Δ of n real numbers in $O(n \log n)$ time.*

In the 2D version of the problem, the input is specified by a collection Δ of vectors in $\mathbb{R}^2$. We first consider the hierarchical version of the problem, namely MLHT. This problem is essentially the same as the *Euclidean Steiner Tree* (EST) problem, in which input consists of a set of points P, and the goal is to compute the shortest tree (measured via Euclidean distance) connecting all points in P, where it is allowed to use points not in P as internal nodes for the tree. It is well known that the EST problem is NP-hard in 2D [12]. We can show that a MLHT of a collection Δ of vectors in $\mathbb{R}^2$ directly corresponds to an EST of the point set $P = \Delta \cup \{(0, 0)\}$. This gives us the following result.

Theorem 8. *The MLHT problem is NP-hard in* $\mathbb{R}^2$.

We next turn our attention to the Free variant of the problem. For this problem variant we do not know whether the problem is NP-hard, or if it can be solved in polynomial time. Instead, we show a very simple and efficient approximation algorithm for the problem.

We first consider a version of the MLFT problem where, instead of using the Euclidean distance to measure the length of a single translation, we use the Manhattan distance. That is, the length of a translation by (x, y) is $|x| + |y|$. We refer to this problem using the Manhattan distance as the *Manhattan MLFT* problem. Now let $(\mathcal{F}, \tau)$ be the optimal solution to the Manhattan MLFT problem for some collection of vectors Δ in $\mathbb{R}^2$. Let $S \in \mathcal{F}$ with $\tau(S) = (x, y)$. We can replace S by two copies S' and S'' of S, and set $\tau(S') = (x, 0)$ and $\tau(S'') = (0, y)$. Since the length of a translation is measured by the Manhattan distance, the resulting solution has exactly the same cost, and is hence still optimal. This implies that the Manhattan MLFT problem can be optimally solved by first solving the problem optimally on x-coordinates, and then solving the problem optimally on y-coordinates independently. By applying Theorem 7 twice, we obtain the following result.

Lemma 9. *The Manhattan MLFT of a collection* Δ *of* n *vectors in* $\mathbb{R}^2$ *can be computed in* $O(n \log n)$ *time.*

Observe that the Manhattan MLFT is also a valid solution for the (Euclidean) MLFT problem. In fact, it can be shown that the Manhattan MLFT is a $\sqrt{2}$-approximation. Instead of proving this result, we present a slight improvement below. Note that we could also use the Manhattan distance for some rotated set of axes. Specifically, let the β-Manhattan MLFT for some collection of vectors Δ be the transformation obtained by first rotating Δ clockwise around the origin by angle β, then computing the Manhattan MLFT $(\mathcal{F}, \tau)$ on the resulting vectors Δ', and then rotating all individual translations in τ counterclockwise by angle β. Observe that the resulting solution $(\mathcal{F}, \tau')$ is again a valid solution for the (Euclidean) MLFT problem.

Theorem 10. *Either the (0-)Manhattan MLFT or the* $\frac{\pi}{4}$*-Manhattan MLFT of some collection* Δ *of vectors in* $\mathbb{R}^2$ *is a c-approximation of the (Euclidean) MLFT of* Δ *with* $c = \sin(\frac{\pi}{8}) + \cos(\frac{\pi}{8}) \approx 1.307$.

Proof. Proof sketch Let $(\mathcal{F}, \tau)$ be the (Euclidean) MLFT of Δ and let L be its total length. Let $S_1, \ldots, S_k$ be the groups in $\mathcal{F}$. For some group S_i let α_i $(0 \leq \alpha_i \leq \frac{\pi}{4})$ be the smallest angle between the vector $\tau(S_i)$ and either the x- or y-axis. If we denote the Euclidean norm using $\|\cdot\|$ and the Manhattan norm using $\|\cdot\|_1$, then, using basic trigonometry, we obtain that $\|\tau(S_i)\|_1 = (\sin(\alpha_i) + \cos(\alpha_i))\|\tau(S_i)\|$. Now let α'_i $(0 \leq \alpha'_i \leq \frac{\pi}{4})$ be the smallest angle between the vector $\tau(S_i)$ and either of the rotated axes of the $\frac{\pi}{4}$-Manhattan

MLFT. Observe that $\alpha_i + \alpha'_i = \frac{\pi}{4}$. Consider the following sum:

$$\sum_{i=1}^{k} \alpha_i \|\tau(S_i)\| + \sum_{i=1}^{k} \alpha'_i \|\tau(S_i)\| =$$
$$\sum_{i=1}^{k} (\alpha_i + \alpha'_i) \|\tau(S_i)\| = \frac{\pi}{4} L$$

Thus either $\sum_{i=1}^{k} \alpha_i \|\tau(S_i)\|$ or $\sum_{i=1}^{k} \alpha'_i \|\tau(S_i)\|$ is at most $\frac{\pi}{8} L$. Without loss of generality, assume that this holds for the first sum. We now consider the sum $\sum_{i=1}^{k} (\sin(\alpha_i) + \cos(\alpha_i)) \|\tau(S_i)\|$, which corresponds to the length of $(\mathcal{F}, \tau)$ in Manhattan distance. As the function $\sin(\alpha) + \cos(\alpha)$ is strictly concave for $0 \leq \alpha \leq \frac{\pi}{4}$ (and the sum of angles is bounded), the worst case is obtained when all angles α_i are equal, that is, $\alpha_i = \frac{\pi}{8}$ for all $1 \leq i \leq k$. Hence, the length of $(\mathcal{F}, \tau)$ in Manhattan distance is at most $\sum_{i=1}^{k} (\sin(\frac{\pi}{8}) + \cos(\frac{\pi}{8})) \|\tau(S_i)\| = (\sin(\frac{\pi}{8}) + \cos(\frac{\pi}{8})) L$. Since the Manhattan MLFT is at least as good as $(\mathcal{F}, \tau)$, we can conclude that the Manhattan MLFT is a c-approximation of the (Euclidean) MLFT with $c = \sin(\frac{\pi}{8}) + \cos(\frac{\pi}{8})$.

Based on Theorem 10, the following natural question arises: can the approximation factor be improved by considering the β-Manhattan MLFT for even more angles? Although this may indeed be the case, we sketch that there is a limit to this approach using a simple example. Consider the collection of vectors/points Δ obtained by densely sampling points along two quarter arcs of the unit circle, as depicted in Fig. 3. For this example it is important that the points on both arcs are sampled such that there is a rotational symmetry around the point $(0.5, 0.5)$. It is easy to verify that, regardless of the angle β, the length of the β-Manhattan MLFT approaches 2 when the arcs are sampled densely enough. However, we can construct a (Euclidean) MLFT $(\mathcal{F}, \tau)$ by choosing appropriate groups S_i along with translations $\tau(S_i)$ corresponding to the difference vectors between all consecutive points along the bottom arc. This allows us to also construct the vectors on the top arc, if we choose the groups S_i carefully. The total length of $(\mathcal{F}, \tau)$ approaches $\frac{\pi}{2}$ when the arcs are sampled densely enough. Thus, even if we consider the β-Manhattan MLFT for all angles β, we can never achieve an approximation ratio better than $\frac{4}{\pi} \approx 1.273$.

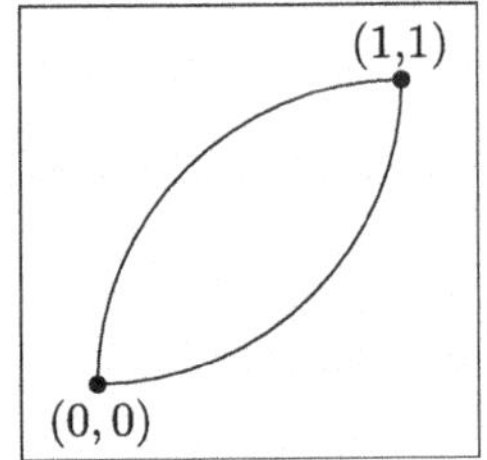

Fig. 3. Two quarter arcs.

Finally, we want to observe that the Manhattan distance does not always make our base problem easier. If we consider the hierarchical version of the problem, then we can simply follow the same reduction as in Theorem 8, but instead reducing from the Rectilinear Steiner Tree Problem. Since this problem is also NP-hard [13], it is also NP-hard to compute the Manhattan MLHT in $\mathbb{R}^2$.

5 Optimizing Cardinality

In this section we consider the variants of the problem that minimize the number of groups in $\mathcal{F}$. Before we discuss the more challenging Free variants, we first briefly discuss the hierarchical variant. Let $(\mathcal{F}, \tau)$ be the MCHT of some collection of vectors Δ in $\mathbb{R}^d$. Assume that there are two groups S and S' in $\mathcal{F}$ such that $S \subset S'$, and there is no $S'' \in \mathcal{F}$ such that $S \subset S'' \subset S'$. We can then replace $\tau(S)$ by $\tau(S) + \tau(S')$ and replace S' by $S' \setminus S$. If $S' \setminus S$ was already in $\mathcal{F}$, then we can simply add $\tau(S')$ to $\tau(S' \setminus S)$ (if $S' \setminus S = \emptyset$, then the group can simply be removed). Note that this operation does not invalidate the solution and also does not increase the number of groups in $\mathcal{F}$. Furthermore, it breaks the subset relation between S and S' and does not introduce any new subset relations between groups. Thus, we can repeat this operation until all groups are disjoint. Finally observe that a disjoint family of groups is also hierarchical. The following result now follows from Theorem 1.

Theorem 11. *We can compute the MCHT for a collection Δ of n vectors in $\mathbb{R}^d$ in $O(dn \log n)$ time.*

We will now consider the Free variant (MCFT) of this problem. Although we are unable to establish the algorithmic complexity of this problem, we will show that a restricted version of the problem, the *monotone* MCFT, is NP-hard. In the monotone version of the problem, all components of vectors in Δ must be non-negative, and the same requirement holds for all vectors in τ in a solution $(\mathcal{F}, \tau)$. In the following we first allow that the dimension of Δ is specified as part of the input. Then we show NP-hardness via a reduction from Vertex Cover [19]. In the Vertex Cover problem we are given a graph $G = (V, E)$ and a number k, and the goal is to determine if there exists a subset $C \subseteq V$ with $|C| \leq k$ such that, for every edge $(u, v) \in E$, either $u \in C$ or $v \in C$. Our reduction is very similar to the NP-hardness proof of the Normal Set Basis problem by Jiang and Ravikumar [18], but requires some extra work.

Theorem 12. *The monotone MCFT problem in $\mathbb{R}^d$ is NP-hard if d is given as part of the input.*

Proof sketch. We use a reduction from Vertex Cover. For a graph $G = (V, E)$ we construct an input Δ for the monotone MCFT problem as follows. We (uniquely) assign 2 dimensions to every vertex of G and 4 dimensions to every edge of G. Let $x_1(v)$, $x_2(v)$ for every $v \in V$ and $x_1(e), x_2(e), x_3(e), x_4(e)$ for every $e \in E$ be the vectors that have a single 1 in the dimension assigned to them, and 0 in all other dimensions. As a result, $d = 2|V| + 4|E|$. For every $v \in V$ we add $\delta(v) = x_1(v) + x_2(v)$ to Δ. For every edge $e = (u, v) \in E$ we add to Δ:

$$\delta_1(e) = x_1(u) + x_1(e) + x_2(e), \qquad \delta_2(e) = x_2(v) + x_2(e) + x_3(e),$$

$$\delta_3(e) = x_2(u) + x_3(e) + x_4(e), \qquad \delta_4(e) = x_1(v) + x_4(e) + x_1(e),$$

$$\delta_5(e) = x_1(e) + x_2(e) + x_3(e) + x_4(e)$$

We can then show that G has a vertex cover of size at most k if and only if Δ has a monotone MCFT $(\mathcal{F}, \tau)$ with $|\mathcal{F}| \leq n + 4|E| + k$. The main idea of the argument is that we need to use at least one group in $\mathcal{F}$ for each vertex of G and at least four groups in $\mathcal{F}$ for each edge of G. Furthermore, we need at least five groups in $\mathcal{F}$ for an edge $e = (u, v)$, unless we can use two groups (instead of one) for either u or v. The vertices that use two groups in $\mathcal{F}$ then directly correspond to a vertex cover in G.

Theorem 12 does not apply to cases where the number of dimensions d of Δ is small. However, we show that the (monotone) d-dimensional version of the problem, even when d is polynomial in n, can be reduced to the 1-dimensional version of the problem under some mild assumptions.

Theorem 13. *There exist a polynomial time reduction of the (monotone) MCFT problem in d dimensions to the (monotone) MCFT problem in 1 dimension, assuming that d is polynomial in n and the input Δ consists of bounded integer vectors in $\mathbb{R}^d$.*

Proof (Sketch). . We construct a linear function $f\colon \mathbb{R}^d \to \mathbb{R}$ to map vectors from the d-dimensional input Δ to the 1-dimensional input Δ'. Due to the linearity of f, any solution $(\mathcal{F}, \tau)$ for Δ can be mapped to a valid solution $(\mathcal{F}, \tau')$ for Δ' by simply using the the function f on the vectors in τ. The main challenge is to show that we can also construct a (linear) inverse mapping f^{-1} that can map an optimal solution for Δ' to a valid solution for Δ of the same cost. Now let $\delta_i \in \Delta$ be denoted as $(\delta_i^1, \ldots, \delta_i^d)$, then f is defined as follows:

$$f(\delta_i) = \sum_{j=1}^{d} \delta_i^j A^{j-1},$$

where $A = 3(n+1)!^2 M$, and M is the maximum absolute value of integers in Δ. This function f also preserves the monotonicity of a solution. Furthermore, the image of f consists of integers with absolute value at most A^d. Hence, the bit complexity of these numbers is at most $\log(A^d) = d\log(3(n+1)!^2 M) = O(d(n \log n + \log M))$. This implies that, even if d depends polynomially on n, the reduction can be computed in polynomial time.

The conditions of Theorem 13 also apply to the instances in the reduction in Theorem 12, and hence we can conclude that the monotone MCFT problem is also NP-hard in 1 dimension. However, the complexity of the non-monotone version of the problem remains an open problem. Intuitively, this version of the problem seems harder than the monotone version, but the reduction in Theorem 12 heavily uses the fact that the problem is monotone. Nonetheless, we believe that the same reduction should also work for the non-monotone version, but that the challenge lies in arguing the correctness of the reduction. We therefore conjecture that the (non-monotone) MCFT problem is also NP-hard.

6 Conclusion and Future Work

We have introduced a new class of problems for measuring the visual complexity of point set mappings, where group translations are treated as single operations. We are able to establish the algorithmic complexity of most of the problems in our problem classification, except for two: the (non-monotone) MCFT problem and the (Euclidean) MLFT problem. We believe that the former problem is NP-hard, but that is not so clear for the MLFT problem, where we believe a polynomial-time algorithm may be possible. Beyond that, our problem classification can be expanded in several natural ways, leading to new open questions:

Stages: Disjoint transformations have the advantage that they can be animated simultaneously, but they are very restricted. We can make them more powerful by considering several disjoint transformations in sequence, as different *stages*. A new optimization criterion may then be to minimize this number of stages in a transformation.

Unlabeled Point Sets: In some cases, the provided point sets may be unlabeled. Additionally identifying a labeling that optimizes the transformation adds another layer of complexity to the described problems.

Transformations: Instead of translations we may also consider other natural transformations like rotations and scaling. As there are multiple ways to scale or rotate one point to another, these new transformations increase the complexity of the problem significantly.

References

1. Alexa, M., Cohen-Or, D., Levin, D.: As-rigid-as-possible shape interpolation. In: Proceedings of the 27th Annual Conference on Computer Graphics and Interactive Techniques, pp. 157 – 164. SIGGRAPH 2000, ACM Press/Addison-Wesley Publishing Co., USA (2000)
2. Alt, H., Behrends, B., Blömer, J.: Approximate matching of polygonal shapes. Ann. Math. Artif. Intell. **13**, 251–265 (1995)
3. Alt, H., Guibas, L.J.: Discrete geometric shapes: Matching, interpolation, and approximation. In: Handbook of Computational Geometry, pp. 121–153. Elsevier (2000)
4. Archambault, D., Purchase, H.C.: Mental map preservation helps user orientation in dynamic graphs. In: Didimo, W., Patrignani, M. (eds.) GD 2012. LNCS, vol. 7704, pp. 475–486. Springer, Heidelberg (2013). https://doi.org/10.1007/978-3-642-36763-2_42
5. Arun, K.S., Huang, T.S., Blostein, S.D.: Least-squares fitting of two 3-D point sets. IEEE Trans. Pattern Anal. Mach. Intell. **PAMI-9**(5), 698–700 (1987)
6. Besl, P.J., McKay, N.D.: A method for registration of 3-D shapes. IEEE Trans. Pattern Anal. Mach. Intell. **14**(2), 239–256 (1992)
7. Cohen, S., Guibas, L.: The Earth Mover's distance under transformation sets. In: Proceedings of the 7th IEEE International Conference on Computer Vision, vol. 2, pp. 1076–1083. IEEE (1999)

8. Du, F., Cao, N., Zhao, J., Lin, Y.R.: Trajectory bundling for animated transitions. In: Proceedings of the 33rd Annual ACM Conference on Human Factors in Computing Systems, pp. 289 – 298. CHI 2015, Association for Computing Machinery, New York, NY, USA (2015)
9. Duckham, M., et al.: Modeling checkpoint-based movement with the earth mover's distance. In: Miller, J.A., O'Sullivan, D., Wiegand, N. (eds.) GIScience 2016. LNCS, vol. 9927, pp. 225–239. Springer, Cham (2016). https://doi.org/10.1007/978-3-319-45738-3_15
10. Eppstein, D., van Kreveld, M., Speckmann, B., Staals, F.: Improved grid map layout by point set matching. Int. J. Comput. Geom. Appl. **25**(02), 101–122 (2015)
11. Franconeri, S., Lin, J., Pylyshyn, Z., Fisher, B., Enns, J.: Evidence against a speed limit in multiple-object tracking. Psychon. Bull. Rev. **15**, 802–808 (2008)
12. Garey, M.R., Graham, R.L., Johnson, D.S.: The complexity of computing Steiner minimal trees. SIAM J. Appl. Math. **32**(4), 835–859 (1977)
13. Garey, M.R., Johnson, D.S.: The rectilinear Steiner tree problem is NP-complete. SIAM J. Appl. Math. **32**(4), 826–834 (1977)
14. Gotsman, C., Surazhsky, V.: Guaranteed intersection-free polygon morphing. Comput. Graph. **25**(1), 67–75 (2001)
15. Heer, J., Robertson, G.: Animated transitions in statistical data graphics. IEEE Trans. Visual Comput. Graphics **13**(6), 1240–1247 (2007)
16. Horn, B.: Closed-form solution of absolute orientation using unit quaternions. J. Opt. Soc. A **4**, 629–642 (1987)
17. Jian, B., Vemuri, B.C.: Robust point set registration using Gaussian mixture models. IEEE Trans. Pattern Anal. Mach. Intell. **33**(8), 1633–1645 (2011)
18. Jiang, T., Ravikumar, B.: Minimal NFA problems are hard. SIAM J. Comput. **22**(6), 1117–1141 (1993)
19. Karp, R.M.: Reducibility among combinatorial problems. In: Miller, R.E., Thatcher, J.W., Bohlinger, J.D. (eds.) Complexity of Computer Computations. The IBM Research Symposia Series, pp. 85–103. Springer, Boston, MA (1972). https://doi.org/10.1007/978-1-4684-2001-2_9
20. Li, L., Yang, M., Wang, C.: Graph correspondence-based point set registration. IEEE Trans. Syst. Man Cybern. Syst. **54**(7), 4101–4112 (2024)
21. Liu, Z., Zhou, L., Leung, H., Shum, H.P.: High-quality compatible triangulations and their application in interactive animation. Comput. Graph. **76**, 60–72 (2018)
22. Meulemans, W.: Efficient optimal overlap removal: algorithms and experiments. Comput. Graph. Forum **38**, 713–723 (2019)
23. Meulemans, W., Dykes, J., Slingsby, A., Turkay, C., Wood, J.: Small multiples with gaps. IEEE Trans. Visual Comput. Graphics **23**(1), 381–390 (2017)
24. Mizuno, K., Wu, H.Y., Takahashi, S., Igarashi, T.: Optimizing stepwise animation in dynamic set diagrams. In: Computer Graphics Forum, vol. 38, pp. 13–24. Wiley Online Library (2019)
25. Munzner, T.: Visualization Analysis and Design. AK Peters/CRC Press (2014)
26. Myronenko, A., Song, X.: Point set registration: coherent point drift. IEEE Trans. Pattern Anal. Mach. Intell. **32**(12), 2262–2275 (2010)
27. Rubner, Y., Tomasi, C., Guibas, L.: The earth mover's distance as a metric for image retrieval. Int. J. Comput. Vis. **40**, 99–121 (2000)
28. Suiker, K.N.: Optimizing staged transition for scatter plots. Master's thesis, TU Eindhoven (2020)
29. Tversky, B.: Cognitive maps, cognitive collages, and spatial mental models. In: Frank, A.U., Campari, I. (eds.) COSIT 1993. LNCS, vol. 716, pp. 14–24. Springer, Heidelberg (1993). https://doi.org/10.1007/3-540-57207-4_2

30. Wang, Y., Archambault, D., Scheidegger, C.E., Qu, H.: A vector field design approach to animated transitions. IEEE Trans. Visual Comput. Graphics **24**(9), 2487–2500 (2018)
31. Wertheimer, M.: Drei Abhandlungen zur Gestalttheorie. Palm and Enke, Erlangen, Germany (1925)

Online and Offline Algorithms for Counting Distinct Closed Factors via Sliding Suffix Trees

Takuya Mieno[1(✉)], Shun Takahashi[2], Kazuhisa Seto[2], and Takashi Horiyama[2]

[1] University of Electro-Communications, Chofu, Japan
tmieno@uec.ac.jp
[2] Hokkaido University, Sapporo, Japan
{seto,horiyama}@ist.hokudai.ac.jp

Abstract. A string is said to be closed if its length is one, or if it has a non-empty factor that occurs both as a prefix and as a suffix of the string, but does not occur elsewhere. The notion of closed words was introduced by [Fici, WORDS 2011]. Recently, the maximum number of distinct closed factors occurring in a string was investigated by [Parshina and Puzynina, Theor. Comput. Sci. 2024], and an asymptotic tight bound was proved. In this paper, we propose two algorithms to count the distinct closed factors in a string T of length n over an alphabet of size σ. The first algorithm runs in $O(n \log \sigma)$ time using $O(n)$ space for string T given in an online manner. The second algorithm runs in $O(n)$ time using $O(n)$ space for string T given in an offline manner. Both algorithms utilize suffix trees for sliding windows.

Keywords: string algorithms · closed words · sliding suffix trees

1 Introduction

String processing is a fundamental area in computer science, with significant importance ranging from theoretical foundations to practical applications. One of the most active areas of this field is the study of repetitive structures within strings, which has driven advances in areas such as pattern matching algorithms [20,23] and compressed string indices [9,22,24]. Understanding repetitive structures in strings is important for the advancement of information processing technology. For surveys on these topics, see [14,34] and [29,30]. The concept of *closed words* [18] is a sort of such repetitive structures of strings. A string is said to be *closed* if its length is one, or if it has a non-empty factor that occurs both as a prefix and as a suffix of the string, but does not occur elsewhere[1]. For example, string abaab is closed because ab occurs both as a prefix and as a suffix, but

[1] The notion of closed words is equivalent to those of return words [15,21] and periodic-like words [13].

R. Královič and V. Kůrková (Eds.): SOFSEM 2025, LNCS 15539, pp. 172–183, 2025.
https://doi.org/10.1007/978-3-031-82697-9_13

does not occur elsewhere in abaab. Closed words have been studied primarily in the field of combinatorics on finite and infinite words [5,6,12,18,19,27,31,32,37]. Regarding the number of closed *factors* (i.e., substrings) appearing in a string, an asymptotic tight bound $\Theta(n^2)$ on the maximum number of distinct closed factors of a string is known [5]. More recently, Parshina and Puzynina refined this bound to $\sim \frac{n^2}{6}$ in 2024 [31]. Despite these progresses on the number of closed factors, there is no non-trivial algorithm for computing the exact number of distinct closed factors of a given string to our knowledge.

In this paper, we present both online and offline algorithms for counting the number of distinct closed factors of a given string T of length n. The first counting algorithm is an online approach performed in $O(n \log \sigma)$ time and $O(n)$ space, where σ is the number of distinct characters in the string. The second counting algorithm is an offline approach performed in $O(n)$ time and $O(n)$ space, assuming T is drawn from an integer alphabet of size $n^{O(1)}$. We begin by characterizing the number of distinct closed factors of T through the *repeating suffixes* of some prefixes and factors of T. Based on this characterization, we design an online algorithm that utilizes Ukkonen's online suffix tree construction [36], as well as suffix trees for a sliding window [17,25,26,33]. We then design a linear-time offline algorithm by simulating sliding-window suffix trees within the static suffix tree of the entire string T. This simulation is of independent interest, as it has the potential to speed up sliding-window algorithms for strings in an offline setting. Furthermore, we explore the enumeration of (distinct) closed factors in a string and propose an algorithm that combines our counting method with a geometric data structure for handling points in the two-dimensional plane [11], resulting a somewhat faster solution.

Related Work. Recent work has highlighted algorithmic advances in the study of closed factors and related problems over the past decade [1,2,4,6–8,28,35]. The line of algorithmic research on closed factors is initialized by Badkobeh et al., [3,4], who addressed various problems related to factorizing a string into a sequence of closed factors and proposed efficient algorithms for these tasks. In the domain of online string algorithms, Alzamel et al. [2] proposed an algorithm that computes closed factorizations in an online settings, and more recently, Sumiyoshi et al. [35] achieved a speedup of $\log \log n$ times in total execution time. Additionally, a new concept, *maximal closed substrings* (MCS), was introduced in [6], along with an $O(n \log n)$-time enumeration algorithm [7].

2 Preliminaries

2.1 Basic Notations

Let Σ be an ordered *alphabet*. An element of Σ is called a *character*. An element of $\Sigma^\star$ is called a *string*. The *empty string*, denoted by ε, is the string of length 0. For a string $T \in \Sigma^\star$, the length of T is denoted by $|T|$. If $T = xyz$ for some strings $x, y, z \in \Sigma^\star$, we call x, y, and z a *prefix*, a *factor*, and a *suffix* of T, respectively. A string $b \neq T$ is said to be a *border* of T if b is both a prefix of

T and a suffix of T. We denote by $bord(T)$ the longest border of T. Note that the longest border always exists since ε is a border of any string. For each i with $1 \leq i \leq |T|$, we denote by $T[i]$ the ith character of T. For each i, j with $1 \leq i \leq j \leq |T|$, we denote by $T[i..j]$ the factor of T that starts at position i and ends at position j. For convenience, we define $T[i..j] = \varepsilon$ for any $i > j$. For a strings T and S, the set $occ_T(S) = \{i \mid T[i..i+|S|-1] = S\}$ of integers is said to be the *occurrences* of S in T. If $occ_T(S) \neq \emptyset$, we say that S *occurs* in T as a factor. A string T is said to be *closed* if there is a border of T that occurs exactly twice in T. We say that a suffix z of T is a closed suffix of T if z is closed. If $|occ_T(S)| = 1$, we say that S is a *unique* factor of T. Also, if $|occ_T(S)| \geq 2$, we say that S is a *repeating* factor of T. We denote by $lrs(T)$ the longest repeating suffix of string T. Further we denote $lrs^2(T) = lrs(lrs(T))$.

In what follows, we fix a non-empty string T of arbitrary length n over an alphabet Σ of size σ. This paper assumes the standard word RAM model with word size $\Omega(\log n)$.

2.2 Suffix Trees

The most important tool of this paper is a *suffix tree* of string T [38]. A suffix tree of T, denoted by $\mathsf{STree}(T)$, is a *compact trie* for the set of suffixes of T. Below, we summarize some known properties of suffix trees and define related notations:

- Each edge of $\mathsf{STree}(T)$ is labeled with a factor of T of length one or more.
- Each internal node u, including the root, has at least two children, and the first characters of the labels of outgoing edges from u are mutually distinct (unless T is a unary string).
- For each node v of $\mathsf{STree}(T)$, we denote by $\mathsf{str}_T(v)$ the string spelled out from the root to v. We also define $\mathsf{strlen}_T(u) = |\mathsf{str}_T(u)|$ for a node u.
- There is a one-to-one correspondence between leaves of $\mathsf{STree}(T)$ and unique suffixes of T. More precisely, for each leaf ℓ, the string $\mathsf{str}_T(\ell)$ equals some unique suffix of T, and vice versa. Note that repeating suffixes of T are not always represented by a node in $\mathsf{STree}(T)$.
- For a substring w of T, the *locus* (v, l) of w is either node v or a location on edge (u, v) such that $\mathsf{str}_T(u)$ is a prefix of w, w is a prefix of $\mathsf{str}_T(v)$, and $|w| = |\mathsf{str}_T(u)| + l$, where u is the parent of v.

Throughout this paper, we assume for convenience that the last character of T is unique. For each $1 \leq i \leq n$, we denote by $\mathsf{leaf}_T(i)$ the leaf ℓ of $\mathsf{STree}(T)$ such that $\mathsf{str}_T(\ell) = T[i..n]$.

It is known that $\mathsf{STree}(T)$ for given string T can be constructed in $O(n)$ time [16] if Σ is *linearly sortable*[2], i.e., any n characters from Σ can be sorted in $O(n)$ time. Additionally, when the input string T is given in an *online* manner,

[2] A typical example of linearly sortable alphabets is an integer alphabet $\{1, 2, \ldots, n^c\}$ for some constant c. Any n characters from the alphabet can be sorted in $O(n)$ time using radix sort.

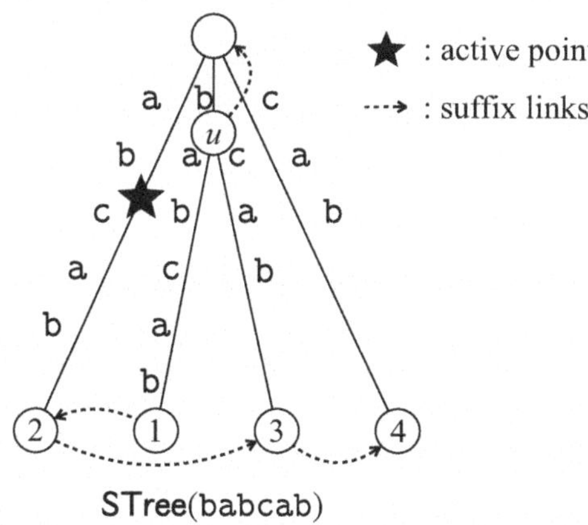

Fig. 1. The suffix tree of string $T = \mathtt{babcab}$. Each leaf of the tree represents a suffix of T, with the integer inside each leaf indicating the starting position of the suffix. Namely, the leaf labeled with number i corresponds to $\mathsf{leaf}_T(i)$. In this suffix tree, $\mathsf{str}_T(u) = \mathtt{b}$, $\mathsf{strlen}_T(u) = 1$, $\mathsf{str}_T(\mathsf{leaf}_T(3)) = \mathtt{bcab}$, and $\mathsf{strlen}_T(\mathsf{leaf}_T(3)) = 4$. The star symbol indicates the locus of the active point, which represents the longest repeating suffix $\mathtt{ab}$ of T. The dotted arrows represent suffix links.

$\mathsf{STree}(T)$ can be constructed in $O(n \log \sigma)$ time using Ukkonen's algorithm [36]. In both algorithms, the resulting suffix trees are edge-sorted, meaning that the first characters of the labels of outgoing edges from each node are (lexicographically) sorted. Thus we assume that all suffix trees in this paper are edge-sorted.

Theorem 1 ([36]). *For incremental $i = 1, 2, \ldots, n$, we can maintain the suffix tree of $T[1..i]$ and the length of $lrs(T[1..i])$ in a total of $O(n \log \sigma)$ time.*

In other words, for every $j = 1, 2, \ldots, n-1$, we can update $\mathsf{STree}(T[1..j])$ to $\mathsf{STree}(T[1..j+1])$ in amortized $O(\log \sigma)$ time. Ukkonen's algorithm maintains the *active point* in the suffix tree, which is the locus of the longest repeating suffix of the string. To maintain the active point efficiently, Ukkonen's algorithm uses auxiliary data structures called *suffix links*: Each internal node u of the suffix tree has a suffix link that points to the node v, such that $\mathsf{str}_T(v)$ is the suffix of $\mathsf{str}_T(u)$ of length $|\mathsf{str}_T(u)| - 1$. See Fig. 1 for examples of a suffix tree and related notations.

Furthermore, based on Ukkonen's algorithm, we can maintain suffix trees for a *sliding window* over T in a total of $O(n \log \sigma)$ time, using space proportional to the window size:

Theorem 2 ([17,25,26,33]). *Using $O(W)$ space, a suffix tree for a variable-width sliding window can be maintained in amortized $O(\log \sigma)$ time per one operation: either (1) appending a character to the right or (2) deleting the leftmost character from the window, where W is the maximum width of the window. Additionally, the length of the longest repeating suffix of the window can be maintained with the same time complexity.*

In other words, for every incremental pair of indices i, j with $i \leq j$, we can update $\mathsf{STree}(T[i..j])$ to either $\mathsf{STree}(T[i..j+1])$ or $\mathsf{STree}(T[i+1..j])$ in amortized $O(\log \sigma)$ time. Later, we use the above algorithmic results as black boxes.

2.3 Weighted Ancestor Queries

For a node-weighted tree $\mathcal{T}$, where the weight of each non-root node is greater than the weight of its parent, a *weighted ancestor query* (WAQ) is defined as follows: given a node u of the tree and an integer t, the query returns the farthest (highest) ancestor of u whose weight is at least t. We denote by $\mathsf{WAQ}_{\mathcal{T}}(u, t)$ the returned node for weighted ancestor query (u, t) on tree $\mathcal{T}$. The subscript $\mathcal{T}$ will be omitted when clear from the context. In general, a WAQ on a tree of size N can be answered in $O(\log \log N)$ time, which is worst-case optimal with linear space. However, if the input is the suffix tree of some string of length n, and the weight function is strlen_T, as defined in the previous subsection, any WAQ can be answered in $O(1)$ time after $O(n)$-time preprocessing [10].

3 Counting Closed Factors Online

In this section, we propose an algorithm to count the distinct closed factors for a string given in an online manner. Since there may be $\Omega(n^2)$ distinct closed factors in a string of length n [5,31], enumerating them requires quadratic time in the worst case. However, for counting the closed factors, we can achieve sub-quadratic time even when the string is given in an online manner, as described below.

3.1 Changes in Number of Closed Factors

First we consider the changes in the number of distinct closed factors when a character is appended. Let $\mathcal{C}(T)$ be the set of distinct closed factors occurring in T. Let $d_j = |\mathcal{C}(T[1..j])| - |\mathcal{C}(T[1..j-1])|$ for $2 \leq j \leq n$. For convenience let $d_1 = 1$.

Observation 1. *For any closed suffixes u and v of the same string, $|bord(u)| \neq |bord(v)|$ holds if $u \neq v$.*

Based on this observation, we count the distinct borders of closed factors instead of the closed factors themselves. We show the following:

Lemma 1. *For each j with $1 \leq j \leq n$, $d_j = |lrs(T[1..j])| - |lrs^2(T[1..j]))|$ holds if $lrs(T[1..j]) \neq \varepsilon$. Also, $d_j = 1$ holds if $lrs(T[1..j]) = \varepsilon$.*

Proof. The latter case is obvious since $lrs(T[1..j]) = \varepsilon$ implies that $T[j]$ is a unique character in $T[1..j]$, making it the only new closed factor. In the following, we assume $lrs(T[1..j]) \neq \varepsilon$ and denote $t_j = lrs(T[1..j])$ and $z_j = lrs^2(T[1..j])$. It can be observed that for any new closed factor $u \in \mathcal{C}(T[1..j]) \setminus \mathcal{C}(T[1..j-1])$,

u is a unique suffix of $T[1..j]$ and the longest border of u repeats in $T[1..j]$. Equivalently, $|u| > |t_j|$ and $|bord(u)| \leq |t_j|$ hold. Additionally, $|u| > |t_j|$ implies $|bord(u)| > |z_j|$. Thus $|z_j| < |bord(u)| \leq |t_j|$, and hence $d_j \leq |t_j| - |z_j|$ by Observation 1. Conversely, for any suffix b of $T[1..j]$ satisfying $|z_j| < |b| \leq |t_j|$, there exists exactly one closed suffix whose longest border is b since b is repeating in $T[1..j]$. Further, since $|b| > |z_j|$, the closed suffix is longer than t_j, and thus the closed suffix is unique in $T[1..j]$. Therefore, $d_j \geq |t_j| - |z_j|$ holds.

We note that the upper bound $d_j \leq |lrs(T[1..j])| - |lrs^2(T[1..j]))|$ of d_j is already claimed in [31], however, no proof was given and no lower bound was claimed. By definition of d_j and Lemma 1, the next corollary immediately follows:

Corollary 1. *The total number of distinct closed factors of T is*

$$\sum_{j \in J} |lrs(T[1..j])| - \sum_{j \in J} |lrs^2(T[1..j]))| + n - |J|$$

where $J = \{j \in [1, n] \mid lrs(T[1..j]) \neq \varepsilon\}$.

3.2 Online Algorithm

By Theorem 1, we can compute $\sum_{j \in J} |lrs(T[1..j])|$ in $O(n \log \sigma)$ time online. Thus, by Corollary 1, our remaining task is to compute $\sum_{j \in J} |lrs^2(T[1..j])|$ online, equivalently, to compute the sequence $\mathcal{Z} = (|z_1|, |z_2| \ldots, |z_n|)$ where $z_j = lrs^2(T[1..j])$ for each j. To compute the sequence, we use sliding suffix trees of Theorem 2. It is known that the starting position of $lrs(T[1..j])$ is not smaller than that of $lrs(T[1..j-1])$. Namely, starting from $t_1 = lrs(T[1..1]) = \varepsilon$, the sequence $\langle t_1, t_2, \ldots, t_n \rangle$ of factors of T can be obtained by $O(n)$ *sliding-operations* on T that consists of (1) appending a character to the right or (2) deleting the leftmost character from the factor, where $t_j = lrs(T[1..j])$ for each j. Thus, by Theorem 2, we can obtain every $|z_i| \in \mathcal{Z}$ in amortized $O(\log \sigma)$ time by appropriately applying sliding-operations to reach the windows t_j for all j with $j \in J$ in order. Finally, we obtain the following:

Theorem 3. *For a string T of length n given in an online manner, we can count the distinct closed factors of T in $O(n \log \sigma)$ time using $O(n)$ space where σ is the alphabet size.*

4 Counting Closed Factors Offline: Shaving Log Factor

In this section, we assume that the alphabet is linearly sortable, allowing us to construct $\mathsf{STree}(T)$ in $O(n)$ time offline [16]. Our goal is to remove the $\log \sigma$ factor from the complexity in Theorem 3, which arises from using dynamic predecessor dictionaries (e.g., AVL trees) at non-leaf nodes. To shave this logarithmic factor, we do not rely on such dynamic dictionaries. Instead, use a WAQ data structure constructed over $\mathsf{STree}(T)$.

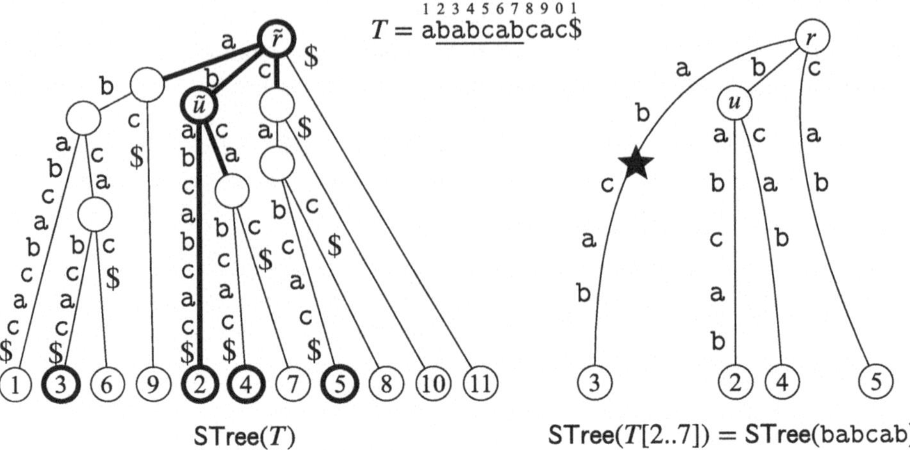

Fig. 2. The suffix tree of string $T =$ ababcabcac\$ is shown on the left. The suffix tree of $T[2..7] =$ babcab, which is the same as the one in Fig. 1, is shown on the right. Suffix links are omitted in this figure. In the tree on the left, each node and edge enclosed in a bold line is connected to a corresponding one in the tree on the right. For clarity, those connections are not drawn.

Simulating Sliding Suffix Tree in Linear Time. The idea is to simulate the online algorithm from Sect. 3 using the suffix tree of the entire string T. In algorithms underlying Theorem 1 and 2, the topology of suffix trees and the loci of the active points change step-by-step. The changes include (i) moving the active point, (ii) adding a node, an edge, or a suffix link, and (iii) removing a node, an edge, or a suffix link.

Our data structure consists of two suffix trees:

(1) The suffix tree $\mathsf{STree}(T)$ of T enhanced with a WAQ data structure.
(2) The suffix tree $\mathsf{STree}(w)$ of the factor $w = T[i..j]$, which represents a sliding window.

Note that $\mathsf{STree}(w)$ includes the active point. All nodes and edges of $\mathsf{STree}(w)$ are connected to their corresponding nodes and edges in $\mathsf{STree}(T)$. Specifically, an internal node u of $\mathsf{STree}(w)$ is connected to the node $\tilde{u}$ of $\mathsf{STree}(T)$ where $\mathsf{str}_w(u) = \mathsf{str}_T(\tilde{u})$. Similarly, an edge (u, v) of $\mathsf{STree}(w)$ is connected to the edge $(\tilde{u}, v')$ of $\mathsf{STree}(T)$ where the first characters of the edge-labels are the same. Such node $\tilde{u}$ and edge $(\tilde{u}, v')$ in $\mathsf{STree}(T)$ always exist: An internal node u implies that, for some distinct characters c_1 and c_2, $\mathsf{str}_w(u)c_1$ and $\mathsf{str}_w(u)c_2$ occur in w, and thus also in T. Furthermore, a leaf ℓ of $\mathsf{STree}(w)$, representing suffix $T[k..j]$ of w, is connected to the leaf of $\mathsf{STree}(T)$ representing the suffix $T[k..n]$ of T. Note that suffix links of $\mathsf{STree}(w)$ do not connect to $\mathsf{STree}(T)$ and are maintained within $\mathsf{STree}(w)$. See Fig. 2 for illustration.

We maintain the active point in $\mathsf{STree}(w)$ as follows:

- When the active point is on an edge and moves down, if the character on the edge following the active point matches the next character $T[j+1]$, the active point simply moves down. Otherwise, if the active point cannot move down, we create a new branching node u at the locus and add a new leaf ℓ and edge $e = (u, \ell)$. The new leaf and edge are connected to the corresponding edge $\tilde{e}$ and leaf of $\mathsf{STree}(T)$ as the above discussion. The edge $\tilde{e}$ of $\mathsf{STree}(T)$ can be found in constant time using a weighted ancestor query on $\mathsf{STree}(T)$.
- Similarly, when the active point is a node u and moves down, if u represents $T[i'..j]$, we query $\mathsf{WAQ}_{\mathsf{STree}(T)}(\mathsf{leaf}_T(i'), j-i'+2)$ and check the resulting edge. The resulting edge of $\mathsf{STree}(T)$ is already connected to an edge of the smaller suffix tree if and only if u has an outgoing edge whose edge label starts with $T[j+1]$. If the active point can move down to the existing edge, we simply move it to the edge (via its corresponding edge in $\mathsf{STree}(T)$). Otherwise, we create a new leaf ℓ and edge $e = (u, \ell)$, connecting them to their corresponding leaf and edge of $\mathsf{STree}(T)$ as described above.

In Ukkonen's algorithm, a new node/edge is created only if the active point reaches it, so every node/edge creation can be simulated as described. Node/edge deletions are straightforward: we simply disconnect the node/edge from $\mathsf{STree}(T)$ and remove them. Therefore, we can simulate the sliding suffix tree over T in amortized $O(1)$ time per sliding-operation. We have shown the next theorem:

Theorem 4. *Given an offline string T over a linearly sortable alphabet, we can simulate a suffix tree for a sliding window in a total of $O(n)$ time.*

By Theorem 4 and the algorithm described in Sect. 3, we obtain the following:

Corollary 2. *We can count the distinct closed factors in T over a linearly sortable alphabet in $O(n)$ time using $O(n)$ space.*

5 Enumerating Closed Factors Offline

In this section, we discuss the enumeration of (distinct) closed factors in an offline string T. We first consider The *open-close-array* OC_T of T, which is the binary sequence of length n where $\mathsf{OC}_T[i] = 1$ iff $T[i..n]$ is closed [28]. Since an open-close-array can be computed in $O(n)$ time, we can enumerate all the occurrences of closed factors of T in $\Theta(n^2)$ time by constructing $\mathsf{OC}_{T[i..n]}$ for all $1 \leq i \leq n$. We then map the occurrences to their corresponding loci on $\mathsf{STree}(T)$ using constant-time WAQs $O(n^2)$ times. Thus we have the following:

Proposition 1. *We can enumerate all the occurrences of closed factors of T and the distinct closed factors of T in $\Theta(n^2)$ time.*

Next, we propose an alternative approach that can achieve subquadratic time when there are few closed factors to output. Using the algorithms described in Sects. 3 and 4, we can enumerate the ending positions and the *longest borders* of distinct closed factors in $O(n + \mathsf{output})$ time. For each such a border b of a closed factor, we need to determine the starting position of the closed factor by finding the nearest occurrence of b to the left. Such occurrences can be computed efficiently by utilizing *range predecessor queries* over the list of (the integer-labels of) the leaves of $\mathsf{STree}(T)$ as follows: Let $b = T[s..t]$ be a border of some closed factor w where b occurs exactly twice in w. We first find the locus of b in $\mathsf{STree}(T)$ by a WAQ. Let v be the nearest descendant of the locus, inclusive. Then, the leaves in the subtree rooted at v represent the occurrences of b in T. Thus the nearest occurrence of b to the left equals the predecessor value of s within the leaves under v. By applying Belazzougui and Puglisi's range predecessor data structure [11], we obtain the following:

Proposition 2. *We can enumerate the distinct closed factors of T in $O(n\sqrt{\log n} + \mathsf{output} \cdot \log^{\epsilon} n)$ time where $\mathsf{output} \in O(n^2)$ is the number of distinct closed factors in T and ϵ is a fixed constant with $0 < \epsilon < 1$.*

6 Conclusions

In this paper, we proposed two algorithms for counting distinct closed factors of a string. The first algorithm runs in $O(n \log \sigma)$ time using $O(n)$ space for a string given in an online manner. The second algorithm runs in $O(n)$ time and space for a static string over a linearly sortable alphabet. Additionally, we discussed how to enumerate the distinct closed factors, and showed a $\Theta(n^2)$-time algorithm using open-close-arrays, as well as an $O(n\sqrt{\log n} + \mathsf{output} \cdot \log^{\epsilon} n)$-time algorithm that combines our counting algorithm with a range predecessor data structure. Since there can be $\Omega(n^2)$ distinct closed factors in a string [5,31], the $\mathsf{output} \cdot \log^{\epsilon} n$ term can be superquadratic in the worst case. This leads to the following open question: For the enumerating problem, can we achieve $O(n\mathsf{polylog}(n) + \mathsf{output})$ time, which is linear in the output size?

Acknowledgments. This work was supported by JSPS KAKENHI Grant Numbers JP23H04381 (TM), JP24K20734 (TM), JP22H00513 (KS), JP20H05964 (TH), and JP22H03549 (TH).

References

1. Alamro, H., Alzamel, M., Iliopoulos, C.S., Pissis, S.P., Sung, W., Watts, S.: Efficient identification of k-closed strings. Int. J. Found. Comput. Sci. **31**(5), 595–610 (2020). https://doi.org/10.1142/S0129054120500288
2. Alzamel, M., Iliopoulos, C.S., Smyth, W.F., Sung, W.: Off-line and on-line algorithms for closed string factorization. Theor. Comput. Sci. **792**, 12–19 (2019). https://doi.org/10.1016/J.TCS.2018.10.033
3. Badkobeh, G., et al.: Closed factorization. In: Proceedings of the Prague Stringology Conference 2014, Prague, Czech Republic, September 1-3, 2014, pp. 162–168. Department of Theoretical Computer Science, Faculty of Information Technology, Czech Technical University in Prague (2014). http://www.stringology.org/event/2014/p15.html

4. Badkobeh, G., et al.: Closed factorization. Discret. Appl. Math. **212**, 23–29 (2016). https://doi.org/10.1016/J.DAM.2016.04.009
5. Badkobeh, G., Fici, G., Lipták, Z.: On the number of closed factors in a word. In: Language and Automata Theory and Applications - 9th International Conference, LATA 2015, Nice, France, March 2-6, 2015, Proceedings. Lecture Notes in Computer Science, vol. 8977, pp. 381–390. Springer (2015). https://doi.org/10.1007/978-3-319-15579-1_29
6. Badkobeh, G., Luca, A.D., Fici, G., Puglisi, S.J.: Maximal closed substrings. In: String Processing and Information Retrieval - 29th International Symposium, SPIRE 2022, Concepción, Chile, November 8-10, 2022, Proceedings. Lecture Notes in Computer Science, vol. 13617, pp. 16–23. Springer (2022). https://doi.org/10.1007/978-3-031-20643-6_2
7. Badkobeh, G., Luca, A.D., Fici, G., Puglisi, S.J.: Maximal closed substrings. CoRR abs/2209.00271 (2022). https://doi.org/10.48550/ARXIV.2209.00271
8. Bannai, H., et al.: Efficient algorithms for longest closed factor array. In: Iliopoulos, C., Puglisi, S., Yilmaz, E. (eds.) SPIRE 2015. LNCS, vol. 9309, pp. 95–102. Springer, Cham (2015). https://doi.org/10.1007/978-3-319-23826-5_10
9. Belazzougui, D., et al.: Block trees. J. Comput. Syst. Sci. **117**, 1–22 (2021). https://doi.org/10.1016/J.JCSS.2020.11.002
10. Belazzougui, D., Kosolobov, D., Puglisi, S.J., Raman, R.: Weighted ancestors in suffix trees revisited. In: 32nd Annual Symposium on Combinatorial Pattern Matching, CPM 2021, July 5-7, 2021, Wrocław, Poland. LIPIcs, vol. 191, pp. 8:1–8:15. Schloss Dagstuhl - Leibniz-Zentrum für Informatik (2021). https://doi.org/10.4230/LIPICS.CPM.2021.8
11. Belazzougui, D., Puglisi, S.J.: Range predecessor and Lempel-Ziv parsing. In: Proceedings of the Twenty-Seventh Annual ACM-SIAM Symposium on Discrete Algorithms, SODA 2016, Arlington, VA, USA, January 10-12, 2016, pp. 2053–2071. SIAM (2016). https://doi.org/10.1137/1.9781611974331.CH143
12. Bucci, M., de Luca, A., De Luca, A.: Rich and periodic-like words. In: Diekert, V., Nowotka, D. (eds.) DLT 2009. LNCS, vol. 5583, pp. 145–155. Springer, Heidelberg (2009). https://doi.org/10.1007/978-3-642-02737-6_11
13. Carpi, A., de Luca, A.: Periodic-like words, periodicity, and boxes. Acta Informatica **37**(8), 597–618 (2001). https://doi.org/10.1007/PL00013314
14. Crochemore, M., Ilie, L., Rytter, W.: Repetitions in strings: algorithms and combinatorics. Theor. Comput. Sci. **410**(50), 5227–5235 (2009). https://doi.org/10.1016/J.TCS.2009.08.024
15. Durand, F.: A characterization of substitutive sequences using return words. Discret. Math. **179**(1–3), 89–101 (1998). https://doi.org/10.1016/S0012-365X(97)00029-0
16. Farach, M.: Optimal suffix tree construction with large alphabets. In: 38th Annual Symposium on Foundations of Computer Science, FOCS '97, Miami Beach, Florida, USA, October 19-22, 1997, pp. 137–143. IEEE Computer Society (1997). https://doi.org/10.1109/SFCS.1997.646102
17. Fiala, E.R., Greene, D.H.: Data compression with finite windows. Commun. ACM **32**(4), 490–505 (1989). https://doi.org/10.1145/63334.63341
18. Fici, G.: A classification of trapezoidal words. In: Proceedings 8th International Conference Words 2011, Prague, Czech Republic, 12-16th September 2011. EPTCS, vol. 63, pp. 129–137 (2011). https://doi.org/10.4204/EPTCS.63.18
19. Fici, G.: Open and closed words. Bulletin of EATCS (123) (2017). http://bulletin.eatcs.org/index.php/beatcs/article/view/508

20. Galil, Z., Seiferas, J.I.: Time-space-optimal string matching. J. Comput. Syst. Sci. **26**(3), 280–294 (1983). https://doi.org/10.1016/0022-0000(83)90002-8
21. Glen, A., Justin, J., Widmer, S., Zamboni, L.Q.: Palindromic richness. Eur. J. Comb. **30**(2), 510–531 (2009). https://doi.org/10.1016/J.EJC.2008.04.006
22. Kempa, D., Prezza, N.: At the roots of dictionary compression: string attractors. In: Proceedings of the 50th Annual ACM SIGACT Symposium on Theory of Computing, STOC 2018, Los Angeles, CA, USA, June 25-29, 2018, pp. 827–840. ACM (2018). https://doi.org/10.1145/3188745.3188814
23. Knuth, D.E., Jr., J.H.M., Pratt, V.R.: Fast pattern matching in strings. SIAM J. Comput. **6**(2), 323–350 (1977). https://doi.org/10.1137/0206024
24. Kociumaka, T., Navarro, G., Olivares, F.: Near-optimal search time in δ-optimal space, and vice versa. Algorithmica **86**(4), 1031–1056 (2024). https://doi.org/10.1007/S00453-023-01186-0
25. Larsson, N.J.: Extended application of suffix trees to data compression. In: Proceedings of the 6th Data Compression Conference (DCC '96), Snowbird, Utah, USA, March 31 - April 3, 1996, pp. 190–199. IEEE Computer Society (1996). https://doi.org/10.1109/DCC.1996.488324
26. Leonard, L., Inenaga, S., Bannai, H., Mieno, T.: Constant-time edge label and leaf pointer maintenance on sliding suffix trees. CoRRabs/2307.01412 (2024). https://doi.org/10.48550/ARXIV.2307.01412
27. De Luca, A., Fici, G.: Open and closed prefixes of sturmian words. In: Karhumäki, J., Lepistö, A., Zamboni, L. (eds.) WORDS 2013. LNCS, vol. 8079, pp. 132–142. Springer, Heidelberg (2013). https://doi.org/10.1007/978-3-642-40579-2_15
28. Luca, A.D., Fici, G., Zamboni, L.Q.: The sequence of open and closed prefixes of a Sturmian word. Adv. Appl. Math. **90**, 27–45 (2017). https://doi.org/10.1016/J.AAM.2017.04.007
29. Navarro, G.: Indexing highly repetitive string collections, part I: repetitiveness measures. ACM Comput. Surv. **54**(2), 29:1–29:31 (2022). https://doi.org/10.1145/3434399
30. Navarro, G.: Indexing highly repetitive string collections, part II: compressed indexes. ACM Comput. Surv. **54**(2), 26:1–26:32 (2022). https://doi.org/10.1145/3432999
31. Parshina, O.G., Puzynina, S.: Finite and infinite closed-rich words. Theor. Comput. Sci. **984**, 114315 (2024). https://doi.org/10.1016/J.TCS.2023.114315
32. Parshina, O.G., Zamboni, L.Q.: Open and closed factors in Arnoux-Rauzy words. Adv. Appl. Math. **107**, 22–31 (2019). https://doi.org/10.1016/J.AAM.2019.02.007
33. Senft, M.: Suffix tree for a sliding window: an overview. In: 14th Annual Conference of Doctoral Students - WDS 2005, pp. 41–46. Matfyzpress (2005). https://physics.mff.cuni.cz/wds/proc/proc-contents.php?year=2005
34. Smyth, W.F.: Computing regularities in strings: a survey. Eur. J. Comb. **34**(1), 3–14 (2013). https://doi.org/10.1016/J.EJC.2012.07.010
35. Sumiyoshi, W., Mieno, T., Inenaga, S.: Faster and simpler online/sliding rightmost lempel-ziv factorizations. In: String Processing and Information Retrieval - 31th International Symposium, SPIRE 2024. Lecture Notes in Computer Science, vol. 14899, pp. 321–335. Springer (2024). https://doi.org/10.1007/978-3-031-72200-4_24

36. Ukkonen, E.: On-line construction of suffix trees. Algorithmica **14**(3), 249–260 (1995). https://doi.org/10.1007/BF01206331
37. Vuillon, L.: A characterization of Sturmian words by return words. Eur. J. Comb. **22**(2), 263–275 (2001). https://doi.org/10.1006/EUJC.2000.0444
38. Weiner, P.: Linear pattern matching algorithms. In: 14th Annual Symposium on Switching and Automata Theory, Iowa City, Iowa, USA, October 15-17, 1973, pp. 1–11. IEEE Computer Society (1973). https://doi.org/10.1109/SWAT.1973.13

Knowledge Neurons in the Knowledge Graph-based Link Prediction Models

Grzegorz P. Mika[1,2](✉), Amel Bouzeghoub[1], Katarzyna Węgrzyn-Wolska[2], and Yessin M. Neggaz[2]

[1] SAMOVAR, Télécom SudParis, Institut Polytechnique de Paris, Palaiseau, France
grzegorz.mika@ip-paris.fr, amel.bouzeghoub@telecom-sudparis.eu
[2] Efrei Research Lab, EFREI Paris, Pantheon Assas University, Villejuif, France
{katarzyna.wegrzyn,yessin.neggaz}@efrei.fr

Abstract. Recent studies have shown that using Graph Transformer and attention weights are often claimed to confer explicability, purportedly helpful in providing insights and explaining why a model makes its decisions. While there is already quite an extensive range of techniques explaining Graph Neural Networks, their explicability and model transparency could be improved. A significant challenge in Explainable AI has recently been correctly interpreting neuron behavior to identify what a deep learning system has internally detected as relevant to the input. To tackle these challenges, we present a knowledge attribution method for the link prediction task to identify the neurons that express the input Knowledge Graph (KG) triples. This method not only enhances the explicability of the model but also improves its transparency, providing a clearer understanding of how specific factual knowledge is stored. It enables human-centric and knowledge attribution explanations by extracting factual knowledge from identified decision drivers. Empirical results on two standard KG-based link prediction datasets shed light on understanding the storage of knowledge within Graph Transformer architecture.

Keywords: Explainable AI · Knowledge Extraction · Link Prediction · Graph Transformer · Knowledge Graphs

1 Introduction

Transformer have revolutionized machine learning, particularly due to their innovative self-attention mechanism. This design enables the model to dynamically focus on pertinent parts of the data, computing attention values for each pair of instances within a layer. Such capability allows Transformer to capture intricate dependencies and relationships in data, leading to superior performance across various tasks. Notably, Transformer architecture has achieved remarkable success in NLP [3,10,35] and computer vision [4,12,21], attributed to its robust feature learning and minimal inductive bias. Building on this success, researchers have extended Transformer architecture to the graph domain, where traditional message-passing schemes are being reimagined. Recent approaches

R. Královič and V. Kůrková (Eds.): SOFSEM 2025, LNCS 15539, pp. 184–197, 2025.
https://doi.org/10.1007/978-3-031-82697-9_14

apply Transformer architecture to entire graphs, treating all nodes as fully connected and leveraging the self-attention mechanism to effectively model complex graph structures for tasks such as node classification [5,20], graph classification [19,41], and link prediction [26,31,36].

This approach, inspired by the selective focus characteristic of human cognition, is promising for developing explainable models. Attention mechanisms enable neural networks to dynamically prioritize specific segments of input data, enhancing both performance and explicability. While attention mechanisms aren't inherently designed to produce human-readable explanations, they can highlight which information flows through a neural network, offering a form of explicability [28]. Beyond merely observing attention for insights, another approach involves explicitly training attention mechanisms to align a network's behavior with desired explanations [30]. Recent studies have shown that large language models, thanks to Transformer architecture, can store factual knowledge, which can be verified using fill-in-the-blank cloze task [16,29] queries. Furthermore, Feed-Forward layers in Transformer function as key-value memory stores [13], facilitating the identification of neurons responsible for encoding specific factual information. These neurons, termed Knowledge Neurons (KNs) [6,7], are responsible for expressing relational facts within large language models.

Despite the recent success of KNs in pre-trained language models [6,7], identifying them within graph-based architectures for link prediction tasks presents two key challenges. First, while masked language modeling is well-established in text data, masked graph modeling is an emerging area that requires further exploration. Developing effective masked graph modeling techniques is essential for identifying KNs in graph-based architectures. Additionally, the knowledge extracted from neurons still requires interpretation to benefit end-users. A further challenge lies in effectively linking the identified neurons with the input data facilitated through the attention matrix.

Inspired by [6,7], we extend this concept to graph-structured data in Graph Transformer. We delve deeper into the role of Transformer architecture within Graph Neural Networks (GNNs) for link prediction task, concentrating on the encoding and storage of factual knowledge. To address the edit sensitivity inherent in GNN models, we employ a Retrieve-and-Read architecture, called KG-R3 [26] that efficiently encodes the structural, semantic, and attribute information of the graph.

This work makes several important contributions to the field:

1. We introduce the extension of the KNs concept to GNNs for link prediction task to identify KNs that express specific factual knowledge, as demonstrated through fill-in-the-blank cloze tasks.
2. We conduct preliminary studies utilizing KNs to edit factual knowledge within the model.
3. We apply the KNs concept, along with the knowledge attribution method, to Knowledge Graph-based Link Prediction Models, enabling the identification of KNs that encode specific factual knowledge within graph-structured data.

4. We perform empirical experiments on standard datasets, FB15K-237 [34] and WN18RR [9], demonstrating a positive correlation between identified KNs and knowledge expression.

The rest of the paper is organized as follows: Sect. 2 presents a critical overview of the related work and state-of-the-art. In Sect. 3, we present some needed basic foundations and describe our methodology, including the Knowledge Graph-based Link Prediction Model, the Masking Construction Process, and Knowledge Neurons Discovery. Section 4 presents the experimental settings, and Sect. 5 reports the study results. Finally, we summarize the paper and propose possibilities for future research in Sect. 6.

2 Related Work

2.1 Explainable Link Predictions

Link prediction task is important in domains like social networks [1], Knowledge Graphs (KGs) [25], and recommender systems [26]. GNNs enhance this task by learning node representations that capture both structural and feature information, achieving state-of-the-art performance [18,27]. Despite these advancements, explaining the inner workings of GNNs in link prediction remains a significant challenge, spurring the development of diverse methods aimed at addressing this issue. HAGERec [40] is an explainable model based on a hierarchical attention graph convolutional networks. KGIN [38] is a framework designed to model users' underlying intents through their interactions with items. Additionally, HGExplainer [24] captures semantic, structural, and attribute information to explain the intricacies of GNN models. Despite advancements, explaining Graph Transformer in link prediction remains challenging, leaving models largely opaque. Our research focuses on analyzing internal neuronal behavior in these models to enhance explicability.

2.2 Knowledge Neurons

Given the considerable scarcity of research analyzing Graph Transformer [35], we devote a brief review to the existing research on Transformer analysis for NLP that inspired us. Transformer, as an important architecture in NLP, and more recently in graph-based models, have attracted renewed attention with regard to the importance of their Feed-Forward Networks (FFNs) modules in the context of explainability. Building on the work of Wu et al. [39] and Dong et al. [11], which highlighted the relevance the FFNs module has for Transformer, Dai et al. [7] and Petroni et al. [29] demonstrated the ability of selected neurons to store factual knowledge - called Knowledge Neurons. The first research on KNs [6,7] comes from the field of model editing, which has so far produced two papers on GNNs, EGNN [22], SEED-GNN [42] and GRE [17]. These works, however, do not study Graph Transformer and focus on less complex architectures of GNNs. However, there is a lack of research for Graph Transformer. To the best of our knowledge, our work is the first to devote space to the analysis of Graph Transformer for link prediction tasks, identifying KNs.

3 Methodology

We begin by defining the notations and formulating the problem, followed by a detailed examination of the proposed framework (see Fig. 1).

Knowledge Graph is formally defined as a set of triples $\mathcal{L} \subseteq \mathcal{E} \times \mathcal{R} \times \mathcal{E}$, where $\mathcal{E}$ denotes the set of entities representing objects, and $\mathcal{R}$ denotes the set of relations, defining the ways in which entities are interconnected. Each triple $< h, r, t > \in \mathcal{L}$ denotes a fact, where h (head) is the subject entity, r (relation) is the directed edge representing the type of relationship, and t (tail) is the object entity [15,37].

Link Prediction involves inferring missing facts within a KG. Given a query in the form of $< h, r, ? >$ or $<?, r, t >$, the model's objective is to rank the correct target entity among a set of candidates. For instance, in the query $< h, r, ? >$, the model aims to predict the appropriate tail entity t that, together with h and r, forms a valid triple in the KG. This process is essential for completing KG by predicting potential links between entities.

Graph Transformer. Each node's feature vector is processed through a self-attention mechanism followed by a FFN. Given an input feature matrix $H \in \mathbb{R}^{n \times d}$, where n represents the number of nodes and d denotes the feature dimensionality, the self-attention mechanism employs three learnable weight matrices:

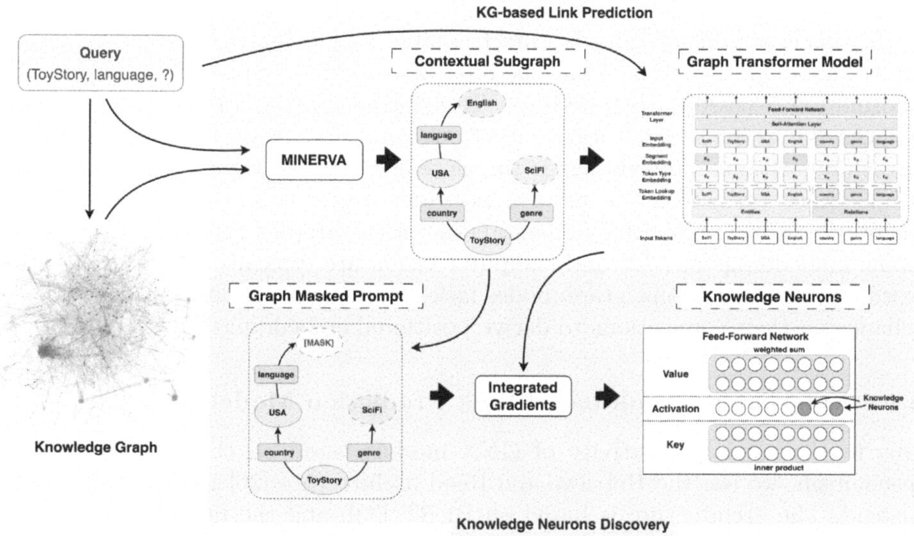

Fig. 1. An overview of our methodology for analyzing KNs within KG-based Link Prediction Models. Initially, we employ the Retrieve-and-Read architecture [26] using contextual subgraphs in Graph Transformer Model. Subsequently, these subgraphs, along with the query, are jointly encoded to utilize Integrated Gradients to identify KNs.

$W_Q \in \mathbb{R}^{d \times d_q}, W_K \in \mathbb{R}^{d \times d_k}$ and $W_V \in \mathbb{R}^{d \times d_v}$, corresponding to queries, keys, and values, respectively. Typically, $d_q = d_k$ to facilitate the computation of attention scores. The self-attention mechanism computes these matrices as follows:

$$Q = HW_Q, K = HW_K, V = HW_V \tag{1}$$

Then, the dot product is utilized to calculate the weights of the values by applying it to the queries with all keys. Subsequently, the final output matrix is computed by

$$Attention(Q, K, V) = softmax\left(\frac{QK^T}{\sqrt{d_k}}V\right) \tag{2}$$

The output of the self-attention mechanism is a matrix of dimensions $n \times d_v$, where each row represents the updated feature vector of a corresponding node. This output is then processed by a two-layer FFNs applied to each node's representation individually, enhancing the model's capacity to capture complex patterns within the graph data. The FFN typically consists of two linear transformations separated by a non-linear activation function, allowing for greater expressiveness in learning node representations. This sequence of self-attention followed by position-wise FFNs enables the Graph Transformer to effectively model both the structural relationships and the features of nodes within the graph.

$$FFN(x) = max(0, xW_1 + b_1)W_2 + b_2 \tag{3}$$

In this equation, W_1 and W_2 are trainable weight matrices, while b_1 and b_2 are trainable bias vectors. While the input feature matrix provides node-specific information accessible to the Graph Transformer during forward propagation, it lacks the graph's structural context, treating each node in isolation. Incorporating positional embeddings is therefore essential, as they enable the Graph Transformer to discern the relative positions and relationships among nodes, facilitating a comprehensive understanding of the graph's topology. While positional embeddings in NLP and computer vision are typically based on absolute word and pixel positions, graph nodes lack such absolute ordering, necessitating reliance on their connections to derive positional embeddings.

3.1 Knowledge Graph-based Link Prediction Model

Due to the editing sensitivity of GNN models based on the message-passing mechanism, we use the Retrieve-and-Read architecture [26] in our experiments instead. The architecture is based on BERT [33], and the input data are contextual subgraphs based on KG and obtained through MINERVA algorithm [8]. The attention structure integrates subgraph information into Transformer representations, allowing nodes and edge tokens to attend to one another. For a triple $< h, r, t >$, token pairs $< h, r >$ and $< r, t >$ share attention, with each token attending to itself. This framework captures factual knowledge by considering both local graph neighborhoods and the global subgraph context.

3.2 The Masking Construction Process

In our research, we utilize a graph-based cloze task to evaluate the capacity of neurons to associate specific factual information, extending the concept of the previous works [7,29] to Graph Transformer. Unlike the approach introduced by Dai et al. [7], our masking procedure requires additional steps, leading to increased time consumption. To enable the calculation of Integrated Gradients [32], it is imperative to prepare a series of analogous contextual subgraphs for each relationship. Each relational fact is represented as a triple $< h, r, t >$, where h denotes the head entity, t the tail entity, and r the relation connecting them. Upon receiving such a fact, the GNN model addresses a cloze-style query x that articulates the fact with the tail entity omitted. For instance, given the fact $< ToyStory, language, english >$, a plausible query could be $< ToyStory, language, ? >$ in the contextual subgraph, serving as a knowledge-expressing prompt. To establish the essential graph prompts for identifying KNs, we generate a collection of 10 subgraphs with similar contexts for selected relations, and at least 10 subgraphs with the same relations, but different h and t. Subsequently, we mask the predictions, specifically the nodes. We ascertain that the set of graph prompts encompasses 10 similar subgraphs for KNs detection. The primary criterion is that the subgraph should consist of the same relational fact and can be linked with different entities.

3.3 Knowledge Neurons Discovery

Similarly to [7,29], we posit that factual knowledge is encoded within FFN memories of Graph Transformer and is expressed through specialized KNs (see Fig. 2). In this section, we present a methodology for attributing knowledge aimed at identifying these KNs. Our methodology is designed to tackle more intricate data structures like graphs and complex architectures featuring three separate Transformer modules. Unlike the study by Dai et al. [7], our work uniquely centers on identifying specific KNs within Graph Transformer that express factual knowledge in the KG-based Link Prediction Models.

In their work, Petroni, et al. [29] posited the notion that a model's knowledge of a fact is contingent upon its ability to accurately predict the correct answer. In our investigation, we extend this concept to encompass graph data structures by introducing the KNs identification method to discern the responsible neurons for predictions within the KG-based Link Prediction Models.

In this study, we extend this analysis to Graph Transformer to locate the neurons responsible for processing factual information in a link prediction task. Mathematically, given a query q, the model's output, denoted as $F(\hat{z}_j^{(l)})$, represents the probability that the pre-trained model assigns to the correct answer, according to the following formula:

$$F(\hat{z}_j^{(l)}) = p(y^* | q, z_j^{(l)} = \hat{z}_j^{(l)}) \tag{4}$$

In this equation, $z_j^{(l)}$ denotes the j-th neuron within the l-th attention head, while y^* signifies the correct answer. $\hat{z}_j^{(l)}$ corresponds to the post-training setting

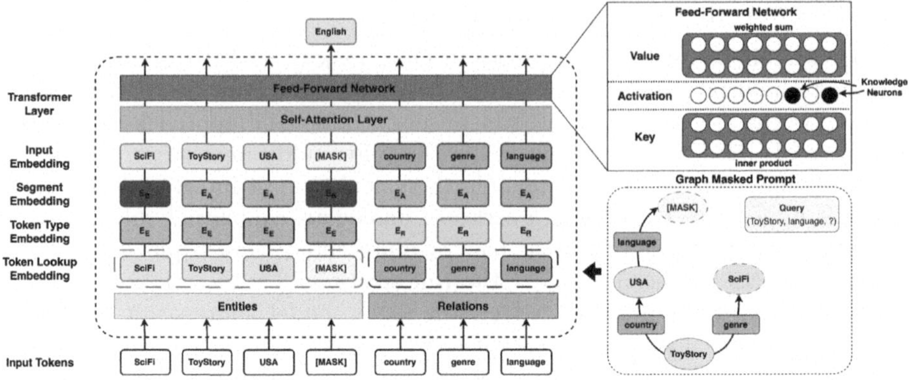

Fig. 2. The schema illustrates the embedding layer of a contextual subgraph within the Graph Transformer. Graph Masked Prompt as input to the Graph Transformer is constructed by masking the fact token. Within the Graph Transformer block, FFN function as a key-value memory system: (1) First linear layer (key) computes intermediate neuron activations through inner product operations; (2) Second linear layer (value) integrates value vectors via a weighted sum, utilizing the activations from the first layer as weights.

of the weight. Integrated Gradients [32] often fall short when applied to graph data, primarily due to their inefficiency in handling the inherent noise and the complex interplay between global and local structures within graphs. To address these challenges, in contrast to [7], we have adopted a modified version of Integrated Gradients, tailored specifically for Graph Transformer. This adaptation enhances the model's ability to effectively capture both global context and local nuances, leading to more accurate and explainable attributions in graph-based tasks. For a given neuron $z_i^{(l)}$, we can calculate the attribution score as follows:

$$IG_i(z_i^{(l)}) = \overline{z}_i^{(l)} \int_0^1 \frac{\varphi F\left(z_i'^{(l)} + \alpha(\overline{z}_i^{(l)} - z_i^{(l)})\right)}{\varphi z_i^{(l)}} d\alpha \tag{5}$$

As we can see, Integrated Gradients defined above are challenging in more advanced and complex deep learning models like KG-R3 [26]. To address this issue, in our setting, we employ the Riemman approximation:

$$IG_i^{approx}(z_i^{(l)}) \approx \frac{\overline{z}_i^{(l)}}{N} \sum_{k=1}^{N} \frac{\varphi F\left(z_i'^{(l)} + \frac{k}{N} \times (\overline{z}_i^{(l)} - z_i'^{(l)})\right)}{\varphi z_i^{(l)}} \tag{6}$$

In this equation, we define the parameter N as the number of the approximation steps.

4 Experimental Setting

4.1 Datasets

We evaluate our model's performance using the established link prediction datasets FB15k-237 [34] and WN18RR [9]. FB15k-237 is a refined version of the FB15k dataset [2], created by removing inverse relations to prevent data leakage between training and test sets. Similarly, WN18RR is a refined version of the WN18 dataset [2], created by removing reversible relations to prevent inference by memorizing training facts. The statistics of these datasets are detailed in Table 1.

Table 1. Datasets used where $|\mathcal{E}|$, $|\mathcal{R}|$ and $|\mathcal{L}|$ respectively stand for the total number of entities, relations, and triples in each split of the two benchmarks.

Dataset	$\|\mathcal{E}\|$	$\|\mathcal{R}\|$	$\|\mathcal{L}\|$ (Train)	$\|\mathcal{L}\|$ (Test)	$\|\mathcal{L}\|$ (Val)
FB15K-237	14,541	237	272,115	20,466	17,535
WN18RR	40,943	11	86,835	3,134	3,034

4.2 Implementation Details

The Retrieve-and-Read architecture, referred to as KG-R3 [26], comprises 9 layers with a hidden size of 320 and an inner FFN hidden size of 1,280. Training was conducted on four GPUs. Our KNs implementation includes an attribution threshold set at 0.2 of the maximum attribution score for each prompt and a refinement threshold of 0.7 for each relation. All experiments were conducted on a Linux system equipped with an NVIDIA V100 GPU featuring 64 GB of memory. The environment utilized CUDA version 12.4 and NVIDIA Driver Version 545.23.06. Python 3.9 and PyTorch 2.4.0 were used for all implementations.

4.3 Baselines

In our evaluation, we assign a neuron's activation value as its attribution score, denoted as $Attr_{base}(z_i^{(l)}) = z_i^{(l)}$. This approach quantifies a neuron's sensitivity to input data, providing a direct measure of its contribution to the model's output. Using this technique, we can effectively interpret the role of individual neurons in the network's decision-making process. Utilizing neuron activation as an attribution method acknowledges its validity as a foundational baseline, inspired by the analogy between feedforward networks and the self-attention mechanism (as discussed in Sect. 3). This rationale is further supported by the prevalent use of self-attention scores as a robust attribution baseline, as noted in [14].

4.4 Metrics

Following the evaluation in [7], we employ two metrics to assess fact updating: (1) Change Rate: the proportion of instances where the original prediction has been altered to a different outcome; (2) Success Rate: the proportion of instances where the updated prediction has become the top-ranked result. To evaluate the impact of these updates on other knowledge within the model, we utilize two additional metrics: (1) Δ Intra-Relation Perplexity: measures the increase in perplexity for graph prompts that share the same relation; (2) Δ Inter-Relation Perplexity: measures the increase in perplexity for graph prompts involving different relations.

5 Results

In this section, we present our experimental evaluation of the method, addressing the following research questions. **RQ1:** What is the distribution of neurons storing factual information in complex models with Graph Transformer? **RQ2:** How effectively are KNs identified in the KG-based Link Prediction Models? **RQ3:** How does the storage of factual knowledge work in Graph Transformer in the context of the real-world industrial applications?

5.1 Analysis of the Identified Knowledge Neurons (RQ1, RQ2)

Our experiments with KG-based Link Prediction Models yielded the results shown in Fig. 3, which illustrates the layered distribution of KNs identified by our attribution method (RQ1). We observe that in both datasets, most fact-related neurons are concentrated in the higher layers of the pre-trained Graph Transformer. This observation aligns with previous studies [6,7,13,33], suggesting that high-level features, such as factual information, are primarily encoded in the final layers of Transformer models. Therefore, we argue that the quality of trained models influences the ability to identify neurons that store specific factual information. Additionally, the FB15K-237 model contains a higher number of identified KNs due to its larger set of triples.

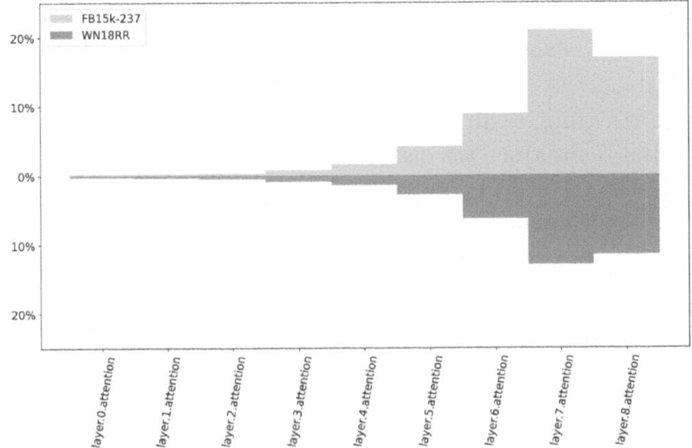

Fig. 3. The distribution of KNs identified by our method across each layer of the KG-based Link Prediction Model, evaluated on two benchmark datasets.

Table 2. Statistics on KNs, with $\cap$ representing the intersection of KNs between fact pairs. The term "rel." abbreviates relation. Our Integrated Gradients method identifies a greater number of exclusive KNs across both benchmarks. Superior results are highlighted in **bold**.

Neurons Type	FB15K-237		WN18RR	
	Integrated Gradients	Baseline	Integrated Gradients	Baseline
Knowledge Neurons	**3.98**	3.72	**3.54**	3.21
$\cap$ of intra-rel. fact pairs	**1.12**	1.38	**1.43**	1.67
$\cap$ of inter-rel. fact pairs	**0.19**	1.04	**0.15**	1.13

Table 2 presents the statistics of KNs (RQ2). Our knowledge attribution method identifies an average of 3.98 KNs per relational fact in FB15K-237 [34] and 3.54 in WN18RR [9]. In comparison, the baseline method identifies 3.72 in FB15K-237 and 3.21 in WN18RR. The similarity in these averages ensures the fairness of subsequent comparisons in this paper. We also compute the intersection of KNs across different relational facts. On average, our method finds that fact pairs sharing the same relation (intra-relation fact pairs) have 1.12 common KNs in FB15K-237 and 1.43 in WN18RR.

Table 3. Sample relational facts from the FB15k-237 dataset [34], highlighting their associated KNs and corresponding activation levels. KNs are indicated in **bold**.

Relational Facts	Neurons	Activation Value
$<ToyStory, Language, English>$	$w^{(8)}_{1235}, w^{(9)}_{1648}$	**5.28**
$<Avatar, Country, USA>$	$w^{(8)}_{1235}, w^{(9)}_{1648}$	-0.45
$<Avatar, Country, USA>$	$w^{(7)}_{1006}, w^{(9)}_{1715}$	**3.67**
$<ToyStory, Language, English>$	$w^{(7)}_{1006}, w^{(9)}_{1715}$	-0.28

Additionally, fact pairs with differing relations (inter-relation fact pairs) share an average of 0.19 KNs in the FB15K-237 dataset [34] and 0.15 in the WN18RR dataset [9]. In contrast, the baseline method identifies a substantial number of neurons shared among intra-relation fact pairs in both datasets, with a significant portion also common among inter-relation fact pairs. This disparity in KN intersections indicates that our method effectively identifies more distinct KNs.

Table 3 presents example prompts that activate specific KNs. Initially, our knowledge attribution method identifies the neurons associated with a given fact. We then calculate the average activation levels of these neurons for each relation in the FB15K-237 dataset [34]. Subsequently, we provide two prompts-one eliciting high average activation and the other low-that engage the same set of neurons. This comparison demonstrates that the activation of identified KNs varies with different queries and contexts, reinforcing that these neurons respond to relevant knowledge-expressing prompts.

5.2 Counterfactual Facts (RQ3)

Given the criticality of transparency in real-world industrial applications, a case study was undertaken to showcase the potential application of the KNs approach within GNN architecture. In our study, we examined the plausibility of altering a query's prediction from correct to incorrect and vice versa. By identifying the KNs associated with a given triple $< h, r, t >$ and retaining only those neurons shared by fewer than 10% of intra-relation facts, we aimed to minimize the impact on other facts sharing the same relation. We conducted experiments on both benchmarks and settings, KNs, and random neurons. Our experimental scope was confined to the top four KNs.

The outcome, demonstrated in Table 4, illustrates the promising success rate of neural modification in updating facts across both benchmarks, while random neurons exhibited failures. These findings align with [7], highlighting the potential influence of altering a select number of neurons on specific facts within GNN models. However, the achieved outcomes were not as favorable as those described in [7], which may be due to the nature of the data structure. KNs in NLP tasks work based on the sequence data. The graph-based data are more complex (non-euclidean structure), including structural, semantic, and attribute information. Moreover, it is conceivable that additional forms of knowledge might prove indispensable for KG-based models. The structure of the subgraphs (set of nodes and edges), and its properties influence the final results, Furthermore, the scale of the knowledge graph affects both the model's performance and the detection of KNs. Notably, the success rate correlates with the number of neurons possessing optimal knowledge.

Table 4. Analyses of fact updates within the model. Arrows indicate the desired direction for each metric: ↑ denotes that higher values are preferable, while ↓ signifies that lower values are favorable. Our approach, involving targeted modifications of KNs, achieves a significant success rate in updating facts, with a controlled impact on other knowledge areas. The most favorable outcomes are highlighted in **bold**.

Metrics	FB15K-237		WN18RR	
	Knowledge Neurons	Random Neurons	Knowledge Neurons	Random Neurons
Change Rate ↑	**32.21%**	5.02%	**23.45%**	2.63%
Success Rate ↑	**25.5%**	0.78%	**17.29%**	1.02%
ΔIntra-relation ↓	9.35	**9.2**	**6.8**	7.4
ΔInter-relation ↓	6.8	**4.0**	5.2	**3.8**

6 Conclusion and Future Work

In this study, we investigate the role of factual Knowledge Neurons (KNs) in Graph Neural Networks (GNNs) for link prediction tasks, utilizing the Integrated Gradients method on a Retrieve-and-Read architecture [26] with Graph Transformer. By applying an attribution technique, we identify KNs that encapsulate factual information on two datasets: FB15K-237 [34] and WN18RR [9]. Our analyses reveal that these neurons are activated by specific knowledge-expressing prompts. Notably, neurons associated with key vectors in the self-attention mechanism play a significant role in knowledge representation, indicating their specialization in processing distinct components of the model in response to particular queries.

Despite the effectiveness of KNs identification, our current research leaves several questions unanswered. First, research is needed to extend experiments to other GNN tasks, such as node classification and graph classification. Identifying KNs is time-consuming, especially on graph-structured data. Therefore, research dedicated to optimizing these processes is needed. Moreover, this research is promising in the Explainable AI field. A key question is whether extending classical GNN architectures to include attention mechanisms and Feed-Forward Networks will also make it possible to identify relevant neurons. The use of Knowledge Graphs in input data offers a promising direction in the further development of neurosymbolic techniques [23], which, when combined with KNs, can serve as a key link between input data and model predictions, generating human-understandable explanations.

Acknowledgments. The contribution has been funded by the Telecom SudParis, Institut Polytechnique de Paris and EFREI Paris, Pantheon Assas University. We thank these institutions for their support.

References

1. Adamic, L.A., Adar, E.: Friends and neighbors on the web. Soc. Netw. **25**(3), 211–230 (2003)
2. Bordes, A., Usunier, N., Garcia-Duran, A., Weston, J., Yakhnenko, O.: Translating embeddings for modeling multi-relational data. Adv. Neural Inf. Process. Syst. **26** (2013)
3. Brown, T., Mann, B., Ryder, N., Subbiah, M., Kaplan, J.D., Dhariwal, P., Neelakantan, A., Shyam, P., Sastry, G., Askell, A., et al.: Language models are few-shot learners. Adv. Neural. Inf. Process. Syst. **33**, 1877–1901 (2020)
4. Carion, N., Massa, F., Synnaeve, G., Usunier, N., Kirillov, A., Zagoruyko, S.: End-to-end object detection with transformers. In: Vedaldi, A., Bischof, H., Brox, T., Frahm, J.-M. (eds.) ECCV 2020. LNCS, vol. 12346, pp. 213–229. Springer, Cham (2020). https://doi.org/10.1007/978-3-030-58452-8_13
5. Chen, J., Gao, K., Li, G., He, K.: NAGphormer: a tokenized graph transformer for node classification in large graphs. In: International Conference on Learning Representations (2023)
6. Chen, Y., Cao, P., Chen, Y., Liu, K., Zhao, J.: Journey to the center of the knowledge neurons: discoveries of language-independent knowledge neurons and degenerate knowledge neurons. In: Proceedings of the 38th AAAI Conference on Artificial Intelligence, vol. 38, pp. 17817–17825 (2024)
7. Dai, D., Dong, L., Hao, Y., Sui, Z., Chang, B., Wei, F.: Knowledge neurons in pretrained transformers. In: Proceedings of the 60th Annual Meeting of the Association for Computational Linguistics (Volume 1: Long Papers), pp. 8493–8502 (2022)
8. Das, R., et al.: Go for a walk and arrive at the answer: reasoning over paths in knowledge bases using reinforcement learning. In: International Conference on Learning Representations (2018)
9. Dettmers, T., Minervini, P., Stenetorp, P., Riedel, S.: Convolutional 2D knowledge graph embeddings. In: Proceedings of the 32nd AAAI Conference on Artificial Intelligence, vol. 32, pp. 1811–1818 (2018)

10. Devlin, J., Chang, M.W., Lee, K., Toutanova, K.: BERT: pre-training of deep bidirectional transformers for language understanding. In: Proceedings of the 2019 Conference of the North American Chapter of the Association for Computational Linguistics: Human Language Technologies, Volume 1 (Long and Short Papers), pp. 4171–4186 (2019)
11. Dong, Y., Cordonnier, J.B., Loukas, A.: Attention is not all you need: pure attention loses rank doubly exponentially with depth. In: Proceedings of the 38th International Conference on Machine Learning, pp. 2793–2803. PMLR (2021)
12. Dosovitskiy, A., et al.: An image is worth 16×16 words: transformers for image recognition at scale. In: International Conference on Learning Representations (2021)
13. Geva, M., Schuster, R., Berant, J., Levy, O.: Transformer feed-forward layers are key-value memories. In: Proceedings of the 2021 Conference on Empirical Methods in Natural Language Processing, pp. 5484–5495 (2021)
14. Hao, Y., Dong, L., Wei, F., Xu, K.: Self-attention attribution: interpreting information interactions inside transformer. In: Proceedings of the 35th AAAI Conference on Artificial Intelligence, vol. 35, pp. 12963–12971 (2021)
15. Hogan, A., et al.: Knowledge graphs. ACM Comput. Surv. **54**(4), 1–37 (2021)
16. Jiang, Z., Xu, F.F., Araki, J., Neubig, G.: How can we know what language models know? Trans. Assoc. Comput. Linguist. **8**, 423–438 (2020)
17. Jiang, Z., et al.: Gradient rewiring for editable graph neural network training. In: Proceedings of the 38th Annual Conference on Neural Information Processing Systems (2024)
18. Kipf, T.N., Welling, M.: Variational graph auto-encoders. arXiv preprint arXiv:1611.07308 (2016)
19. Li, J., Zhang, K., Pu, X., Kong, Y.: Graph attention mixup transformer for graph classification. In: Tanveer, M., Agarwal, S., Ozawa, S., Ekbal, A., Jatowt, A. (eds.) Neural Information Processing: 29th International Conference, ICONIP 2022, Virtual Event, November 22–26, 2022, Proceedings, Part VI, pp. 188–199. Springer Nature Singapore, Singapore (2023). https://doi.org/10.1007/978-981-99-1645-0_16
20. Liu, C., et al.: Gapformer: graph transformer with graph pooling for node classification. In: Proceedings of the 32nd International Joint Conferences on Artificial Intelligence Organization, pp. 2196–2205 (2023)
21. Liu, Z., et al.: Swin transformer: hierarchical vision transformer using shifted windows. In: Proceedings of the 2021 IEEE/CVF International Conference on Computer Vision, pp. 9992–10002 (2021)
22. Liu, Z., et al.: Editable graph neural network for node classifications. In: Proceedings of the 19th International Workshop on Mining and Learning with Graphs (2023)
23. Mika, G.P.: Toward a transparent recommender system. In: Proceedings of the 14th International Rule Challenge, 4th Doctoral Consortium, and 6th Industry Track @ RuleML+RR, vol. 2644, pp. 111–119 (2020)
24. Mika, G.P., Bouzeghoub, A., Węgrzyn-Wolska, K., Neggaz, Y.M.: HGExplainer: explainable graph neural network. In: Proceedings of the 22nd IEEE/WIC International Conference on Web Intelligence and Intelligent Agent Technology, pp. 221–229 (2023)
25. Nickel, M., Murphy, K., Tresp, V., Gabrilovich, E.: A review of relational machine learning for knowledge graphs. Proc. IEEE **104**(1), 11–33 (2015)

26. Pahuja, V., Wang, B., Latapie, H., Srinivasa, J., Su, Y.: A retrieve-and-read framework for knowledge graph link prediction. In: Proceedings of the 32nd ACM International Conference on Information and Knowledge Management, pp. 1992–2002 (2023)
27. Pan, L., Shi, C., Dokmanić, I.: Neural link prediction with walk pooling. In: International Conference on Learning Representations (2022)
28. Park, D.H., et al.: Multimodal explanations: justifying decisions and pointing to the evidence. In: Proceedings of the IEEE Conference on Computer Vision and Pattern Recognition, pp. 8779–8788 (2018)
29. Petroni, F., et al.: Language models as knowledge bases? In: Proceedings of the 2019 Conference on Empirical Methods in Natural Language Processing and the 9th International Joint Conference on Natural Language Processing, pp. 2463–2473 (2019)
30. Ross, A.S., Hughes, M.C., Doshi-Velez, F.: Right for the right reasons: training differentiable models by constraining their explanations. In: Proceedings of the 26th International Joint Conference on Artificial Intelligence, pp. 2662–2670 (2017)
31. Shomer, H., Ma, Y., Mao, H., Li, J., Wu, B., Tang, J.: LPFormer: an adaptive graph transformer for link prediction. In: Proceedings of the 30th ACM SIGKDD Conference on Knowledge Discovery and Data Mining, pp. 2686–2698 (2024)
32. Sundararajan, M., Taly, A., Yan, Q.: Axiomatic attribution for deep networks. In: Proceedings of the 34th International Conference on Machine Learning, pp. 3319–3328 (2017)
33. Tenney, I., Das, D., Pavlick, E.: BERT rediscovers the classical NLP pipeline. In: Proceedings of the 57th Annual Meeting of the Association for Computational Linguistics, pp. 4593–4601 (2019)
34. Toutanova, K., Chen, D.: Observed versus latent features for knowledge base and text inference. In: Proceedings of the 3rd Workshop on Continuous Vector Space Models and Their Compositionality, pp. 57–66 (2015)
35. Vaswani, A., et al.: Attention is all you need. Adv. Neural Inf. Process. Syst. **30** (2017)
36. Wang, D., et al.: MM-transformer: a transformer-based knowledge graph link prediction model that fuses multimodal features. Symmetry **16**(8), 961 (2024)
37. Wang, Q., Mao, Z., Wang, B., Guo, L.: Knowledge graph embedding: a survey of approaches and applications. IEEE Trans. Knowl. Data Eng. **29**(12), 2724–2743 (2017)
38. Wang, X., et al.: Learning intents behind interactions with knowledge graph for recommendation. In: Proceedings of the 30th Web Conference 2021, pp. 878–887 (2021)
39. Wu, F., Fan, A., Baevski, A., Dauphin, Y.N., Auli, M.: Pay less attention with lightweight and dynamic convolutions. In: International Conference on Learning Representations (2019)
40. Yang, Z., Dong, S.: HAGERec: hierarchical attention graph convolutional network incorporating knowledge graph for explainable recommendation. Knowl. Based Syst. **204**, 106194 (2020)
41. Zhang, Z., et al.: GBCA: graph convolution network and BERT combined with co-attention for fake news detection. Pattern Recogn. Lett. **180**, 26–32 (2024)
42. Zhong, S., et al.: GNNs also deserve editing, and they need it more than once. In: Proceedings of the 41st International Conference on Machine Learning, vol. 235, pp. 61727–61746 (2024)

Disjoint Covering of Bipartite Graphs with s-clubs

Angelo Monti[1] and Blerina Sinaimeri[2(✉)]

[1] Sapienza University, via Salaria 113, 00139 Rome, Italy
monti@di.uniroma1.it
[2] Luiss University, Viale Romania 32,00197 Rome, Italy
bsinaimeri@luiss.it

Abstract. For a positive integer s, an s-club in a graph G is a set of vertices inducing a subgraph with diameter at most s. As generalizations of cliques, s-clubs offer a flexible model for real-world networks. This paper addresses the problems of partitioning and disjoint covering of vertices with s-clubs on bipartite graphs. First we prove that for any fixed $s \geq 3$ and fixed $k \geq 3$, determining whether the vertices of G can be partitioned into at most k disjoint s-clubs is NP-complete even for bipartite graphs. Note that our NP-completeness result is stronger than the one in Abbas and Stewart (1999), as we assume that both s and k are constants and not part of the input.

Additionally, we study the Maximum Disjoint (t, s)-Club Covering problem (MAX-DCC(t, s)), which aims to find a collection of vertex-disjoint (t, s)-clubs (i.e. s-clubs with at least t vertices) that covers the maximum number of vertices in G. We prove that it is NP-hard to achieve an approximation factor of $\frac{95}{94}$ for MAX-DCC$(t, 3)$ for any fixed $t \geq 8$ and for MAX-DCC$(t, 2)$ for any fixed $t \geq 5$ even for bipartite graphs. Previously, results were known only for MAX-DCC$(3, 2)$. Finally, we provide a polynomial-time algorithm for MAX-DCC$(2, 2)$ resolving an open problem from Dondi *et al.* (2019).

Keywords: s-club · graph covering · bipartite graph

1 Introduction

For a positive integer s, an s-club in a graph G is a set of vertices that induces a subgraph of G of diameter at most s. Clubs are generalizations of cliques (1-clubs are exactly cliques) and offer a wider and more practical way to model real-world interactions [13–16,18]. Partitioning a graph into cliques is important for clustering and community detection. Consequently, there has been research into partitioning graphs into s-clubs as a way to extend these methods to more flexible and realistic groupings. In this paper we focus exclusively on bipartite graphs. Bipartite graphs are of particular interest due to their wide range of applications in various fields such as scheduling, matching problems, and network flow optimization [2]. We examine two closely related problems involving

R. Královič and V. Kůrková (Eds.): SOFSEM 2025, LNCS 15539, pp. 198–210, 2025.
https://doi.org/10.1007/978-3-031-82697-9_15

the partitioning and covering of a graph's vertices using s-clubs. The first problem is the Minimum Partition s-Club problem, where the objective is to partition the vertices of a graph G into the minimum number of disjoint s-clubs. This problem is NP-hard for $s \geq 2$ [7], even when restricted to bipartite or chordal graphs [1,3]. Clearly from these results we have that the decision version of this problem - where for a fixed s, given a graph G and an integer k, we determine whether it is possible to partition the vertices of G into at most k disjoint s-clubs - is NP-complete. However, this does not address the complexity of the decision problem when k is also fixed (that is both s and k are not part of the input). We thus, consider the problem below.

k-Partition s-Club problem (PC(k, s))

Instance: A graph $G = (V, E)$.
Question: Is there a partition of V into at most k vertex disjoint s-clubs?

Previous studies have explored the complexity of this problem for some fixed values of s and k. For $k = 1$, the problem is equivalent to determining the diameter of a graph and thus is trivially solvable in polynomial time. For $s = 1$ the problem is equivalent to determine whether there exists a partition of the vertices into k cliques, that is equivalent to determine whether there exists a k-coloring of the complement graph. This problem is NP-complete for any $k \geq 3$ [11] and polynomial for $k = 2$. When we restrict to bipartite graphs, $s = 1$ corresponds to the problem of determining whether there exists a perfect matching of size k in a graph and is thus polynomial. For $s = 2$ and $k = 2$ the problem is polynomial[1] [10]. In this paper we show that for bipartite graphs and any fixed $s, k \geq 3$ the PC(k, s) problem is NP-complete. Note that our NP-completeness result is stronger than the one in [1] as we assume that both s and k are constants and not part of the input.

In some real-world applications, it may not be feasible to partition all vertices into s-clubs. To address such cases, a variant of this problem, known as the Maximum Disjoint (t, s)-Club Covering problem (Max-DCC(t, s)), was introduced by Dondi *et al.* in [9]. This problem seeks to find, given a graph G, a collection of disjoint (t, s)-clubs that covers the maximum number of vertices in G. A (t, s)-club is an s-club with at least t vertices. The concept of (t, s)-clubs extends that of s-clubs by adding a minimum size constraint, and is motivated by applications where identifying large, well-connected subgraphs is important (see, e.g., [9,14]). Hence, the second problem we consider is the following:

Maximum Disjoint. (t, s)-Club Covering problem (MAX-DCC(t, s))

Instance: A graph $G = (V, E)$.
Required: A collection of vertex disjoint (t, s)-clubs that covers the maximum number of vertices in V.

[1] Notice that if we do not require the s-clubs to be disjoint, then the case $s = 2$, $k = 2$ is shown to be NP-complete for general graphs [8].

To the best of our knowledge the only cases considered in literature are the cases $s = 2$ and $s = 3$ [9]. In particular, for $s = 2$ in [9] it is proved that Max-DCC(3, 2) is APX-hard and the case Max-DCC(2, 2) is left open. Here we show that Max-DCC(t, 2) is APX-hard for any fixed $t \geq 5$ even for bipartite graphs and Max-DCC(2, 2) can be solved in polynomial time for general graphs. For $s = 3$ in [9] it is shown that Max-DCC(2, 3) is polynomial and the case Max-DCC(3, 3) is left as an open problem. We show that $MAX\text{-}DCC(t, 3)$ is APX-hard for any fixed $t \geq 8$, even for bipartite graphs.

The paper is organized as follows: In Sect. 2 we introduce the definitions we need for the paper. In Sect. 3 we show that for any fixed $s \geq 3$ and for any fixed $k \geq 3$ the PC(k, s) problem is NP-complete. In Sect. 4 we prove that it is NP-hard to achieve an approximation factor of $\frac{95}{94}$ for MAX-DCC(t, 3) for any fixed $t \geq 8$ and for MAX-DCC(t, 2) for any fixed $t \geq 5$ even for bipartite graphs. On the positive side we provide a polynomial-time algorithm for MAX-DCC(2, 2). Finally, in Sect. 5 we conclude with some open problems.

2 Preliminaries

All the graphs we consider here are undirected and simple (with no loops or multiple edges). For a graph $G = (V, E)$ and a subset $V' \subseteq V$ we denote by $G[V']$ the subgraph induced by the vertices in V'. For any two vertices $u, v \in V$ we denote by $u - - - -v$ a path connecting u and v in G. To simplify notation and avoid confusion, we will slightly abuse notation by writing (u, v) for the edge between u and v, instead of $\{u, v\}$. A subset V' of vertices is called an *s-club* if the diameter of $G[V']$ is at most s. In other words, every pair of vertices in the s-club can be connected by a path of length at most s within the subgraph. An (t, s)-*club* is an s-club of at least t vertices.

The *closed neighborhood* of a vertex v in a graph $G = (V, E)$, denoted by $N[v]$, is the set consisting of the vertex v itself and all vertices adjacent to v. Formally, $N[v] = \{v\} \cup \{u \in V \mid (u, v) \in E\}$.

A graph $G = (V, E)$ is *bipartite* if its vertices can be partitioned into two independent sets. We denote it as $G = (V_1, V_2, E)$, where V_1 and V_2 are the independent sets. A tree T_G is said to be a *spanning tree* of a connected graph G if T_G is a subgraph of G (not necessarily induced) and T_G contains all vertices of G. A *rooted tree* is a tree with a special vertex labelled as the *root* of the tree. In a tree, a vertex v is said to be at *level* l if v is at a distance l from the root. The *height* of a tree is the maximum level which occurs in the tree. The *parent* of a vertex v is the vertex connected to v on the path to the root.

For any integer k we denote by $[k]$ the set $\{1, 2, \ldots, k\}$. We denote by $2^{[k]}$ the family of all possible subsets of $[k]$.

3 NP-Completeness of PC(k, s) problem

We reduce from the k-List Coloring problem, which determines whether a graph admits a proper coloring compatible with a given list assignment to its vertices with colors in $[k]$. Specifically, for a graph $G = (V, E)$ and a list assignment $L : V \to 2^{[k]}$, the task is to determine whether there exists a proper coloring c such that each vertex is assigned a color from its list. Recall that a proper coloring is an assignment of colors to the vertices of a graph so that no two adjacent vertices have the same color. If c exists, it is called an L-coloring of G. It is known that k-List Coloring is NP-complete, even for bipartite graphs when $k = 3$ [12]. In this paper, we consider a slightly different variant of this problem, which we will define and prove to be NP-complete below.

Balanced Bipartite. k-List Coloring problem

Instance: A bipartite graph $G = (V_1, V_2, E)$ and a list assignment $L : V_1 \cup V_2 \to 2^{[k]}$.

Question: Is there an L-coloring such that for each color used, either there is only one vertex of that color, or at least one vertex from each of the partitions V_1 and V_2 is assigned that color?

Due to space constraints, we state the result without providing the proof.

Lemma 1. *The Balanced Bipartite k-List Coloring problem is NP-complete for any fixed $k \geq 3$.*

We prove now that the PC(k, s) problem is NP-complete for any fixed $k \geq 3$, $s \geq 3$ and even for bipartite graphs. We will reduce from the Balanced Bipartite k-List-Coloring problem. Let $G = (V_1, V_2, E)$, together with a list assignment L, be an instance of this problem. We construct a new bipartite graph $G' = (V', E')$ using the following process:

- First, we take the bipartite complement of G, which is the graph obtained by complementing the edges between V_1 and V_2.
- Next, for each color $i \in [k]$, we add the color vertices c_i and $\hat{c}_i$, along with the edge $\{c_i, \hat{c}_i\}$.
- Finally, for each vertex $x \in V_1$ and each color $i \in L(x)$, we add a path of length $s-1$ between x and $\hat{c}_i$. This means we introduce new *auxiliary* vertices $a^1_{x,i}, a^2_{x,i}, \ldots, a^{s-2}_{x,i}$, and connect them as follows: x is connected to $a^1_{x,i}$, $a^1_{x,i}$ is connected to $a^2_{x,i}$, and so on, until $a^{s-2}_{x,i}$ is connected to $\hat{c}_i$. We call this path $x - - - - \hat{c}_i$ an *auxiliary* path. We repeat this construction similarly for the vertices in V_2 and color vertices c_i.

See Fig. 1(a)-(b) for an example of the construction. More formally,

$$
\begin{aligned}
V' = & V_1 \cup V_2 \cup \{c_i \mid i \in [k]\} \cup \{\hat{c}_i \mid i \in [k]\} \\
& \cup \{a^j_{x,i} \mid x \in V_1, i \in L(x), j \in [s-2]\} \\
& \cup \{a^j_{y,i} \mid y \in V_2, i \in L(y), j \in [s-2]\}, \\
E' = & \overline{E} \cup \{\{c_i, \hat{c}_i\} \mid i \in [k]\} \\
& \cup \{(x, a^1_{x,i}), (a^1_{x,i}, a^2_{x,i}), \ldots, (a^{s-2}_{x,i}, \hat{c}_i) \mid x \in V_1, i \in L(x)\} \\
& \cup \{(y, a^1_{y,i}), (a^1_{y,i}, a^2_{y,i}), \ldots, (a^{s-2}_{y,i}, c_i) \mid y \in V_2, i \in L(y)\}.
\end{aligned}
$$

where $\overline{E}$ are the edges that are not present in G.

Lemma 2. *The graph G' is bipartite.*

Proof. We provide now the bipartition V_1', V_2' for G'. We need to consider two cases based on the parity of s.

Case 1: Let s be odd. We begin by considering the bipartition V_1, V_2 of G, and we assign colors and auxiliary vertices to extend these two sets. To achieve this, observe that each auxiliary path consists of an odd number of vertices (specifically s vertices). This allows us to assign the endpoints of each path to the same set in the bipartition, while alternating the assignment of the auxiliary vertices between the two sets. More formally, we define:

$$\begin{aligned} V_1' =& V_1 \cup \{\hat{c}_i \mid i \in [k]\} \cup \{a^j_{x,i} \mid x \in V_1, i \in L(x), j \in [s-2] \wedge j \text{ even}\} \\ &\cup \{a^j_{y,i} \mid y \in V_2, i \in L(y), j \in [s-2] \wedge j \text{ odd}\}. \\ V_2' =& V_2 \cup \{c_i \mid i \in [k]\} \cup \{a^j_{x,i} \mid x \in V_1, i \in L(x), j \in [s-2] \wedge j \text{ odd}\} \\ &\cup \{a^j_{y,i} \mid y \in V_2, i \in L(y), j \in [s-2] \wedge j \text{ even}\}. \end{aligned}$$

Refer to Fig. 1(c) for an example of how the bipartition is defined.

Case 2: s is even. This case follows in the same way as the previous one, by noticing that in this case each auxiliary path consists of an even number of vertices and thus the two endpoints have to be assigned to different sets in the bipartition. Thus we define:

$$\begin{aligned} V_1' =& V_1 \cup \{c_i \mid i \in [k]\} \cup \{a^j_{x,i} \mid x \in V_1, i \in L(x), j \in [s-2] \wedge j \text{ even}\} \\ &\cup \{a^j_{y,i} \mid y \in V_2, i \in L(y), j \in [s-2] \wedge j \text{ odd}\}. \\ V_2' =& V_2 \cup \{\hat{c}_i \mid i \in [k]\} \cup \{a^j_{x,i} \mid x \in V_1, i \in L(x), j \in [s-2] \wedge j \text{ odd}\} \\ &\cup \{a^j_{y,i} \mid y \in V_2, i \in L(y), j \in [s-2] \wedge j \text{ even}\}. \end{aligned}$$

It is straightforward to verify that in both cases we have a bipartite graph. □

Lemma 3. *G has a balanced L-colouring using colors in $[k]$ if and only if G' has a partition into at most k vertex-disjoint s-clubs.*

Proof. ⇒ Let c be a balanced L-coloring of G, and let $S_1, S_2, \dots, S_k$ denote the sets of vertices assigned colors $1, 2, \dots, k$, respectively. Notice that for some i, it may be that $S_i = \emptyset$. We define in G' the following sets for all $i \in [k]$,

$$S_i' = S_i \cup C_i \cup A_i \text{ where } C_i = \{c_i, \hat{c}_i\} \text{ and } A_i = \{a^l_{v,i} \mid v \in S_i, l \in [s-2]\}$$

In other words, S_i' consists of the vertices in S_i, along with the corresponding color vertices c_i and $\hat{c}_i$, as well as all the auxiliary vertices on the paths connecting the vertices of S_i to the color vertices.

Clearly the S_i's form a partition of V'. It remains to show that for all $i \in [k]$, S_i' is an s-club. To this purpose we determine the distance of any two vertices u, v in S_i'. For simplicity we denote by $V_1^i = S_i' \cap V_1$ and $V_2^i = S_i' \cap V_2$ the two bipartitions of S_i' restricted to the vertices of the graph G. The following cases need to be considered::

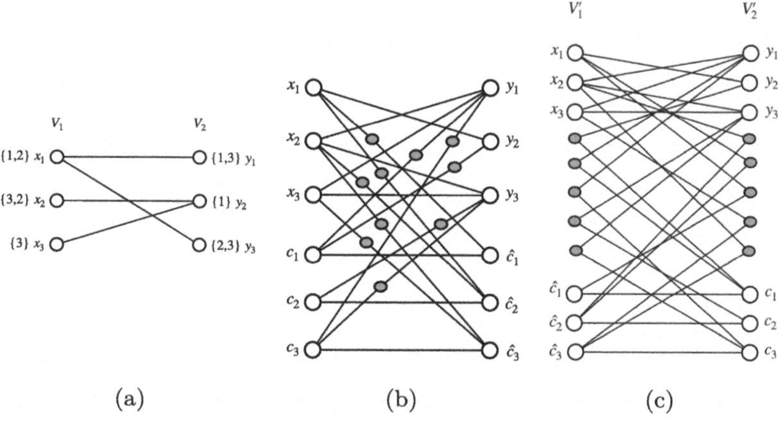

Fig. 1. (a) A bipartite graph G with the list coloring L and (b) the corresponding graph G' for $s = 3$ and $k = 3$. The gray vertices represent the auxiliary vertices added in the construction. (c) The bipartite representation of G'.

1. u, v are in S_i:
   - $u \in V_1^i$ and $v \in V_2^i$ then $d(u, v) = 1$ as S_i induces an independent set in G and thus a biclique in G'.
   - $u, v \in V_1^i$, by the definition of Balanced k-list coloring, there exists a vertex $y \in V_2^i$ such that both u and v are connected to y, which implies $d(u, v) = 2$. A similar argument applies for $u, v \in V_2^i$.
2. $u \in S_i$, $v \in C_i$: We assume $u \in V_1^i$ (the case $u \in V_2^i$ is symmetric). There are two cases to consider:
   - $v = \hat{c}_i$: then $d(u, v) = s - 1$ thanks to the auxiliary path $u - - - - \hat{c}_i$.
   - $v = c_i$: then $d(u, v) = s$ thanks to the path $u - - - - \hat{c}_i - c_i$.
3. $u \in A_i$, $v \in S_i$: Let $u = a^j_{x,i}$ for $x \in V_1^i, j \in [s - 2]$ (the case $u = a^j_{y,i}$ for $y \in V_2^i$ follows in an identical way).
   (a) $v = x' \in V_1^i$; Then if $x = x'$ we have $d(a^j_{x,i}, x') \leq s - 2$, by construction. Otherwise by definition of the balanced k-list coloring, there exists $y \in V_2^i$ such that x and x' are both connected to y, so we consider the cycle $C = x' - y - x - - - - a^j_{x,i} - - - - \hat{c}_i - - - - x'$. We have that the total length of the cycle is exactly $1 + 1 + s - 1 + s - 1 = 2s$. Then any two vertices in this cycle are at distance at most s. As both x' and $a^j_{x,i}$ are in C, we have $d(a^j_{x,i}, x') \leq s$.
   (b) $v = y \in V_2^i$. Then as the vertices in $V_1^i \cap V_2^i$ form a biclique we have the path $y - x - - - - a^j_{x,i}$ which has length at most $1 + s - 1$. Hence $d(a^j_{x,i}, y) \leq s$.
4. u, v in C_i: $d(u, v) = 1$ by construction of G'.
5. $u \in A_i$, $v \in C_i$: we have $d(u, v) \leq s - 2$ if u and v belong to the same part of the bipartition in G' (due to the existence of an auxiliary path between them). Otherwise, $d(u, v) \leq s - 1$, owing to the auxiliary path and the edge between c_i and $\hat{c}_i$.

6. $u, v \in A_i$: Let $u = a^j_{x,i}$ for $x \in V^i_1, j \in [s-2]$ (the case $u = a^j_{y,i}$ for $y \in V^i_2$ follows in an identical way).
   - $v = a^{j'}_{x',i}$ for $x' \in V^i_1, j' \in [s-2]$. This case is proven in an identical way to the case 3.(a).
   - $v = a^{j'}_{y,i}$ for $y \in V^i_2, j' \in [s-2]$. Then we consider the cycle $C = x - y - - - - a^{j'}_{y,i} - - - - c_i - \hat{c}_i - - - - a^j_{x,i} - - - - x$. We have that the total length of the cycle is exactly $1 + s - 1 + 1 + s - 1 = 2s$. Then any two vertices in this cycle are at distance at most s and thus we have $d(a^j_{x,i}, a^{j'}_{y,i}) \leq s$.

$\Leftarrow$ Suppose that $S'_1, S'_2, \dots, S'_k$ form a partition of G' in k s-clubs. Notice that in G' for any $i \neq j$ with $i, j \in [k]$, we have $d(c_i, c_j) \geq 2(s-1) > s$ (recall that $s \geq 3$). Indeed the shortest path between c_i and c_j occurs when there exists a node y in G' with two auxiliary paths to c_i and c_j respectively (i.e. $c_i - - - - y - - - - c_j$). Similarly, $d(c_i, \hat{c_j}) \geq 2(s-1)+1 > s$ as the shortest path between them takes places when there exists a node y with auxiliary paths to both c_i and c_j (i.e. $c_i - - - - y - - - - c_j - \hat{c_j}$) or a node x with auxiliary paths to both $\hat{c}_i$ and $\hat{c}_i$ (i.e. $c_i - \hat{c}_i - - - - x - - - - \hat{c_j}$). Thus, since there are k pairs $(c_i, \hat{c}_i)$, each pair must belong to exactly one s-club. Therefore, without loss of generality, we can assume that for any $i \in [k]$, both c_i and $\hat{c}_i$ are contained in S_i. We will now prove the following three properties

1. for each $x \in S'_i \cap V_1$ and for each $y \in S'_i \cap V_2$ the auxiliary paths $x - - - - \hat{c}_i$ and $y - - - - c_i$ are in G'.
2. for each $x \in S'_i \cap V_1$ and $y \in S'_i \cap V_2$ it must hold that $(x, y) \in E(G')$
3. if $|S'_i \cap (V_1 \cup V_2)| > 1$ we have at least one vertex in $S'_i \cap V_1$ and at least one vertex in $S'_i \cap V_2$

The first property, guarantees that $i \in L(u)$ for each $u \in S_i \cap (V_1 \cup V_2)$. The second property, guarantees that the vertices in $S'_i \cap (V_1 \cup V_2)$ form an independent set in G. These two properties imply that $C_1, C_2, \dots C_k$ where $C_i = S'_i \cap (V_1 \cup V_2)$ is an L-coloring for G. Moreover, the third property ensures that the L-coloring is balanced. It remains to prove the three properties.

1. Let $x \in S'_i \cap V_1$. In absence of the auxiliary path $x - - - - \hat{c}_i$ the shortest path between x and $\hat{c}_i$ in G' has length at least $s+1$ (i.e. $x - y - x' - - - - \hat{c}_i$ or $x - y - x' - - - - c_i - \hat{c}_i$ for $y \in S_i \cap V_2$ and $x' \in S_i \cap V_1$). However since both x and $\hat{c}_i$ are in the s-club S_i, it must hold that $d(x, \hat{c}_i) \leq s$. Therefore the auxiliary path is in G' and all of its vertices are in S'_i.
2. From the previous point, we know that $a^{s-2}_{x,i}$ is in S'_i. We show that if $(x, y) \notin E(G')$ then $d(y, a^{s-2}_{x,i}) > s$ in G'. Notice that the only vertices adjacent to $a^{s-2}_{x,i}$ are $\hat{c}_i$ and $a^{s-3}_{x,i}$. Thus, we can reach $a^{s-2}_{x,i}$ either through $\hat{c}_i$ and hence through c_i, or through $a^{s-3}_{x,i}$ and hence through x. The shortest path passing through c_i is $y - - - - c_i - \hat{c_j} - a^{s-2}_{x,i}$ which is at least $s+1$ long. The shortest path passing through x is $y - x' - y' - x - - - - a^{s-2}_{x,i}$ for some $x' \in S'_i \cap V_1$ and $y' \in S'_i \cap V_1$, which is at least $s+2$ long. Thus, in both cases we reach a contradiction as y and $a^{s-2}_{x,i}$ belong to the s-club S'_i.

3. We will prove that if $x, x' \in S_i \cap V_1$ and $S_i \cap V_2 = \emptyset$ then $d(x, x') > s$ in $G'[S_i]$ (the proof for $y, y' \in S_i \cap V_1$ and $S_i \cap V_1 = \emptyset$ is analogous). The shortest path between x and x' passes through c_i (i.e. $x - - - - c_i - - - - x'$) and hence $d(x, x') > 2(s-1) > s$.

This concludes the proof. □

Theorem 1. *The PC(k, s) problem is NP-complete for any fixed $k \geq 3$, $s \geq 3$ and even for bipartite graphs.*

Proof. The proof follows from Lemmas 1, 2 and 3. □

4 Hardness of MAX-DCC(t, s) problem

4.1 The MAX-DCC($t, 3$) problem

In [9] it was proven that $MAX\text{-}DCC(2,3)$ can be solved in polynomial time and the complexity of $MAX\text{-}DCC(3,3)$ was posed as an open problem. In this section we prove that $MAX\text{-}DCC(t,3)$ is APX-hard for any fixed $t \geq 8$, even in bipartite graphs. We will use the following problem, which was proven to be NP-hard to approximate in [4].

Maximum 2 Bounded 3-Dimensional Matching problem (Max-2B3DM)

Instance: A set $M \subseteq X \times Y \times Z$ of ordered triples where X, Y and Z are disjoint sets and the number of occurrences in M of an element in X, Y or Z is bounded by constant 2.

Required: The largest matching $M' \subseteq M$, that is, the largest subset such that no two elements of M agree in any coordinate.

In order to prove that MAX-DCC($t, 3$) with $t \geq 8$ is NP-hard we give an L-reduction from (Max-2B3DM). For the definition of an L-reduction see [17].

We begin by describing the reduction. Let $M = \{C_1, C_2, \ldots, C_m\}$ be an instance of the Max-2B3DM problem, where each C_i is an ordered triple in $X \times Y \times Z$. Fix a constant $t \geq 8$. We construct a bipartite graph $G_{M,t,3} = (V_1, V_2, E)$, as follows:

$$\begin{aligned}
V_1 &= X \cup Y \cup Z \cup \{a_i \mid i \in [m]\} \\
V_2 &= \{c_i \mid i \in [m]\} \cup \{h_{i,j} \mid i \in [m], j \in [t-5]\} \\
E &= \Big\{(c_i, a_i) \mid i \in [m]\Big\} \bigcup \Big\{(a_i, h_{i,j}) \mid i \in [m], j \in [t-5]\Big\} \bigcup_{i \in [m]} E_{C_i}
\end{aligned}$$

where $E_{C_i} = \{(c_i, x), (c_i, y), (c_i, z)\}$ for each triple $C_i = (x, y, z)$ in M. As an example see Figure 2.

Claim 1. *The graph $G_{M,t,3}$ can be constructed in polynomial time, is bipartite and has maximum degree $t - 4$.*

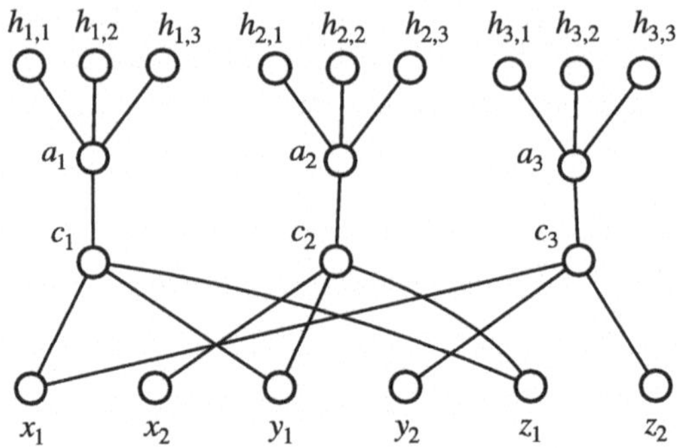

Fig. 2. The graph $G_{M,8,3}$ obtained when the instance of Max-2B3DM problem is the set $M = \{(x_1, y_1, z_1), (x_2, y_1, z_1)\ (x_1, y_2, z_2)\}$.

Proof. It is easy to see that $G_{M,t,3}$ can be constructed in polynomial time. Then by construction V_1 and V_2 form a bipartition as there are no edges within the sets V_1 and V_2. Furthermore, for every $i \in [m]$, the vertex c_i has degree 4, a_i has degree $t-4$, and for every $j \in [t-5]$, the vertex $h_{i,j}$ has degree 1. Finally, every vertex $u \in X \cup Y \cup Z$ has degree 2, by the definition of the Max-2B3DM problem. □

We need the following lemma.

Lemma 4. *In the graph $G_{M,t,3}$, all 3-clubs are of size at most t, and the only 3-clubs of size exactly t are of the form $N[c_i] \cup \{h_{i,j} \mid j \in [t-5]\}$ for $1 \le i \le m$.*

Proof. To prove the claim, we will show that any 3-club in $G_{M,t,3}$ either contains exactly t vertices or has at most 7 vertices (with $7 < t$). Clearly the subgraphs of $G_{M,t,3}$ induced by $N[c_i] \cup \{h_{i,j} | \ j \in [t-5]\}$ for $1 \le i \le m$ are 3-clubs of size t.

To simplify the analysis, we classify the vertices in $G_{M,t,3}$ into four distinct sets: the vertices in $W = X \cup Y \cup Z$, the vertices in $C = \{c_1, \ldots, c_m\}$, the vertices in $A = \{a_1, \ldots, a_m\}$, and the vertices in $H = \{h_{i,j} \mid i \in [m],\ j \in [t-5]\}$.

Let S be a 3-club in $G_{M,t,3}$. We will now show that if S contains at least two vertices from C then $|S| \le 7 < t$. We make the following considerations regarding the composition of the vertices in S.

- *the number of vertices in $S \cap H$ is zero.* Note that such 3-club cannot include any vertices from H, as a vertex in H would be at a distance of at least 4 from one of the two vertices in C.
- *the number of vertices in $S \cap C$ is at most* 3. Suppose on the contrary there are 4 vertices from C, say c_1, c_2, c_3 and c_4, For S to be a 3-club, there must be a vertex from $S \cap W$ adjacent to each pair of the c_i ($i \in [4]$) vertices. Let u be the vertex $S \cap W$ adjacent to c_1 and c_2, and v be the vertex from $S \cap W$ adjacent c_3 and c_4. The vertices u and v are at distance at least 4 in

$G_{M,t,3}[S]$ because $N[u] \cap N[v] = \emptyset$ as all vertices from W have degree 2. This contradicts the assumption that S is a 3-club.

- *The number of vertices in* $S \cap W$ *is at most* 4. Suppose on the contrary that there are 5 such vertices, $U = \{u_1, u_2, u_3, u_4, u_5\}$. At least one vertex $u \in U$ must have degree 1 in $G_{M,t,3}[S]$. Otherwise, if all vertices in U had degree 2, there would be 10 edges between these 5 vertices and the vertices in $S \cap C$. However, as shown in the previous point, there can be at most 3 vertices in $S \cap C$, and since each of these vertices has degree 3, this leads to a total of only 9 edges between U and $S \cap C$. Finally notice that u is at distance at least four from one of the vertices $S \cap C$ (recall that in C there are at least two vertices) contradicting the fact that S is a 3-club.

Given the above, if $S \cap A = \emptyset$ then trivially the 3-club cannot have more than 7 vertices. Now assume that $S \cap A \neq \emptyset$. Notice that it must hold $|S \cap A| = 1$, since any two vertices in $S \cap A$ would be at distance at least 4 from each other. Let $a_i \in S \cap A$, then the vertices in W that a_i can reach with a path of length at most three are the ones adjacent to c_i. Thus $|S \cap W| \leq 3$ and the total number of vertices in the 3-club remains bounded by $1 + 3 + 3 = 7$. □

Theorem 2. *Let t be a constant with $t \geq 8$. It is NP-hard to approximate the solution of MAX-DCC(t, 3) within a factor of $\frac{95}{94}$, even for bipartite graphs with a constant maximum degree of $t - 4$.*

Proof. Let $G_{M,t,3}$ be the graph obtained from an instance M of Max-2B3DM, for a given fixed t, with $t \geq 8$. By Claim 1 we have that $G_{M,t,3}$ can be constructed in polynomial time, has maximum degree of $t - 4$ and is bipartite. We prove now that from a matching of $k \geq 1$ triples in M, we can always obtain, in polynomial time, a disjoint cover with $(t, 3)$-clubs that covers $t \cdot k$ vertices in $G_{M,t,3}$. Moreover, for every disjoint cover of $G_{M,t,3}$ with $(t, 3)$-clubs that covers k vertices in $G_{M,t,3}$ it is possible to obtain in polynomial time a matching of $\frac{k}{t}$ triples in M (note that by Lemma 4 we have that k is a multiple of t).

Let $S = \{C_1, \ldots C_k\}$ be a subset of k triples from M that correspond to a matching. We define for each $i \in [k]$ the following set in $G_{M,t,3}$.

$$S'_i = \{x, y, z, c_i, a_i, h_{i,1} \ldots h_{i,t-5}\}$$

Notice that S'_i is obviously a 3-club of size t and, since the triples in S form a matching, it follows that $S'_i \cap S'_j = \emptyset$ for any $i \neq j$. Thus, $S'_1, S'_2, \ldots, S'_k$ form a disjoint cover with $(t, 3)$-clubs that covers $t \cdot k$ vertices of $G_{M,t,3}$.

Let now $S_1, S_2, \ldots, S_r$ be a disjoint cover with $(t, 3)$-clubs that covers k vertices in $G_{M,t,3}$. By Lemma 4, we know that this cover is the union of $\frac{k}{t}$ disjoint 3-clubs, each containing t vertices of the form $N[c_i] \cup \{h_{i,j} | \ j \in [t - 5]\}$ for some i, $1 \leq i \leq |M|$ where c_i corresponds to a triple (x, y, z) in M. To obtain the $\frac{k}{t}$ triples in M corresponding to a matching, it is enough to take the triples corresponding to the $\frac{k}{t}$ disjoint 3-clubs from S.

We therefore have an L-reduction with $a = t$ and $b = \frac{1}{t}$ from Max-2B3DM to Max-DCC(t, 3). In [5], it was proven that for Max-2B3DM, achieving an approximation factor better than $\frac{95}{94}$ is NP-hard. Therefore we deduce the same result for Max-DCC(t, 3) (*see* [17] *for more information about L-reductions and in particular Theorem* 16.5*).* □

4.2 The MAX-DCC(t, 2) problem

In [9] it was proven that $MAX\text{-}DCC(3, 2)$ is APX-hard and the complexity of $Max\text{-}DCC(2, 2)$ was posed as an open problem. In this section we prove that $MAX\text{-}DCC(t, 2)$ is APX-hard for any fixed $t \geq 5$, even in bipartite graphs and that $MAX\text{-}DCC(2, 2)$ can be solved in polynomial time.

The following theorem can be proved using similar ideas as in Theorem 2.

Theorem 3. *Let t a constant with , $t \geq 5$. It is NP-hard to approximate the solution of Max-DCC(t, 2) to within $\frac{95}{94}$ even for bipartite graphs of degree at most $t - 1$.*

In cite [9], it is shown that for any $s \geq 3$ Max-DCC(2, s) is solvable in linear time. The authors leave the problem of determining the complexity of Max-DCC(2, 2) as an open problem. We prove the following.

Theorem 4. *Max-DCC(2, 2) can be solved in linear time.*

Proof. Let G be the input graph for Max-DCC(2, 2). We will prove the claim by describing a linear algorithm that produces a cover of disjoint (2, 2)-clubs that covers the maximum number of vertices of G. Consider the graph G', obtained by removing the isolated vertices from G. Note that isolated vertices cannot be covered by a (2, 2)-club, so they can be disregarded. We can assume that G' is connected as otherwise, we apply the following procedure to each connected component.

From G', construct a rooted spanning tree $T_{G'}$. This can be done in linear time see [6]. Notice that $T_{G'}$ has a height of at least one as G' has no isolated vertices. Thus, let x be a vertex of $T_{G'}$ at maximum level and let y be the parent of x. Consider the subtree T_y rooted at y. Clearly, T_y contains at least 2 vertices. Moreover, as x has maximum level then T_y is of height 1 and all vertices in this subtree are at a distance at most 2 from each other. Hence, the set S_1 of vertices in T_y form a (2, 2)-club. We add S_1 to the solution and remove T_y from $T_{G'}$. The remaining tree is still connected, so we can iterate the process until either we reach an empty tree or we reach a tree consisting of a single vertex. In the latter case, let s be this vertex and let S_i be the (2, 2)-club added to the solution by the last iteration. Notice that s is at distance 1 from the root of this subtree and at distance 2 from any vertex of S_i. Therefore, we can add s to S_i and still have a (2, 2)-club. In the end the solution obtained is a cover of *all* the vertices of G' with disjoint (2, 2)-clubs and thus is a solution for Max-DCC(2, 2) on G. □

5 Open Problems

Several problems remain open in the context of bipartite graphs. Specifically, the cases of PC$(t, 2)$ for any $t \geq 3$ and PC$(2, s)$ for $s \geq 3$ remain unsolved.

Regarding the second problem, for $s = 3$, the complexity of the Max-DCC$(t, 3)$ problem is still open for $3 \leq t \leq 7$. Similarly, for $s = 2$, the case of Max-DCC$(4, 2)$ remains unsolved.

Acknowledgments. This research was supported in part by MUR PRIN Project EXPAND, grant number 2022TS4Y3N.

Disclosure of Interests. The authors have no competing interests to declare that are relevant to the content of this article.

References

1. Abbas, N., Stewart, L.: Clustering bipartite and chordal graphs: Complexity, sequential and parallel algorithms. Discret. Appl. Math. **91**(13), 123 (1999). https://doi.org/10.1016/s0166-218x(98)00094-8
2. Asratian, A.S., Denley, T.M.J., Häggkvist, R.: Bipartite Graphs and their Applications. Cambridge University Press (1998)
3. Chang, J.M., Yang, J.S., Peng, S.L.: On the complexity of graph clustering with bounded diameter. In: 2014 International Computer Science and Engineering Conference (ICSEC), vol. 10, p. 1822. IEEE (Jul 2014).https://doi.org/10.1109/icsec.2014.6978122
4. Chlebík, M., Chlebíková, J.: Approximation hardness for small occurrence instances of NP-Hard problems. In: Petreschi, R., Persiano, G., Silvestri, R. (eds.) Algorithms and Complexity, pp. 152–164. Springer Berlin Heidelberg, Berlin, Heidelberg (2003). https://doi.org/10.1007/3-540-44849-7_21
5. Chlebík, M., Chlebíková, J.: Complexity of approximating bounded variants of optimization problems. Theoret. Comput. Sci. **354**(3), 320–338 (2006). https://doi.org/10.1016/j.tcs.2005.11.029, foundations of Computation Theory (FCT 2003)
6. Cormen, T.H., Leiserson, C.E., Rivest, R.L., Stein, C.: Introduction to Algorithms. The MIT Press, 2nd edn. (2001), http://www.amazon.com/Introduction-Algorithms-Thomas-H-Cormen/dp/0262032937%3FSubscriptionId%3D13CT5CVB80YFWJEPWS02%26tag%3Dws%26linkCode%3Dxm2%26camp%3D2025%26creative%3D165953%26creativeASIN%3D0262032937
7. Deogun, J.S., Kratsch, D., Steiner, G.: An approximation algorithm for clustering graphs with dominating diametral path. Inf. Process. Lett. **61**(3), 121127 (1997). https://doi.org/10.1016/s0020-0190(97)81663-8
8. Dondi, R., Lafond, M.: On the tractability of covering a graph with 2-clubs. Algorithmica **85**(4), 992–1028 (2023). https://doi.org/10.1007/S00453-022-01062-3
9. Dondi, R., Mauri, G., Zoppis, I.: On the tractability of finding disjoint clubs in a network. Theor. Comput. Sci. **777**, 243251 (Jul 2019). https://doi.org/10.1016/j.tcs.2019.03.045
10. Fleischner, H., Mujuni, E., Paulusma, D., Szeider, S.: Covering graphs with few complete bipartite subgraphs. Theoret. Comput. Sci. **410**(2123), 20452053 (2009). https://doi.org/10.1016/j.tcs.2008.12.059

11. Garey, M.R., Johnson, D.S.: Computers and Intractability: A Guide to the Theory of NP-Completeness. Freeman, W. H (1979)
12. Gravier, S., Kobler, D., Kubiak, W.: Complexity of list coloring problems with a fixed total number of colors. Discret. Appl. Math. **117**(13), 6579 (2002). https://doi.org/10.1016/s0166-218x(01)00179-2
13. Komusiewicz, C.: Multivariate algorithmics for finding cohesive subnetworks. Algorithms **9**(1), 21 (2016)
14. Laan, S., Marx, M., Mokken, R.J.: Close communities in social networks: boroughs and 2-clubs. Social Network Anal. Mining **6**(1) (Apr 2016). https://doi.org/10.1007/s13278-016-0326-0
15. . Mokken, R.J.: Cliques, clubs and clans. Quality amp; Quantity 13(2), 161173 (Apr 1979). https://doi.org/10.1007/bf00139635 ¡error l="308" c="Invalid command: paragraph not started." /¿
16. Mokken, R.J., Heemskerk, E.M., Laan, S.: Close communication and 2-clubs in corporate networks: Europe 2010. Social Netw. Anal. Mining **6**(1) (Jun 2016). https://doi.org/10.1007/s13278-016-0345-x
17. Williamson, D.P., Shmoys, D.B.: The design of approximation algorithms. Cambridge university press (2011)
18. Zoppis, I., Dondi, R., Santoro, E., Castelnuovo, G., Sicurello, F., Mauri, G.: Optimizing social interaction. In: Proceedings of the 11th International Joint Conference on Biomedical Engineering Systems and Technologies, pp. 651657. SCITEPRESS - Science and Technology Publications (2018). https://doi.org/10.5220/0006730606510657

Query Learning of Context-Deterministic and Congruential Context-Free Languages over Infinite Alphabets

Yutaro Numaya(✉), Yoshito Kawasaki, Ryo Yoshinaka, and Ayumi Shinohara

Tohoku University, Sendai, Japan
yutaro.numaya.q3@dc.tohoku.ac.jp

Abstract. This paper is concerned with learning some context-free languages (CFLs) over infinite alphabets from membership and equivalence queries. Argyros and D'Antoni (2018) proposed a query learning algorithm for regular languages over infinite alphabets using deterministic symbolic automata. Our algorithms learn context-deterministic CFLs and congruential CFLs over infinite alphabets. We enhance the existing algorithms for learning those CFL classes over finite alphabets by adapting Argyros and D'Antoni's technique. Our result shows that their technique can be extended to learn richer classes beyond regular languages.

Keywords: Active learning · Query learning · Distributional learning · Context-free languages · Infinite alphabets

1 Introduction

Learning formal languages involves inferring automata or grammars from observed strings over an alphabet and has a wide range of applications in fields such as pattern recognition, model checking, data mining, computational linguistics, and more [10]. This paper is concerned with the *query learning* (a.k.a. active learning) of context-free languages (CFLs) over large or infinite alphabets.

In her seminal paper, Angluin [1] presented the first query learning algorithm for regular languages in polynomial time, using a teacher who answers *membership queries* (MQs) and *equivalence queries* (EQs). Since then, several variants have been proposed in the literature (e.g., [5,11,14,18]). Most existing learning algorithms assume that the alphabet is finite and typically small. However, sequential data in the real world, such as time-series data, often come from large or infinite alphabets. Recently, several attempts have been made to extend existing algorithms to learn languages over infinite alphabets. In order to represent regular languages over infinite alphabets, finite-state automata need to be extended. One of the most popular extensions is *symbolic finite-state automata* (SFA) [21]. They have *predicates* as edge labels, which represent potentially infinite sets of input symbols in a compact manner. Particularly for the

R. Královič and V. Kůrková (Eds.): SOFSEM 2025, LNCS 15539, pp. 211–224, 2025.
https://doi.org/10.1007/978-3-031-82697-9_16

needs in model checking, the learning of different types of SFAs has been discussed [4,9,15,16]. Among those, Argyros and D'Antoni's work [3] is prominent due to its generality. They proposed an algorithm framework for learning deterministic SFAs over any kind of predicates that are query-learnable. Their technique has been applied to other settings, for example, learning nondeterministic SFAs [6] and weighted SFAs [20], and learning with a teacher who sometimes does not answer MQs [13].

On the other hand, the learning of CFLs is considered to be quite a challenging task. Only a limited number of context-free subclasses are known to be efficiently query-learnable. *Simple deterministic* CFLs [12], *context-deterministic* CFLs [19], and *congruential* CFLs are representative examples. Those are superclasses of regular languages and incomparable with each other.

The main focus of this paper is the query learning of CFLs over large or infinite alphabets. Indeed, much of the data in the real world, such as XML documents and program traces, have context-free structures over large or virtually infinite alphabets [8]. Numaya et al. [17] define an extension of context-free grammars called Ψ-CFGs with a predicate set Ψ, and show that *substitutable* Ψ-CFLs over infinite alphabets are *identifiable in the limit from positive data*. The learning model differs from the query learning and their learning target class is rather small. In fact, the class of substitutable CFLs is incomparable with that of regular languages. As far as we know, except for their work, there are no results on learning CFLs over infinite alphabets, particularly under the query learning model. In this paper, we show that context-deterministic Ψ-CFLs and congruential Ψ-CFLs over infinite alphabets are efficiently learnable using MQs and EQs provided that the predicate set Ψ is also query-learnable. Our result can be seen as a context-free counterpart of Argyros and D'Antoni's framework [3] for the query learning of SFAs.

2 Preliminaries

2.1 Languages and Grammars

Let Σ be an *alphabet*, a nonempty finite or countably infinite set of characters. The set of finite strings over Σ is denoted by Σ^*. A *language* is any subset of Σ^*. We write ε for the empty string. A *context* is just a pair of strings $u_1, u_2 \in \Sigma^*$, for which we write $u_1 \Box u_2$. The composition of a context $u_1 \Box u_2 \in \Sigma^* \times \Sigma^*$ and a string $v \in \Sigma^*$ is $u_1 \Box u_2 \odot v = u_1 v u_2 \in \Sigma^*$. This operation is naturally extended to sets $\Pi \subseteq \Sigma^* \times \Sigma^*$ and $V \subseteq \Sigma^*$ as $\Pi \odot V = \{\pi \odot v \,|\, \pi \in \Pi,\ v \in V\}$. We extract substrings and contexts from a string $w \in \Sigma^*$ as

$$\mathrm{Sub}(w) = \{v \in \Sigma^* \mid \pi \odot v = w \text{ for some } \pi \in \Sigma^* \times \Sigma^*\},$$
$$\mathrm{Con}(w) = \{\pi \in \Sigma^* \times \Sigma^* \mid \pi \odot v = w \text{ for some } v \in \Sigma^*\}.$$

These functions are naturally extended to the functions from sets of strings. We denote the set of contexts that accept a string $v \in \Sigma^*$ in a language L by

$$L \oslash v = \{\pi \in \Sigma^* \times \Sigma^* \mid \pi \odot v \in L\}.$$

Similarly, the set of strings that a context $\pi \in \Sigma^* \times \Sigma^*$ accepts is

$$L \oslash \pi = \{v \in \Sigma^* \mid \pi \odot v \in L\}.$$

When L is understood, particularly when our learning target is L, we use $v^{\triangleright}$ for $L \oslash v$ and use $\pi^{\triangleleft}$ for $L \oslash \pi$, where $v \in \Sigma^*$ and $\pi \in \Sigma^* \times \Sigma^*$. Two strings $u, v \in \Sigma^*$ are said to be *congruent*, denoted as $u \equiv_L v$, if $u^{\triangleright} = v^{\triangleright}$. The congruence class of v is denoted by $[v]_L = \{u \in \Sigma^* \mid u \equiv_L v\}$. The subscript L is omitted when it is understood. We also write $u \sim_F v$ if $u^{\triangleright} \cap F = v^{\triangleright} \cap F$ for any $F \subseteq \Sigma^* \times \Sigma^*$.

A *predicate system* is a tuple $\langle \Sigma, \Psi, [\![\cdot]\!] \rangle$, where Σ is a possibly infinite alphabet, Ψ is a recursive set of finite descriptions called *predicates*, and $[\![\cdot]\!]$ is a denotation function from Ψ to subsets of Σ. We assume that it is decidable whether $a \in [\![\psi]\!]$ for any $a \in \Sigma$. The description size of a predicate $\psi \in \Psi$ is denoted by $|\psi|$. Abusing the notation, we often represent the predicate system by Ψ, where Σ and $[\![\cdot]\!]$ are implicitly understood.

A Ψ-CFG is a tuple $G = \langle \Psi, N, P, S \rangle$, consisting of a predicate system Ψ, a finite set N of *nonterminal symbols*, a finite set P of *production rules* of the form $A \to \alpha$ where $A \in N$ is a nonterminal symbol and $\alpha \in (N \cup \Psi)^*$ is a sequence of nonterminal symbols and predicates, and a *start symbol* $S \in N$. We write a derivation as $\beta A \gamma \Rightarrow_G \beta \alpha \gamma$ when $A \to \alpha \in P$, and $\beta \psi \gamma \Rightarrow_G \beta a \gamma$ when $a \in [\![\psi]\!]$. The reflexive transitive closure of $\Rightarrow_G$ is $\overset{*}{\Rightarrow}_G$. The subscript G of $\Rightarrow_G$ is omitted if it is understood from the context. The language generated by each nonterminal symbol A is $\mathcal{L}(G, A) = \{u \in \Sigma^* \mid A \overset{*}{\Rightarrow}_G u\}$ and the language of G is $\mathcal{L}(G) = \mathcal{L}(G, S)$, called a Ψ-CFL. A Ψ-CFG is said to be in the *branching normal form* if all of its production rules are of the form $A \to BC$ or $A \to \psi$ or $A \to \varepsilon$, where $A, B, C \in N$ and $\psi \in \Psi$. We call each type of rules *branching-rules*, *predicate-rules*, and *ε-rules*, respectively. Obviously every Ψ-CFL admits a grammar in the branching normal form. A classic CFG over a finite alphabet Σ can be seen as a special case, where each terminal symbol $a \in \Sigma$ has a unique predicate $a \in \Psi = \Sigma$ that represents $[\![a]\!] = \{a\}$. We call such a CFG a Σ-CFG.

2.2 Query Learning Model

We review the query learning model introduced by Angluin [1]. This paper is concerned with the query learning of character sets as well as languages.

For a countably infinite set $\mathcal{D}$ called the *domain*, a *concept* C is any subset of $\mathcal{D}$. Let $\mathcal{R}$ be any recursive set of finite descriptions and f be a function mapping descriptions to concepts. A learner can make membership queries (MQs) and equivalence queries (EQs) concerning a target concept C_*. To an MQ on $c \in \mathcal{D}$, the teacher answers "YES" if $c \in C_*$ and "NO" otherwise. To an EQ on $R \in \mathcal{R}$, if $f(R) \neq C_*$, the teacher gives a counterexample in $(C_* \setminus f(R)) \cup (f(R) \setminus C_*)$. If $f(R) = C_*$, the teacher returns "YES", and the learning process finishes. A class $\mathcal{C}$ of concepts is said to be learnable using $\mathcal{R}$ if there exists an algorithm that learns any concept $C \in \mathcal{C}$ from MQs and EQs. We sometimes say that $\mathcal{R}$ is learnable when $\mathcal{C} = \{f(R) \mid R \in \mathcal{R}\}$. Note that the notion of the teacher is an abstract idealization of information that the learner can obtain from the

Algorithm 1: The Ψ-CDET learner

Require: a predicate learner Λ for Ψ

1 $F := \{\square\}$; $K := \{\varepsilon\}$; Construct a conjecture grammar $\hat{G}$ from F, K;
2 **while** the answer to an EQ on $\hat{G}$ is not "YES" **do**
3   Let w be the counterexample for the EQ;
4   **if** $w \notin \mathcal{L}(\hat{G})$ **then**
5   **foreach** predicate-rule $\langle\!\langle \pi \rangle\!\rangle \to \varphi$ of $\hat{G}$ **do**
6   **if** $a \notin [\![\varphi]\!]$ and $\pi \odot a \in L_*$ for some $a \in \Sigma \cap \mathrm{Sub}(w)$ **then**
7   Give a to Λ_π as a positive counterexample;
8   **if** we did not give counterexamples to any instance of Λ **then**
9   Spawn an instance Λ_π for each $\pi \in \mathrm{Con}(w) \setminus F$;
10   Add $\mathrm{Con}(w)$ to F;
11   **else**
12   Add CDETWITNESS($\langle\!\langle \square \rangle\!\rangle, w$) to K;
13   Construct a conjecture grammar $\hat{G}$ from F, K;
14 **return** $\hat{G}$;

learning environment. We do not require the equivalence of two descriptions to be computable. It is known that random sampling probabilistically simulates the EQ protocol. For more details, see [2].

Throughout this paper, we fix a predicate system $\langle \Sigma, \Psi, [\![\cdot]\!] \rangle$ and assume Ψ is learnable in the above sense. Moreover, for technical convenience, we assume that Ψ has a predicate $\bot$ representing the empty set. This does not affect the learnability of Ψ. If $\Psi \setminus \{\bot\}$ is learnable, one may raise $\bot$ as the first EQ and then start learning $\Psi \setminus \{\bot\}$ if the teacher does not say "YES" to the first EQ.

Example 1. Let the alphabet Σ be the set $\mathbb{Z}$ of integers and the predicate set Ψ be the union closure of the atomic predicates of intervals of the form $[a, b)$ for $a \in \mathbb{Z} \cup \{-\infty\}$ and $b \in \mathbb{Z} \cup \{\infty\}$, where $[\![[a, b)]\!] = \{\, z \in \Sigma \mid a \leq z < b \,\}$. The union of two predicates ψ_1 and ψ_2 is denoted by $\psi_1 \vee \psi_2$. The description size of a predicate $\psi = [a_1, b_1) \vee \cdots \vee [a_k, b_k)$ is $|\psi| = \sum_{i=1}^{k} (\log_2\lceil |a_i| + 1 \rceil + \log_2\lceil |b_i| + 1 \rceil)$, assuming $|-\infty| = |\infty| = 1$. Those predicates are learnable with $\mathcal{O}(|\psi|)$ MQs and $\mathcal{O}(|\psi|)$ EQs.

3 Learning Context-Deterministic Ψ-CFLs

Shirakawa and Yokomori [19] proposed a query learning algorithm for learning context-deterministic Σ-CFLs. We extend their algorithm to Ψ-CFLs with the aid of a predicate learner. Our learning target class Ψ-CDET is given as follows.

Definition 1. Let $G = \langle \Psi, N, P, S \rangle$ be a Ψ-CFG. A *context of a nonterminal symbol* $A \in N$ is any π such that $S \overset{*}{\Rightarrow}_G \pi \odot A$. The grammar G is said to be *context-deterministic* if $\mathcal{L}(G, A) = \mathcal{L}(G) \oslash \pi$ for any context π of every nonterminal symbol A. By Ψ-CDET, we denote the class of languages generated by context-deterministic Ψ-CFGs in the branching normal form such that for any context π, there is $\psi \in \Psi$ such that $(\mathcal{L}(G) \oslash \pi) \cap \Sigma = [\![\psi]\!]$.

We will discuss in Sect. 6 how to lift the assumption that generating grammars are in the branching normal form in Definition 1. The last requirement might appear strong, but actually it is weaker than that Ψ is closed under union with the assumption that G is in the branching normal form. We remark that Argyros and D'Antoni's algorithm for deterministic SFAs [3] requires the predicates to be closed under the Boolean operations. If Ψ is closed under union, for every nonterminal symbol A, there exists a predicate $\psi_A \in \Psi$ such that $[\![\psi_A]\!] = \mathcal{L}(G, A) \cap \Sigma$. Then, for any context π, if the set $\pi^{\lhd} \cap \Sigma$ is not empty, say $a \in \pi^{\lhd} \cap \Sigma$, there is a derivation $S \overset{*}{\Rightarrow} \pi \odot A \Rightarrow \pi \odot \psi_A \Rightarrow \pi \odot a$ for some nonterminal symbol A, for which $\pi^{\lhd} \cap \Sigma = [\![\psi_A]\!]$ holds.

Our learning algorithm for Ψ-CDET is shown in Algorithm 1. A sample run can be found in Sect. 5. Let Λ be a predicate learner for Ψ. Hereafter, fix a target language L_* in Ψ-CDET and a grammar $G_* = \langle \Psi, N_*, P_*, S_* \rangle$ for L_* satisfying the requirement in Definition 1. We write $\pi^{\lhd} = L_* \oslash \pi$ for $\pi \in \Sigma^* \times \Sigma^*$.

Our conjecture $\hat{G} = \langle \Psi, \hat{N}, \hat{P}, \hat{S} \rangle$ is constructed from a finite set $F \subseteq \Sigma^* \times \Sigma^*$ of contexts and a finite set $K \subseteq \Sigma^*$ of strings. The set F grows our grammar, and the set K restrains the growth. The initial F and K are $\{\Box\}$ and $\{\varepsilon\}$, respectively, and we expand these sets monotonically when getting counterexamples by EQs.

The nonterminal symbol set $\hat{N}$ is constructed as

$$\hat{N} := \{\langle\!\langle \pi \rangle\!\rangle \mid \pi \in F\},$$

where $\langle\!\langle \pi \rangle\!\rangle$ denotes a nonterminal symbol indexed by π. The start symbol $\hat{S}$ is $\langle\!\langle \Box \rangle\!\rangle$.

For every nonterminal symbol $A \in N_*$, we would like to obtain a corresponding nonterminal symbol $\langle\!\langle \pi_A \rangle\!\rangle$ with some context π_A of A and intend $\langle\!\langle \pi_A \rangle\!\rangle$ to generate $\pi_A^{\lhd}$. In particular, each $\langle\!\langle \pi_A \rangle\!\rangle$ should have a predicate-rule $\langle\!\langle \pi_A \rangle\!\rangle \to \varphi$ satisfying $[\![\varphi]\!] = \pi_A^{\lhd} \cap \Sigma$. Accordingly, our learner spawns an instance Λ_π of Λ for each $\pi \in F$ and acts as a teacher so that these instances learn desirable predicates. For any $a \in \Sigma$ and $\pi \in \Sigma^* \times \Sigma^*$, one can decide whether $a \in \pi^{\lhd} \cap \Sigma$ using an MQ on $\pi \odot a$. When Λ_π asks an MQ on $a \in \Sigma$, we answer "YES" if $a \in \pi^{\lhd} \cap \Sigma$, and "NO" otherwise. After some MQs, Λ_π raises an EQ on some $\varphi \in \Psi$.

Our production rules are constructed as

$$\begin{aligned}\hat{P} :=& \{\langle\!\langle \pi_0 \rangle\!\rangle \to \langle\!\langle \pi_1 \rangle\!\rangle\langle\!\langle \pi_2 \rangle\!\rangle \mid \pi_0 \odot (\pi_1^{\lhd} \cap K)(\pi_2^{\lhd} \cap K) \subseteq L_*,\ \langle\!\langle \pi_0 \rangle\!\rangle, \langle\!\langle \pi_1 \rangle\!\rangle, \langle\!\langle \pi_2 \rangle\!\rangle \in \hat{N}\} \\ &\cup \{\langle\!\langle \pi \rangle\!\rangle \to \varphi \mid \varphi \text{ is the predicate } \Lambda_\pi \text{ gave us in the last EQ}, \langle\!\langle \pi \rangle\!\rangle \in \hat{N}\} \\ &\cup \{\langle\!\langle \pi \rangle\!\rangle \to \varepsilon \mid \pi \odot \varepsilon \in L_*,\ \langle\!\langle \pi \rangle\!\rangle \in \hat{N}\}.\end{aligned}$$

These rules are constructed using polynomially many MQs in $|F|$ and $|K|$.

Lemma 1. *Assume that for every $A \in N_*$, the set F has a context π_A such that $S_* \overset{*}{\Rightarrow}_{G_*} \pi_A \odot A$, and for every predicate-rule $\langle\!\langle \pi \rangle\!\rangle \to \varphi$ of $\hat{G}$, we have $[\![\varphi]\!] \supseteq \pi^{\lhd} \cap \Sigma$. Then, $\mathcal{L}(\hat{G}) \supseteq L_*$.*

Proof. One can show by induction that $A \overset{*}{\Rightarrow}_{G_*} u$ implies $\langle\!\langle \pi_A \rangle\!\rangle \overset{*}{\Rightarrow}_{\hat{G}} u$. □

We say that a predicate-rule $\langle\!\langle \pi \rangle\!\rangle \to \varphi$ is *insufficient* if $[\![\varphi]\!] \not\supseteq \pi^{\lhd} \cap \Sigma$ holds. (The proof of) Lemma 1 shows that given a positive counterexample $w \in L_* \setminus \mathcal{L}(\hat{G})$,

- our conjecture $\hat{G}$ has some insufficient predicate-rule, or
- there is $A \in N_*$ used to derive w in G_* such that F has no context of A.

Our learner first checks whether $\hat{G}$ has some insufficient predicate-rule. For each predicate-rule $\langle\!\langle \pi \rangle\!\rangle \to \varphi$, we look for $a \in \Sigma \cap \mathrm{Sub}(w)$ satisfying $a \in \pi^{\lhd} \setminus [\![\varphi]\!]$. If such terminal symbol a is found, the rule $\langle\!\langle \pi \rangle\!\rangle \to \varphi$ is insufficient. We thus provide Λ_π with a as a positive counterexample.

If no insufficient predicate-rules are found, there should be a nonterminal symbol $A \in N_*$ such that $S_* \overset{*}{\Rightarrow}_{G_*} \pi_A \odot A \overset{*}{\Rightarrow}_{G_*} w$. Our learner adds all contexts extracted from w to F in order to get a nonterminal symbol $\langle\!\langle \pi_A \rangle\!\rangle$ corresponding to A. In addition, we spawn an instance of Λ for each newly added context.

Our production rules may contradict the intention that each nonterminal symbol $\langle\!\langle \pi \rangle\!\rangle$ should generate $\pi^{\lhd}$. Let us say that

- a branching-rule $\langle\!\langle \pi_0 \rangle\!\rangle \to \langle\!\langle \pi_1 \rangle\!\rangle\langle\!\langle \pi_2 \rangle\!\rangle$ is *incorrect* if $\pi_0 \odot \pi_1^{\lhd}\pi_2^{\lhd} \not\subseteq L_*$ holds,
- a predicate-rule $\langle\!\langle \pi \rangle\!\rangle \to \varphi$ is *excessive* if $[\![\varphi]\!] \not\subseteq \pi^{\lhd} \cap \Sigma$ holds.

Lemma 2. *If $\hat{G}$ has neither incorrect rules nor excessive rules, then $\mathcal{L}(\hat{G}) \subseteq L_*$.*

Proof. One can show by induction that $\langle\!\langle \pi \rangle\!\rangle \overset{*}{\Rightarrow}_{\hat{G}} u$ implies $\pi \odot u \in L_*$. □

When learning a context-deterministic Σ-CFL, the learner should care only about incorrect branching-rules because it constructs terminal-rules $\langle\!\langle \pi \rangle\!\rangle \to a$ just in the case when $a \in \pi^{\lhd} \cap \Sigma$. On the other hand, our Ψ-CDET learner needs to cope with excessive predicate-rules as well as incorrect branching-rules.

(The proof of) Lemma 2 shows if $\langle\!\langle \pi \rangle\!\rangle \overset{*}{\Rightarrow}_{\hat{G}} u$ and $u \notin \pi^{\lhd}$, there are some incorrect or excessive rules used in the derivation. Incorrect branching-rules $\langle\!\langle \pi_0 \rangle\!\rangle \to \langle\!\langle \pi_1 \rangle\!\rangle\langle\!\langle \pi_2 \rangle\!\rangle$ are created when $\pi_1^{\lhd}\pi_2^{\lhd} \not\subseteq \pi_0^{\lhd}$ but $(\pi_1^{\lhd} \cap K)(\pi_2^{\lhd} \cap K) \subseteq \pi_0^{\lhd}$. Expanding K with appropriate strings will remove such incorrect rules. On the other hand, predicate learner instances Λ_π are responsible for excessive predicate-rules $\langle\!\langle \pi \rangle\!\rangle \to \varphi$. Based on this fact, the procedure CdetWitness($\langle\!\langle \pi \rangle\!\rangle, u$) with $\langle\!\langle \pi \rangle\!\rangle \overset{*}{\Rightarrow}_{\hat{G}} u$ and $u \notin \pi^{\lhd}$ searches the derivation tree in a top-down manner for an incorrect or excessive rule. If an incorrect branching-rule is found, it outputs witness strings. If an excessive predicate-rule is found, it gives a negative counterexample to the instance. Note that the derivation cannot be by a single ε-rule, $\langle\!\langle \pi \rangle\!\rangle \Rightarrow \varepsilon = u$. We have two cases.

Case 1. Suppose $\langle\!\langle \pi \rangle\!\rangle \Rightarrow \langle\!\langle \pi_1 \rangle\!\rangle\langle\!\langle \pi_2 \rangle\!\rangle \overset{*}{\Rightarrow} u_1u_2 = u$. If $\pi_i \odot u_i \notin L_*$ for some $i \in \{1, 2\}$, some incorrect rule is used in $\langle\!\langle \pi_i \rangle\!\rangle \overset{*}{\Rightarrow} u_i$, and we recursively call CdetWitness($\langle\!\langle \pi_i \rangle\!\rangle, u_i$). Otherwise, the branching-rule is incorrect. Return $\{u_1, u_2\}$ to add them to K and halt the recursive procedure.

Case 2. Suppose $\langle\!\langle \pi \rangle\!\rangle \Rightarrow \varphi \Rightarrow u \in \Sigma$. The predicate-rule $\langle\!\langle \pi \rangle\!\rangle \to \varphi$ is excessive because $u \notin \pi^{\lhd} \cap \Sigma$. We give $u \in \Sigma$ to the instance Λ_π as a negative counterexample. Return the empty set and halt the recursive procedure.

For a fixed set F of contexts, all incorrect or excessive rules will eventually be removed from our conjecture $\hat{G}$ and then $\mathcal{L}(\hat{G}) \subseteq L_*$ holds.

The efficiency of our algorithm depends on the predicate learner Λ. Let $\mathcal{M}(n)$ and $\mathcal{E}(n)$ be the upper bounds on the numbers of MQs and EQs, respectively, that Λ raises for learning a predicate of size at most n. In addition, let

$$k_\pi := \min\{\, |\psi| \mid [\![\psi]\!] = \pi^{\lhd} \cap \Sigma,\ \psi \in \Psi \,\} \text{ for } \pi \in \Sigma^* \times \Sigma^*,$$
$$k_* := \max\{\, k_\pi \mid \pi \in \Sigma^* \times \Sigma^* \,\}.$$

The parameter k_* is the maximum size of a predicate that Λ may try to learn in the whole learning process. Every instance Λ_π finally hypothesizes the desirable predicate $\psi \in \Psi$ with $[\![\psi]\!] = \pi^{\lhd} \cap \Sigma$ unless our algorithm terminates before that.

Theorem 1. *The class Ψ-CDET is learnable from membership and equivalence queries, the numbers of which are bounded by polynomials in $|N_*|$, $\mathcal{M}(k_*)$, $\mathcal{E}(k_*)$ and ℓ, where ℓ is the length of a longest counterexample given by a teacher.*

Proof. For each counterexample, our algorithm performs one of the following:

- adding contexts to F to create more nonterminal symbols,
- adding two strings to K to remove an incorrect branching-rule,
- giving a counterexample to at least one instance of Λ.

The set F is expanded only when a given positive counterexample is generated by G_* using a nonterminal symbol of which F has no context. This happens at most $|N_*|$ times, and each time it adds at most ℓ^2 contexts. Hence $|F| = |\hat{N}| = \mathcal{O}(|N_*|\ell^2)$. Our grammar has $|\hat{N}|$ predicate-rules, at most $|\hat{N}|$ ε-rules and at most $|\hat{N}|^3$ branching-rules. The algorithm adds two strings to K only to remove an incorrect branching-rule. This implies $|K| = \mathcal{O}(|\hat{N}|^3) = \mathcal{O}(|N_*|^3\ell^6)$. Every instance Λ_π returns a predicate in Ψ representing $\pi^{\lhd} \cap \Sigma$ after our learner answers MQs and EQs of Λ_π at most $\mathcal{M}(k_*)$ and $\mathcal{E}(k_*)$ times, respectively. Hence, by Lemmas 1 and 2, the number of EQs our algorithm asks is bounded by a polynomial in $|N_*|$, $\mathcal{M}(k_*)$, $\mathcal{E}(k_*)$ and ℓ.

Our algorithm uses MQs

- at most $\mathcal{O}(|K|^2)$ times to construct a branching-rule,
- one time to construct an ε-rule,
- one time to answer each MQ of an instance,
- at most $|\hat{N}|\ell$ times to look for insufficient rules for each positive example,
- at most $\mathcal{O}(\ell)$ times to find incorrect or excessive rules for each negative example.

Consequently, the number of MQs our algorithm makes is bounded by a polynomial in $|N_*|$, $\mathcal{M}(k_*)$, $\mathcal{E}(k_*)$ and ℓ.

□

Algorithm 2: The Ψ-CONG learner

Require: a predicate learner Λ for Ψ

1 $K := \{\varepsilon\}$; $F := \{\square\}$; Construct a conjecture grammar $\hat{G}$ from K, F;
2 **while** the answer to an EQ on $\hat{G}$ is not "YES" **do**
3 Let w be the counterexample for the EQ;
4 **if** $w \notin \mathcal{L}(\hat{G})$ **then**
5 **foreach** predicate-rule $\langle\!\langle v \rangle\!\rangle \to \varphi$ of $\hat{G}$ **do**
6 **if** $a \notin [\![\varphi]\!]$ and $v \sim_F a$ for some $a \in \Sigma \cap \mathrm{Sub}(w)$ **then**
7 Give a to Λ_v as a positive counterexample;
8 **if** we did not give counterexamples to any instance of Λ **then**
9 Spawn an instance Λ_v for each $v \in \mathrm{Sub}(w) \setminus K$;
10 Add $\mathrm{Sub}(w)$ to K; Add $\mathrm{Con}(w)$ to F; Initialize Λ_v for all $v \in K$;
11 **else**
12 $\pi :=$ CONGWITNESS$(\langle\!\langle v \rangle\!\rangle, w, \square)$ where $\langle\!\langle v \rangle\!\rangle \in \hat{I}$ and $\langle\!\langle v \rangle\!\rangle \overset{*}{\Rightarrow} w$;
13 **if** $\pi \neq$ null **then**
14 Add π to F; Initialize Λ_v for all $v \in K$;
15 Construct a conjecture grammar $\hat{G}$ from K, F;
16 **return** $\hat{G}$;

4 Learning Congruential Ψ-CFLs

Clark [7] has a learning algorithm for congruential Σ-CFLs. Following [7], this section allows a Ψ-CFG G to have multiple start symbols and defines the language of G to be the union of the languages of those start symbols.

Definition 2. A Ψ-CFG G is said to be *congruential* if $u, v \in \mathcal{L}(G, A)$ implies $u \equiv_{\mathcal{L}(G)} v$ for every nonterminal symbol A. By Ψ-CONG, we denote the class of languages generated by congruential Ψ-CFGs in the branching normal form.

We present a learning algorithm (Algorithm 2) for Ψ-CONG, which works under the assumption that Ψ is closed under union. We embed the predicate learner Λ into Clark's algorithm. In contrast with the Ψ-CDET learner in Sect. 3, the Ψ-CONG learner uses a finite string set K for building nonterminal symbols and rules, while using a finite context set F for filtering branching-rules, and similarly constructs predicate-rules with the aid of predicate learner instances. However, unlike the Ψ-CDET learner, it often returns "wrong" answers to the MQs posed by instances. Whenever we are aware of the risk of wrong answers, all instances are initialized.

Let L_* be the target language generated by a grammar $G_* = \langle \Psi, N_*, P_*, I_* \rangle$. Our conjecture $\hat{G} = \langle \Psi, \hat{N}, \hat{P}, \hat{I} \rangle$ is constructed from K and F. The nonterminal symbol set $\hat{N}$ and the start symbol set $\hat{I}$ are constructed as

$$\hat{N} := \{\langle\!\langle v \rangle\!\rangle \mid v \in K\}, \qquad \hat{I} := \{\langle\!\langle v \rangle\!\rangle \mid v \in K \cap L_*\}.$$

We intend each nonterminal symbol $\langle\!\langle v \rangle\!\rangle \in \hat{N}$ to generate $[v]_{L_*}$ so that we can use $\langle\!\langle v_A \rangle\!\rangle$ to simulate a nonterminal symbol A of G_* if $v_A \in \mathcal{L}(G_*, A)$. Particularly, each $\langle\!\langle v \rangle\!\rangle$ should have a predicate-rule $\langle\!\langle v \rangle\!\rangle \to \varphi$ with $[\![\varphi]\!] = [v]_{L_*} \cap \Sigma$. To this end, the learner spawns an instance Λ_v of Λ for each $v \in K$ and lets them propose predicates. To each MQ by Λ_v on $a \in \Sigma$, we should answer "Yes" if $v \equiv_{L_*} a$, but it is not computable. Instead, our algorithm behaves as if the learning target of Λ_v is the set $[v]_{/\sim_F} \cap \Sigma = \{\, a \in \Sigma \mid v \sim_F a \,\}$. That is, we answer "Yes" if $v \sim_F a$ and "No" otherwise. This is computable with $2|F|$ MQs.

Lemma 3. *For any $v \in \mathrm{Sub}(L_*)$ and $F \subseteq \Sigma^* \times \Sigma^*$, if $F \cap v^{\triangleright} \neq \emptyset$,*

$$[v]_{/\sim_F} \cap \Sigma = [\![\textstyle\bigvee_{\psi \in \Psi'} \psi]\!]$$

for some $\Psi' \subseteq \Psi_{G_*}$, where $\Psi_{G_*} \subseteq \Psi$ is the set of the predicates appearing in G_*.

Since Ψ is closed under union, Ψ has a predicate representing $[v]_{/\sim_F} \cap \Sigma$, so Λ_v works well. After some MQs, Λ_v gives us a predicate $\varphi \in \Psi$ as an EQ. Our production rules are given as

$$\begin{aligned}\hat{P} := \{ & \langle\!\langle v_0 \rangle\!\rangle \to \langle\!\langle v_1 \rangle\!\rangle \langle\!\langle v_2 \rangle\!\rangle \mid (v_0^{\triangleright} \cap F) \odot v_1 v_2 \subseteq L_*,\ \langle\!\langle v_0 \rangle\!\rangle, \langle\!\langle v_1 \rangle\!\rangle, \langle\!\langle v_2 \rangle\!\rangle \in \hat{N} \} \\ & \cup \{ \langle\!\langle v \rangle\!\rangle \to \varphi \mid \varphi \text{ is the predicate} \Lambda_v \text{ gave us in the last EQ}, \langle\!\langle v \rangle\!\rangle \in \hat{N} \} \\ & \cup \{ \langle\!\langle \varepsilon \rangle\!\rangle \to \varepsilon \}.\end{aligned}$$

Lemma 4. *Assume that for every $A \in N_*$, the set K has a string v_A such that $A \stackrel{*}{\Rightarrow}_{G_*} v_A$, and for every predicate-rule $\langle\!\langle v \rangle\!\rangle \to \varphi$ of $\hat{G}$, we have $[\![\varphi]\!] \supseteq [v]_{L_*} \cap \Sigma$. Then, $\mathcal{L}(\hat{G}) \supseteq L_*$.*

We say that a predicate-rule $\langle\!\langle v \rangle\!\rangle \to \varphi$ is *insufficient* if $[\![\varphi]\!] \not\supseteq [v]_{L_*} \cap \Sigma$ holds. Lemma 4 implies that given a positive counterexample $w \in L_* \setminus \mathcal{L}(\hat{G})$,

- our conjecture $\hat{G}$ has some insufficient predicate-rules, or
- there is $A \in N_*$ such that $S_* \stackrel{*}{\Rightarrow} xAy \stackrel{*}{\Rightarrow} xv_Ay = w$ and $K \cap \mathcal{L}(G_*, A) = \emptyset$.

To find an insufficient predicate-rule, we check whether there exist $\langle\!\langle v \rangle\!\rangle \to \varphi \in \hat{P}$ and $a \in \Sigma \cap \mathrm{Sub}(w)$ such that $a \notin [\![\varphi]\!]$ and $v \sim_F a$. If such $\langle\!\langle v \rangle\!\rangle$ and a are found, we give a to Λ_v as a positive counterexample. On the other hand, if we cannot find such an insufficient predicate-rule using the characters in the positive counterexample w, there must be a nonterminal symbol A of G_* used to generate w but our grammar has no corresponding nonterminal symbol. In this case, by adding all substrings of w to K, we get a nonterminal symbol $\langle\!\langle v_A \rangle\!\rangle$ corresponding to A. In addition, we add all contexts extracted from w to F so that the condition of Lemma 3 holds for any new nonterminal symbol $\langle\!\langle v \rangle\!\rangle$. Our algorithm then spawns an instance of the predicate learner Λ for each newly added element of K. In this case, we initialize all instances, since an expansion of F may make our past and future answers to MQs raised by instances inconsistent with any predicates in Ψ.

We now explain how to process negative counterexamples. Our grammar generates undesirable strings due to the following two types of rules. Let us say

- a branching-rule $\langle\!\langle v_0 \rangle\!\rangle \to \langle\!\langle v_1 \rangle\!\rangle\langle\!\langle v_2 \rangle\!\rangle$ is *incorrect* if $v_0^{\triangleright} \odot v_1 v_2 \not\subseteq L_*$ holds,
- a predicate-rule $\langle\!\langle v \rangle\!\rangle \to \varphi$ is *excessive* if $[\![\varphi]\!] \not\subseteq [v]_{L_*} \cap \Sigma$ holds.

Lemma 5. *If $\hat{G}$ has neither incorrect rules nor excessive rules, then $\mathcal{L}(\hat{G}) \subseteq L_*$.*

Recall that to Λ_v's MQ on $a \in \Sigma$, our learner should answer "YES" just if $v \equiv_{L_*} a$, but in fact it returns "YES" if $v \sim_F a$. Our learner may return wrong answers. Thus, there can be two reasons for an excessive predicate-rule. One is that the instance did not get enough information to guess the target predicate. The other is that we gave wrong answers to some MQs from the instance.

By Lemma 5, given a negative counterexample $w \in \mathcal{L}(\hat{G}) \setminus L_*$, some incorrect or excessive rules are used to derive w in $\hat{G}$. For any derivation $\langle\!\langle v \rangle\!\rangle \overset{*}{\Rightarrow} u$, if $v \not\equiv_{L_*} u$, there are some incorrect or excessive rules in the derivation. For $\langle\!\langle v \rangle\!\rangle \in \hat{I}$ with $\langle\!\langle v \rangle\!\rangle \overset{*}{\Rightarrow} w$, we call CONGWITNESS($\langle\!\langle v \rangle\!\rangle, w, \Box$) to search the derivation tree for an incorrect or excessive rule and return a context if it is useful to remove the found rule.

Suppose we call CONGWITNESS($\langle\!\langle v \rangle\!\rangle, u, x\Box y$) with $xvy \in L_*$ and $xuy \notin L_*$, i.e., $v \not\equiv_{L_*} u$. We have two cases.

CASE 1. Suppose $\langle\!\langle v \rangle\!\rangle \Rightarrow \langle\!\langle v_1 \rangle\!\rangle\langle\!\langle v_2 \rangle\!\rangle \overset{*}{\Rightarrow} u_1 u_2 = u$. If $xv_1v_2y \notin L_*$, the rule $\langle\!\langle v \rangle\!\rangle \to \langle\!\langle v_1 \rangle\!\rangle\langle\!\langle v_2 \rangle\!\rangle$ is incorrect, which is witnessed by the context $x\Box y$. We thus return $x\Box y$. Otherwise, if $xv_1u_2y \in L_*$, then $v_1 \not\equiv_{L_*} u_1$ due to $x\Box u_2 y \in v_1^{\triangleright} \setminus u_1^{\triangleright}$, and we recurse with $(\langle\!\langle v_1 \rangle\!\rangle, u_1, x\Box u_2 y)$. If not, $v_2 \not\equiv_{L_*} u_2$ due to $xv_1\Box y \in v_2^{\triangleright} \setminus u_2^{\triangleright}$, and we recurse with $(\langle\!\langle v_2 \rangle\!\rangle, u_2, xv_1\Box y)$.

CASE 2. Suppose $\langle\!\langle v \rangle\!\rangle \Rightarrow \varphi \Rightarrow u \in \Sigma$. The predicate-rule $\langle\!\langle v \rangle\!\rangle \to \varphi$ is excessive because $u \in [\![\varphi]\!] \setminus [v]_{L_*}$, which is witnessed by $x\Box y$. We have two subcases. If $v \sim_F u$, no contexts of F but the new context $x\Box y$ separates $[v]_{L_*} \cap \Sigma$ and $[u]_{L_*} \cap \Sigma$. The instance Λ_v may have raised an MQ on $u \in \Sigma$ in the past, in which case we should have wrongly answered "YES". Therefore, we return $x\Box y$ and add $x\Box y$ to F so that we will not repeat the same mistake thereafter. On the other hand, if $v \not\sim_F u$, we give $u \in \Sigma$ to Λ_v as a negative counterexample. The witness context $x\Box y$ brings no new information on the congruence classes of characters. We return null.

Now, we will discuss the correctness and the query complexity of our algorithm. Let $\mathcal{M}(n)$ and $\mathcal{E}(n)$ be the upper bounds on the numbers of MQs and EQs, respectively, raised by Λ for learning a predicate of size at most n. Let

$$k_\Delta := \min\{\, |\psi| \mid [\![\psi]\!] = \bigcup_{a \in \Delta} [a]_{L_*} \cap \Sigma,\ \psi \in \Psi \,\} \text{ for } \Delta \subseteq \Sigma \cap \mathrm{Sub}(L_*),$$
$$k_* := \max\{\, k_\Delta \mid \Delta \subseteq \Sigma \cap \mathrm{Sub}(L_*)\}.$$

For any $v \in \mathrm{Sub}(L_*)$ and a context set F, we have $[v]_{/\sim_F} \cap \Sigma = \bigcup_{a \in \Delta} [a]_{L_*} \cap \Sigma$ for $\Delta = [v]_{/\sim_F} \cap \Sigma$. The positive integer k_* bounds the maximum size of a predicate that Λ may try to learn in the whole process of learning L_*. The predicate learner learns those character sets using at most $\mathcal{M}(k_*)$ MQs and $\mathcal{E}(k_*)$ EQs.

Theorem 2. *If Ψ is closed under union, the class Ψ-CONG is learnable from membership and equivalence queries, the numbers of which are bounded by polynomials in $|N_*|$, $\mathcal{M}(k_*)$, $\mathcal{E}(k_*)$ and ℓ, where ℓ is the length of a longest counterexample given by a teacher.*

Proof. Unlike the Ψ-CDET learner, the Ψ-CONG learner initializes predicate learner instances. We show that we initialize them at most $|N_*|$ times. Other than this point, the rest of the proof is much the same as Theorem 1.

Whenever our algorithm initializes instances, it adds a context to F, which splits at least one equivalence class over Σ with respect to $\sim_F$ into two. It follows that if the algorithm initializes them at most as many times as the number of congruence classes over Σ with respect to $\equiv_{L_*}$, then for all $a, b \in \Sigma$, $a \sim_F b$ implies $a \equiv_{L_*} b$. Thereafter, our algorithm will never return wrong answers to instances and thus never initialize them. The number of congruence classes over Σ with respect to $\equiv_{L_*}$ is bounded by $|N_*|$. Hence, the number of initialization of instances is also bounded by $|N_*|$.

□

5 Running Example

We provide a sample run of the Ψ-CDET learner in Sect. 3. Consider a Ψ-CFG whose interval predicate system Ψ is that of Example 1, production rules are

$$S \to AX,\ X \to SC,\ A \to [0,2) \vee [6,7),\ S \to [2,5),\ C \to [5,6) \vee [7,9),$$

and start symbol is S. This generates a language in the class Ψ-CDET:

$$L_* = \{\, a_1 \cdots a_n\, b\, c_1 \cdots c_n \mid a_i \in \{0,1,6\},\ b \in \{2,3,4\},\ c_i \in \{5,7,8\},\ n \geq 0 \,\}.$$

For simplicity, we restrict the alphabet to $\Sigma = \{0, 1, \ldots, 9\}$ and write, for example, 23 for the string "two · three", not "twenty-three".

The Ψ-CDET learner starts by setting $F = \{\Box\}$ and $K = \{\varepsilon\}$ and constructing a grammar that generates the empty set. The learner makes an EQ and then gets a positive counterexample $2 \in L_*$. We add all contexts extracted from 2 to F, so $F = \{\Box, \Box 2, 2\Box\}$. Then three instances $\Lambda_\Box, \Lambda_{\Box 2}, \Lambda_{2\Box}$ of the predicate learner are spawned. At first, they all raise an EQ on $\bot$ without MQs, so our conjecture is still a grammar generating the empty set since all predicate-rules are useless. Again, we are given a counterexample 2 for an EQ.

Our learner finds the predicate-rule $\langle\!\langle \Box \rangle\!\rangle \to \bot$ insufficient by $2 \in \Box^{\triangleleft} \setminus [\![\bot]\!]$, and so gives 2 to $\Lambda_\Box$ as a positive counterexample. The instance $\Lambda_\Box$ thus raises some MQs and proposes $[2,5)$ as an EQ, which is the desirable predicate. On the other hand, we will never give any counterexample to $\Lambda_{2\Box}$ and $\Lambda_{\Box 2}$, which implies their conjectures will forever be $\bot$. We construct $\hat{G}$ that has the predicate-rule $\langle\!\langle \Box \rangle\!\rangle \to [2,5)$ and the branching-rule $\langle\!\langle \Box \rangle\!\rangle \to \langle\!\langle \Box \rangle\!\rangle \langle\!\langle \Box \rangle\!\rangle$, among others. This generates $\{c_1 \cdots c_n \mid c_i \in \{2,3,4\}, n > 0\} \neq L_*$. We make an EQ on $\hat{G}$ and get a negative counterexample $23 \in \mathcal{L}(\hat{G}) \setminus L_*$.

To find an incorrect or excessive rule, we call CDETWITNESS($\langle\!\langle\Box\rangle\!\rangle$, 23). This finds that $\langle\!\langle\Box\rangle\!\rangle \to \langle\!\langle\Box\rangle\!\rangle\langle\!\langle\Box\rangle\!\rangle$ is an incorrect branching-rule and adds 2 and 3 to K to remove it. We update $\hat{G}$, so that $\mathcal{L}(\hat{G}) = \{c \mid c \in \{2,3,4\}\} \subseteq L_*$. We receive a positive counterexample $637 \in L_* \setminus \mathcal{L}(\hat{G})$ for an EQ.

There is no insufficient predicate-rule in $\hat{G}$. We thus add all contexts extracted from 637 to F, spawn some instances, and answer their queries. Our grammar $\hat{G}$ still has some undesirable rules, but as described above, our learner will remove all incorrect branching-rules and correct all predicate-rules by providing counterexamples for instances. Therefore, our leaner will finally gain a grammar that generates L_* and then terminate.

6 Discussion

Throughout this paper, we have assumed all target Ψ-CFGs are in the branching normal form. It is known that every congruential Σ-CFG can be converted into another congruential Σ-CFG in the branching normal form. It is the case for congruential Ψ-CFGs if Ψ is closed under the Boolean operations. A congruential Ψ-CFG G in general may have "non-congruential" predicates ψ, in the sense that there are $a, b \in [\![\psi]\!]$ such that $a \not\equiv_{\mathcal{L}(G)} b$, but one can easily see that every congruence class of characters can be represented by a Boolean combination of predicates used in the grammar G. Hence, we can transform G so that all predicates used there are "congruential", and then apply the standard technique for converting a CFG into the branching normal form. Recall that Argyros and D'Antoni's algorithm works under the assumption that the predicate set is closed under the Boolean operations [3]. Therefore, our Ψ-CONG learner is no weaker than theirs. On the other hand, there is a congruential Ψ-CFL that cannot be represented by a congruential Ψ-CFG in the branching normal form when Ψ is not closed under the Boolean operations. An example is $L_0 = \{ac, ad, bc\}$, which can be generated by a congruential Ψ-CFG with rules $S \to \psi_{a,b}\psi_c \mid \psi_a\psi_d$ where $[\![\psi_{a,b}]\!] = \{a, b\}$ and $[\![\psi_x]\!] = \{x\}$ for $x \in \{a, c, d\}$, and Ψ misses ψ_b.

On the other hand, the language L_0 cannot be generated by a context-deterministic Ψ-CFG in the branching normal form regardless of the richness of Ψ. In this regard, our result on learning Ψ-CDET might appear weak. One can overcome this weakness by allowing grammars to include "regular rules" of the form $A \to \psi B$ and applying Argyros and D'Antoni's technique for identifying ψ. Our learner should construct rules of the form $\langle\!\langle\pi_1\rangle\!\rangle \to \varphi_{\pi_1/\pi_2}\langle\!\langle\pi_2\rangle\!\rangle$ where φ_{π_1/π_2} is the hypothesis by a predicate learner instance Λ_{π_1/π_2}, whose target set should be $\psi_{\pi_1/\pi_2} = \{\, a \in \Sigma \mid a\pi_2^{\lhd} \subseteq \pi_1^{\lhd} \,\}$. While the membership of the target set $[\![\psi_\pi]\!] = \pi^{\lhd} \cap \Sigma$ of Λ_π in Algorithm 1 is computable with an MQ, it is not the case for ψ_{π_1/π_2}. One can only approximate $[\![\psi_{\pi_1/\pi_2}]\!]$ by $\{\, a \in \Sigma \mid a(\pi_2^{\lhd} \cap K) \subseteq \pi_1^{\lhd} \,\}$. Due to the uncertainty of the answers, we need to initialize instances in our learning algorithm. This also requires Ψ to satisfy an additional property.

Furthermore, Yoshinaka [22] proposed a query learning algorithm unifying the two learning algorithms for context-deterministic Σ-CFLs and congruential Σ-CFLs where generating grammars are not limited to the branching normal form. This type of generalization of our algorithms should be tackled.

Another challenge is to generalize Ishizaka's work [12] on the query learning of simple deterministic CFLs. His algorithm also observes and exploits the context–substring relation determined by the learning target language. This should also be an interesting direction for future work.

Acknowledgments. The authors are grateful to the anonymous reviewers for their helpful comments.This work is supported in part by JSPS KAKENHI Grant Numbers JP18K11150 (RY), JP20H05703 (RY), JP23K11325 (RY) and JP21K11745 (AS).

References

1. Angluin, D.: Learning regular sets from queries and counterexamples. Inf. Comput. **75**(2), 87–106 (1987)
2. Angluin, D.: Queries and concept learning. Mach. Learn. **2**(4), 319–342 (1988)
3. Argyros, G., D'Antoni, L.: The learnability of symbolic automata. In: 30th International Conference on Computer Aided Verification, pp. 427–445 (2018)
4. Argyros, G., Stais, I., Kiayias, A., Keromytis, A.D.: Back in black: Towards formal, black box analysis of sanitizers and filters. In: IEEE Symposium on Security and Privacy (SP 2016), pp. 91–109 (2016)
5. Bollig, B., Habermehl, P., Kern, C., Leucker, M.: Angluin-style learning of NFA. In: Proceedings of the 21st International Joint Conference on Artificial Intelligence, pp. 1004–1009 (2009)
6. Chubachi, K., Hendrian, D., Yoshinaka, R., Shinohara, A.: Query learning algorithm for residual symbolic finite automata. In: 10th International Symposium on Games, Automata, Logics, and Formal Verification, pp. 140–153 (2019)
7. Clark, A.: Distributional learning of some context-free languages with a minimally adequate teacher. In: Grammatical Inference: Theoretical Results and Applications, pp. 24–37 (2010)
8. D'Antoni, L., Alur, R.: Symbolic visibly pushdown automata. In: Biere, A., Bloem, R. (eds.) CAV 2014. LNCS, vol. 8559, pp. 209–225. Springer, Cham (2014). https://doi.org/10.1007/978-3-319-08867-9_14
9. Drews, S., D'Antoni, L.: Learning symbolic automata. In: Legay, A., Margaria, T. (eds.) TACAS 2017. LNCS, vol. 10205, pp. 173–189. Springer, Heidelberg (2017). https://doi.org/10.1007/978-3-662-54577-5_10
10. de la Higuera, C.: Grammatical Inference: Learning Automata and Grammars. Cambridge University Press (2010)
11. Isberner, M., Howar, F., Steffen, B.: The TTT algorithm: A redundancy-free approach to active automata learning. In: Runtime Verification, pp. 307–322 (2014)
12. Ishizaka, H.: Polynomial time learnability of simple deterministic languages. Mach. Learn. **5**, 151–164 (1990)
13. Kawasaki, Y., Hendrian, D., Yoshinaka, R., Shinohara, A.: Query learning of minimal deterministic symbolic finite automata separating regular languages. In: SOFSEM 2024: Theory and Practice of Computer Science, pp. 340–354 (2024)
14. Kearns, M., Vazirani, U.: An introduction to computational learning theory. The MIT Press (1994)
15. Maler, O., Mens, I.: A generic algorithm for learning symbolic automata from membership queries. In: Models, Algorithms, Logics and Tools - Essays Dedicated to Kim Guldstrand Larsen on the Occasion of His 60th Birthday, pp. 146–169 (2017)

16. Mens, I., Maler, O.: Learning regular languages over large ordered alphabets. Logical Methods Comput. Sci. **11**(3) (2015)
17. Numaya, Y., Hendrian, D., Yoshinaka, R., Shinohara, A.: Identification of substitutable context-free languages over infinite alphabets from positive data. In: Proceedings of 16th edition of the International Conference on Grammatical Inference, pp. 23–34 (2023)
18. Rivest, R.L., Schapire, R.E.: Inference of finite automata using homing sequences. Inf. Comput. **103**(2), 299–347 (1993)
19. Shirakawa, H., Yokomori, T.: Polynomial-time MAT learning of c-deterministic context-free grammars. Transac. Inform. Process. Society Japan **34**(3), 380–390 (1993)
20. Suzuki, K., Hendrian, D., Yoshinaka, R., Shinohara, A.: Query learning algorithm for symbolic weighted finite automata. In: 15th International Conference on Grammatical Inference, pp. 202–216 (2021)
21. Veanes, M., de Halleux, P., Tillmann, N.: Rex: Symbolic regular expression explorer. In: 2010 Third International Conference on Software Testing, Verification and Validation, pp. 498–507 (2010)
22. Yoshinaka, R.: Integration of the dual approaches in the distributional learning of context-free grammars. In: Language and Automata Theory and Applications, pp. 538–550 (2012)

Minimal Schnyder Woods and Long Induced Paths in 3-Connected Planar Graphs

Christian Ortlieb(✉)

Institute of Computer Science, University of Rostock, Rostock, Germany

Abstract. We investigate a new structural property of Schnyder woods: every minimal Schnyder wood of a 3-connected planar graph of order n has a tree of depth at least $\log_2(n)/(3\log_2(3))$. This bound is tight. Our result directly implies that such a graph has an induced path of length at least $\log_2(n)/(3\log_2(3))$, improving the previous best lower bound on the length of such a path.

Keywords: Induced path · 3-connected planar graph · Schnyder wood · depth

1 Introduction

Already in 1986, Erdős et al. [5] investigated the problem of finding long induced paths. Let $p(G)$ be the size, i.e., the number of vertices, of a longest induced path of G. For a connected graph G with radius $r(G)$, Erdős et al. [5] showed that $p(G) \geq 2r(G) - 1$. Fourteen years later, Arocha and Valencia [1] gave the lower bound $\log_{\Delta}(n)$ on the diameter (and hence on $p(G)$) for a 3-connected planar graph G of order n with bounded maximum degree Δ. For unbounded Δ, they gave an induced path of size $\sqrt{\log_3(\Delta)}$. In 2016, Di Giacomo et al. [4] showed that $p(G) \geq \frac{\log_2(n)}{12\log_2\log_2(n)}$ for 3-connected planar graphs G. And they gave an upper bound showing $p(G) \leq 1.3\log_2(n) + 5$ for a family of specific 3-connected planar graphs. The same year, Esperet et al. [6] improved the lower bound to $(\log_2(n) - 3\log_2\log_2(n))/6$ with a similar approach. Recently, we [12] improved the lower bound to $\log_2(n)/6$ using a new technique based on deep trees in Schnyder woods.

In this paper, we give a better lower bound of $p(G) \geq \log_2(n)/(3\log_2(3)) \geq \log_2(n)/4.76$. We approach the problem via deep trees in minimal Schnyder woods.

Given a planar embedding of a 3-connected planar graph and a minimal Schnyder wood on this embedding (Formal definitions are given in Sect. 2.), we show that at least one of the three trees has depth at least $\log_2(n)/(3\log_2(3))$.

This research is supported by the grant SCHM 3186/2-1 (401348462) from the Deutsche Forschungsgemeinschaft (DFG, German Research Foundation).

R. Královič and V. Kůrková (Eds.): SOFSEM 2025, LNCS 15539, pp. 225–237, 2025.
https://doi.org/10.1007/978-3-031-82697-9_17

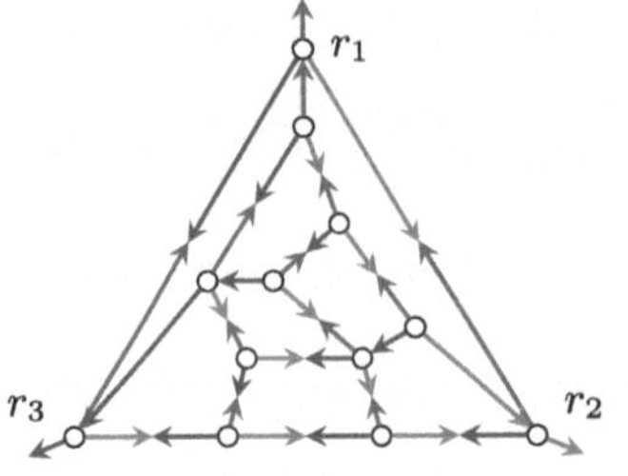

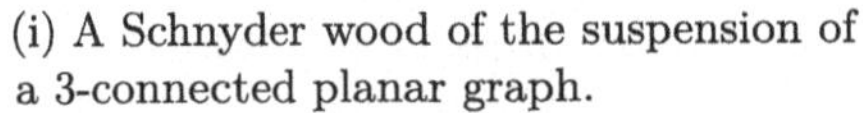
(i) A Schnyder wood of the suspension of a 3-connected planar graph.

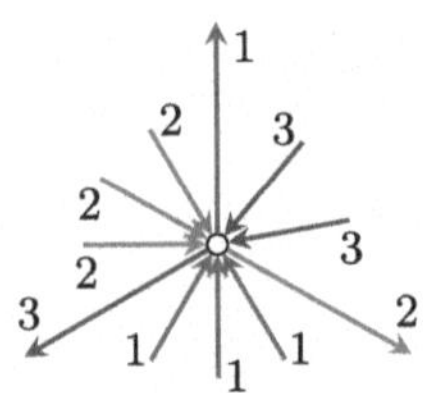

(ii) Example for Definition 1(iii) at a vertex in a Schnyder wood.

Fig. 1. Illustrations for the definition of Schnyder woods.

We also show that this bound is tight, i.e., for every $0 < \varepsilon < 1$ there exists a 3-connected planar graph with a minimal Schnyder wood such that every tree of this Schnyder wood has depth at most $\log_2(n)/(3(1-\varepsilon)\log_2(3)) + 1$. Actually, the 3-connected planar graph that we need for our lower bound has a unique Schnyder wood for our choice of the outer face. Thus, our bound is tight not only for minimal but also for arbitrary Schnyder woods of 3-connected plane graphs, that is, planar graphs with a fixed embedding. Since every leaf-to-root-path in a tree of a Schnyder wood is an induced path (Corollary 1), the lower bound directly implies that $p(G) \geq \log_2(n)/(3\log_2(3))$. As stated in [12], this new structural property of Schnyder woods is not only of theoretical interest, but also comes with the following additional benefits.

We have an easy linear time algorithm that computes those long induced paths. Furthermore, we obtain that there are at least $f/(2\Delta)$ different such paths, where f is the number of faces and Δ the maximum degree. And, for every such path, there exists a planar grid drawing such that this path is monotone in both coordinates.

The paper is organized as follows. In Sect. 2, we give basic definitions and lemmas. In Sect. 3, we define the graph G^k and use it to give an upper bound on the depth of a tree in a minimal Schnyder wood. In Sect. 4, we give a procedure that transforms any 3-connected planar graph to G^k for a suitable k. This we use to give our lower bound.

2 Schnyder Woods

We only consider simple undirected graphs. A graph is *plane* if it is planar and embedded into the Euclidean plane. Although parts of this paper use orientation on edges, we will always let vw denote the undirected edge $\{v, w\}$.

Let $\sigma := \{r_1, r_2, r_3\}$ be a set of three vertices of the outer face boundary of a plane graph G in clockwise order (but not necessarily consecutive). We call r_1, r_2 and r_3 *roots*. With a little abuse of notation, we define a *half-edge* as an arc that has a start vertex but no end vertex. The *suspension* G^σ of G is the graph obtained from G by adding at each root of σ a half-edge pointing into the outer face.

Definition 1 (Felsner [7]). *Let $\sigma = \{r_1, r_2, r_3\}$ and G^σ be the suspension of a 3-connected plane graph G. A* Schnyder wood *of G^σ is an orientation and coloring of the edges of G^σ (including the half-edges) with the colors 1,2,3 (red, green, blue) such that*

(i) Every edge e is oriented in one direction (we say e is unidirected*) or in two opposite directions (we say e is* bidirected*). Every direction of an edge is colored with one of the three colors 1,2,3 (we say an edge is i-*colored *if one of its directions has color i) such that the two colors i and j of every bidirected edge are distinct (we call such an edge i-j-*colored*). Throughout the paper, we assume modular arithmetic on the colors 1,2,3 in such a way that $i+1$ and $i-1$ for a color i are defined as $(i \mod 3)+1$ and $(i+1 \mod 3)+1$. For a vertex v, a uni- or bidirected edge is* incoming *(i-colored)* in *v if it has a direction (of color i) that is directed toward v, and* outgoing *(i-colored)* of *v if it has a direction (of color i) that is directed away from v.*

(ii) For every color i, the half-edge at r_i is unidirected, outgoing and i-colored.

(iii) Every vertex v has exactly one outgoing edge of every color. The outgoing 1-, 2-, 3-colored edges e_1, e_2, e_3 of v occur in clockwise order around v. For every color i, every incoming i-colored edge of v is contained in the clockwise sector around v from e_{i+1} to e_{i-1} (Fig. 1ii). This clockwise sector includes e_{i+1} and e_{i-1}.

(iv) No inner face boundary contains a directed cycle in one color.

For an illustration of Definition 1 see Fig. 1i.

For a Schnyder wood and color i, let T_i be the directed graph that is induced by the directed edges of color i. The following result justifies the name of Schnyder woods.

Lemma 1 ([9,14]). *For every color i of a Schnyder wood of G^σ, T_i is a directed spanning tree of G in which all edges are oriented to the root r_i.*

For a vertex v, we denote by $\text{depth}_i(v)$ the length of the v-r_i-path in the tree T_i. For a directed graph H, we denote by H^{-1} the graph obtained from H by reversing the direction of all its edges.

Lemma 2 (Felsner [8]). *For every $i \in \{1,2,3\}$, $T_i^{-1} \cup T_{i+1}^{-1} \cup T_{i+2}$ is acyclic.*

Using results on orientations with prescribed outdegrees on the respective completions, Felsner and Ossona de Mendez [9,10] showed that the set of Schnyder woods of a planar suspension G^σ forms a distributive lattice. The order relation of this lattice is defined on the superposition of the dual and the primal graph and also requires a Schnyder wood on the dual graph. We refer the interested reader to [13] for a definition of the minimal Schnyder wood for 3-connected planar graphs whose notation coincides with our notation.

But, as we are mostly working on planar triangulations, we work with the following simpler statement. One can easily deduce from the result of Felsner and Ossona de Mendez [9,10] that for triangulations the order relation of this lattice relates a Schnyder wood of G^σ to a second Schnyder wood if the former can be obtained from the latter by reversing the orientation of a directed clockwise cycle. This yields the following lemma.

Lemma 3 ([9,10]). *Let G be a triangulated planar graph. The minimal element of the lattice of all Schnyder woods of G^σ contains no clockwise directed cycle.*

We call the minimal element of the lattice of all Schnyder woods of G^σ also the *minimal Schnyder wood* of G^σ. If the lattice has only one element, we say that the Schnyder wood is *unique*.

3 Upper Bound on the Maximum Depth of a Tree

In this section, we define a sequence of graphs with a minimal Schnyder wood such that for each $0 < \varepsilon < 1$ there exists an N such that in each of those graphs of order $n \geq N$ each tree of the Schnyder wood has depth at most $1/(3(1-\varepsilon)\log_2(3))\log_2(n)+1$.

Those graphs are specific planar 3-trees. A *planar 3-tree* is a graph that can be constructed by the following procedure. Starting with a triangle, we iteratively select an inner face, add a new vertex v in its interior and connect this vertex with the three vertices of that face. During this construction, we assign a *level* to each newly added vertex v as follows. Every vertex on the outer face has level 0. And for v, we define $\text{level}(v) := \max\{\text{level}(w) \mid w \text{ is adjacent to } v\} + 1$. Observe that planar 3-trees are triangulated. Define the *complete* planar 3-tree of level k to be the 3-tree with the maximum number of vertices such that every vertex has level at most k. Let an *inner leaf* be a leaf of a tree of the Schnyder wood that is not on the boundary of the outer face. This construction procedure motivates the following lemma.

Lemma 4. *Let G be a triangulated plane graph and S be a Schnyder wood of G^σ. Let v be an inner leaf of the tree T_i for some $i \in \{1,2,3\}$. Let vp and vq be the outgoing $(i+1)$-colored edge and the outgoing $(i+2)$-colored edge at v, respectively. Let f be the inner face that has vp and vq on its boundary.*

If we add a vertex w in f and connect it to the vertices v, p and q, then there is exactly one way to augment S to a Schnyder wood of the suspension of $G+w$. If S is minimal or unique, then the resulting Schnyder wood is minimal or unique (w.r.t. the choice of the roots), respectively. Also, $\text{depth}_i(w) = \text{depth}_i(v) + 1$, $\text{depth}_{i+1}(w) = \text{depth}_{i+1}(v)$, $\text{depth}_{i+2}(w) = \text{depth}_{i+2}(v)$ *and v is not a leaf of T_i in the resulting Schnyder wood.*

For every $k \geq 1$, define G^k to be the planar 3-tree non-isomorphic to the triangle (as the triangle does not have inner leaves) such that in every tree of the Schnyder wood of its suspension every inner leaf has depth k. Observe that this is a valid definition. Let v be a leaf of w.l.o.g. T_3 in the Schnyder wood of the suspension of a planar 3-tree. Then, there exists a face that has v, its outgoing 1-colored edge vp and its outgoing 2-colored edge vq on its boundary. By Lemma 4, we can now add a vertex w in that face, connect it to p, q and v and color the new edges such that we obtain a Schnyder wood. In the resulting Schnyder wood, we have $\text{depth}_1(w) = \text{depth}_1(v)$, $\text{depth}_2(w) = \text{depth}_2(v)$, $\text{depth}_3(w) = \text{depth}_3(v)+1$ and v is not a leaf in T_3 anymore. If we iterate this for every leaf of depth smaller than k, we eventually arrive at the graph G^k (Fig. 2 and 3). Observe that the number of vertices in G^k rapidly increases. This in turn yields that the depth of the deepest tree in the Schnyder wood is small in terms of the number of vertices.

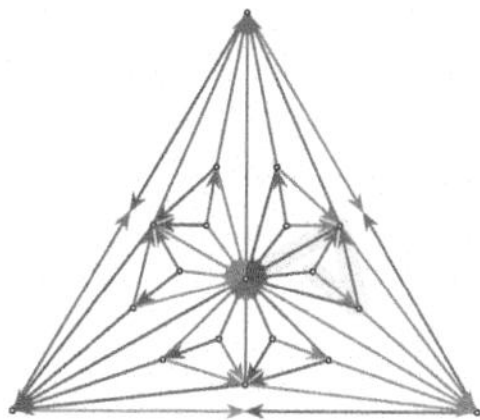

Fig. 2. Illustration for the definition of G^k. G^2 together with its Schnyder wood. G^2 has 19 vertices. The path marked in yellow maps to the sequence $(3, 1, 2)$.

Lemma 5. *G^k has $3 + \sum_{s,t,r=0,\dots,k-1} \binom{s+t+r}{s}\binom{t+r}{t}$ vertices. And for every $0 < \varepsilon < 1$ and $c > 0$ there exists a $K \geq 0$ such that $|V(G^k)| \geq c \cdot 3^{3(1-\varepsilon)(k-1)}$ for every $k \geq K$.*

Proof (Sketch). We give a bijection between the interior vertices of G^k and the sequences of the numbers 1, 2 and 3 in which each number appears at most $k-1$ times. Counting those sequences then shows the claimed statement.

By the definition of planar 3-trees, we have that every inner vertex v of G^k is adjacent to at least one vertex u such that $\text{level}(v) = \text{level}(u) + 1$. It is possible to show that for every inner vertex v except for the one vertex x that is adjacent to the three vertices on the outer face there exists exactly one such vertex u. Also, x is the only vertex of level 1. Hence, for every such vertex v, there is exactly one path $P_v = (v_0, \dots, v_l)$ with $x = v_0$ and $v = v_l$ such that $\text{level}(v_s) = \text{level}(v_{s-1}) + 1$ for every $s = 1, \dots, l$. This path maps to a sequence via $f(P_v) = (c(v_0v_1), \dots, c(v_{l-1}v_l))$ where $c(e)$ refers to the color of the edge e (Fig. 2). It is possible to show that f is a bijection between the interior vertices of G^k and the sequences of the numbers 1, 2 and 3 in which each number appears at most $k-1$ times.

Thus, we are left to count those sequences. We also count the vertices on the outer face, we use Stirling's formula and some $0 < \varepsilon < 1$ and obtain that

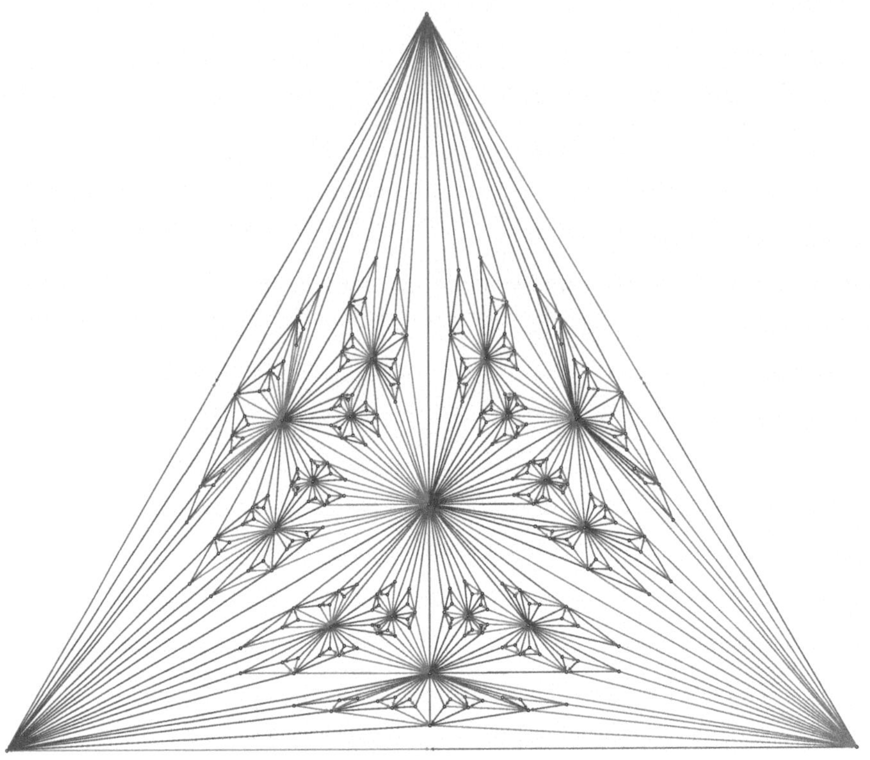

Fig. 3. Illustration for the definition of G^k. G^3 together with its Schnyder wood. G^3 has 274 vertices.

$$|V(G^k)| = 3 + \sum_{s,t,r=0,\dots,k-1} \binom{s+t+r}{s}\binom{t+r}{t} \geq \frac{(3k-3)!}{((k-1)!)^3}$$

$$\geq exp\left(-\frac{96k-21}{432k^2-852k+420}\right) \cdot \frac{\sqrt{3}}{2\pi} \cdot \frac{1}{k-1} \cdot 3^{3\varepsilon(k-1)} \cdot 3^{3(1-\varepsilon)(k-1)}.$$

Observe that

$$exp\left(-\frac{96k-21}{432k^2-852k+420}\right) \cdot \frac{\sqrt{3}}{2\pi} \cdot \frac{1}{k-1} \cdot 3^{3\varepsilon(k-1)} \longrightarrow \infty, \text{ for } k \longrightarrow \infty.$$

Hence, for every $0 < \varepsilon < 1$ and $c > 0$, there exists a $K \geq 0$ such that

$$exp\left(-\frac{96k-21}{432k^2-852k+420}\right) \cdot \frac{\sqrt{3}}{2\pi} \cdot \frac{1}{k-1} \cdot 3^{3\varepsilon(k-1)} \geq c$$

for every $k \geq K$. This concludes the proof.

Theorem 1. *For every $0 < \varepsilon < 1$ and n sufficiently large (n depends on ε), there exists a planar graph of order n with a unique, and thus, minimal Schnyder wood such that every tree of the Schnyder wood has depth at most $\log_2(n)/(3(1-\varepsilon)\log_2(3)) + 1$. For ε small enough, we obtain $\log_2(n)/(3(1-\varepsilon)\log_2(3)) + 1 < \log_2(n)/4.75 + 1$.*

Proof (Sketch). Remember that G^k and the Schnyder wood of its suspension are defined such that every tree of the Schnyder wood has depth k. It is possible to show that this Schnyder wood is also unique and thus minimal. By Lemma 5, for every $0 < \varepsilon < 1$ there exists a $K > 0$ such that $n := |V(G^k)| \geq 1 \cdot 3^{3(1-\varepsilon)(k-1)}$ for every $k \geq K$. Thus, we obtain

$$3^{3(1-\varepsilon)(k-1)} \leq n \Leftrightarrow k \leq \frac{1}{3(1-\varepsilon)\log_2(3)} \log_2(n) + 1.$$

Since $3\log_2(3) > 4.75$, we can choose ε such that $3(1-\varepsilon)\log_2(3) > 4.75$. Hence, $k \leq 1/4.75 \log_2(n) + 1$ for n sufficiently large.

4 Lower Bound

In this section, G always refers to a 3-connected plane graph such that S is the minimal Schnyder wood of G^σ. G might have additional structural properties if explicitly stated. We show that S has a tree of depth at least $1/(3\log_2(3)) \cdot \log_2(n)$. Define the depth of G to be the maximum depth of a tree of S and denote it by $\text{depth}(G)$. In our proof, we essentially show that G^k is indeed the worst case example. We give a procedure that transforms every given graph G to G^k for some $k \leq \text{depth}(G)$. We realize this by defining a graph $H(G)$ and iteratively applying this construction, i.e., we form the graphs $H(H(G)), \ldots$ until we obtain the graph G^k. Throughout this procedure, we only increase the number of vertices and decrease the depth of the deepest tree. In the end, we obtain that $\text{depth}(G) \geq \text{depth}(G^k) \geq 1/(3\log_2(3)) \cdot \log_2(|V(G^k)|) \geq 1/(3\log_2(3)) \cdot \log_2(|V(G)|)$. Hence, we need a lower bound on the depth of G^k.

Lemma 6. *We have that $n := |V(G^k)| \leq 1/2 \cdot 3^{3k-2} - 1/2$, and thus, $\text{depth}(G^k) = k > 1/(3\log_2(3)) \cdot \log_2(n)$.*

Proof. By Lemma 5, $n - 3$ equals the number of sequences of the colors 1, 2 and 3 such that each color appears at most $k-1$ times. This is clearly upper bounded by the number of sequences of length at most $3k-3$ such that the colors appear an arbitrary number of times. Hence, we obtain that

$$n - 3 \leq \sum_{l=0}^{3k-3} 3^l = \frac{3^{3k-2} - 1}{3 - 1} \Rightarrow \frac{1}{3\log_2(3)} \log_2(n) < k.$$

Lemma 7 (Di Battista et al. [3]). *The boundary of every inner face f of G can be partitioned into six paths $P_{1,3}$, $p_{2,3}$, $P_{2,1}$, $p_{3,1}$, $P_{3,2}$ and $p_{1,2}$ which appear in that clockwise order. For those paths the following holds (Fig. 4).*

(i) $P_{i,j}$ *consists of one edge which is either unidirected i-colored, unidirected j-colored or i-j-colored. Color i is directed in clockwise direction and color j in counterclockwise direction around f.*
(ii) $p_{i,j}$ *consists of a possibly empty sequence of i-j-colored edges such that color i is directed clockwise around f.*

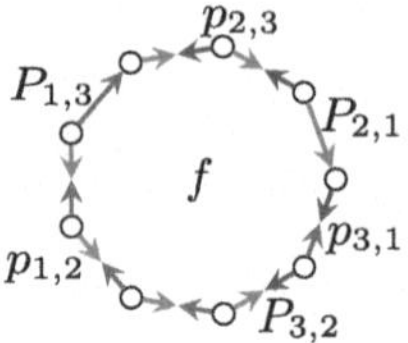

Fig. 4. Illustration for Lemma 7. A face f and the paths on its boundary.

The following definition is similar to the definition of $\tau(G)$ in [11]. The two graphs only differ on the outer face. A similar construction, but in the reverse direction, is used by Bonichon et al. [2]. Let P be the counterclockwise 3-colored path on the boundary of some inner face. By Lemma 7, P consists of $p_{2,3}$ (a possibly empty sequence of 2-3-colored edges) and possibly $P_{1,3}$ (an edge which is either unidirected 1-colored, unidirected 3-colored or 1-3-colored). Since S is minimal, we we obtain that if $p_{2,3}$ is non-empty, then $P_{1,3}$ is either unidirected 3-colored or 3-1-colored [11], and we might define $\tau(G)$ as follows.

Definition 2. *Define* $\tau(G)$ *to be a triangulation of* G *obtained as follows. First, add the edges* r_1r_2, r_2r_3 *and* r_3r_1 *if they are not yet in* $E(G)$*. Change the coloring of the edges incident to the roots such that the resulting orientation and coloring is still a Schnyder wood. If we add* r_ir_{i+1}, $i \in \{1, 2, 3\}$, *then* r_ir_{i+1} *is incoming i-colored and outgoing* $(i+1)$*-colored at* r_i*. And hence, the edge incident to* r_i *that is* i-$(i+1)$*-colored before we add* r_ir_{i+1} *becomes unidirected i-colored and incoming at* r_i*. Similarly, the edge incident to* r_{i+1} *that is* i-$(i+1)$*-colored before we add* r_ir_{i+1} *becomes unidirected* $(i+1)$*-colored and incoming at* r_{i+1} *(Fig. 5i).*

Let f *be an inner face of* G*. Let* P *be the counterclockwise 3-colored path on the boundary of* f *and let* $v_1, \ldots, v_k$ *be its vertices in counterclockwise order around* f*. If* $k \geq 3$*, proceed as follows. Add 3-colored edges* $v_1v_k, \ldots, v_{k-2}v_k$ *directed towards* v_k *and for* $j = 2, \ldots, k-1$ *change the color and orientation of* v_jv_{j+1} *such that* v_jv_{j+1} *is 2-colored and directed towards* v_j *(Fig. 5ii and 5iii). Proceed the same way for the counterclockwise 1-colored path and the counterclockwise 2-colored path on the boundary of* f*.*

Lemma 8. *The orientation and coloring* S' *of the suspension of* $\tau(G)$ *is a minimal Schnyder wood. And we have* $\text{depth}(\tau(G)) \leq \text{depth}(G)$.

For the subsequent proofs to work, we need that every inner leaf has the same depth. Hence, we give the following definition. By Lemma 4, this can easily be achieved.

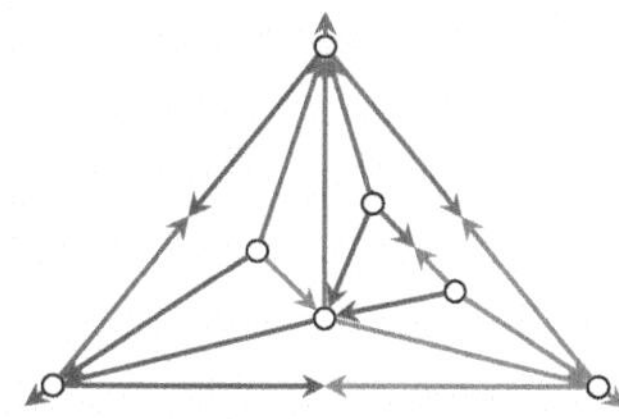

(i) The graph we obtain after the addition of r_1r_2 and r_3r_1 if we start with a wheel graph.

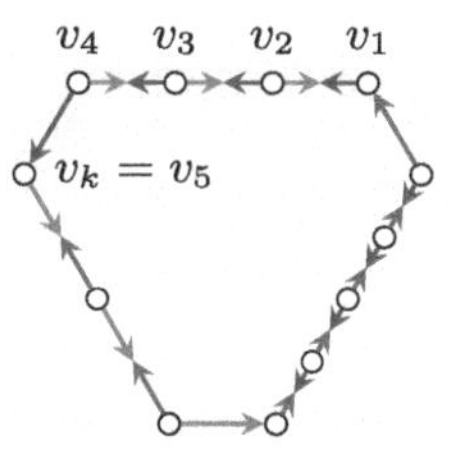

(ii) An inner face of G.

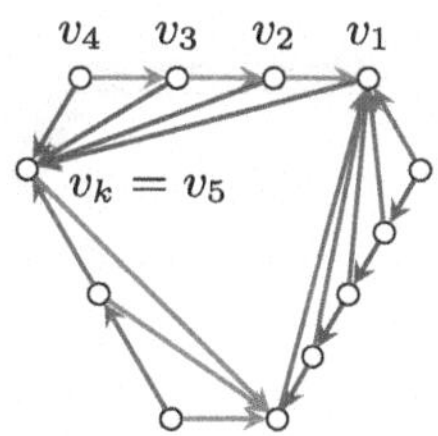

(iii) The corresponding subgraph of $\tau(G)$.

Fig. 5. Illustration for the definition of $\tau(G)$. The counterclockwise 3-colored path P on the boundary of the face of G is highlighted in yellow.

Definition 3. *Let G be triangulated and of depth k. Define $\overline{G}$ to be the graph obtained from G by the following iterative process. For all $i \in \{1, 2, 3\}$, whenever there is an inner leaf v in T_i that does not have depth k, we add a vertex u in the face delimited by the outgoing edge vp of v in color $i + 1$, the outgoing edge vq of v in color $i + 2$ and pq. We orient and color the edges incident to u such that we obtain a Schnyder wood, i.e., the edges uv, up and uq are outgoing at u and i-, $(i + 1)$- and $(i + 2)$-colored, respectively.*

Remark 1. Observe that in the setting of Definition 3, we obtain the following. By Lemma 4, $\text{depth}_i(u) = \text{depth}_i(v) + 1$, $\text{depth}_{i+1}(u) = \text{depth}_{i+1}(v)$, $\text{depth}_{i+2}(u) = \text{depth}_{i+2}(v)$ and v is not a leaf of T_i anymore. Hence, in $\overline{G}$, every inner leaf has depth k. Also, by Lemma 4, the resulting Schnyder wood is still minimal.

For $i \in \{1, 2, 3\}$ and a triangulated graph G, let $L_i(G)$ be the set of inner leaves of T_i and $l_i^\circ(G) := |L_i(G)|$. Also, we denote by $f^\circ(G)$ the number of inner faces of G.

Lemma 9. *Let $i \in \{1, 2, 3\}$ and C be a facial cycle of $\overline{G} - L_i(\overline{G})$, i.e., a cycle that forms the boundary of a face. Then, there is no $(i - 1)$-colored edge e in $\overline{G}$ with head in the interior of C and tail on C.*

Lemma 10 (folklore). *Let G be triangulated and of order n. Then, $\sum_{i=1}^{3} l_i^\circ(G) \leq f^\circ(G) = 2n - 5$.*

Lemma 9 allows for the following definition.

Definition 4. *Let G be of depth k. Define G_1 to be the graph obtained by the following process. First, triangulate G as described in Definition 2 obtaining $\tau(G)$. Then, add vertices as described in Definition 3 obtaining $\overline{\tau(G)}$.*

Let $G_1 = \overline{\tau(G)} - L_1(\overline{\tau(G)})$. Now, for every facial cycle C of G_1 that is not a triangle do the following. For every vertex z on C with an outgoing 2-colored edge zy such that $zy \in E(\overline{\tau(G)})$ and $y \in L_1(\overline{\tau(G)})$ is in the interior of C, let v_z

be the vertex where the 2-colored path in $\overline{\tau(G)}$ *from* z *to the root* r_2 *first meets* C*. Add the edge* zv_z*, color it with color 2 and orient it from* z *to* v_z*. In order to guarantee that all inner leaves of* T_2 *and* T_3 *in* G_1 *have depth* k *and all inner leaves of* T_1 *have depth* $k-1$*, we iteratively add vertices as in Definition 3.* G_2 *and* G_3 *are defined symmetrically.*

Lemma 11. G_i *is a planar triangulated graph and the Schnyder wood we obtain for* G_i^σ *is a minimal Schnyder wood for all* $i \in \{1, 2, 3\}$*.*

Proof (Sketch). We argued in Remark 1 and Lemma 4 that adding vertices in the manner of Definition 3 preserves the minimality of the Schnyder wood. Hence, in the following, we only consider the graph before we add those vertices. W.l.o.g. let $i = 1$. As described in Remark 1 and Lemma 8, the orientation and coloring of $\overline{\tau(G)}$ is a minimal Schnyder wood. First, observe that adding the 2-colored edges to $\overline{\tau(G)} - L_1(\overline{\tau(G)})$ does not create multi-edges by Lemma 2, i.e., for every pair of vertices there is at most one edge incident to both.

Second, we show that G_1 is indeed planar. Assume that G_1 is not planar. Then, there exists a facial cycle C in $\overline{\tau(G)} - L_1(\overline{\tau(G)})$ with four vertices $x, v, y, u \in C$ in that clockwise order such that there are 2-colored edges $xy, vu \in E(G_1)$. This implies that there are 2-colored paths P_{xy} and P_{vu} in $\overline{\tau(G)}$ in the interior of C connecting x with y and v with u, respectively. Those paths need to intersect. Let p be the last vertex of this intersection in the direction of color 2. At p Definition 1(iii) is violated, a contradiction. This implies that G_1 is planar.

Lemma 9 yields that our construction does not split 3-colored paths from a vertex to the root. Since we do only delete leaves of T_1, we also do not split 1-colored paths. And if we split a 2-colored path, then we patch this path with a 2-colored edge. These are the key observations in order to show that our construction yields a Schnyder wood. Finally, it is also possible to show that this Schnyder wood of G_1^σ is minimal.

In the following, we need to deal with multiple graphs and their Schnyder woods. If needed, we add a specifier. For $i, j \in \{1, 2, 3\}$, we refer by $r_i(G_j)$ to the root of the i-colored tree of the Schnyder wood of G_j^σ.

Definition 5. *Define* $H(G)$ *to be the graph with a Schnyder wood obtained by the following procedure. Take* G_1*,* G_2 *and* G_3*. Identify the edges on the outer face, recolor and reorient those edges as follows. Identify* $r_1(G_3)r_3(G_3)$ *with* $r_1(G_2)r_2(G_2)$*, color it with color 1 and orient it towards* $r_1(G_2) = r_1(G_3)$*. Identify* $r_2(G_3)r_3(G_3)$ *with* $r_1(G_1)r_2(G_1)$*, color it with color 2 and orient it towards* $r_2(G_1) = r_2(G_3)$*. Identify* $r_1(G_1)r_3(G_1)$ *with* $r_2(G_2)r_3(G_2)$*, color it with color 3 and orient it towards* $r_3(G_1) = r_3(G_2)$ *(Fig. 6). Delete redundant half-edges.*

Lemma 12. $H(G)$ *is triangulated and its orientation and coloring yields a minimal Schnyder wood of its suspension. Furthermore,* $\text{depth}(G) \geq \text{depth}(\overline{\tau(G)}) = \text{depth}(H(G))$ *and* $|V(G)| \leq |V(H(G))|$*.*

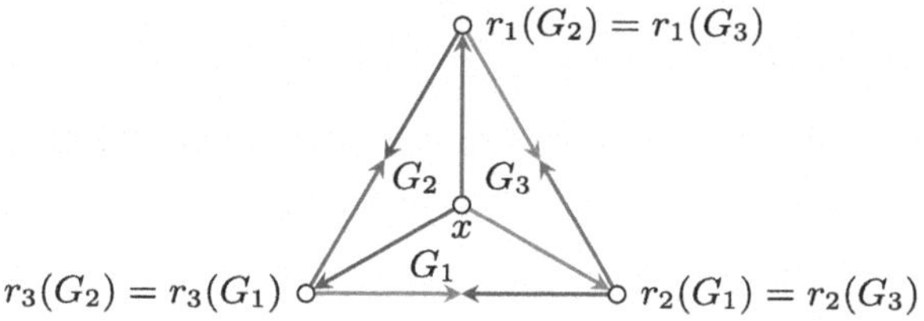

Fig. 6. Illustration for Definition 5. Here $x = r_1(G_1) = r_2(G_2) = r_3(G_3)$.

Proof. $H(G)$ is triangulated by construction. As observed in Lemma 11, the orientations and colorings of G_1, G_2 and G_3 are minimal Schnyder woods. Hence, by construction the orientation and coloring of $H(G)$ is a Schnyder wood. Assume, for the sake of contradiction, that there is a clockwise cycle C in $H(G)$. Observe that C cannot contain a vertex on the outer face of $H(G)$. Since all the outgoing edges of the vertex $x = r_1(G_1) = r_2(G_2) = r_3(G_3)$ end at vertices that are on the outer face of $H(G)$, C cannot contain x. Thus, C is completely contained in w.l.o.g. G_1, contradicting the minimality of the Schnyder wood of G_1^σ. And hence, the Schnyder wood of the suspension of $H(G)$ is minimal.

Let us consider the depth. By Lemma 8, $\text{depth}(G) \geq \text{depth}(\tau(G))$. By Definition 3, $\text{depth}(\tau(G)) = \text{depth}(\overline{\tau(G)})$. For $i \in \{2,3\}$, by Definition 4, the depth of the i-colored tree of $\overline{\tau(G)}$ equals the depth of the i-colored tree in G_1 and, by Definition 5, the root of the i-colored tree of G_1 becomes the root of the i-colored tree of $H(G)$. Also, by Definition 4, the depth of the 1-colored tree of G_1 is by one smaller than the depth of the 1-colored tree of $\overline{\tau(G)}$. And, by Definition 5, the root of the 1-colored tree of G_1 has depth one in the 1-colored tree of $H(G)$. This holds symmetrically for G_2 and G_3. This yields that $\text{depth}(\overline{\tau(G)}) = \text{depth}(H(G))$, and altogether, $\text{depth}(G) \geq \text{depth}(H(G))$.

It remains to show that $|V(G)| \leq |V(H(G))|$. Using Lemma 10, we obtain that

$$\begin{aligned}|V(H(G))| &\geq 1 + 3 + 3(|V(\overline{\tau(G)})| - 3) - \sum_{i=1}^{3} l_i^\circ(\overline{\tau(G)}) \\ &\geq 3|V(\overline{\tau(G)})| - 5 - 2|V(\overline{\tau(G)})| + 5 \\ &= |V(\overline{\tau(G)})|.\end{aligned}$$

Obviously, we have $|V(\overline{\tau(G)})| \geq |V(G)|$, and thus, $|V(H(G))| \geq |V(G)|$.

Let A be a subgraph of G. Observe that we delete edges and vertices going from G to G_1, G_2 and G_3. Hence, there only remain subgraphs of A in G_1, G_2 and G_3. Let A' be the subgraph of $H(G)$ given by the union of the three subgraphs of A in G_1, G_2 and G_3. We say that A' *originates* from A. We define this relation to be transitive, i.e., if A originates from B and B originates from C, then A originates from C.

Let D be the subgraph of $\overline{\tau(G)}$ that is induced by its triangular outer face. Observe that the subgraph in $H(G)$ that originates from D is the complete graph K_4. We iterate this idea and finally obtain the following theorem.

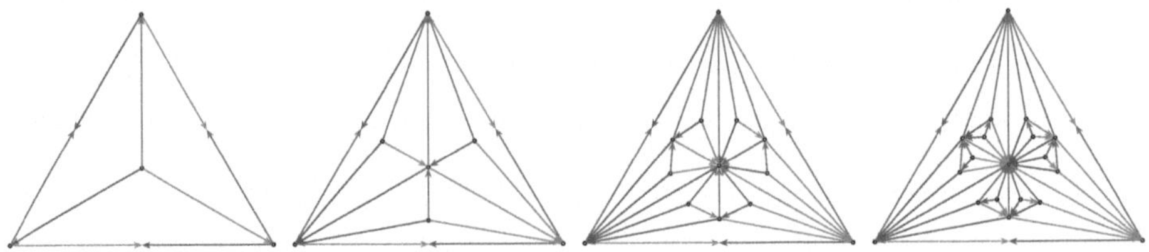

Fig. 7. Let $\text{depth}(\overline{\tau(G)}) = 2$. From left to right we have the subgraphs of $H(G)$, $H^2(G)$, $H^3(G)$ and $H^4(G)$ that originate from the triangular outer face of $\overline{\tau(G)}$. Observe that the rightmost graph corresponds to G^2, compare Fig. 2.

Theorem 2. *Let G be a 3-connected planar graph with a minimal Schnyder wood. Then,*

$$\text{depth}(G) \geq \frac{1}{3\log_2(3)} \log_2(|V(G)|).$$

Proof. Define $H^s(G) := H(H^{s-1}(G))$ for $s \geq 2$ with $H^1(G) := H(G)$. The Schnyder wood we obtain for $H(G)$ is the minimal Schnyder wood, by Lemma 12. Thus, $H^s(G)$ is well-defined. Let $t = \text{depth}(\tau(G))$. Since $H(G)$ is already triangulated and all leaves of the three trees of the Schnyder wood have depth t, we have $\overline{\tau(H(G))} = H(G)$ and thus, by Lemma 12 and Definition 3, $\text{depth}(H^{3t-2}(G)) = \text{depth}(\tau(G)) = t$. One can show that $H^{3t-2}(G) = G^t$ (Fig. 7). And hence, Lemma 6 and 12 yield

$$\begin{aligned}
\text{depth}(G) &\geq \text{depth}(H(G)) \geq \ldots \geq \text{depth}(H^{3t-2}(G)) = \text{depth}(G^t) \\
&\geq \frac{1}{3\log_2(3)} \log_2(|V(G^t)|) = \frac{1}{3\log_2(3)} \log_2(|V(H^{3t-2}(G))|) \geq \ldots \\
&\geq \frac{1}{3\log_2(3)} \log_2(|V(G)|).
\end{aligned}$$

Corollary 1. *Every 3-connected planar graph G on n vertices has an induced path of size at least $\lfloor 1/(3\log_2(3)) \log_2(n) \rfloor + 1$.*

Proof. Take a minimal Schnyder wood of G^σ. For a vertex $v \in V(G)$ and $i \in \{1, 2, 3\}$, the v-r_i-path $P_i(v)$ in the tree T_i is always induced. Assume, for the sake of contradiction, that there exists a vertex $v \in V(G)$ for which this does not hold. Then, there is a j-colored edge $e = xy$ in G with $x, y \in P_i(v)$, $i, j \in \{1, 2, 3\}$. By Lemma 1, T_i is a tree and hence $i \neq j$. Now, either $T_i^{-1} \cup T_j$ or $T_i \cup T_j$ has an oriented cycle, contradicting Lemma 2. Hence, Theorem 2 directly yields an induced path of size $\lfloor 1/(3\log_2(3)) \log_2(n) \rfloor + 1$.

5 Conclusion

In this paper, we gave a tight bound on the depth of a minimal Schnyder wood and used this bound to give a lower bound on the length of an induced path. Observe that the suspension of the graph G^k which we used to show that our lower bound is tight has a unique Schnyder wood for our choice of the outer face. Hence, if we want to exploit this method further, then we need to allow to choose the outer face.

References

1. Arocha, J.L., Valencia, P.: Long induced paths in 3-connected planar graphs. Discuss. Math. Graph Theory **20**(1), 105–107 (2000). https://doi.org/10.7151/dmgt.1110
2. Bonichon, N., Felsner, S., Mosbah, M.: Convex drawings of 3-connected plane graphs. Algorithmica **47**(4), 399–420 (2007)
3. Di Battista, G., Tamassia, R., Vismara, L.: Output-sensitive reporting of disjoint paths. Algorithmica **23**(4), 302–340 (1999). https://doi.org/10.1007/PL00009264
4. Di Giacomo, E., Liotta, G., Mchedlidze, T.: Lower and upper bounds for long induced paths in 3-connected planar graphs. Theoret. Comput. Sci. **636**, 47–55 (2016). https://doi.org/10.1016/j.tcs.2016.04.034
5. Erdős, P., Saks, M., Sós, V.T.: Maximum induced trees in graphs. J. Combin. Theory Ser. B **41**(1), 61–79 (1986). https://doi.org/10.1016/0097-3165(86)90115-9
6. Esperet, L., Lemoine, L., Maffray, F.: Long induced paths in graphs. European J. Combin. **62**, 1–14 (2017). https://doi.org/10.1016/j.ejc.2016.11.011
7. Felsner, S.: Geodesic embeddings and planar graphs. Order **20**, 135–150 (2003)
8. Felsner, S.: Convex drawings of planar graphs and the order dimension of 3-polytopes. Order **18**(1), 19–37 (2001). https://doi.org/10.1023/A:1010604726900
9. Felsner, S.: Geometric graphs and arrangements. Advanced Lectures in Mathematics, Friedr. Vieweg & Sohn, Wiesbaden (2004). https://doi.org/10.1007/978-3-322-80303-0
10. Ossona de Mendez, P.: Orientations bipolaires. Ph.D. thesis, École des Hautes Études en Sciences Sociales, Paris (1994)
11. Ortlieb, C.: Toward Grünbaum's conjecture for 4-connected graphs. In: Proceedings of the 49th International Symposium on Mathematical Foundations of Computer Science (MFCS'24). Leibniz Int. Proc. Inform. **306**, 77:1–77:13 (2024). https://doi.org/10.4230/LIPICS.MFCS.2024.77
12. Ortlieb, C.: Schnyder woods and long induced paths in 3-connected planar graphs. In: Proceedings of Latin American Theoretical Informatics Conference (2024), to appear
13. Ortlieb, C., Schmidt, J.M.: Toward Grünbaum's Conjecture. In: 19th Scandinavian Symposium and Workshops on Algorithm Theory (SWAT 2024). Leibniz International Proceedings in Informatics (LIPIcs), vol. 294, pp. 37:1–37:17. Schloss Dagstuhl – Leibniz-Zentrum für Informatik, Dagstuhl, Germany (2024). https://doi.org/10.4230/LIPIcs.SWAT.2024.37
14. Schnyder, W.: Embedding planar graphs on the grid. In: Proceedings of the first annual ACM-SIAM symposium on Discrete algorithms, pp. 138–148 (1990)

DAG Scheduling in the BSP Model

Pál András Papp[(✉)], Georg Anegg, and Albert-Jan N. Yzelman

Computing Systems Lab, Huawei Zurich Research Center, Zurich, Switzerland
{pal.andras.papp,albertjan.yzelman}@huawei.com

Abstract. We study the problem of scheduling an arbitrary computational DAG on a fixed number of processors while minimizing the makespan. While previous works have mostly studied this problem in fairly restricted models, we define and analyze DAG scheduling in the Bulk Synchronous Parallel (BSP) model, which is a well-established parallel computing model that captures the communication cost between processors much more accurately. We provide a taxonomy of simpler scheduling models that can be understood as variants or special cases of BSP, and discuss how the properties and optimum cost of these models relate to BSP. This essentially allows us to dissect the different building blocks of the BSP model, and gain insight into how these influence the scheduling problem.

We then analyze the hardness of DAG scheduling in BSP in detail. We show that the problem is solvable in polynomial time for some very simple classes of DAGs, but it is already NP-hard for in-trees or DAGs of height 2. We also prove that in general DAGs, the problem is APX-hard: it cannot be approximated to a $(1+\epsilon)$-factor in polynomial time for some specific $\epsilon > 0$. We then separately study the subproblem of scheduling communication steps, and we show that the NP-hardness of this problem depends on the problem parameters and the communication rules within the BSP model. Finally, we present and analyze a natural formulation of our scheduling task as an Integer Linear Program.

Keywords: Bulk synchronous parallel · NP-hard · APX-hard · ILP

1 Introduction

The optimal scheduling of complex workloads is a fundamental problem not only in computer science, but also in other areas like logistics or operations research. In a computational context, the most natural application of scheduling is when we have a complex computation consisting of many different subtasks, and we want to execute this on a parallel (multi-processor or multi-core) architecture, while minimizing the total time required for this. Unsurprisingly, this topic has been extensively studied since the 1960s, and has gained even more importance recently with the widespread use of manycore architectures.

In these scheduling problems, a computational task is represented as a Directed Acyclic Graph (DAG), where each node corresponds to an operation

R. Královič and V. Kůrková (Eds.): SOFSEM 2025, LNCS 15539, pp. 238–253, 2025.
https://doi.org/10.1007/978-3-031-82697-9_18

or subtask, and each directed edge (u, v) indicates a dependency relation, i.e. that the processing of node u has to be finished before the processing of node v begins, since the output of operation u is needed as an input for operation v. This DAG model of general computations is not only prominent in scheduling, but also in further topics such as pebble games.

However, from a complexity-theoretic perspective, the scheduling of general DAGs is already a hard problem even in very simple settings, e.g. even in models that heavily simplify or completely ignore the communication costs between processors, which is the main bottleneck in many computational tasks in practice. Due to this, previous theoretical works have mostly focused on analyzing the complexity of scheduling in these rather simple models, and while more realistic models were sometimes introduced, the theoretical properties of scheduling in these more realistic models received little attention.

On the other hand, the parallel computing community has developed far more advanced models to accurately quantify the real cost of parallel algorithms in practice. One of the most notable is the Bulk Synchronous Parallel (BSP) model, which is still relatively simple, but provides a delicate cost function that captures the volume of communicated data and the synchronization costs in a given parallel schedule. BSP (and similar models) are fundamental tools for evaluating and comparing concrete practical implementations of parallel algorithms. However, previous theoretical works on BSP only focus on finding and analyzing parallel schedules for specific algorithms, and do not study BSP as a model for scheduling general DAGs, i.e. an arbitrary computational task.

Our goal in this paper is to bridge this gap between theory and practice to some extent, and understand the fundamental theoretical properties of DAG scheduling in the BSP model. This more detailed model results in a more complex scheduling problem with some entirely new aspects compared to classical models; as such, understanding its key properties is a crucial step towards designing efficient parallel schedules for computations in practice. Our hope is that our insights inspire a new line of work on the theoretical study of scheduling in more advanced parallel computing models that are developed and applied on the practical side. More specifically, our main contributions are as follows:

(i) We first define DAG scheduling in the BSP model. We then provide a taxonomy of scheduling models from previous works, and show that many of these can be understood as a special case of BSP. Analyzing the relations between these models essentially allows us to gain insight into how the different aspects of BSP affect the scheduling problem.
(ii) We then analyze the complexity of BSP scheduling, with the goal of understanding when (i.e. for which kind of DAGs) it becomes NP-hard. We show that the problem is still solvable in polynomial time for DAGs that consist of several connected chains, but it is already NP-hard for slightly more complex classes of DAGs, such as in-trees or DAGs of height 2.
(iii) We show that for general DAGs, the problem is NP-hard to even approximate to a $(1 + \epsilon)$ factor, for some $\epsilon > 0$. In order words, BSP scheduling

is APX-hard: it allows no polynomial-time approximation scheme unless P=NP.

(iv) We separately analyze the subproblem of scheduling communication steps, assuming that the rest of the schedule is already fixed. We discuss the complexity of this subproblem for several different variants of BSP.

(v) Finally, we present and analyze a natural formulation of the BSP scheduling task as an Integer Linear Programming (ILP) problem.

Our main technical contributions are points (ii)–(iv) above; however, the remaining points also provide valuable insight into the problem. The proof details and some further model discussion are deferred to the full version of the paper [42].

2 Related Work

DAG scheduling is a fundamental problem in computer science, and has been studied extensively since the 1970s. The first papers considered a simple setting where communication between processors is free, which essentially corresponds to the PRAM model. The numerous results in this model include polynomial algorithms for $P = 2$ processors [5,14,45], polynomial algorithms for special classes of DAGs [10,11,15], hardness results for $P = \infty$ [3,25,50], and results for weighted DAGs [4,24,37]. The results on some of these topics, e.g. approximation algorithms, are still rapidly improving in recent years [16,28,30,47]. On the other hand, some basic questions are still open even in this fundamental model: e.g. it is still not known whether scheduling for some fixed $P > 2$ is NP-hard.

A more realistic version of this model was introduced in the late 1980s [40,54], where there is a fixed communication delay between processors. There are also numerous algorithms and hardness results for this setting, in particular for unit-length delays [19,21] or infinitely many processors [21,33,38]. The approximability of the optimal solution is also a central question in this model that receives significant attention even in recent years [7,8,23,31,32].

This communication delay model is still unrealistic, as it allows an unlimited amount of data to be sent in a single time unit. To our knowledge, the only more sophisticated DAG scheduling model which measures communication volume is the recently introduced single-port duplex model [39]. However, this model has not been studied from a theoretical perspective, and in contrast to BSP, it cannot be extended by some real-world aspects such as synchronization costs.

There are also numerous extensions of these models with further aspects, e.g. heterogeneous processors or deadlines for each task [9,26,27,29,43,48].

On the other hand, BSP has been introduced as a prominent model of parallel computing in 1990 [51], and studied extensively ever since [2,34,46]. The model has also found its way into various applications, most notably through the BSPlib standard library [20] and its different implementations [55,56]. Fundamental results on BSP include the analysis of prominent algorithms in this model [35,36] or the extension of BSP to multi-level architectures [52,53]. However, the BSP model (and similar models with even more parameters, such as LogP [6]) have mostly been used so far to analyze the computational costs of specific parallel implementations of concrete algorithms. In contrast to this, in our work, we apply BSP as a general model to evaluate the scheduling of any DAG.

3 Model and Background

Computational tasks are modelled as a *Directed Acyclic Graph* G, with the set of nodes (subtasks) denoted by V, and the set of directed edges (dependencies) by E. We use u and v to denote individual nodes, and n to denote the number of nodes $|V|$. An edge (u, v) indicates that subtask u has to be finished before the computation of subtask v begins. We also use $[k]$ as a shorthand notation for the integer set $\{1, ..., k\}$.

We assume that we have $P \in O(1)$ identical processors, and our goal is to execute all nodes of G on these, while minimizing the total time this takes.

The BSP model. BSP is a very popular model for the design and evaluation of parallel algorithms. To our knowledge, general DAG scheduling has not been studied in this model before, but extending the interpretation of BSP to this setting is rather straightforward.

In contrast to classical models where nodes are assigned to concrete points in time, the BSP model instead divides the execution of nodes into larger batches, so-called *supersteps*. Each superstep consists of two phases, in the following order:

1. *Computation phase*: each processor may execute an arbitrary number of computation steps, but no communication between the processors is allowed.
2. *Communication phase*: processors can communicate an arbitrary number of values to each other, but no computation is executed.

The motivation behind supersteps is to encourage sending data in large batches, since in practice, communication often has a large fixed cost (e.g. synchronization, network initialization) that is independent of data volume.

Formal Definition. Let us denote the number of supersteps in our schedule by S. While S can be freely chosen in a schedule, we provide our definitions for a fixed S for simplicity. A *BSP schedule* with S supersteps consists of:

- An assignment of nodes to processors $\pi : V \rightarrow [P]$ and to supersteps $\tau : V \rightarrow [S]$. For simplicity, we introduce the notation $H^{(s,p)} = \{v \in V \mid \pi(v) = p, \tau(v) = s\}$ for the set of nodes assigned to processor p and superstep s. We can imagine the nodes of $H^{(s,p)}$ to be executed in an arbitrary (but topologically correct) order on p in superstep s.
- A set Γ of 4-tuples $(v, p_1, p_2, s) \in V \times [P] \times [P] \times [S]$, indicating that the output of node v is sent from processor p_1 to processor p_2 in the communication phase of superstep s. In this base variant of BSP, we only include p_1 in these 4-tuples for clarity, but we always assume $p_1 = \pi(v)$, i.e. the value is sent from the processor where it was computed.

A valid BSP schedule must satisfy the following conditions:

(i) A node v can only be computed if all of its predecessors are available, i.e. they were computed on processor $\pi(v)$ in an earlier (or the same) superstep, or sent to $\pi(v)$ before the given superstep. That is, for all $(u, v) \in E$, if $\pi(u) = \pi(v)$ then we must have $\tau(u) \leq \tau(v)$, and if $\pi(u) \neq \pi(v)$ then we must have $(u, \pi(u), \pi(v), s) \in \Gamma$ for some $s < \tau(v)$.
(ii) We only communicate values that are already computed: if $(v, p_1, p_2, s) \in \Gamma$, then $p_1 = \pi(v)$, and $\tau(v) \leq s$.

Cost Function. The computation phase can be executed in parallel on the different processors, so its cost (the amount of time it takes) in superstep $s \in [S]$ is the largest amount of computation executed on any of the processors. More formally, the *work cost* of superstep s (first for a given processor $p \in [P]$, and then in general) is defined as

$$C_{work}{}^{(s,p)} = |H^{(s,p)}| \qquad \text{and} \qquad C_{work}{}^{(s)} = \max_{p \in [P]} C_{work}{}^{(s,p)} .$$

Communication costs, on the other hand, are governed by two further problem parameters: $g \in \mathbb{N}$ is the cost of communicating a single unit of data, and $L \in \mathbb{N}$ is the fixed latency cost incurred by each superstep. BSP assumes that different values can be communicated in parallel in general, but any processor can only send and receive a single value in any time unit. As such, BSP considers the number of values sent and received by processor p in superstep s, and then defines the *communication cost* of a superstep (for p, or in general) as the maximum of these; this cost function is called *h-relation.* More formally, let

$$C_{sent}{}^{(s,p)} = |\{(v, p, p', s) \in \Gamma\}| \qquad \text{and} \qquad C_{rec}{}^{(s,p)} = |\{(v, p', p, s) \in \Gamma\}|$$

for some fixed $s \in [S]$ and $p \in [P]$ (over all $v \in V$ and $p' \in [P]$), and then let

$$C_{comm}{}^{(s,p)} = \max(C_{sent}{}^{(s,p)}, C_{rec}{}^{(s,p)}) \quad \text{and} \quad C_{comm}{}^{(s)} = \max_{p \in [P]} C_{comm}{}^{(s,p)} .$$

The cost $C^{(s)}$ of superstep s and the cost C of the entire schedule is defined as:

$$C^{(s)} = C_{work}{}^{(s)} + g \cdot C_{comm}{}^{(s)} + L \qquad \text{and} \qquad C = \sum_{s \in [S]} C^{(s)} .$$

For an example, consider the BSP schedule shown in Fig. 1, and let $s = 1$, $p_1 = 1$, $p_2 = 2$. Here processor p_1 computes 4 nodes, and processor p_2 computes 5 nodes, so $C_{work}{}^{(s,p_1)} = 4$, $C_{work}{}^{(s,p_2)} = 5$, and $C_{work}{}^{(s)} = \max(4, 5) = 5$ in the computation phase. In the communication phase, p_1 must send a single value to p_2 (so $C_{sent}{}^{(s,p_1)} = C_{rec}{}^{(s,p_2)} = 1$), while p_2 must send two values to p_1 ($C_{sent}{}^{(s,p_2)} = C_{rec}{}^{(s,p_1)} = 2$). This implies $C_{comm}{}^{(s,p_1)} = C_{comm}{}^{(s,p_2)} = \max(2, 1) = 2$, and hence $C_{comm}{}^{(s)} = 2$. The total cost of the superstep is $C^{(s)} = 5 + 2 \cdot g + L$. For more details on BSP, we refer the reader to [2,34].

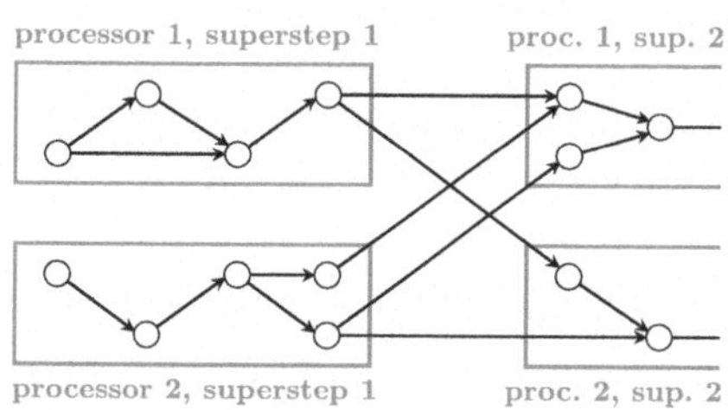

Fig. 1. Example BSP schedule for a DAG. The labelled boxes only represent the computation phases; the supersteps itself also consists of the communication phase that follows.

Table 1. Taxonomy of DAG scheduling models. The horizontal axis shows communication cost models, and the vertical axis shows whether simultaneous computation and communication is allowed.

	free comm. (no cost)	simplified comm. cost (any amount of data in a single step)	exact comm. cost (depends on data volume)
comp. & comm. simultaneously	classical scheduling [5,45]	commdelay [23,40]	single-port duplex [39]
comp. & comm. in separate phases	classical scheduling [5,45]	commdelay with phase [13]	**BSP [novel for DAGs]**

As a Scheduling Problem. We can now formally define our problem.

Definition 1. Given an input DAG, the goal of *BSP scheduling* is to find a feasible BSP schedule (π, τ, Γ) as described above, with minimal cost C.

In the decision version of the problem, we also have a maximal cost parameter C_0, and we need to decide if there is a BSP schedule with cost $C \leq C_0$. For simplicity, we will often focus on the simplest case of $L = 0$.

From a complexity perspective, it is important to note that we consider the parameters P, g, L to be small fixed constants (properties of our computing architecture), and not parts of the problem input. We especially emphasize this for P, since in contrast to our work, some others assume that P is an input variable that can be up to linear in n; however, this is unrealistic in most applications, and also makes the problem unreasonably hard even for trivial DAGs. In general, both settings (fixed P and variable P) have been extensively studied before, and are distinguished by "Pm" and "P" in the classical 3-field notation [18].

4 Comparison to Other Models

4.1 Taxonomy of Scheduling Models

In the most basic *classical scheduling* model, nodes are assigned to processors $\pi : V \rightarrow [P]$ and time steps $t : V \rightarrow \mathbb{Z}^+$, with two simple conditions: $\nexists u, v \in V$ with $\pi(u) = \pi(v)$, $t(u) = t(v)$), and $\forall (u, v) \in E$ we have $t(u) < t(v)$. The cost of a schedule here is simply $\max_{v \in V} t(v)$. A more realistic version of this model also adds a fixed communication delay g between processors: $\forall (u, v) \in E$, in case if $\pi(u) \neq \pi(v)$, we now need to have $t(v) > t(u) + g$.

BSP has two major differences from this latter *commdelay* model. Firstly, commdelay allows a processor to execute computations and communications simultaneously, while in BSP, these must happen in separate phases. Moreover, commdelay implicitly allows any amount of data to be sent from p_1 to p_2 simultaneously, whereas BSP also considers data volume: sending k values from p_1 to p_2 takes k times as long as a single value. Previous work has briefly considered the extension of commdelay with both of these modifications separately:

- The work of [13] considers a variant of commdelay where computation and communication can only happen in separate phases.
- The work of [39] introduces a *single-port duplex* (SPD) model with communication volume: a schedule here must also specifically assign the send/recieve steps to concrete (disjoint) time intervals for each processor.

The summary of these models in Table 1 shows that BSP scheduling indeed fills a natural place in this taxonomy. Note that in the last column, the difference between SPD and BSP is in fact twofold. Firstly, BSP assumes that at any time, a processor can either compute or communicate, but not both. Secondly, BSP divides the schedule into supersteps, which can be understood as a *barrier synchronization* requirement: to send a value v, we first need a point in time (after computing v) when no processor is computing, to initiate the process of sending v (and possibly other values). By separating these two properties, we can extend our taxonomy into Table 2 to include further model variants.

Table 2. Extended table of DAG scheduling models. The vertical axis is split according to (i) whether computation and communication are allowed simultaneously, and (ii) whether barrier synchronization is required for communication.

		free communication (no cost)	simplified comm. cost (any amount of data in a single step)	exact comm. cost (cost depends on data volume)
comp.&comm. simultaneously	no sync	classical scheduling	commdelay	single-port duplex
	sync	classical scheduling	commdelay with sync points	maxBSP
comp.&comm. separately	no sync	classical scheduling	$\alpha-\beta$ with $\beta=0$ (or subset-CD)	$\alpha-\beta$ with $\alpha=0$ (or subset-BSP)
	sync	classical scheduling	commdelay with phases	BSP

- If global synchronization is required (so we have supersteps), but processors can compute and communicate simultaneously: the first paper on BSP [51] implicitly assumes this *maxBSP* model variant, defining $C^{(s)}$ as the maximum of ${C_{work}}^{(s)}$ and $g \cdot {C_{comm}}^{(s)} + L$. In contrast, recent textbooks [2] apply our definition from Sect. 3 with a sum. Note that defining maxBSP for general DAGs requires further consideration, to ensure that the computation and communication phases of a superstep are indeed parallelizable.

- If processors can only either compute or communicate at a given time, but no synchronization is needed: one can interpret this as the $\alpha-\beta$ model [44] with a choice of $\alpha=0$, although the definition of this model varies (see Appendix A.4 of [42]). This allows e.g. p_1 and p_2 to stop computing and exchange values, while the rest of the processors keep computing in the meantime.

One natural question regarding this taxonomy is how the optimum costs in these models (denoted by OPT) relate to each other for a given DAG. Firstly, note that in any of the models, one of the processors must have a work cost of $\frac{n}{P}$ at least; this provides a lower bound. Moreover, executing the entire DAG on a single processor (without communication) always yields a valid solution of cost n, and hence it always gives a factor P approximation of the optimum.

Proposition 1. *We have* $\frac{n}{P} \leq \mathrm{OPT} \leq n$ *in any of these models.*

Next we compare the optimum cost in the two fundamental models, classical scheduling and commdelay, to the optimum in BSP (denoted by OPT_{class}, OPT_{CD} and OPT_{BSP}, respectively, assuming $L=0$ in BSP). These clearly satisfy $\mathrm{OPT}_{class} \leq \mathrm{OPT}_{CD} \leq \mathrm{OPT}_{BSP}$. Finding the maximal difference is more involved; however, note that e.g. Proposition 1 already implies that the optimum costs in any two models differ by at most a factor P.

Lemma 1. *We have* $\mathrm{OPT}_{BSP} \leq P \cdot \mathrm{OPT}_{class}$ *for any DAG and parameters* P *and* g. *Moreover,* $\mathrm{OPT}_{CD} \leq (1+g) \cdot \mathrm{OPT}_{class}$. *These bounds are essentially tight: there are DAG constructions with* $\frac{\mathrm{OPT}_{CD}}{\mathrm{OPT}_{class}} = P$, $\frac{\mathrm{OPT}_{BSP}}{\mathrm{OPT}_{CD}} = P$, *and* $\frac{\mathrm{OPT}_{CD}}{\mathrm{OPT}_{class}} = (1+g-\varepsilon)$ *for any* $\varepsilon > 0$.

Due to our focus on BSP, we also analyze the relation between the models in the last column of Table 2. Let us denote their optimum costs by OPT_{SPD}, OPT_{mBSP}, OPT_{β} and OPT_{BSP} from top to bottom, again for $L=0$. The restrictiveness of the models implies $\mathrm{OPT}_{SPD} \leq \mathrm{OPT}_{mBSP} \leq \mathrm{OPT}_{BSP}$ and $\mathrm{OPT}_{SPD} \leq \mathrm{OPT}_{\beta} \leq \mathrm{OPT}_{BSP}$; we complement with some further observations.

Theorem 1. *For any DAG and parameters* P, g, *the optimal costs in the models of the last column can differ by a factor* 2 *at most, i.e.* $\mathrm{OPT}_{BSP} \leq 2 \cdot \mathrm{OPT}_{SPD}$. *Moreover, we show DAG constructions that prove (for any* $\varepsilon > 0$*)*

i) a lower bound of $(2-\varepsilon)$ *for* $\frac{\mathrm{OPT}_{\beta}}{\mathrm{OPT}_{SPD}}$, $\frac{\mathrm{OPT}_{BSP}}{\mathrm{OPT}_{SPD}}$, $\frac{\mathrm{OPT}_{\beta}}{\mathrm{OPT}_{mBSP}}$ *and* $\frac{\mathrm{OPT}_{BSP}}{\mathrm{OPT}_{mBSP}}$,
ii) a looser lower bound of $(\frac{3}{2}-\varepsilon)$ *for* $\frac{\mathrm{OPT}_{mBSP}}{\mathrm{OPT}_{SPD}}$, $\frac{\mathrm{OPT}_{BSP}}{\mathrm{OPT}_{\beta}}$ *and* $\frac{\mathrm{OPT}_{mBSP}}{\mathrm{OPT}_{\beta}}$.

On a high level, $\mathrm{OPT}_{BSP} \leq 2 \cdot \mathrm{OPT}_{SPD}$ follows from a simple conversion of an SPD schedule to BSP. The lower bounds in i) use a construction that allows to parallelize computation and communication almost perfectly, while ii) uses a DAG where different processors would ideally always send data at different times, so requiring barrier synchronization notably increases the optimum.

4.2 Communication Models Within BSP

Even within BSP, there are different options to model the communication rules, and these seemingly small changes can have a significant effect on the model.

For instance, our base assumed that $\pi(v) = p_1$ for all $(v, p_1, p_2, s) \in \Gamma$ for simplicity, i.e. values are sent from the processor where they were computed. However, to transfer a value from p_1 to p_2, one might as well send it from p_1 to a third processor p_3 first, and then from p_3 to p_2. Moreover, there are simple examples where such *free data movement* between processors can indeed result in a lower communication cost altogether. Consider the schedule in Fig. 2 with $P = 3$ processors; this has a node that is computed on p_3 in superstep 1, but later only needed on p_1 in superstep 3. With direct transfer, we can send this data from p_3 to p_1 in either superstep 1 or 2; however, p_1 must already receive a value in superstep 1, and p_3 must send a value in superstep 2, so both options increase the cost in one of the supersteps. On the other hand, with free data movement, we can send the value from p_3 to p_2 in superstep 1, and then from p_2 to p_1 in superstep 2, without increasing the cost in either superstep.

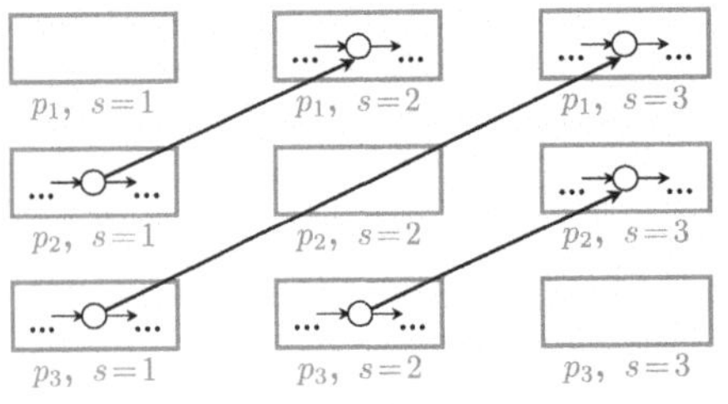

Fig. 2. An example BSP schedule where free data movement allows for a lower communication cost than direct data transfer.

Table 3. Communication models *within BSP*, and their properties shown in Sects. 7 and 8. Our main results (Theorems 2–5) hold in all of these models.

	Singlecast	Broadcast
Direct transfer	textbfDS model CS: *open problem* ILP: $O(n \cdot P \cdot S)$ *vars*	**DB model** CS: *NP-hard* ILP: $O(n \cdot P \cdot S)$ *vars*
Free data movement	**FS model** CS: *NP-hard* ILP: $O(n \cdot P^2 \cdot S)$ *vars*	**FB model** CS: *NP-hard* ILP: $O(n \cdot P \cdot S)$ *vars*

Another interesting question occurs when processor p wants to send a single value v to multiple other processors $p_1, ..., p_k$ in the same superstep. In our base model, this requires a separate entry (v, p, p_i, s) for all $i \in [k]$, and hence contributes k units to the send cost $C_{send}{}^{(s,p)}$. This is a reasonable assumption e.g. if the communication topology is a clique, and thus p needs to send this value over k distinct network links. However, in e.g. a star-shaped communication topology, it can be more reasonable to only charge a single unit of send cost for this, i.e. to assume that data transfers are *broadcast operations*, and their values can be received by any number of processors.

These options can be combined to form 4 different *communication models* within BSP. We name these models in Table 3. Our results in Sects. 7 and 8 show that these models can indeed influence some properties of the problem.

5 NP-Hardness

Regarding the complexity of BSP scheduling, it is not surprising that the problem is NP-hard in general DAGs. However, this raises a natural follow-up question, which has also been studied in simpler models: in which subclasses of DAGs is the problem still solvable in polynomial time, and when does it become NP-hard?

The simplest non-trivial subclass is *chain DAGs*, where both the indegree and outdegree of nodes is at most 1. This subclass has been analyzed in different scheduling models [12,48], including a proof of NP-hardness in BSP [17] if we assume a compressed input representation (see [42] for a discussion). We also consider an extension of this subclass: we say that a DAG is a *connected chain DAG* if it can be obtained by adding an extra source node v_0 to a chain DAG, and drawing an edge from v_0 to the first node in every chain. We show that for these simple classes of DAGs, the optimal BSP schedule can still be found in polynomial time. The key observation is that the optimal BSP schedule in chain DAGs always consists of at most P supersteps; this allows us to find the optimum through a rather complex dynamic programming approach.

Theorem 2. *The BSP scheduling problem can be solved in polynomial time in n for chain DAGs and connected chain DAGs.*

On the other hand, it turns out that BSP scheduling already becomes NP-hard for slightly more complex DAGs. In particular, we consider (i) DAGs with height only 2, where *height* is the number of nodes in the longest directed path, and (ii) *in-trees*, which are DAGs where every node has outdegree at most 1. We prove that in these (still relatively simple) classes, the problem is already NP-hard. Since the scheduling problem is known to be polynomially solvable for these classes in simpler models (specifically, for in-trees in classical scheduling [15], and for height-2 DAGs in commdelay [33]), this shows that our more advanced model indeed comes at the cost of increased complexity.

Theorem 3. *BSP scheduling is NP-hard if restricted to DAGs of height* 2.

Theorem 4. *BSP scheduling is NP-hard if restricted to in-trees.*

Proof. The proof of Theorem 3 uses a reduction from k-clique, and can be understood as an adaptation of the reduction approach for partitioning in [41] to our setting. Intuitively, the second level of the DAG consists of gadgets representing each node of the input graph, while the first level consists of gadgets representing the edges; each edge gadget is connected to the two incident node gadgets. Our construction ensures that at most k of the node gadgets can be assigned to a given processor p without incurring a too large work cost, and that each edge gadget incurs a communication step exactly if one of its incident node gadgets is not assigned to p. With the appropriate cost limit C_0, the DAG only admits a valid schedule if there are k original nodes that induce $\binom{k}{2}$ edges.

Theorem 4 is our most technical proof, requiring $P = 16$ processors. Firstly, our choice of C_0 ensures that we can only have $\frac{n}{P}$ computations on any processor, and only d communication steps altogether (for some d). Then on the one hand, our construction contains a large gadget that occupies a single processor p for the entire schedule (otherwise the communication cost is too high), and d further gadgets that each need to send a value to p. Any schedule needs to manage these gadgets carefully to ensure that one of these d values is already available every time we do a communication step; this ensures that the communications can only happen at specific times at the earliest. On the other hand, we use critical paths to ensure that specific nodes need to be computed by a given time step. We then attach further gadgets to given segments of these paths, which forces us to split the work between multiple processors (and thus communicate) to satisfy these deadlines; hence communications must happen at specific times at the latest. Together, these define the exact times when we need to communicate, otherwise the cost exceeds C_0. Due to this, there is essentially only one way to develop a valid schedule in our DAG. The underlying reduction then uses the 3-partition problem with gadgets representing each input number a_i, and ensures that it is only possible to satisfy each communication deadline if the numbers a_i are sorted into triplets that each sum up to a given value.

6 Inapproximability

Given the NP-hardness of the problem, another follow-up question is whether we can at least approximate the optimal solution in polynomial time. However, we show that on general DAGs, BSP scheduling is APX-hard: there is a constant $\epsilon > 0$ such that it is already NP-hard to approximate the optimum cost to a $(1+\epsilon)$ factor. To our knowledge, for the case of constant P, no similar hardness results are known for other DAG scheduling problems.

Theorem 5. *BSP scheduling is APX-hard: there is a constant $\epsilon > 0$ such that it is NP-hard to approximate the optimum to a $(1+\epsilon)$ factor, already for $P = 2$.*

Proof. We provide a reduction from MAX-3SAT(B), i.e. maximizing the satisfied clauses in a 3SAT formula where each variable occurs at most B times; this is already APX-hard for $B \in O(1)$ [1,49]. Our construction consists of a separate gadget for each clause and each variable of the given 3SAT formula. On a high level, the variable gadgets contain two separate subgadgets that we need to assign to the two different processors; this choice corresponds to setting the variable to true or false in the underlying formula. The clause gadgets then contain subgadgets corresponding to the 3 literals in the clause, which need to be assigned according to the truth value chosen in the corresponding variable gadget. Whenever a clause is unsatisfied, the corresponding gadgets in the schedule incur a slightly higher work cost than a satisfied clause. With this, the total cost of the BSP schedule has a $\Theta(n)$ term that is proportional to the number of unsatisfied clauses, which allows us to complete the reduction.

The technical part of the proof is to show that any reasonable schedule in our DAG is indeed structured as described above, representing a solution of MAX-3SAT(B). Intuitively, our construction ensures that using any other kind of subschedule in our gadgets results in a higher work or communication cost. More formally, we can always apply a sequence of transformation steps to convert any other BSP schedule into a schedule that represents a valid 3SAT assignment, while only decreasing the cost of the solution.

7 Problem Within a Problem: Communication Scheduling

Since h-relations are a complex communication metric, it is natural to wonder if the hardness of BSP scheduling lies only within the assignment of nodes to processors and steps (as in simpler models), or if the scheduling of communication steps also adds another layer of complexity on this. More specifically, assume that the nodes are already assigned to processors and supersteps, but we still need to decide when the values are communicated between the chosen processors; is this subproblem easier to solve?

Definition 2. In the *communication scheduling (CS) problem*, π and τ are already fixed, and we need to find a Γ that minimizes the resulting BSP cost.

We assume that π and τ allow a feasible communication schedule, i.e. $\forall\, (u, v) \in E$, we have $\tau(u) \leq \tau(v)$ if $\pi(u) = \pi(v)$, and $\tau(u) < \tau(v)$ if $\pi(u) \neq \pi(v)$.

As a heuristic approach, one might consider an eager or lazy communication policy, i.e. to communicate each value immediately after it is computed (eager), or only in the last possible superstep before it is needed (lazy). However, a both of these can easily be suboptimal; see the full version for a concrete example [42].

This shows that CS is indeed an interesting theoretical problem. Furthermore, from the practical side, CS has significantly less degree of freedom, so a heuristic approach could aim to first find a good initial schedule, and then try to improve this by fixing π and τ, and reducing communication cost separately.

For the simplest case of $P = 2$, a greedy approach already finds the optimal solution (in any model of Table 3, since these are all equivalent for $P = 2$).

Lemma 2. *The CS problem can be solved in polynomial time for $P = 2$.*

Interestingly, for general P, the hardness of CS may depend on the communication model: we prove that CS is already NP-hard in the DB, FS or FB models, but we leave it as an open question whether it is also NP-hard in DS.

Theorem 6. *The CS problem is NP-hard in the DB, FS and FB models.*

Proof. The proof uses a reduction from 3D-matching: given sets X, Y, Z of size N, and M triplets from $X{\times}Y{\times}Z$, we need to select N triplets that form a disjoint cover. We convert this into a CS problem where each triplet is represented by a value which needs to be sent from p_0 to several other processors, with the concrete deadlines depending on the given triplet. The construction consists of two parts. The first part allows us to send exactly $(M - N)$ values from p_0 to all other processors (the triplets that are not chosen) within the allowed cost. The second part then allows us to send N further values from p_0 to the other processors, but only if the corresponding triplets are disjoint.

We also note that if we consider a natural extension of the problem with node weights (see the full version for details), then it becomes relatively simple to reduce CS to standard packing problems, already for $P{=}\,2$ processors.

Lemma 3. *With communication weights, CS is already NP-hard for* $P{=}\,2$.

8 Formulation as an ILP Problem

Finally, we discuss a straightforward approach to formulate BSP scheduling as an Integer Linear Programming (ILP) problem. Since today's ILP solvers can often solve even fairly large instances in reasonable time, this naive ILP formulation may already be a viable approach in several applications, especially if the computation is modelled as a DAG at relatively coarse granularity. A similar approach with ILP solvers has already provided remarkable empirical results for various combinatorial problems [22]. Our ILP formulation also generalizes very naturally to several model extensions, e.g. with node weights.

Proposition 2. *BSP scheduling can be formulated as an ILP on* $O(n \cdot P \cdot S)$ *variables in models DS, DB and FB, and on* $O(n \cdot P^2 \cdot S)$ *variables in model FS.*

References

1. Berman, P.R., Scott, A., Karpinski, M.: Approximation hardness and satisfiability of bounded occurrence instances of sat. Tech. rep., SIS-2003-269 (2003)
2. Bisseling, R.H.: Parallel Scientific Computation: A Structured Approach Using BSP. Oxford University Press, USA (2020)
3. Bodlaender, H.L., Fellows, M.R.: W [2]-hardness of precedence constrained k-processor scheduling. Oper. Res. Lett. **18**(2), 93–97 (1995)
4. Chen, L., Jansen, K., Zhang, G.: On the optimality of approximation schemes for the classical scheduling problem. In: Proceedings of the 25th Annual ACM-SIAM Symposium on Discrete Algorithms (SODA), pp. 657–668. SIAM (2014)
5. Coffman, E.G., Graham, R.L.: Optimal scheduling for two-processor systems. Acta Informatica **1**(3), 200–213 (1972)
6. Culler, D., et al.: LogP: towards a realistic model of parallel computation. In: Proceedings of the fourth ACM SIGPLAN symposium on Principles and practice of parallel programming (PPoPP), pp. 1–12 (1993)

7. Davies, S., Kulkarni, J., Rothvoss, T., Sandeep, S., Tarnawski, J., Zhang, Y.: On the hardness of scheduling with non-uniform communication delays. In: ACM-SIAM Symposium on Discrete Algorithms (SODA), pp. 316–328. SIAM (2022)
8. Davies, S., Kulkarni, J., Rothvoss, T., Tarnawski, J., Zhang, Y.: Scheduling with communication delays via LP hierarchies and clustering. In: 61st Annual Symposium on Foundations of Computer Science (FOCS), pp. 822–833. IEEE (2020)
9. Davies, S., Kulkarni, J., Rothvoss, T., Tarnawski, J., Zhang, Y.: Scheduling with communication delays via LP hierarchies and clustering II: weighted completion times on related machines. In: ACM-SIAM Symposium on Discrete Algorithms (SODA), pp. 2958–2977. SIAM (2021)
10. Dolev, D., Warmuth, M.K.: Scheduling precedence graphs of bounded height. J. Algorithms **5**(1), 48–59 (1984)
11. Dolev, D., Warmuth, M.K.: Profile scheduling of opposing forests and level orders. SIAM J. Algebraic Discrete Methods **6**(4), 665–687 (1985)
12. Du, J., Leung, J.Y., Young, G.H.: Scheduling chain-structured tasks to minimize makespan and mean flow time. Inf. Comput. **92**(2), 219–236 (1991)
13. Fujimoto, N., Hagihara, K.: On approximation of the bulk synchronous task scheduling problem. IEEE Trans. Parallel Distrib. Syst. **14**(11), 1191–1199 (2003)
14. Gabow, H.N.: An almost-linear algorithm for two-processor scheduling. J. ACM **29**(3), 766–780 (1982)
15. Garey, M., Johnson, D., Tarjan, R., Yannakakis, M.: Scheduling opposing forests. SIAM J. Algebraic Discrete Methods **4**(1), 72–93 (1983)
16. Garg, S.: Quasi-ptas for scheduling with precedences using LP hierarchies. In: 45th International Colloquium on Automata, Languages, and Programming (ICALP 2018). Schloss Dagstuhl-Leibniz-Zentrum fuer Informatik (2018)
17. Goldman, A., Mounié, G., Trystram, D.: Near optimal algorithms for scheduling independent chains in BSP. In: Proceedings. 5th International Conference on High Performance Computing, pp. 310–317. IEEE (1998)
18. Graham, R.L., Lawler, E.L., Lenstra, J.K., Kan, A.R.: Optimization and approximation in deterministic sequencing and scheduling: a survey. In: Annals of Discrete Mathematics, vol. 5, pp. 287–326. Elsevier (1979)
19. Hanen, C., Munier, A.: An approximation algorithm for scheduling dependent tasks on m processors with small communication delays. In: Proceedings 1995 INRIA/IEEE Symposium on Emerging Technologies and Factory Automation. ETFA 1995. vol. 1, pp. 167–189. IEEE (1995)
20. Hill, J.M., et al.: BSPlib: the BSP programming library. Parallel Comput. **24**(14), 1947–1980 (1998)
21. Hoogeveen, J., Lenstra, J.K., Veltman, B.: Three, four, five, six, or the complexity of scheduling with communication delays. Oper. Res. Lett. **16**(3), 129–137 (1994)
22. Jenneskens, E.L., Bisseling, R.H.: Exact k-way sparse matrix partitioning. In: 2022 IEEE International Parallel and Distributed Processing Symposium Workshops (IPDPSW), pp. 754–763. IEEE (2022)
23. Kulkarni, J., Li, S., Tarnawski, J., Ye, M.: Hierarchy-based algorithms for minimizing makespan under precedence and communication constraints. In: Proceedings of the 14th Annual ACM-SIAM Symposium on Discrete Algorithms (SODA), pp. 2770–2789. SIAM (2020)
24. Lenstra, J.K., Kan, A.R., Brucker, P.: Complexity of machine scheduling problems. In: Annals of Discrete Mathematics, vol. 1, pp. 343–362. Elsevier (1977)
25. Lenstra, J.K., Rinnooy Kan, A.: Complexity of scheduling under precedence constraints. Oper. Res. **26**(1), 22–35 (1978)

26. Lenstra, J.K., Shmoys, D.B., Tardos, É.: Approximation algorithms for scheduling unrelated parallel machines. Math. Program. **46**(1), 259–271 (1990)
27. Leung, J.Y.T., Young, G.H.: Minimizing total tardiness on a single machine with precedence constraints. ORSA J. Comput. **2**(4), 346–352 (1990)
28. Levey, E., Rothvoss, T.: A (1+ epsilon)-approximation for Makespan scheduling with precedence constraints using LP hierarchies. In: Proceedings of the 48th Annual ACM Symposium on Theory of Computing (STOC), pp. 168–177 (2016)
29. Li, S.: Scheduling to minimize total weighted completion time via time-indexed linear programming relaxations. In: IEEE 58th Annual Symposium on Foundations of Computer Science (FOCS), pp. 283–294. IEEE Computer Society (2017)
30. Li, S.: Towards PTAS for precedence constrained scheduling via combinatorial algorithms. In: Proceedings of the ACM-SIAM Symposium on Discrete Algorithms (SODA), pp. 2991–3010. SIAM (2021)
31. Liu, Q.C., Purohit, M., Svitkina, Z., Vee, E., Wang, J.R.: Scheduling with communication delay in near-linear time. In: 39th International Symposium on Theoretical Aspects of Computer Science (STACS 2022). Schloss Dagstuhl-Leibniz-Zentrum für Informatik (2022)
32. Maiti, B., Rajaraman, R., Stalfa, D., Svitkina, Z., Vijayaraghavan, A.: Scheduling precedence-constrained jobs on related machines with communication delay. In: 61st Annual Symposium on Foundations of Computer Science (FOCS), pp. 834–845. IEEE (2020)
33. Markenscoff, P., Li, Y.Y.: Scheduling a computational DAG on a parallel system with communication delays and replication of node execution. In: Proceedings of the 7th International Parallel Processing Symposium, pp. 113–117. IEEE (1993)
34. McColl, B.: Mathematics, Models and Architectures, p. 653. Cambridge University Press (2021)
35. McColl, W.F.: Scalable computing. In: van Leeuwen, J. (ed.) Computer Science Today. LNCS, vol. 1000, pp. 46–61. Springer, Heidelberg (1995). https://doi.org/10.1007/BFb0015236
36. McColl, W.F., Tiskin, A.: Memory-efficient matrix multiplication in the BSP model. Algorithmica **24**(3), 287–297 (1999)
37. Mnich, M., Wiese, A.: Scheduling and fixed-parameter tractability. Math. Program. **154**(1), 533–562 (2015)
38. Munier, A., König, J.C.: A heuristic for a scheduling problem with communication delays. Oper. Res. **45**(1), 145–147 (1997)
39. Özkaya, M.Y., Benoit, A., Uçar, B., Herrmann, J., Çatalyürek, Ü.V.: A scalable clustering-based task scheduler for homogeneous processors using DAG partitioning. In: IEEE International Parallel and Distributed Processing Symposium (IPDPS), pp. 155–165. IEEE (2019)
40. Papadimitriou, C., Yannakakis, M.: Towards an architecture-independent analysis of parallel algorithms. In: Proceedings of the 20th Annual ACM Symposium on Theory of Computing (STOC), pp. 510–513 (1988)
41. Papp, P.A., Anegg, G., Yzelman, A.J.N.: Partitioning hypergraphs is hard: Models, inapproximability, and applications. In: Proceedings of the 35th ACM Symposium on Parallelism in Algorithms and Architectures (SPAA), pp. 415–425 (2023)
42. Papp, P.A., Anegg, G., Yzelman, A.N.: DAG scheduling in the BSP model. arXiv preprint arXiv:2303.05989 (2024)
43. Rajaraman, R., Stalfa, D., Yang, S.: Approximation algorithms for scheduling under non-uniform machine-dependent delays. arXiv preprint arXiv:2207.13121 (2022)

44. Sequential and Parallel Algorithms and Data Structures. Springer, Cham (2019). https://doi.org/10.1007/978-3-030-25209-0_9
45. Sethi, R.: Scheduling graphs on two processors. SIAM J. Comput. **5**(1), 73–82 (1976)
46. Skillicorn, D.B., Hill, J., McColl, W.F.: Questions and answers about BSP. Sci. Program. **6**(3), 249–274 (1997)
47. Svensson, O.: Conditional hardness of precedence constrained scheduling on identical machines. In: Proceedings of the 42nd ACM Symposium on Theory of Computing (STOC), pp. 745–754 (2010)
48. Timkovsky, V.G.: Identical parallel machines vs. unit-time shops and preemptions vs. chains in scheduling complexity. Eur. J. Oper. Res. **149**(2), 355–376 (2003)
49. Trevisan, L.: Non-approximability results for optimization problems on bounded degree instances. In: Proceedings of the Thirty-Third Annual ACM Symposium on Theory of Computing, pp. 453–461 (2001)
50. Ullman, J.D.: Np-complete scheduling problems. J. Comput. Syst. Sci. **10**(3), 384–393 (1975)
51. Valiant, L.G.: A bridging model for parallel computation. Commun. ACM **33**(8), 103–111 (1990)
52. Valiant, L.G.: A bridging model for multi-core computing. In: Halperin, D., Mehlhorn, K. (eds.) ESA 2008. LNCS, vol. 5193, pp. 13–28. Springer, Heidelberg (2008). https://doi.org/10.1007/978-3-540-87744-8_2
53. Valiant, L.G.: A bridging model for multi-core computing. J. Comput. Syst. Sci. **77**(1), 154–166 (2011)
54. Veltman, B., Lageweg, B., Lenstra, J.K.: Multiprocessor scheduling with communication delays. Parallel Comput. **16**(2–3), 173–182 (1990)
55. Yzelman, A., Bisseling, R.H.: An object-oriented bulk synchronous parallel library for multicore programming. Concurr. Comput. Pract. Exp. **24**(5), 533–553 (2012)
56. Yzelman, A., Bisseling, R.H., Roose, D., Meerbergen, K.: MulticoreBSP for C: a high-performance library for shared-memory parallel programming. Int. J. Parallel Prog. **42**(4), 619–642 (2014)

Incremental Computation of the Set of Period Sets

Eric Rivals(✉)

LIRMM, Université Montpellier CNRS, Montpellier, France
rivals@lirmm.fr

Abstract. Overlaps between words are crucial in many areas of computer science, such as code design, stringology, and bioinformatics. A self overlapping word is characterized by its periods and borders. A period of a word u is the starting position of a suffix of u that is also a prefix u, and such a suffix is called a border. Each word of length, say $n > 0$, has a set of periods, but not all combinations of integers are sets of periods. Computing the period set of a word u takes linear time in the length of u. We address the question of computing, the set, denoted Γ_n, of all period sets of words of length n. Although period sets have been characterized, there is no formula to compute the cardinality of Γ_n (which is exponential in n), and the known dynamic programming algorithm to enumerate Γ_n suffers from its space complexity. We present an incremental approach to compute Γ_n from Γ_{n-1}, which reduces the space complexity, and then a constructive certification algorithm useful for verification purposes. The incremental approach defines a parental relation between sets in Γ_{n-1} and Γ_n, enabling one to investigate the dynamics of period sets, and their intriguing statistical properties. Moreover, the period set of a word u is key for computing the absence probability of u in random texts. Thus, knowing Γ_n is useful to assess the significance of word statistics, such as the number of missing words in a random text.

Keywords: string · overlap · period · autocorrelation · combinatorics · algorithm

1 Introduction

Considering finite words, we say that a word u overlaps a word v if a suffix of u equals a prefix of v of the same length. A pair of words u, v can have several overlaps of different lengths. For instance, over the alphabet $\{a, b\}$, let $u := ababba$ and $v := abbabb$, then u overlaps v with the suffix-prefix $abba$, and with a. It appears that the longest overlap contains all other overlaps: to find all overlaps from u to v, it suffices to study the overlaps of $abba$ with itself.

For a word u, a suffix that equals a prefix of u is called a *border*, and the length of u minus the length of a border, is called a *period*. Computing all self-overlaps of a word u is computing all its borders or all its periods, which can be done in linear time (see the algorithm for computing the border array in [25], [chap.

R. Královič and V. Kůrková (Eds.): SOFSEM 2025, LNCS 15539, pp. 254–268, 2025.
https://doi.org/10.1007/978-3-031-82697-9_19

1]).This well-studied problem was also solved for preprocessing the pattern in the seminal Knuth-Morris-Pratt pattern matching algorithm [10]): the borders serve to optimally shift the window along the text when seeking the pattern. For instance the word *ababaaba* has length $n = 8$ and set of periods: $\{0, 5, 7\}$ (zero being the trivial period - the whole word matches itself).

One can easily see that distinct words of the same length can share the same set of periods, even if one forbids a permutation of the alphabet. For a word u, let us denote by $P(u)$ its period set (which we abbreviate by PS). In this work, we investigate algorithms to enumerate all possible period sets for any words of a given length n. This set is denoted Γ_n for $n > 0$ and is non trivial if the alphabet contains at least two symbols. Brute force enumeration can consider all possible words of length n and compute their period set, but this approach becomes computationally unaffordable for $n > 30$. Currently, only a dynamic programming algorithm exists to enumerate Γ_n, but it suffers from high space complexity [23].

The notion of overlap in strings is crucial in many areas and applications, among others: combinatorics, bioinformatics, code design, or string algorithms. Interest in Γ_n sparkled mostly in the 80's, when researchers started to evaluate the average behavior of pattern matching algorithms, or that of filtering strategies for sequence alignment, text comparisons or clustering. A powerful filtering when comparing two texts, is to list their k-long substrings (a.k.a. k-mers), for appropriate values of k, and then compute e.g., a Jaccard distance between their set of k-mers, to see whether the two texts are similar enough to warrant a costly alignment procedure [28].

In a different area, testing Pseudo-Random Number Generators can also be translated into a question on vocabulary statistics. Indeed, in random real numbers written as sequence of digits, all substrings of a given length, say k, should ideally have an almost equal number of occurrences (i.e., should not significantly deviate from the theoretical expectation). Empirical tests, named *Monkey Tests*, were developed for such generators [11,14,17]. It turns out that the absence probability of a word/string in a random text is essentially controlled by the period set of the word [7]. Hence, the need for enumerating Γ_n appears in diverse domains of the literature [18,19].

In network communication, the so-called prefix-free, bifix-free, and cross-bifix-free codes are used for synchronization purposes. Their design require to select a set of words that are mutually non overlapping – a long studied topic. Nielsen published in 1973 a construction algorithm [15] together with a note on the expected duration of a search for a fixed pattern in random data [16], thereby linking explicitly pattern matching and code design. Improved algorithms for such code design were recently published [1,2].

Many aspects of periodicity of words were and are still extensively studied in combinatorics, e.g. [4,6]. For instance, the periods of random words were investigated in [9] and the Fine and Wilf (FW) Theorem was generalized to three or more words [3]. Algorithms for constructing extremal FW-words relative to a subset of periods were gradually improved in [26,27], a question that is related to the ones we investigate in Sect. 5. Last, the sequence formed by the cardinality of Γ_n, which is denoted κ_n, has an entry in the OEIS, and some known lower and upper bounds [7,23], which helped closing a conjecture related to κ_n in 2023 [24].

Plan. In Sect. 2, we introduce a notation, definitions, and review some known results. In Sect. 3, we propose an incremental approach to compute Γ_n from Γ_{n-1}, which uses $O(n)$ memory. In Sect. 4, we give an algorithm to build a word that is a witness for a period set. In Sect. 5, we propose to explore the dynamics of period sets when n increases, before exploring the distribution of period sets according to their basic period and concluding in Sect. 6. Appendices providing additional algorithms, proofs and explanations can be read in [20].

2 Related Works, Notation and Preliminary Results

2.1 Notation

Here we introduce a notation and basic definitions.

For two integers $p, q \in \mathbb{N}_{>0}$, we denote the fact that p divides q by $p \mid q$ and the opposite by $p \nmid q$. We consider that strings and arrays are indexed from 0. We use $=$ to denote equality, and $:=$ to denote a definition. The cardinality of a discrete set X is denoted by $\sharp(X)$.

An *alphabet* Σ is a finite set of *letters*. A finite sequence of elements of Σ is called a *word* or a *string*. The set of all words over Σ is denoted by $\Sigma^\star$, and ε denotes the empty word. For a word x, $|x|$ denotes the *length* of x. Let n be an integer. The set of all words of length n is denoted by Σ^n. Given two words x and y, we denote by $x.y$ the *concatenation* of x and y. For a word w, we define the powers of w inductively: $w^0 := \epsilon$ and for any $n > 0$, $w^n := w.w^{n-1}$. A word u is *primitive* if there exists no word v such that $u = v^k$ with $k \geq 2$.

Let $u := u[0\,..\,n-1] \in \Sigma^n$. For any $0 \leq i \leq j \leq n-1$, we denote by $u[i-1]$ the ith symbol in u, and by $u[i\,..\,j]$, the substring starting at position i and ending at position j. In particular, $u[0\,..\,j]$ denotes a prefix and $u[i\,..\,n-1]$ a suffix of u.

Let $x, y \in \Sigma^*$ and let j be an integer such that $0 \leq j \leq |x|$ and $j \leq |y|$. If $x[n-j\,..\,|x|-1] = y[0\,..\,j-1]$, then the *merge* of x and y with offset j, which is denoted by $x \oplus_j y$, is defined as the concatenation of $x[0\,..\,n-j-1]$ with y. I.e., $x \oplus_j y := x[0\,..\,n-j-1].y$.

2.2 Periods, Period Set, and the Set of Period Sets

Let us first define the concepts of period, period set, basic period, and set of period sets, and then recall some useful results.

Definition 1 (Period/border). The string $u = u[0\,..\,n-1]$ has period $p \in \{0, 1, \ldots, n-1\}$ if and only if $u[0\,..\,n-p-1] = u[p\,..\,n-1]$, i.e., for all $0 \leq i \leq n-p-1$, we have $u[i] = u[i+p]$. Moreover, we consider that $p = 0$ is a period of any string of length n. The substring $u[p\,..\,n-1]$ is called a *border*.

Zero is also called the trivial period. The *period set* of a string u is the set of all its periods and is denoted by $P(u)$. Formally, for any length $n > 0$, any word u in Σ^n, $P(u) := \{p \in \mathbb{N} : p \text{ is a period in } u\}$. The *weight* of a period set is its cardinality. The smallest non-zero period of u is called its *basic period*. When $P(u) = \{0\}$, we consider that its basic period is the string length n.

Note that two definitions of period set exist in the literature: the one above, also used in [7] (even if the set was encoded in a binary string called autocorrelation), where $P(u)$ is a subset of $\{0, 1, \ldots, n-1\}$, and the one from [13] in which it equals $P(u) \cup \{n\}$ (where the length n is included in $P(u)$). We will use the first one.

Let $\Gamma_n := \{Q \subset \{0, 1, \ldots, n-1\} \mid \exists u \in \Sigma^n : Q = P(u)\}$ be the set of all period sets of strings of length n. We denote its cardinality by κ_n. The period sets in Γ_n can be partitioned according to their basic period; thus, for $0 \leq p < n$, we denote by $\Gamma_{n,p}$ the subset of period sets whose basic period is p, and by $\kappa_{n,p}$ the cardinality of this subset. A surprising result from Guibas and Odlyzko's characterization of period sets [7] is the *alphabet independence* of Γ_n: Any alphabet of size at least two gives rise to the same set of period sets, i.e., to Γ_n. In the sequel, we assume $\sharp(\Sigma) > 1$. Circa 20 years later, Halava et al. gave a simpler proof of the alphabet independence result by solving the following question [8]. For any period set Q of Γ_n, let v be a word over an alphabet Σ with $\sharp(\Sigma) > 1$ such that $P(v) = Q$; then compute a binary word u such that $P(u) = Q$. Indeed, they gave a linear time algorithm to compute such a word u from v.

For the most important properties on periods, we refer the reader to [7,12, 13,20]. Below we recall the two characterizations of period sets from [7] and from [13], and state the version for strings of the famous Fine and Wilf (FW) Theorem [5]. We refer the reader to [23,24] for more properties on period sets and on Γ_n.

2.3 Characterizations of Period Sets

For a given word length $n > 0$, the question of characterization is: among all possibles subsets of $\{0, 1, \ldots, n-1\}$, recognize those that are period sets for at least one word of length n.

In 1981, Guibas and Odlyzko (GO for short) proved a double characterization of period sets: one is based on two rules, the Forward Propagation Rule (FPR) and the Backward Propagation Rule (BPR), the second is the recursive predicate Ξ (which is given in extenso in [7,23].

First, let us formulate the FPR and BPR in terms of sets (rather than in term of binary vector as in [7]). Let P a subset of $\{0, 1, \ldots, n-1\}$.

Definition 2 (FPR). *P satisfies the FPR iff for all pairs p, q in P satisfying $0 \leq p < q < n$, it follows that $p + i(q-p) \in P$ for all $i = 2, \ldots, \lfloor (n-p)/(q-p) \rfloor$.*

Definition 3 (BPR). *P satisfies the BPR iff for all pairs p, q in P satisfying $0 \leq p < q < 2p$, and $(2p - q) \notin P$, it follows that $p - i(q-p) \notin P$ for all $i = 2, \ldots, \min(\lfloor p/(q-p) \rfloor, \lfloor (n-p)/(q-p) \rfloor)$.*

We can now state GO's characterization theorem

Theorem 1. *Let P a subset of $\{0, 1, \ldots, n-1\}$. The four following statements are equivalent: (1) P is the period set of a binary word of length n.*
(2) P is the period set of a word of length n.
(3) Zero belongs to P and P satisfies the forward and backward propagation rules.
(4) P satisfies predicate Ξ.

The equivalence of (1) and (2) yields the abovementioned alphabet independence of Γ_n. The authors also noticed that the FPR and BPR are local properties [7, Lemma 3.1], which we use later on. Finally, predicate Ξ is a recursive procedure to check whether P belongs to Γ_n in $O(n)$ time: it considers two cases depending on whether the basic period is $\leq \lfloor n/2 \rfloor$ (case a) or not (case b) [7].

Besides this characterization, [7] studied the set of distinct strings over a given alphabet that share the same period set (a.k.a. its *population*). They also reported values of κ_n, the cardinality of Γ_n, for $n < 55$, exhibited lower and upper bounds for $\log(\kappa_n)/\log(n)^2$, and conjectured its convergence when $n \to \infty$. Much later, improved lower and upper bounds were provided and served to close the conjecture [23,24].

An alternative characterization of PS appears in 2002 as Theorem 8.1.11 in [13],[Chap. 8]; it comprises four equivalent statements, of which the first three "are proved in [7]" (see notes in [13, p. 310]; the first two are identical to (1) and (2) in Theorem 1). We thus recall only statement (iv) of this characterization in the next theorem.

Theorem 2 ([13]). *Let $P := \{0 = p_0 < p_1 < \ldots < p_s = n\}$ be a set of integers and let $d_h := p_h - p_{h-1}$ for $1 \leq h \leq s$. Then P is a period set (i.e., $P \in \Gamma_n$) if and only if for each h such that $d_h + p_h \leq n$, one has: (a) $p_h + d_h \in P$ and (b) if $d_h = kd_{h+1}$ for some integer k, then $k = 1$.*

Note that in this formulation n belongs to P (see our remark about two definitions of period sets), but as when $h = s$ one has $d_s + p_s > n$, one sees that conditions (a) and (b) are in fact only required for $1 \leq h < s$.

Given a period set $P(w)$ of some word w, the proof shows that there exists a binary word having the period set $P(w)$. It is a proof of existence, it is not constructive, but the authors also give an example on how to build a binary string for $Q := \{0, 11, 14, 17, 18\}$ knowing that Q is the period set of the word $w = abcabcadefgabcabca$. Knowing that $Q \in \Gamma_n$, an algorithm for solving this problem was given by [8].

In this work, we address a different question, termed *constructive certification*: Given Q a subset of $\{0, 1, \ldots, n-1\}$, build a (binary) string u such that $P(u) = Q$ iff $Q \in \Gamma_n$, or return the empty string otherwise.

2.4 Additional Properties of Periods

In [7], the authors give a version for strings of the famous Fine and Wilf Theorem [5], a.k.a. the periodicity lemma. A nice proof was provided by Halava and colleagues [8].

Theorem 3 (Fine and Wilf). *Let p, q be periods of $u \in \Sigma^n$. If $n \geq p + q - \gcd(p, q)$, then $\gcd(p, q)$ is a period of u.*

We can reformulate Theorem 3 as a condition that must be satisfied by a period set P.

Theorem 4. *Let $P \in \Gamma_n$. Let p, q be periods of P such that $\gcd(p, q) \notin P$, and define $FW(p, q) := p + q - \gcd(p, q)$. Then $FW(p, q)$ must be strictly larger than n.*

We call $FW(p, q)$ the *Fine and Wilf (FW) limit* of (p, q). We say (p, q) violates the *FW condition* at length n, if $FW(p, q) \leq n$.

2.5 Dynamic Programming Algorithm to Enumerate Γ_n

In 2001, further investigation of Γ_n led to a dynamic programming programming algorithm to enumerate all period sets in Γ_n: it converts the recursive approach of predicate Ξ into a dynamic program that stores all Γ_i for $0 < i \leq \lfloor 2n/3 \rfloor$ [22,23]. With some practical improvements, the range was reduced to $0 < i \leq \lfloor n/2 \rfloor$. However, as κ_n is exponential in n, this induces a large memory usage, which remains a serious drawback. Hence, the quest for memory efficient algorithms. The authors also demonstrated that Γ_n equipped with set inclusion is a lattice, but this did not help to improve Γ_n enumeration.

3 Incremental Enumeration of Γ_n

Rationale of the incremental approach

In their seminal work [7], GO manipulate period sets, not as sets, but as a binary strings of length n, where position i is set to 1 if i is a period. For example, the binary encoding of period set $\{0, 3, 5\}$ for length $n = 6$ is `100101`. They call these binary strings correlations, and even autocorrelations to emphasize that it encodes all self-overlaps a string [7, p. 21]. The rule based characterization (statement (3) of Theorem 1) implies a *special substring* property [7]. Indeed, noting the locality of the forward and backward propagation rules, the authors state in Lemma 3.1 in [7]: If a binary string v satisfies the forward and backward propagation rules, then so does any prefix or suffix of v.

As the rule based characterization of period sets – statement (3) of Theorem 1 – also requires that an autocorrelation has its first bit equal to one, or equivalently that zero belongs to a period set, one gets the following theorem ([23, Thm 1.3]):

Theorem 5. *Let v be an autocorrelation of length n. Any substring $v_i \dots v_j$ of v with $0 \leq i \leq j < n$ such that $v_i = 1$ is an autocorrelation of length $j - i + 1$.*

Applying Theorem 5 to a prefix of v, one gets for any $n > 0$: The prefix of length $(n-1)$ from an autocorrelation of length n is an autocorrelation of length $(n-1)$. In terms of period sets, this statement can be reformulated as:

Corollary 1. *If P is a period set of Γ_n, then $P \setminus \{n-1\}$ belongs to Γ_{n-1}.*

First, this means that, knowing Γ_n, it is easy to compute Γ_{n-1}. It suffices to consider each element of Γ_n in turn (or in parallel) and to possibly remove the period $(n-1)$ from it (i.e., if $(n-1)$ belongs to it) to obtain an element of Γ_{n-1}. With this procedure one can obtain the same element of Γ_{n-1} twice, and one must keep track of this to avoid redundancy.
Conversely, we get the Lemma that underlies the incremental approach for computing Γ_n:

Lemma 1. *Let Q be a period set of Γ_n. Then Q can only be of two alternative forms: either P or $P \cup \{n-1\}$, for some P in Γ_{n-1}.*

Incremental Algorithm Framework.
Lemma 1 suggests an approach for computing Γ_n using Γ_{n-1}. Consider each P from Γ_{n-1}, and check whether the candidate sets P and $P \cup \{n-1\}$ are period sets of Γ_n. Algorithm 1 presents a generic incremental algorithm for Γ_n, where `certify` denotes the certification function used. In general, a certification function takes as input n and any subset Q of $\{0, 1, \dots, n-1\}$, returns True if and only if Q belongs to Γ_n.

Algorithm 1: IncrementalGamma(length $n > 1$; set Γ_{n-1})

```
  Output: Γn: the set of period sets for length n;
1 G := ∅;                                          // G: variable to store Γn
2 for all P ∈ Γ(n − 1) do
3     if certify( P, n ) then insert P in G ;
4     Q := P ∪ {n − 1} // build extension P with period n − 1;
5     if certify( Q, n ) then insert Q in G ;
6 return G;
```

The recursive predicate Ξ from [7] is indeed a certification function: it does exactly what is required for any subset of $\{0, 1, \dots, n-1\}$, in $O(n)$ time [7]. Thus, using predicate Ξ, Algorithm 1 correctly computes Γ_n from Γ_{n-1}. We will discuss alternative certification functions below.

Besides its simplicity, the main advantage of Algorithm 1, compared to the dynamic programming enumeration algorithm of [22], is its space complexity. Here, the computation considers each period set P from Γ_{n-1} in turn (and independently from the others), executes twice the certification function for P and Q; this implies that the memory required, besides storage of P, Q, is the one used by the certification function. With the predicate Ξ, it is linear in n, so $O(n)$ space. The time complexity is proportional to κ_n (i.e., the cardinality of Γ_n) times the running time of the certification function, which yields the following theorem. Of course, the set Γ_{n-1} must be available on disk space before hand.

Theorem 6. *Provided that Γ_{n-1} is available on external memory, then*

1. *Algorithm 1 using any certification function correctly computes Γ_n from Γ_{n-1}.*
2. *Using the predicate Ξ as certification function, it runs in $O(n\kappa_n)$ time and $O(n)$ space.*

Moreover, it is worth noticing that Algorithm 1 is embarrassingly parallelizable. Note that the output contains κ_n period sets, whose cardinality is bounded by n and sometimes equal n. For known bounds on κ_n see [23,24]. So, the output size is bounded by $n\kappa_n$.

Alternative certification functions

In our incremental setup, that is when computing Γ_n using Γ_{n-1}, we know that P belongs to Γ_{n-1} (line 2 in Algorithm 1). Hence, the candidate sets, denoted P and Q, are not **any subset** of $\{0, 1, \ldots, n-1\}$, but already satisfy some constraints for length $(n-1)$. Therefore, finding alternative certification functions is interesting. Here, we discuss two alternative functions, and in Sect. 4, we exhibit a constructive certification algorithm, which not only certifies a candidate set, but also computes a witness, i.e., a word whose period set is the candidate set, only if the answer is positive.

To simplify Algorithm 1 by improving how certifications are done for the candidate sets P and $Q := P \cup \{n-1\}$, we can take advantage of two facts. First, the only period that can be added is $n-1$. This limits the cases for which we need to check some conditions. Second, P and Q are not independent, and if $n-1$ is compulsory at length n, because it is generated by the FPR from smaller periods, then only candidate set $Q := P \cup \{n-1\}$ may belong to Γ_n, but not P.

Beides Predicate Ξ, a second certification function, we call it *rule based*, exploits statement (3) of Theorem 1 and uses procedures to verify if the forward and backward propagation rules are satisfied. Algorithm 3 in [20] gives the code for computing Γ_n using the rule based certification.

A third certification approach, which exploits the characterization of Theorem 2, can perform the certification of both P and Q in $O(\sharp(P))$ time. Some details are also given in [20].

4 Constructive Certification of a Period Set

Let Q be subset of $\{0, 1, \ldots, n-1\}$. We say that a word u *realizes* Q if $P(u) = Q$.

The certification functions used in Sect. 3 yield a True/False answer, but no witness for a period set. As there is no resources providing Γ_n for many word lengths, checking the output of an enumeration algorithm, remains difficult. Hence, we need a constructive certification function, which given length

$n > 0$ and set Q, provides a word realizing Q if only if $Q \in \Gamma_n$, and the empty string otherwise. Given the alphabet independence of Γ_n, we restrict the search to binary words. We present an algorithm called *binary realization* (see Algorithm 2) solving this question, and demonstrate its linear complexity.

Using Algorithm 2 as a certification function in Algorithm 1, for each $R \in \Gamma_n$, we get a word u realizing R. Then, computing the period set of u allows us to check that $P(u) = R$. Let us define the notion of *nested set.*

Definition 4. Let $n > 0$, P be a subset of $\{0, 1, \ldots, n-1\}$, and q be an element of P. We denote by P_q, the *nested set of P starting at period q*:

$$P_q := \{(r - q) \text{ for each } r \in P \text{ such that } r \geq q\}.$$

By construction P_q starts with 0; moreover, if we choose $q = 0$ then $P_q = P$.

4.1 Binary Realization of a Subset of {0, 1, ..., n-1}

Algorithm 2 computes a word u that realizes a set P for length $n > 0$, or returns the empty word ϵ if P is not a period set of Γ_n. For legibility, the preliminary checks on P are not written in Algorithm 2: they include checking that P is a subset of $[0, \ldots, n-1]$, is ordered, and has zero as first period. The word u is written over the alphabet $\{\mathtt{a}, \mathtt{b}\}$.

The algorithm considers elements of P backwards, starting with largest integer first, since P is ordered. At each execution of the for loop, it considers the current integer $P[i]$ as a period and builds a suffix of u of length $n - P[i]$ (variable *lg*). In fact, it considers a potentially larger and larger nested sets, and computes a suffix of u for this length. At the end of the for loop, the variable *suffix* contains a string of length *lg* realizing the nested set. Note that algorithm uses three variables (whose names start with *prev*) to store the length, the inner period, and the suffix obtained with the previous period.

The base case is processed before the loop and consider the nested set for $P[k-1] = max(P)$ for the length $n - max(P)$ without any period. Hence, the suffix $\mathtt{a}.\mathtt{b}^{(\text{prevLg -1})}$ is a realization for nested set $\{0\}$ for length $n - max(P)$.

In the for loop the key variable is the *innerPeriod*, which equals the offset $P[i+1]-P[i]$, which is the basic period of the current nested set. If innerPeriod < prevIP then the FPR is violated and the algorithm returns ϵ (line 7). Two cases are considered depending on whether the current length is smaller twice the previous length (case 1) or not (case 2). Because of the notion of period, the suffix must start and end with a copy of *prevSuffix.* The construction of the suffix depends on the case. In case 1, the two copies of *prevSuffix* are concatenated or overlap themselves, and some additional conditions are required (line 9). These conditions are dictated by the characterization of period set from [7] (see their predicate Ξ). Whenever one is not satisfied, Algorithm 2 returns the empty word as expected. In case 2, the two copies of *prevSuffix* must be separated by m additional symbols (to be determined). One builds a *newPrefix* that starts with *prevSuffix* followed by $\mathtt{a}^m$, and one checks whether *newPrefix* is primitive. This *newPrefix* is the part that ensures the suffix will have *innerPeriod* as a

Algorithm 2: Binary Realization

```
Input: n > 0: integer; P: a subset of [0, 1, ..., n − 1] including 0, in a sorted
       array
Output: a binary string realizing P at length n xor the empty string
        otherwise;
1  k := ♯(P) ;                                        // k: cardinality of P
2  if k = 1 then return a.b^(n−1)                      // trivial case where P = {0};
   // processing the largest period and init. variables
3  prevLg := n − P[k − 1]; prevIP := prevLg; prevSuffix := a.b^(prevLg−1);
4  for i going from k − 2 to 0 do
5      lg := n − P[i];
6      innerPeriod := P[i + 1] − P[i];
7      if innerPeriod < prevIP then return ϵ;
8      if lg ≤ 2× prevLg then                          // condition for case 1
9          if (innerPeriod = prevIP ) OR (( prevIP ∤ innerPeriod ) AND (
             (innerPeriod = prevLg) OR ( prevSuffix has period innerPeriod )))
            then
               // suffix := a prefix of prevSuffix concat. with
                  prevSuffix
10             suffix := prevSuffix[0..innerPeriod−1] . prevSuffix;
11         else return ϵ                               // invalid case for length lg;
12     else                                            // condition for case 2
           // suffix := prevSuffix newsymbols prevSuffix
13         m := lg −2× prevLg;
14         newPrefix := prevSuffix .a^m;
15         if newPrefix is not primitive then
16             newPrefix := prevSuffix .a^(m−1)b;
           // Invariant: newPrefix is primitive
17         suffix := newPrefix . prevSuffix;
       // update variables
18     prevLg := lg; prevIP := innerPeriod; prevSuffix := suffix;
19 return suffix;
```

period. The primitivity is required, since *newPrefix* may have a proper period, but this period shall not divide *innerPeriod*. If primitivity is not satisfied, then changing the last symbol a of *newPrefix* by b will make it primitive. This is enforced by Lemma 3 from [8], which states that for any binary word w, wa or wb is primitive. So, we know that at least one of the two forms of *newPrefix* is primitive as necessary. It can be that both are primitive and suitable. Finally, we build the current *suffix* by concatenating *newPrefix* with *prevSuffix*.

Complexity. First, in case 1, checking the condition "*prevSuffix has period innerPeriod*" can be done in linear time in $|prevSuffix|$ (which is $\leq n$). Overall, this can be executed $\sharp(P)$ times. Second, the primitivity test performed in case 2 takes a time proportional to the length of the string *newPrefix*. However, the sum of these lengths, for all iterations of the loop, is bounded by n. Other instructions of the for loop take constant time. Overall, the time complexity of Algorithm 2 by $O(\sharp(P) \times n)$. However, when Algorithm 2 is plugged in Algorithm 1 it processes special instances: either P or $Q := P \cup \{n-1\}$, with $P \in \Gamma_{n-1}$. Then, the time taken by all verifications of condition "*prevSuffix has period innerPeriod*" for all cases 1 is bounded by n, due to the properties of periods that generate more than two repetitions in a string (see [20, Lemma 3] and [8, Lemma 2]). For the instances processed in Algorithm 1, the time complexity of Algorithm 2 is $O(\sharp(P) + n)$ or $O(n)$.

Remark we can modify Algorithm 2 to build, instead of a binary word, a realizing word that maximizes the number of distinct symbols used in it.

4.2 Examples of Binary Realization

We consider the case of the period set $P := \{0, 3, 6, 8\}$ from Γ_9, which does not belong to Γ_{10}, and show the traces of execution for both lengths $n := 9$ and $n := 10$. The table below shows the trace for $n := 9$. The operator $\oplus_j$ merges the two strings with an offset of length j if the corresponding prefix and suffix are equal, for any appropriate integer j. So when $n = 9$ and $i = 0$, the merge $v := w \oplus_3 w$ with $w = abaaba$ is feasible since w has period 3. When $n = 10$, the merge $w := y \oplus_3 y$ with $y = abab$ is not possible since $a \neq b$. The trace for $n = 10$ is given in [20].

period	length	inner period	case	suffix	valid
8	9-8 = 1	9-8 = 1	2	$z := a$	true
6	9-6 = 3	8-6 = 2	2	$y := z.b.z = aba$	true
3	9-3 = 6	6-3 = 3	2	$w := y.y = abaaba$	true
0	9-0 = 9	3-0 = 3	1	$v := w \oplus_3 w = (aba)^3$	true

5 Fate and Dynamics of Period Sets

The incremental algorithm and Lemma 1 induces a *parental* relationship between sets in Γ_{n-1} and Γ_n. Any period set P occurs first in $\Gamma_{max(P)+1}$. When the length increases from $n-1$ to n, a period set P in Γ_{n-1} faces three cases: (1) P may remain as is in Γ_n, (2) P has an *extension* with period $n-1$ (i.e., $P \cup \{n-1\} \in \Gamma_n$), or (3) P *dies*, i.e., is neither in case (1) nor case (2). Note that cases (1) and (2) are not exclusive from each other. Thus, P at length $n-1$ can be the parent of at most two period sets in in Γ_n. One can thus investigate the dynamics of period sets when the word length n increases starting with $n = 1$.

Example 1. For instance, $\{0, 3, 6\}$ is born in Γ_7 and still belongs to Γ_8 and Γ_9; its extension $\{0, 3, 6, 7\}$ also belongs to Γ_8. From a dynamic view point, $\{0, 3, 6\}$ is the *parent* of both $\{0, 3, 6\}$ and $\{0, 3, 6, 7\}$ in Γ_8. On the contrary, $\{0, 4, 6\}$

belongs Γ_7, but dies at $n = 8$, since the pair $(4, 6)$ violates the FW condition at $n = 8$ (which would require to add $gcd(4, 6)$ as a new period). Last, $\{0, 2, 4, 6\}$ belongs to Γ_7, Γ_8, and generates $\{0, 2, 4, 6, 8\}$ in Γ_9 because the extension with period 8 is required by the FPR, but it never dies.

With these definitions at hand, consider the parental relation when the word length n increases starting from $\Gamma_1 := \{\{0\}\}$: it forms an infinite directed tree whose nodes are period sets and arcs represent the parental relation. The tree is rooted with period set $\{0\}$, is structured in successive layers corresponding to PS for successive word lengths, and the outdegree of each node can be 0, 1, or 2. When a period set dies, a branch of the tree becomes a dead end.

For each period set P, one may ask at how many consecutive word lengths it exists. We show that its fate depends only on the periods in P, and define below two variables that give the limit of its existence, and give algorithms to compute these.

Definition 5 (Recursive FW limit and next extension). Let $P := \{0 < p_1 < \ldots < p_k\}$ with $P \in \Gamma_{max(P)+1}$. The recursive FW limit of P, denoted $\text{rfw}(P)$, is defined by $\text{rfw}(P) := +\infty$ if $p_1 | p_i$ for all $0 < i \leq k$, and $\text{rfw}(P) := min_{0<i<j\leq k}\{2p_i - p_j : p_i, p_j \in P \text{ and } 2p_i - p_j > p_k + 1\}$. The next extension of P, denoted $e(P)$, is $e(P) := min_{0<i<j\leq k}\{FW(p_i, p_j) : p_i, p_j \in P \text{ and } \gcd(p_i, p_j) \notin P\}$ if $\sharp(P) > 1$, and $e(\{0\}) = +\infty$ otherwise.

If $\sharp(P) > 1$, as the word length increases, the current periods of P will induce, by the FPR, new compulsory periods larger than $max(P)$; the minimum among those is $e(P)$, meaning that when the word length reaches $e(P) + 1$ then P will necessarily be extended (i.e., possiblity (2) but not (1)), unless P dies. When the word length reaches $\text{rfw}(P)$, then at least one pair of periods will violate the FW condition, and thus P must die at that length. Of course, if $\sharp(P) = 1$ or all non zero periods in P are multiple of its basic period, then $\text{rfw}(P) = +\infty$. Thus, any given period set P ceases to belong to Γ_l if $l := min(e(P)+1, \text{rfw}(P))$ (i.e., case (1) is forbidden), it dies at length $\text{rfw}(P)$, and it is extended at length $e(P) + 1$ if $e(P) + 1 < \text{rfw}(P)$.

A challenging open question is to compute how many sets dies at length n, without enumerating Γ_n. The sequence of the numbers of dying period sets at length n is $0, 0, 0, 0, 0, 1, 1, 2, 1, 3, 2, 8$ for $n := [1, 12]$.

6 Conclusion and Exploration of Γ_n: distributions of period sets with respect to basic period and to weight

The key element of a period set is its basic period, which defines the first level of periodicity in a word. How period sets in Γ_n are distributed according to their basic period is non trivial. Enumerating Γ_n allows inspecting this distribution. The left plot in Fig. 1 displays $\kappa_{n,p}$, the counts of period sets for all possible basic periods p, in Γ_{60}. In predicate Ξ [7], one separates period sets depending on the basic period being $\leq \lfloor n/2 \rfloor$ (case **a**) or larger than $\lfloor n/2 \rfloor$ (case **b**). The smooth decrease of counts beyond $\lfloor n/2 \rfloor$ is explained by the combinatorial property that links number of period sets in case **b** and the number of binary partitions of an integer (see Lemma 5.8 in [23]). However, the distribution of counts for period sets in case **a**, still requires some investigation and statistical modeling. Here, we observe that between basic period 1 and 30, $\kappa_{n,p}$ reaches local maxima when p divides the string length n (e.g. see the peaks at $p = 10, 12, 15, 20$, or 30, which all correspond to period sets of case **a**).

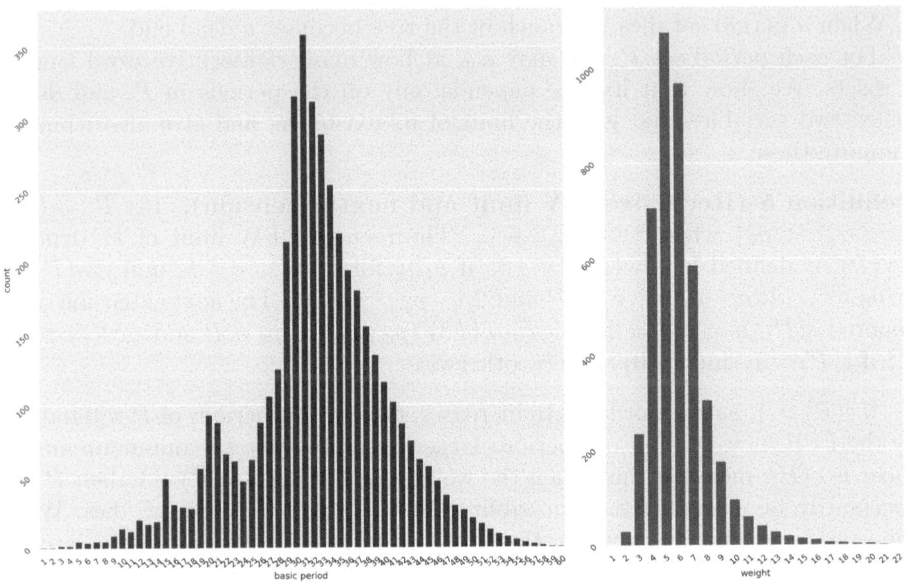

Fig. 1. Distribution in Γ_{60} of the number of period sets by basic period (left) and by weight (right), for string length of $n := 60$. Beyond basic period 30, the counts decrease smoothly with the basic period. Between basic period 1 and 30 the counts increase to a local maximum when the basic period reaches $\lfloor n/x \rfloor$ for $1 < x \leq 12 =$ (e.g. basic periods 10, 12, 15, 20, 30). The distribution by weight (right) is limited to weight below 22; it is unimodal and right skewed towards low weights.

Other works have investigated combinatorial parameters that control the number of periods of a word [6]. Thanks to enumeration of Γ_n, one can study the distribution of period sets with respect to weight and how it evolves with n. The right plot of Fig. 1 displays the number of period sets with equal weight (i.e., same number of periods) for $n = 60$. This distribution is right skewed and illustrates the constraints imposed by multiple periods. Similar figures to Fig. 1 for other string lengths are shown in [20].

Conclusion. We provide algorithms to enumerate Γ_n incrementally with low space requirement, with several certification functions, and an algorithm for binary realization of a period set. We define a parental relation between period

sets for distinct word lengths (which makes up a tree), and propose a way to study their dynamics as n increases. Many questions remain open – besides that on the number of dying PS mentioned in Sect. 5. Can one speed up enumeration by exploiting the tree and a dynamic update of the recursive FW limit and next extension of a PS? Can we exploit the tree to unravel how population sizes of PS evolve with n? Extending the notions presented here to generalizations of words, like partial words, degenerate strings, or multidimensional words opens avenues of future work. As seen in Sect. 5, the number of dying PS is not monotonically increasing in function of n; thus understanding the sequences of κ_n and $\kappa_{n,p}$ is both stimulating and challenging (see Fig. 1).
Resource: we provide files containing the period sets of Γ_n for $n = 1, \ldots, 100$ at [21].

Acknowledgements. This work is part of a project that has received funding from the European Unions Horizon 2020 research and innovation programme under the Marie Skodowska-Curie grant agreement No 956229. We thank the anonymous reviewers for their suggestions and comments.

References

1. Bajic, D., Loncar-Turukalo, T.: A simple suboptimal construction of cross-bifix-free codes. Cryptogr. Commun. **6**(1), 27–37 (2013). https://doi.org/10.1007/s12095-013-0088-8
2. Bilotta, S., Pergola, E., Pinzani, R.: A new approach to cross-bifix-free sets. IEEE Trans. Inf. Theory **58**(6), 4058–4063 (2012). https://doi.org/10.1109/TIT.2012.2189479
3. Castelli, M., Mignosi, F., Restivo, A.: Fine and Wilfs theorem for three periods and a generalization of sturmian words. Theoret. Comput. Sci. **218**(1), 83–94 (1999). https://doi.org/10.1016/s0304-3975(98)00251-5
4. Ehrenfeucht, A., Silberger, D.: Periodicity and unbordered segments of words. Discret. Math. **26**(2), 101–109 (1979). https://doi.org/10.1016/0012-365x(79)90116-x
5. Fine, N.J., Wilf, H.S.: Uniqueness theorems for periodic functions. Proc. Amer. Math. Soc. **16**, 109–114 (1965)
6. Gabric, D., Rampersad, N., Shallit, J.: An inequality for the number of periods in a word. Int. J. Found. Comput. Sci. **32**(05), 597–614 (2021). https://doi.org/10.1142/s0129054121410094
7. Guibas, L.J., Odlyzko, A.M.: Periods in strings. J. of Combinatorial Theory series A **30**, 19–42 (1981). https://doi.org/10.1016/0097-3165(81)90038-8
8. Halava, V., Harju, T., Ilie, L.: Periods and binary words. J. Comb. Theory, Ser. A **89**(2), 298–303 (2000). https://doi.org/10.1006/JCTA.1999.3014,
9. Holub, S., Shallit, J.O.: Periods and borders of random words. In: Ollinger, N., Vollmer, H. (eds.) 33rd Symposium on Theoretical Aspects of Computer Science, STACS 2016, February 17-20, 2016, Orléans, France. LIPIcs, vol. 47, pp. 44:1–44:10. Schloss Dagstuhl - Leibniz-Zentrum für Informatik (2016). https://doi.org/10.4230/LIPIcs.STACS.2016.44
10. Knuth, D., Morris, J., Pratt, V.: Fast pattern matching in strings. SIAM J. Comput. **6**, 323–350 (1977)

11. Leopardi, P.: Testing the tests: using random number generators to improve empirical tests. In: L' Ecuyer, P., Owen, A.B. (eds.) Monte Carlo and Quasi-Monte Carlo Methods 2008, pp. 501–512. Springer Berlin Heidelberg, Berlin, Heidelberg (2009). https://doi.org/10.1007/978-3-642-04107-5_32
12. Lothaire, M. (ed.): Combinatorics on Words. Cambridge University Press, second edn. (1997)
13. Lothaire, M.: Algebraic Combinatorics on Words. Cambridge University Press, Cambridge (2005). http://www-igm.univ-mlv.fr/~berstel/Lothaire/index.html
14. Marsaglia, G., Zaman, A.: Monkey tests for random number generators. Comput. Math. Appl. **26**(9), 1–10 (1993)
15. Nielsen, P.: A note on bifix-free sequences (corresp.). IEEE Trans. Info. Theory **19**(5), 704-706 (1973). https://doi.org/10.1109/tit.1973.1055065
16. Nielsen, P.: On the expected duration of a search for a fixed pattern in random data (corresp.). IEEE Trans. Info. Theory **19**(5), 702–704 (1973). https://doi.org/10.1109/tit.1973.1055064
17. Percus, O.E., Whitlock, P.A.: Theory and application of Marsaglia's monkey test for pseudorandom number generators. ACM Trans. Model. Comput. Simul. **5**(2), 87–100 (1995)
18. Rahmann, S., Rivals, E.: Exact and efficient computation of the expected number of missing and common words in random texts. In: Giancarlo, R., Sankoff, D. (eds.) CPM 2000. LNCS, vol. 1848, pp. 375–387. Springer, Heidelberg (2000). https://doi.org/10.1007/3-540-45123-4_31
19. Rahmann, S., Rivals, E.: On the distribution of the number of missing words in random texts. Comb. Probab. Comput. **12**(01) (2003). https://doi.org/10.1017/s0963548302005473
20. Rivals, E.: Incremental computation of the set of period sets. arXiv (2024). https://doi.org/10.48550/arXiv.2410.12077, arXiv:2410.12077, 21 pages
21. Rivals, E.: Sets of period sets for words of length n. Zenodo (2024). https://doi.org/10.5281/zenodo.13826260, data set
22. Rivals, E., Rahmann, S.: Combinatorics of Periods in Strings. In: Orejas, F., Spirakis, P.G., van Leeuwen, J. (eds.) Automata, Languages and Programming: 28th International Colloquium, ICALP 2001 Crete, Greece, July 8–12, 2001 Proceedings, pp. 615–626. Springer Berlin Heidelberg, Berlin, Heidelberg (2001). https://doi.org/10.1007/3-540-48224-5_51
23. Rivals, E., Rahmann, S.: Combinatorics of periods in strings. J. Combinatorial Theory Ser. A **104**(1), 95–113 (2003). https://doi.org/10.1016/s0097-3165(03)00123-7
24. Rivals, E., Sweering, M., Wang, P.: Convergence of the number of period sets in strings. In: Etessami, K., Feige, U., Puppis, G. (eds.) 50th International Colloquium on Automata, Languages, and Programming (ICALP 2023). Leibniz International Proceedings in Informatics (LIPIcs), vol. 261, pp. 100:1–100:14. Schloss Dagstuhl – Leibniz-Zentrum für Informatik, Dagstuhl, Germany (2023). https://doi.org/10.4230/LIPIcs.ICALP.2023.100
25. Smyth, W.F.: Computating Pattern in Strings. Pearson - Addison Wesley (2003)
26. Tijdeman, R., Zamboni, L.: Fine and Wilf words for any periods. Indag. Math. **14**(1), 135–147 (2003). https://doi.org/10.1016/s0019-3577(03)90076-0
27. Tijdeman, R., Zamboni, L.: Fine and Wilf words for any periods ii. Theoret. Comput. Sci. **410**(30–32), 3027–3034 (2009). https://doi.org/10.1016/j.tcs.2009.02.004
28. Ukkonen, E.: Approximate string-matching with q-grams and maximal matches. Theor. Comp. Sci. **92**(1), 191–211 (1992)

Tolerant Testing and Distance Estimation for Distributions Under Memory Constraints

Sampriti Roy(✉) and Yadu Vasudev

Department of Computer Science and Engineering Indian Institute of Technology Madras, Chennai, India
cs18d200@smail.iitm.ac.in

Abstract. We investigate tolerant testing and distance estimation problems in distribution testing when the amount of memory is limited. In particular, our aim is to provide a good estimate of the distance measure (in total variation distance) using constant memory words. Though the problems are well understood in the standard access model with an $\Omega(\frac{n}{\log n})$ lower bound, it has not been explored under memory constraints.

We propose algorithms that require $O(\frac{n}{\epsilon^7})$ queries for estimating distance to a known distribution and $O(\frac{n}{\epsilon^6})$ queries (where $\epsilon \in (0, 1)$ is an error parameter) for estimating distance to uniformity that use only $O(1)$ memory words. Our algorithms match the $\Omega(\frac{n}{\epsilon^4})$ lower bound established by [19] in terms of the dependence on n. Additionally, we present an $(\epsilon, c\epsilon)$ tolerant closeness tester in terms of ℓ_2 distance for all $c \geq 3$, which requires $O(\frac{n^3}{m\epsilon^2})$ samples and $O(m)$ bits of memory for $\log n/\epsilon^4 \leq m \leq n/\epsilon^2$. Furthermore, we outline an $O(\frac{n^2\epsilon_1^2}{m\epsilon_2^4})$ trade-off for (ϵ_1, ϵ_2) tolerant closeness testing problem $(0 < \epsilon_1 < \epsilon_2 < 1)$ using ℓ_1 distance, where $\log n \leq m \leq n \log n$ is the available memory in bits.

Keywords: Property testing · Distribution testing · Streaming

1 Introduction

In the realm of statistical analysis and data science, understanding the underlying distribution of data is a fundamental problem. In distribution testing, which is a subfield of property testing, we study this problem with goal of checking if an unknown distribution has a property or is ϵ-far from having a property. The distribution is accessible using samples and the complexity measure of interest is the number of samples required to test the property. The surveys of Rubinfeld [27], Canonne [5] provide an overview of the results in this area. A closely related question is the problem of estimating the distance of a distribution to a property. For instance, an algorithm that can estimate the distance of an unknown distribution to the uniform distribution gives a testing algorithm. It is known that the

S. Roy—Supported by IITM Pravartak Technologies Foundation.

R. Královič and V. Kůrková (Eds.): SOFSEM 2025, LNCS 15539, pp. 269–283, 2025.
https://doi.org/10.1007/978-3-031-82697-9_20

distance estimation problem is provable harder (in terms of sample complexity) than the problem of testing.

An intermediate problem that lies between testing and estimation is the question of tolerant testing [25], which takes parameters ϵ_1 and ϵ_2 and tests if the distribution is ϵ_1-close to the property or ϵ_2-far. It is easy to see that if there is a tolerant (ϵ_1, ϵ_2)-tester for every setting of the parameter, then this gives an algorithm for estimating the distance with similar sample complexity. The focus of this paper is on tolerant testing and estimation of probability distributions.

While the number of samples has been the natural measure of complexity in distribution testing, a recent line of work [9,19,26] has studied the trade-offs between sample complexity and space complexity for such problems. The algorithms for testing distributions assume that the algorithm has sufficient space to store all the samples that it has received, whereas in practice it is possible that there are memory constraints on how many samples can be stored at any given time. If the distribution is supported on the set $[n] = \{1, 2, \ldots, n\}$, then storing one sample from the distribution requires $O(\log n)$ bits, which we will denote by a memory word. If the memory available to the algorithm is only $O(1)$ words (or $O(\log n)$ bits), then this means that at any point in time the algorithm can perform computation on only a constant number of samples.

In this work, we study tolerant testing and distance estimation under such memory conditions. In particular, we design algorithms for estimating the distance (in ℓ_1-norm) of an unknown distribution to the uniform distribution, and extend the idea to also estimate the distance of an unknown distribution to a known fixed distribution. Our algorithms store only a constant number of memory words at any point of time. We show that the number of samples that we use matches the known lower bound for this problem; we leave the technical aspects of the results to the end of this section.

Background and Related Work. Distribution testing was introduced by the pioneering work of Goldreich and Ron [23] who gave a sample-efficient algorithm for testing if an unknown distribution is uniform or ϵ-far from uniform. Shortly after, a study by Batu et al. [2] addressed the problem of testing whether two unknown random variables share the same distribution (known as closeness or equivalence testing). This initiated a line of research focusing on testing uniformity, identity, closeness [4,13,20,21,24,30,31], structure of distributions (monotonicity, k-modality, decomposibility etc.) [3,6,12,16,17,22] and many more using the standard access model when i.i.d samples are drawn from an unknown distribution. Other works by [28,29] investigated the tolerant testing problems in the standard setting where the aim is to test whether an unknown distribution is ϵ_1 close to a property or ϵ_2 far from it in total variance distance. Though the sample complexity is strongly sublinear $[(\Theta(\sqrt{n}))]$ on the domain size when $\epsilon_1 = 0$, the authors showed that the sample complexity jumps to barely sublinear $[\Theta(\frac{n}{\log n})]$ for $\epsilon_1 = \epsilon_2/2$. Recently, Canonne et al. [7]) studied the tolerant testing problem for different regimes of ϵ_1, ϵ_2 and fully characterized the sample complexity in terms of $n, \epsilon_1, \epsilon_2$. Parnas et al. [25] studied the problem of estimating distance to a property and showed its connection to tolerant testing. Daskalakis et al. [18] further studied the tolerant testing problems using different distance measures. Another work by Chakraborty et al. [10] investigated the connection between the sample complexities of tolerant and non-tolerant version of distribution testing. In particular, they proved that for any label invariant property if testing requires s samples, tolerant testing requires $min\{s^2, n\}$ samples.

Distribution testing under strict memory constraints has gained considerable attention in recent times. Chien et al. [14] put forth a sample-space trade-off for testing (ϵ, δ)-weakly continuous properties, as defined by Valiant [31]. Diakonikolas et al. [19] explore distribution testing under memory limitations, focusing on uniformity testing, identity testing, and closeness testing, and proposed sample-space trade-offs for these problems. Canonne et al. [9] refine the results from [19] by presenting simpler algorithms for the same problems. Additionally, [26] investigate sample-space trade-offs for identity testing using conditional sampling access as proposed by [8,11] and offer a trade-off for monotonicity testing, extending the approach to a broader class of decomposable distributions. Another work by Acharya et al. [1] focuses on estimating the entropy of distributions with constant memory words when samples appear as a stream. While these studies concentrate exclusively on classical distribution testing with memory constraints, our work addresses the tolerant testing and distance estimation for probability distributions.

1.1 Our Contributions

We investigate the problem of tolerant testing and distance estimation under limited memory constraints. Specifically, we consider a model where samples from an unknown distribution D are received sequentially, and the available memory is insufficient to store all the samples. Within this framework, our goal is to provide an accurate estimate of the distance (total variation distance) to a given property while using only a constant number of memory words (or $O(\log n)$ bits). Our key contributions are as follows:

- We provide an algorithm for estimating the distance to a known distribution D^* using constant space. In particular, we present an $O(n/\epsilon^7)$ query algorithm for this problem.
- We show that algorithm for estimating distance to a known distribution can be used for estimating distance to uniformity with a little modification in the analysis which leads to an $O(\frac{n}{\epsilon^6})$-query algorithm where only constant amount of space is used. This matches the lower bound for the uniformity testing problem obtained by Diakonikolas et al. [19].

Additionally, we outline an $O(\frac{n^2\epsilon_1^2}{m\epsilon_2^4})$ trade-off for (ϵ_1, ϵ_2) tolerant closeness testing problem using ℓ_1 distance, where $\log n \leq m \leq n \log n$ is the available memory in bits. We also study tolerant closeness testing in ℓ_2 distance measure and present an $O(\frac{n^3}{m\epsilon^2})$ trade-off for the same. The memory requirement of our algorithm is $O(m)$ bits where $\log n/\epsilon^4 \leq m \leq n/\epsilon^2$.

2 Notation and Preliminaries

We use D as a discrete probability distribution supported over $[n]$ and $\mathcal{U}_n$ denotes a uniform distribution over $[n]$. Let $\mathcal{D}$ be the set of all probability distributions supported on $[n]$. A property $\mathcal{P}$ is a subset of $\mathcal{D}$. The total variation distance, or ℓ_1 distance between two distributions D_1, D_2 is defined by $d_{TV}(D_1, D_2) = (1/2)\sum_{i=1}^{n} |D_1(i) - D_2(i)|$ and the ℓ_2 distance is defined by $d_{\ell_2}(D_1, D_2) = \sqrt{\sum_{i=1}^{n}(D_1(i) - D_2(i))^2}$. We say that a distribution D is ϵ far from $\mathcal{P}$ if D is ϵ far from all the distributions having the property $\mathcal{P}$, i.e. $d_{TV}(D, D') > \epsilon$ for every $D' \in \mathcal{P}$. While testing property of a distribution D, the goal is to check whether D is satisfying the property or ϵ-far from it. The tolerant testing problems need additional restriction, where the aim is to check whether D is ϵ_1-close to a property or ϵ_2-far from it, for $\epsilon_1 < \epsilon_2$. Distance estimation is another aspect of tolerant testing where we need to approximate the distance of a distribution to a property. Let $x_1, ...x_s$ be sample from D, we define the probability of self-collision of D by $||D||_2 = Pr[x_i = x_j]$. Similarly, let $x_1, ..., x_s$ and $y_1, ..., y_s$ are samples from D_1 and D_2 respectively, then the collision probability between samples from D_1, D_2 is denoted by $\langle D_1 D_2 \rangle = Pr[x_i = y_j]$. Consider S samples are drawn from a distribution D, where $S_1, S_2 \subset S$, the *bipartite collision* of D with respect to S (defined by $\mathsf{Collision}\ (S_1, S_2)$) is the number of collisions between S_1 and S_2. We denote SAMP to define standard access model when i.i.d samples can be drawn according to a distribution D. In our case, we consider the samples appear from D as a stream while there is a memory constraint, and the task is to provide space efficient algorithms for distribution testing problems. We state the following inequalities that will be used throughout the paper,

Lemma 1 (Cauchy-Schwartz inequality). *If $a_1, ..., a_n$ and $b_1, ..., b_n$ are real numbers, then $\left(\sum_{i=1}^{n} a_i b_i\right)^2 \leq \sum_{i=1}^{n} a_i^2 \cdot \sum_{i=1}^{n} b_i^2$.*

Lemma 2 (Hoeffding's inequality). *Let $X_1, ..., X_m \in [a_i, b_i]$ be independent random variables, and $X = (X_1 + X_2 + ... + X_m)/m$. Then $Pr[|X - \mathbb{E}[X]| \geq t] \leq 2exp(\frac{-2(mt)^2}{\sum_i (b_i - a_i)^2})$.*

3 Estimating Distance to a Known Distribution Using Constant Memory Words

In this section, we present an algorithm for estimating distance to a known distribution using constant memory words. Let D be an unknown distribution over $[n]$, and D^* be a distribution over $[n]$ which is explicitly given where $d = d_{TV}(D, D^*)$.

The goal is to come up with a good estimate of d, (say $\hat{d}$) while sampling according to SAMP oracle and using only constant amount of space. Observe that $d_{TV}(D, D^*) = \sum_{i=1}^{n} D(i)|1 - \frac{D^*(i)}{D(i)}|$, which can be written as $\mathbb{E}_{i \sim D}[|1 - \frac{D^*(i)}{D(i)}|]$. At a very high level, the core idea behind our algorithm is to estimate the quantity $|1 - \frac{D^*(i)}{D(i)}|$ using samples from D. In order to make the algorithm sample-efficient in terms of ϵ, we mix the distribution D with D^* to get D', such that for all $i \in [n]$, $D'(i) = (1 - \frac{\epsilon}{2})D(i) + \frac{\epsilon}{2} \cdot D^*(i)$. This approach ensures that $D'(i)$ has the probability weight at least $\frac{\epsilon}{2}D^*(i)$ for every $i \in [n]$ which will be beneficial in our analysis later on. Now, let $d' = d_{TV}(D', D^*)$. We show that d' is not significantly far from d.

Lemma 3. $d - \frac{\epsilon}{2} \leq d' \leq d + \frac{\epsilon}{2}$

Proof. Observe that $d' = d_{TV}(D', D^*) = \frac{1}{2}\sum_{i=1}^{n} |D(i) - \frac{\epsilon}{2}D(i) + \frac{\epsilon}{2}D^*(i) - D^*(i)|$. Using triangle inequality, we get $d' \leq \frac{1}{2}\sum_{i=1}^{n} |D(i) - D^*(i)| + \frac{1}{2} \cdot \frac{\epsilon}{2}\sum_{i=1}^{n} |D(i) - D^*(i)|$ which implies that $d' \leq d + \frac{\epsilon}{2}d$. Considering $d \leq 1$, we obtain $d' \leq d + \frac{\epsilon}{2}$. Similarly, using reverse triangle inequality, we get $d' \geq \frac{1}{2}\sum_{i=1}^{n} ||D(i) - D^*(i)| - \frac{\epsilon}{2}|D^*(i) - D(i)|| \geq \frac{1}{2}[\sum_{i=1}^{n} |D(i) - D^*(i)| - \frac{\epsilon}{2}\sum_{i=1}^{n} |D^*(i) - D(i)|]$ which implies $d' \geq d - \frac{\epsilon}{2}d$. Considering $d \leq 1$, we obtain $d' \geq d - \frac{\epsilon}{2}$.

For the rest of this section, we estimate d' by taking samples from D' and later use the Lemma 3 to conclude that the estimate is good for estimating d as well. Observe that while sampling from D' a point i from $[n]$ is returned with probability $D'(i) = (1 - \frac{\epsilon}{2})D(i) + \frac{\epsilon}{2}D^*(i)$ where $D'(i) \geq (\epsilon/2) \cdot D^*(i)$. The algorithm is formalized below.

Algorithm 1: Estimating $d_{TV}(D', D^*)$

Input : A data stream where elements are sampled from D', error parameter $\epsilon \in (0, 1]$, an explicitly given distribution D^*

Output: An estimate of $d_{TV}(D', D^*)$

1 Initialize $\tilde{d} = 0$

2 **for** *each* $i = 1$ *to* $R = \frac{8}{\epsilon^4}$ **do**

3   Let x be the next element in the stream

4   N_x = Number of times x occurs in the next $s = n/\epsilon^3$ samples of D'

5     $d_i = |1 - \frac{(s+1)D^*(x)}{(N_x+1)}|$

6     $\tilde{d} = \tilde{d} + d_i$

7 $\hat{d} = \frac{\tilde{d}}{R}$

Theorem 1. *The algorithm takes* $O(\frac{n}{\epsilon^7})$ *samples from* D' *and with high probability returns an estimate of* $d = d_{TV}(D', D^*)$ *to be* $\hat{d}$ *such that* $d' - \epsilon \leq \hat{d} \leq 2d' + 4\epsilon$*. The memory requirement of the algorithm is* $O(1)$ *words.*

Proof. Observe that the random variable $N_x \sim Bin(s, D'(x))$ and so, $\mathbb{E}[N_x] = s \cdot D'(x)$. Also, $\mathbb{E}[N_x^2] = sD'(x) + s(s-1)D'(x)^2$. The following lemma calculates the expected value of $\frac{1}{(N_x+1)}$, and $\frac{1}{(N_x+1)^2}$.

Lemma 4. *Let N_x be the number of times an element x occurs in a stream of length s, where $N_x \sim Bin(s, D'(x))$,then*

- $\mathbb{E}[\frac{1}{N_x+1}] = \frac{1-\delta_x}{D'(x)(s+1)}$; *for* $\delta_x = (1 - D'(x))^{s+1}$.
- $\mathbb{E}[\frac{1}{(N_x+1)^2}] \leq \frac{1}{(s+1)^2 D'(x)^2}$.

Proof.

$$\begin{aligned}
\mathbb{E}[\frac{1}{(N_x+1)}] &= \frac{1}{D'(x)(s+1)} \sum_{l=0}^{s} \frac{(s+1)}{l+1} \binom{s}{l} D'(x)^{l+1} \cdot (1 - D'(x))^{s-l} \\
&= \frac{1}{D'(x)(s+1)} \sum_{l=0}^{s} \binom{s+1}{l+1} D'(x)^{l+1} \cdot (1 - D'(x))^{s-l} \\
&= \frac{1 - (1 - D'(x))^{s+1}}{D'(x)(s+1)} = \frac{1 - \delta_x}{D'(x)(s+1)}
\end{aligned}$$

The first equation is obtained by writing the expectation explicitly when N_x is $Bin(s, D'(x))$. The second equality comes from the fact that $\frac{(s+1)}{l+1}\binom{s}{l} = \binom{s+1}{l+1}$. Now, observe that $\sum_{l=0}^{s} \binom{s+1}{l+1} D'(x)^{l+1} \cdot (1 - D'(x))^{s-l} + \binom{s+1}{0} D'(x)^0 (1 - D'(x))^{s+1} = 1$, subtracting the term $(1 - D'(x))^{s+1}$ from 1, we get the result. To bound the $\mathbb{E}[\frac{1}{(N_x+1)^2}]$ we use the following fact from [15],

FACT 1 ([15]). *Let $X \sim Bin(n,p)$, then $\mathbb{E}[(1 + X)^{-\alpha}] \leq (np)^{-\alpha}$ for $\alpha \geq 1$.*

Hence, using the above fact, we can deduce that $\mathbb{E}[\frac{1}{(N_x+1)^2}] \leq \frac{1}{(sD'(x))^2}$. When s is very large, the length of the stream s is of same order as that of $s + 1$. [In our case, $s = O(\frac{n}{\epsilon^2})$, and so, $s + 1 = O(\frac{n}{\epsilon^4})$]. Hence, we can write the expected value as $\mathbb{E}[\frac{1}{(N_x+1)^2}] \leq \frac{1}{(s+1)^2 D'(x)^2}$.

The following lemma depicts the fact that the expected value of $\hat{d}$ is a good estimate of d'.

Lemma 5. $d' \leq \mathbb{E}[\hat{d}] \leq 2d' + 3\epsilon$.

Proof. We know, $\mathbb{E}[\hat{d}] = \frac{1}{R} \sum_{i \in [R]} \mathbb{E}[d_i]$.

$$\begin{aligned}
\mathbb{E}[d_i] &= \sum_{x=1}^{n} D'(x) \mathbb{E}[|1 - \frac{(s+1)D^*(x)}{(N_x+1)}|] \geq \sum_{x=1}^{n} D'(x) |\mathbb{E}[1 - \frac{(s+1)D^*(x)}{(N_x+1)}]| \\
&= \sum_{x=1}^{n} D'(x) |1 - \frac{(s+1)(1-\delta_x)D^*(x)}{D'(x)(s+1)}| \geq \sum_{x;D'(x)>(1-\delta_x)D^*(x)} D'(x) |1 - \frac{(1-\delta_x)D^*(x)}{D'(x)}| \\
&\geq \sum_{x;D'(x)>D^*(x)} D'(x) - (1-\delta_x)D^*(x) > \sum_{x;D'(x)>D^*(x)} (D'(x) - D^*(x)) = d'
\end{aligned}$$

The first inequality in obtained by the Jensen's inequality of convex function. The second inequality comes by taking sum over the elements for which $D'(x) > (1 - \delta_x)D^*(x)$. The third inequality follows from the fact that

$\{x; D'(x) > D^*(x)\} \subset \{x; D'(x) > (1-\delta_x)D^*(x)\}$. The last equality comes from the fact that $d' = (1/2)\sum_{x=1}^{n}|D'(x)-D^*(x)| = \sum_{x;D'(x)>D^*(x)}(D'(x)-D^*(x))$.

Now, we upper bound the expectation $\mathbb{E}[d_i] = \sum_{x=1}^{n} D'(x)\mathbb{E}[|1-\frac{(s+1)D^*(x)}{(N_x+1)}|]$. In order to do that we use the fact that $\mathbb{E}\Big[|X-Y|\Big] \leq \sqrt{\mathbb{E}\Big[(X-Y)^2\Big]}$.

$$\begin{aligned}\mathbb{E}\Big[|1-\frac{(s+1)D^*(x)}{(N_x+1)}|\Big]^2 &\leq \mathbb{E}[1-\frac{2(s+1)D^*(x)}{(N_x+1)}+\frac{(s+1)^2D^*(x)^2}{(N_x+1)^2}] \\ &< 1-\frac{2(1-\delta_x)D^*(x)}{D'(x)}+\frac{D^*(x)^2}{D'(x)^2} \\ &= (1-\frac{D^*(x)}{D'(x)})^2+(\sqrt{\frac{2\delta_x D^*(x)}{D'(x)}})^2\end{aligned}$$

The second inequality is obtained by substituting the value of $\mathbb{E}[\frac{1}{(N_x+1)^2}] \leq \frac{1}{(s+1)^2D'(x)^2}$ from the Lemma 4. Now, observe that $(1-\frac{D^*(x)}{D'(x)})$ can be positive or negative. So, the expectation can be written as,

$$\mathbb{E}\Big[|1-\frac{(s+1)D^*(x)}{(N_x+1)}|\Big]^2 < \begin{cases}\left(1-\frac{D^*(x)}{D'(x)}+\sqrt{\frac{2\delta_x D^*(x)}{D'(x)}}\right)^2 & ;\ \text{when} D'(x) > D^*(x) \\ \left(\frac{D^*(x)}{D'(x)}-1+\sqrt{\frac{2\delta_x D^*(x)}{D'(x)}}\right)^2 & ;\ \text{when} D'(x) \leq D^*(x)\end{cases}$$

Now, substituting the expected value, we get,

$$\begin{aligned}\mathbb{E}[d_i] < &\sum_{x;D'(x)>D^*(x)} D'(x)(1-\frac{D^*(x)}{D'(x)}) + \sum_{x;D'(x)\leq D^*(x)} D'(x)(\frac{D^*(x)}{D'(x)}-1) \\ &+\sum_{x=1}^{n} D'(x)\sqrt{\frac{2\delta_x D^*(x)}{D'(x)}} \\ = &\sum_{x;D'(x)>D^*(x)} (D'(x)-D^*(x)) + \sum_{x;D'(x)<D^*(x)} (D^*(x)-D'(x)) \\ &+\sqrt{2}\sum_{x;D'(x)>\epsilon^2/n}\sqrt{\delta_x D'(x)D^*(x)}+\sqrt{2}\sum_{x;D'(x)\leq\epsilon^2/n}\sqrt{\delta_x D'(x)D^*(x)} \quad (1)\end{aligned}$$

When $D(x) > \epsilon^2/n$, $\delta_x = (1-D(x))^{s+1} < e^{-1/\epsilon} < \epsilon^2$, for $s = n/\epsilon^3$. Also, we use the Cauchy-Schwartz inequality to bound $\sum_{x;D'(x)>\epsilon^2/n}\sqrt{D'(x)D^*(x)}$. Substituting $a_i = \sqrt{D'(x)}$, and $b_i = \sqrt{D^*(x)}$ in Lemma 1, we observe that $\sum_{x=1}^{n}\sqrt{D'(x)D^*(x)} \leq 1$. So, $\sum_{x;D'(x)>\epsilon^2/n}\sqrt{\delta_x}\sqrt{D'(x)D^*(x)}$ becomes at most ϵ. For the next part of the sum, substituting $D'(x) \leq \epsilon^2/n$, and applying Lemma 1 for $\sum_{x=1}^{n}\sqrt{D^*(x)} < \sqrt{n}$, we thus write the equation as, $\mathbb{E}[d_i] < 2d' + \sqrt{2}\epsilon + \sqrt{2}\frac{\epsilon}{\sqrt{n}}\cdot\sqrt{n} < 2d' + 3\epsilon$.

The minimum value of the estimator d_i can be 0. The maximum value of d_i can be $max_x\{\frac{(s+1)D^*(x)}{N_x+1}-1\}$. Given that $D'(x) \geq (\epsilon/2)\cdot D^*(x)$, the expected value of N_x is at least $(\epsilon/2)\cdot D^*(x)\cdot s$. Applying an additive Chernoff bound, with high probability, we obtain that $N_x \geq [(\epsilon/2) \cdot D^*(x) \cdot s] - 1$. Hence, $max\ d_i \leq \frac{(s+1)D^*(x)}{(\epsilon/2)\cdot D^*(x)\cdot s} \approx 2/\epsilon$. Hence, d_i can take the value in the range $[0, 2/\epsilon]$. By applying Hoeffding bound, we get, $Pr[|\hat{d} - \mathbb{E}[\hat{d}]| \geq \epsilon] \leq e^{-\frac{\epsilon^4 R}{4}}$. For, $R = 8/\epsilon^4$, with probability at least 3/4, $\mathbb{E}[\hat{d}] - \epsilon \leq \hat{d} \leq \mathbb{E}[\hat{d}] + \epsilon$. Together with Lemma 5 we have, $d' - \epsilon \leq \hat{d} \leq 2d' + 4\epsilon$.

Furthermore, we know that $d - \frac{\epsilon}{2} \leq d' \leq d + \frac{\epsilon}{2}$, combining this with Theorem 1 we formalize the following,

Theorem 2. *There exists an algorithm that takes $O(\frac{n}{\epsilon^7})$ samples from D' which is a mixture of D and D^* and with high probability returns an estimate of $d = d_{TV}(D, D^*)$ to be $\hat{d}$ such that $d - 3\epsilon/2 \leq \hat{d} \leq 2d + 5\epsilon$. The memory requirement of the algorithm is $O(1)$ words.*

Proof. By the Theorem 1, we get $d' - \epsilon \leq \hat{d} \leq 2d' + 4\epsilon$. In order to argue that $\hat{d}$ is a good estimate of d, we use the Lemma 3 and obtain $d - 3\epsilon/2 \leq \hat{d} \leq 2d + 4\epsilon$. Also, the memory requirement for the algorithm is $O(1)$ words or, $O(\log n)$ bits.

3.1 Estimating Distance to Uniformity Using Constant Memory Words

Our algorithm for estimating distance to a known distribution can be used to estimate distance to a uniform distribution when $D^*(i) = 1/n$ for all $i \in [n]$. Similar to the Algorithm 1 we sample from the distribution D', where D' is a mixture of D and $\mathcal{U}_n$ such that for all $i \in [n]$, $D'(i) = (1 - \frac{\epsilon}{2})D(i) + \frac{\epsilon}{2n}$. This guarantees the fact that for all $i \in [n]$, $D'(i) \geq \epsilon/2n$. However, the sample complexity becomes $O(n/\epsilon^6)$ instead of $O(n/\epsilon^7)$. We formalize the theorem below and describe how the extra $(1/\epsilon)$ can be saved in case of estimating distance to uniformity. We leave out many of the technical details in this version of the paper.

Theorem 3. *There exists an algorithm that takes $O(\frac{n}{\epsilon^6})$ samples from D' which is a mixture of D and $\mathcal{U}_n$ and with high probability returns an estimate of $d = d_{TV}(D, \mathcal{U}_n)$ to be $\hat{d}$ such that $d - 3\epsilon/2 \leq \hat{d} \leq 2d + 4\epsilon$. The memory requirement of the algorithm is $O(1)$ words.*

Proof. Let $d = d_{TV}(D, \mathcal{U}_n)$ and $d' = d_{TV}(D', \mathcal{U}_n)$, then analogous to the Lemma 3, we have $d - \frac{\epsilon}{2} \leq d' \leq d + \frac{\epsilon}{2}$. Next, we can show that $\hat{d}$ is a good estimate of d'. In particular, we can prove that $d' \leq \mathbb{E}[\hat{d}] \leq 2d' + 3\epsilon$. We follow the similar proof of the Lemma 5 by substituting $D^*(x) = 1/n$ and lower bound $\hat{d} \geq d'$. To upper bound $\mathbb{E}[\hat{d}]$, we again follow the same proof as that of the Lemma 5 by replacing $D^*(x) = 1/n$ except from the Eq. 1. Observe that in the previous section, when $D'(x) = (1 - \frac{\epsilon}{2})D(x) + \frac{\epsilon}{2}D^*(x)$, there are two possibilities that $D'(x) > \epsilon^2/n$, and $D'(x) \leq \epsilon^2/n$ depending on the value of $D^*(x)$. Therefore, the sum has been divided into two parts in the Eq. 1. When $D^*(x) = 1/n$ for all $x \in [n]$, the mixture distribution becomes $D'(x) = (1 - \frac{\epsilon}{2})D(x) + \frac{\epsilon}{2n}$ with a guarantee that $D'(x) \geq \epsilon/2n$ for all $x \in [n]$. Hence, we can rewrite the equation as $\mathbb{E}[d_i] < \sum_{x;D'(x)>1/n}(D'(x) - \frac{1}{n}) + \sum_{x;D'(x)<1/n}(\frac{1}{n} - D'(x)) + \sqrt{2}\sum_{x;D'(x)>\epsilon/2n}\sqrt{\frac{\delta_x D'(x)}{n}}$.

As $D'(x) > \epsilon/2n$, $\delta_x = (1 - D'(x))^{s+1} < e^{-1/\epsilon} < \epsilon^2$ for $s = n/\epsilon^2$. Specifically, in our case of estimating distance to uniformity by the definition of mixture distribution D', $s = n/\epsilon^2$ will suffice to go through the analysis. Finally, we use the Cauchy-Schwartz inequality to bound $\sum_{x;D'(x)>\epsilon/2n}\sqrt{D'(x)}$. Substituting $a_i = \sqrt{D'(x)}$, and $b_i = 1$ in the Lemma 1, we observe that $\sum_{x=1}^{n}\sqrt{D'(x)} \leq \sqrt{n}$. So, $\sum_{x;D'(x)>\epsilon/n}\sqrt{\delta_x}\sqrt{\frac{D'(x)}{n}}$ becomes at most ϵ. Together, we get $\mathbb{E}[\hat{d}] < 2d + 2\epsilon$. Similar to the previous section, using Hoeffding inequality 2 with $R = 8/\epsilon^4$ and combining with the relation between d and d', we argue that $d - 3\epsilon/2 \leq \hat{d} \leq 2d + 4\epsilon$. As $s = n/\epsilon^2$, and $R = 8/\epsilon^4$, the sample complexity of the algorithm becomes $O(n/\epsilon^6)$ with a memory requirement of $O(1)$ words.

Memory Lower Bound for Tolerant Uniformity Testing. Diakonikolas et al. [19] prove a memory lower bound for uniformity testing problem when samples appear sequentially and available memory is $O(m)$ bits. We state their result formally below.

Theorem 4 (Uniformity testing lower bound [19]). *Let $\mathcal{A}$ be an algorithm which uses S samples from a distribution D with m bits of available memory, tests if a distribution D is uniform versus ϵ far from uniform with error probability 1/3, then $S.m = \Omega(n/\epsilon^4)$. Furthermore, if $S < n^{0.9}$ and $m > S^2/n^{0.9}$ then $S \cdot m = \Omega(n \log n/\epsilon^4)$.*

Observe that our algorithm for estimating distance to uniformity requires $O(\frac{n}{\epsilon^6})$ samples while using only $O(1)$ memory words or $O(\log n)$ bits of memory. As, distance estimation is harder than the classical property testing, the above lower bound is trivially applied for estimating distance to uniformity. Hence, for $m = O(\log n)$, our algorithm matches the lower bound with regard to the dependence on n. The dependence of ϵ in the algorithm is sub-optimal, but we do not address that question in this paper.

4 Sample-Space Trade-Off for Tolerant Testing in Various Distance Measures

In this section, we present sample-space trade-off for tolerant testing problems. In particular, we investigate the problems of tolerant closeness testing in ℓ_1 or total variation distance and ℓ_2 distance when both the distributions are unknown and there is a memory constraint of m bits.

4.1 Tolerant Closeness Testing in Total Variation Distance

In this section, we explore the sample-space trade-off for tolerant closeness testing in terms of total variation distance. The core component of our algorithm is the tolerant closeness tester introduced by [7]. We demonstrate how to adapt their algorithm to be more space-efficient by sequentializing its process. Essentially, their algorithm involves taking samples from two distributions D_1 and D_2, and then computing a statistic based on the frequency of each index $i \in [n]$ within these samples. The algorithm uses this statistic to make decisions about acceptance or rejection. With a memory constraint of up to m bits, our approach suggests storing frequencies for at most $m/\log n$ indices at a time, and then reusing this space for $n \log n/m$ blocks sequentially. Before delving into our algorithm, we first outline the results from [7].

Theorem 5. *[7] Let D_1 and D_2 be two unknown distributions over $[n]$. There exists an algorithm such that for an absolute constant $c > 0$ and $0 \leq \epsilon_2 \leq 1$, $0 \leq \epsilon_1 \leq c\epsilon_2$ distinguishes between $d_{TV}(D_1, D_2) \leq \epsilon_1$ and $d_{TV}(D_1, D_2) > \epsilon_2$ with probability at least 4/5. The algorithm takes $O(n(\frac{\epsilon_1}{\epsilon_2^2})^2 + n(\frac{\epsilon_1}{\epsilon_2^2}) + \frac{\sqrt{n}}{\epsilon_2^2} + \frac{n^{2/3}}{\epsilon_2^{4/3}})$ samples from each D_1 and D_2.*

Similar to [7], we sample two sets of $Poi(S)$ points from D_1 and D_2 respectively, where $Poi(S)$ is poisson samples. Let $\tilde{X}_i$ and X_i be the count of occurrences of symbol $i \in [n]$ in the first and the second set of the samples from D_1. Also, let $\tilde{Y}_i$ and Y_i be the count of symbol i in the first and the second set of samples from D_2. Define f_i in the following way,

$$f_i = \begin{cases} max\{\sqrt{Sn} \cdot |D_1(i) - D_2(i)|, n \cdot (D_1(i) + D_2(i)), 1\}; & \text{if } S \geq n \\ max\{S \cdot (D_1(i) + D_2(i)), 1\}; & \text{if } S < n \end{cases}$$

In order to define an estimator $\hat{f}_i$, $\tilde{X}_i$ and $\tilde{Y}_i$ will be used.

$$\hat{f}_i = \begin{cases} max\{\frac{|\tilde{X}_i - \tilde{Y}_i|}{\sqrt{S/n}}, \frac{\tilde{X}_i + \tilde{Y}_i}{S/n}, 1\}; & \text{if } S \geq n \\ max\{\tilde{X}_i + \tilde{Y}_i, 1\}; & \text{if } S < n \end{cases}$$

Also $Z_i = (X_i - Y_i)^2 - X_i - Y_i$. We outline the tester as follows,

---

Algorithm 2: Tolerant ℓ_1 Closeness

Input : SAMP access from both D_1 and D_2, error parameter $0 \leq \epsilon_1 < \epsilon_2 \leq 1$, $c > 0$, threshold $\tau = c \cdot min(\frac{S^{3/2}\epsilon_2}{n^{1/2}}, \frac{S^2\epsilon_2^2}{n})$ memory constraint $\log n \leq m \leq n \log n$, initialize $Z = 0$

Output: Accept if $d_{TV}(D_1, D_2) \leq \epsilon_1$, Reject if $d_{TV}(D_1, D_2) \geq \epsilon_2$

1 Divide $[n]$ into $\frac{n \log n}{m}$ consecutive blocks with each block $B_j = \{\frac{(j-1)m}{\log n} + 1, ..., \frac{jm}{\log n}\}$ set of points
2 **for** *each block* B_j; $1 \leq j \leq \frac{n \log n}{m}$ **do**
3 Sample two sets of $Poi(S)$ points from both D_1 and D_2
4 Compute Z_i and $\hat{f}_i$ for each $i \in B_j$
5 Calculate $Z = Z + \frac{Z_i}{\hat{f}_i}$
6 **if** $Z \geq \tau$ **then**
7 Reject
8 **else**
9 Accept

Theorem 6. *The algorithm Tolerant ℓ_1 Closeness uses $O[\frac{n \log n}{m}(n(\frac{\epsilon_1}{\epsilon_2^2})^2 + n(\frac{\epsilon_1}{\epsilon_2^2}) + \frac{\sqrt{n}}{\epsilon_2^2} + \frac{n^{2/3}}{\epsilon_2^{4/3}})]$ samples from each D_1 and D_2 and distinguishes between*

$d_{TV}(D_1, D_2) \leq \epsilon_1$ and $d_{TV}(D_1, D_2) \geq \epsilon_2$ with probability at least 4/5 for $0 \leq \epsilon_1 < \epsilon_2 \leq 1$. The memory requirement of the algorithm is $O(m)$ bits for $\log n \leq m \leq n \log n$.

Proof. The key concept of the algorithm is similar to that of [7] apart from the fact that we sequentialize their algorithm in order to trade-off between sample and space complexity. In the Line 5 we keep a running sum Z instead of calculating and storing it in one shot. However, Z calculates $\sum_{i=1}^{n} \frac{Z_i}{\hat{f}_i}$ after coming out of the for loop. Hence, our statistics is essentially same as that of [7] and the correctness of our algorithm directly follows from 5.

4.2 Tolerant Closeness Testing in ℓ_2 Distance

Given sample access from two unknown distributions D_1 and D_2 respectively, the task is to check if D_1 is ϵ close to D_2 in ℓ_2 distance, or $c\epsilon$ far from each other in ℓ_2 distance, for all $c \geq 3$. The bipartite collision tester defined by [19] is the main building block of our algorithm. We use the following lemma from [26] to prove the correctness of our algorithm.

Lemma 6 ([26]). *Let D be an unknown distribution over $[n]$ and S be the set of samples drawn according to SAMP . Let $I \subset [n]$ be an interval and S_I be the set of points lying in the interval I. Let S_I be divided into two subsets S_1 and S_2; $S_1 \cup S_2 = S_I$ such that $|S_1| \cdot |S_2| \geq O(|S_I|/\epsilon^4)$, then with probability at least 2/3, $||D_I||_2^2 - \frac{\epsilon^2}{64|I|} \leq \frac{coll(S_1,S_2)}{|S_1|\cdot|S_2|} \leq ||D_I||_2^2 + \frac{\epsilon^2}{64|I|}$. where $coll(S_1, S_2)$ is the number of collisions between the two set S_1 and S_2.*

Instead of considering the small intervals I, we use the entire domain $[n]$ in our case. Also, an ϵ^2-approximation of $||D||_2^2$ suffices for our analysis. Following the similar proof as that of the above lemma, to guarantee $||D||_2^2 - \epsilon^2 \leq \frac{coll(S_1,S_2)}{|S_1||S_2|} \leq ||D||_2^2 + \epsilon^2$, $|S_1|.|S_2|$ should be at least $O(\frac{S}{n^2\epsilon^4})$. Therefore, S should be at least $O(n^2)$ to see at least one collision. We present the algorithm below.

Algorithm 3: Tolerant ℓ_2 Closeness

Input : SAMP access to D_1 and D_2, error parameter $\epsilon \in (0, 1]$, $c \geq 3$, memory constraint $\log n/\epsilon^4 \leq m \leq n/\epsilon^2$

Output: Accept if $d_{\ell_2}(D_1, D_2) \leq \epsilon$, Reject if $d_{\ell_2}(D_1, D_2) \geq c\epsilon$

1. Draw $S_1 = O(\frac{m}{\log n})$ points from D_1 and store it in memory
2. Draw $S_2 = O(\frac{n^3}{\epsilon^2 m})$ points from D_1 and count $coll(S_1, S_2)$, let $X = \frac{coll(S_1,S_2)}{|S_1||S_2|}$
3. Draw $S_1' = O(\frac{m}{\log n})$ points from D_2 and store in memory
4. Draw another $S_2' = O(\frac{n^3}{m\epsilon^2})$ points from D_2 and count $coll(S_1', S_2')$ and $coll(S_1, S_2')$. Let $Y = \frac{coll(S_1,S_2')}{|S_1||S_2'|}$ and $Z = \frac{coll(S_1',S_2')}{|S_1'||S_2'|}$
5. **if** $X - 2Y + Z \geq (c^2 - 4)\epsilon^2$ **then**
6. | Reject
7. **else**
8. | Accept

Theorem 7. *The algorithm takes $O(\frac{n^3}{m\epsilon^2})$ samples for $\log n/\epsilon^4 \leq m \leq n/\epsilon^2$ and does the following,*

- *If $d_{\ell_2}(D_1, D_2) < \epsilon$, returns Accept with probability 2/3*
- *If $d_{\ell_2}(D_1, D_2) \geq c\epsilon$, returns Reject with probability 2/3 for every $c \geq 3$*

The algorithm takes $O(m)$ bits of memory.

Proof. Given the size of S_1 and S_2 in the Algorithm 3, $|S_1| + |S_2| = O(\frac{n^3}{m\epsilon^2})$ for $m \leq n/\epsilon^2$, $S \geq O(n^2)$, and we observe that $|S_1| \cdot |S_2| = \frac{n^3}{\epsilon^2 \log n} \geq O(\frac{n^3}{m\epsilon^6})$ for $m \geq \frac{\log n}{\epsilon^4}$. In other words, we can say that $|S_1| \cdot |S_2| \geq \frac{|S_1|+|S_2|}{\epsilon^4} > \frac{|S_1|+|S_2|}{n^2\epsilon^4}$. Similarly, $|S_1||S_2'| \geq \frac{|S_1|+|S_2'|}{n^2\epsilon^4}$. Hence, we have the following observations, with probability at least 2/3,

- $||D_1||_2^2 - \epsilon^2 \leq \frac{coll(S_1,S_2)}{|S_1||S_2|} \leq ||D_1||_2^2 + \epsilon^2$ or, $||D_1||_2^2 - \epsilon^2 \leq X \leq ||D_1||_2^2 + \epsilon^2$
- $\langle D_1 D_2 \rangle_2 - \epsilon^2 \leq \frac{coll(S_1,S_2')}{|S_1||S_2'|} \leq \langle D_1 D_2 \rangle_2 + \epsilon^2$ or, $\langle D_1 D_2 \rangle_2 - \epsilon^2 \leq Y \leq \langle D_1 D_2 \rangle_2 + \epsilon^2$
- $||D_2||_2^2 - \epsilon^2 \leq \frac{coll(S_1',S_2')}{|S_1'||S_2'|} \leq ||D_2||_2^2 + \epsilon^2$ or, $||D_2||_2^2 - \epsilon^2 \leq Z \leq ||D_2||_2^2 + \epsilon^2$

Where $\langle D_1 D_2 \rangle_2$ is the collision probability between D_1 and D_2. Also, notice that $d_{\ell_2}^2(D_1, D_2) = ||D_1||_2^2 - 2\langle D_1 D_2 \rangle_2 + ||D_2||_2^2$.

Completeness : Let $d_{\ell_2}(D_1, D_2) < \epsilon$. It implies $||D_1||_2^2 - 2\langle D_1 D_2 \rangle_2 + ||D_2||_2^2 < \epsilon^2$. Writing the equation in terms of X, Y, Z, we get, $X - \epsilon^2 - 2Y - 2\epsilon^2 + Z - \epsilon^2 < \epsilon^2$ or, $X - 2Y + Z < 5\epsilon^2 < (c^2 - 4)\epsilon^2$. Hence, the algorithm returns Accept for $c \geq 3$.

Soundness : Let $d_{\ell_2}(D_1, D_2) \geq c\epsilon$. It implies $||D_1||_2^2 - 2\langle D_1 D_2 \rangle_2 + ||D_2||_2^2 \geq c^2\epsilon^2$. Writing it in terms of X, Y, Z, we get, $X + \epsilon^2 - 2Y + 2\epsilon^2 + Z + \epsilon^2 \geq c^2\epsilon^2$, or, $X - 2Y + Z \geq (c^2 - 4)\epsilon^2$. Hence, the algorithm returns Reject .

We store the sets $S_1 = O(\frac{m}{\log n})$, and $S_1' = O(\frac{m}{\log n})$ in memory which needs $O(m)$ bits of storage.

5 Conclusion

We present algorithms aimed at estimating the distance to uniformity, distance to a known distribution using only constant amount of memory. For estimating distance to uniformity our algorithm achieves near optimal bound. We also discuss smooth sample-space trade-offs for tolerant testing problems considering different distance measures. However, our analysis focuses solely on SAMP access. It would be interesting to explore how sample-space trade-offs might change for estimation problems when COND access is available.

Acknowledgments. We thank the anonymous reviewers for their insightful feedback which helped us refine the ϵ factors in the sample complexities of the distance estimation algorithms presented in Sect. 3.

References

1. Acharya, J., Bhadane, S., Indyk, P., Sun, Z.: Estimating entropy of distributions in constant space. In: Wallach, H., Larochelle, H., Beygelzimer, A., d'Alché-Buc, F., Fox, E., Garnett, R. (eds.) Advances in Neural Information Processing Systems, vol. 32. Curran Associates, Inc. (2019)
2. Batu, T., Fortnow, L., Fischer, E., Kumar, R., Rubinfeld, R., White, P.: Testing random variables for independence and identity. In: FOCS '01, Proceedings of the 42Nd IEEE Symposium on Foundations of Computer Science, pp. 442–. IEEE Computer Society, Washington, DC, USA (2001)
3. Batu, T., Kumar, R., Rubinfeld, R.: Sublinear algorithms for testing monotone and unimodal distributions. In: STOC '04, Proceedings of the Thirty-sixth Annual ACM Symposium on Theory of Computing, pp. 381–390. ACM, New York, NY, USA (2004)
4. Batu, T., Fortnow, L., Rubinfeld, R., Smith, W.D., White, P.: Testing closeness of discrete distributions. J. ACM **60**(1) (2013)
5. Canonne, C.L.: Topics and techniques in distribution testing: a biased but representative sample. Found. Trends Commun. Inf. Theory **19**(6), 1032–1198 (2022). https://doi.org/10.1561/0100000114
6. Canonne, C., Diakonikolas, I., Gouleakis, T., Rubinfeld, R.: Testing shape restrictions of discrete distributions. In: Proceedings of the 33rd International Symposium on Theoretical Aspects of Computer Science (STACS 2016), pp. 1–14. Leibniz International Proceedings in Informatics (LIPIcs), Schloss Dagstuhl - Leibniz-Zentrum fuer Informatik, Germany (2016)
7. Canonne, C.L., Jain, A., Kamath, G., Li, J.: The price of tolerance in distribution testing. In: Loh, P., Raginsky, M. (eds.) Conference on Learning Theory, 2–5 July 2022, London, UK. Proceedings of Machine Learning Research, vol. 178, pp. 573–624. PMLR (2022). https://proceedings.mlr.press/v178/canonne22a.html
8. Canonne, C.L., Ron, D., Servedio, R.A.: Testing probability distributions using conditional samples. SIAM J. Comput. **44**(3), 540–616 (2015)
9. Canonne, C.L., Yang, J.Q.: Simpler distribution testing with little memory. In: 2024 Symposium on Simplicity in Algorithms (SOSA), pp. 406–416 (2024)
10. Chakraborty, S., Fischer, E., Ghosh, A., Mishra, G., Sen, S.: Exploring the gap between tolerant and non-tolerant distribution testing. In: Chakrabarti, A., Swamy, C. (eds.) Approximation, Randomization, and Combinatorial Optimization. Algorithms and Techniques (APPROX/RANDOM 2022). Leibniz International Proceedings in Informatics (LIPIcs), vol. 245, pp. 27:1–27:23. Schloss Dagstuhl – Leibniz-Zentrum für Informatik, Dagstuhl, Germany (2022). https://doi.org/10.4230/LIPIcs.APPROX/RANDOM.2022.27
11. Chakraborty, S., Fischer, E., Goldhirsh, Y., Matsliah, A.: On the power of conditional samples in distribution testing. SIAM J. Comput. **45**(4), 1261–1296 (2016)
12. Chan, S.O., Diakonikolas, I., Sun, X., Servedio, R.A.: Learning mixtures of structured distributions over discrete domains. In: Proceedings of the 2013 Annual ACM-SIAM Symposium on Discrete Algorithms (SODA), pp. 1380–1394 (2013). https://doi.org/10.1137/1.9781611973105.100
13. Chan, S.O., Diakonikolas, I., Valiant, P., Valiant, G.: Optimal algorithms for testing closeness of discrete distributions. In: Proceedings of the 2014 Annual ACM-SIAM Symposium on Discrete Algorithms (SODA), pp. 1193–1203 (2014). https://doi.org/10.1137/1.9781611973402.88

14. Chien, S., Ligett, K., McGregor, A.: Space-efficient estimation of robust statistics and distribution testing. In: Yao, A.C. (ed.) Innovations in Computer Science - ICS 2010, Tsinghua University, Beijing, China, January 5–7, 2010. Proceedings, pp. 251–265. Tsinghua University Press (2010)
15. Cribari-Neto, F., Garcia, N.L., Vasconcellos, K.L.P.: A note on inverse moments of binomial variates. Braz. Rev. Econometrics **20**(2) (2000). https://EconPapers.repec.org/RePEc:sbe:breart:v:20:y:2000:i:2:a:2760
16. Daskalakis, C., Diakonikolas, I., Servedio, R.A.: Learning k-modal distributions via testing. Theory Comput. **10**, 535–570 (2012). https://api.semanticscholar.org/CorpusID:7739848
17. Daskalakis, C., Diakonikolas, I., Servedio, R., Valiant, G., Valiant, P.: Testing k-modal distributions: optimal algorithms via reductions. In: Conference, 24th Annual ACM-SIAM Symposium on Discrete Algorithms, SODA (2013)
18. Daskalakis, C., Kamath, G., Wright, J.: Which distribution distances are sublinearly testable? CoRR arXiv:1708.00002 (2017)
19. Diakonikolas, I., Gouleakis, T., Kane, D.M., Rao, S.: Communication and memory efficient testing of discrete distributions. In: Annual Conference Computational Learning Theory (2019)
20. Diakonikolas, I., Kane, D.M., Nikishkin, V.: Optimal algorithms and lower bounds for testing closeness of structured distributions. In: 2015 IEEE 56th Annual Symposium on Foundations of Computer Science, pp. 1183–1202 (2015). https://doi.org/10.1109/FOCS.2015.76
21. Diakonikolas, I., Kane, D.M., Nikishkin, V.: Testing identity of structured distributions. In: SODA '15, Proceedings of the Twenty-Sixth Annual ACM-SIAM Symposium on Discrete Algorithms, p. 18411854. Society for Industrial and Applied Mathematics, USA (2015)
22. Fischer, E., Lachish, O., Vasudev, Y.: Improving and extending the testing of distributions for shape-restricted properties. Algorithmica **81**, 37653802 (2019)
23. Goldreich, O., Ron, D.: On testing expansion in bounded-degree graphs. Electron. Colloq. Comput. Complexity **7** (2000)
24. Paninski, L.: A coincidence-based test for uniformity given very sparsely sampled discrete data. IEEE Trans. Inf. Theor. **54**(10), 4750–4755 (2008)
25. Parnas, M., Ron, D., Rubinfeld, R.: Tolerant property testing and distance approximation. J. Comput. Syst. Sci. **72**(6), 1012–1042 (2006). https://doi.org/10.1016/J.JCSS.2006.03.002
26. Roy, S., Vasudev, Y.: Testing properties of distributions in the streaming model. In: 34th International Symposium on Algorithms and Computation, ISAAC 2023. LIPIcs, vol. 283, pp. 56:1–56:17. Schloss Dagstuhl - Leibniz-Zentrum für Informatik (2023). https://doi.org/10.4230/LIPICS.ISAAC.2023.56
27. Rubinfeld, R.: Taming big probability distributions. XRDS **19**(1), 2428 (2012). https://doi.org/10.1145/2331042.2331052
28. Valiant, G., Valiant, P.: A CLT and tight lower bounds for estimating entropy. Electron. Colloq. Comput. Complexity (ECCC) **17**, 183 (2010)

29. Valiant, G., Valiant, P.: The power of linear estimators. In: FOCS '11, Proceedings of the 2011 IEEE 52Nd Annual Symposium on Foundations of Computer Science, pp. 403–412. IEEE Computer Society, Washington, DC, USA (2011)
30. Valiant, G., Valiant, P.: An automatic inequality prover and instance optimal identity testing. In: FOCS '14, Proceedings of the 2014 IEEE 55th Annual Symposium on Foundations of Computer Science, pp. 51–60. IEEE Computer Society, Washington, DC, USA (2014)
31. Valiant, P.: Testing symmetric properties of distributions. SIAM J. Comput. **40**(6), 1927–1968 (2011)

Packed Acyclic Deterministic Finite Automata

Hiroki Shibata[1]( ), Masakazu Ishihata[2], and Shunsuke Inenaga[3]

[1] Joint Graduate School of Mathematics for Innovation, Kyushu University, Fukuoka, Japan
shibata.hiroki.753@s.kyushu-u.ac.jp
[2] NTT Communication Science Laboratories, Tokyo, Japan
[3] Department of Informatics, Kyushu University, Fukuoka, Japan

Abstract. An acyclic deterministic finite automaton (ADFA) is a data structure that represents a set of strings (i.e., a dictionary) and facilitates a pattern-searching problem of determining whether a given pattern string is present in the dictionary. We introduce the *packed ADFA* (PADFA), a compact variant of ADFA designed to achieve more efficient pattern searching by encoding specific paths as packed strings stored in contiguous memory. We theoretically demonstrate that pattern searching in PADFA is near time-optimal with a small additional overhead and becomes fully time-optimal for sufficiently long patterns. Moreover, we prove that a PADFA requires fewer bits than a trie when the dictionary size is relatively smaller than the number of states in the PADFA. Lastly, we empirically show that PADFAs improve both the space and time efficiency of pattern searching on real-world datasets.

Keywords: pattern searching · acyclic DFA · packed string

1 Introduction

Text indexing is a central problem in string processing with many real-world applications, including information retrieval, natural language processing, and bioinformatics. In this paper, we focus on the *pattern-searching problem*, which involves preprocessing a given *dictionary* (a set of distinct strings) into an indexing structure and checking whether a pattern string is contained within the dictionary. An *acyclic deterministic finite automaton* (ADFA) [4,10], also known as a *deterministic acyclic finite state automaton* (DAFSA), is a fundamental indexing structure for pattern searching. A *trie* [17] for a dictionary $\mathcal{S}$ of k strings is an ADFA that forms a rooted tree, and the k paths from its root to an accepting state correspond to strings in $\mathcal{S}$. A minimal ADFA (minADFA) for the same $\mathcal{S}$ can be obtained by merging all isomorphic subtrees of the trie. Given a pattern string P of length m, both the trie and the minADFA allow us to determine $P \in \mathcal{S}$ in $\mathcal{O}(m \log \sigma)$ time, where σ represents the alphabet size, while the minADFA is smaller than the trie. However, in practice, tries are considered to operate more efficiently than minADFAs. This is because a trie has a simple tree structure, whereas the minADFA is a directed acyclic graph (DAG), which introduces additional processing overhead. Therefore, determining which structure offers superior memory efficiency in practical usage remains an open question.

R. Královič and V. Kůrková (Eds.): SOFSEM 2025, LNCS 15539, pp. 284–297, 2025.
https://doi.org/10.1007/978-3-031-82697-9_21

Modern computers can process a single word of data in one operation. Therefore, when the alphabet size σ is sufficiently smaller than the length of a machine word, computers can process strings more efficiently by packing multiple characters into a single machine word and processing them in a single operation. Whereas an ordinary string stores each character in a separate machine word, a *packed string* stores characters in consecutive memory locations. This packing technique enhances both the space and time efficiency of text processing. For instance, comparing two packed strings is α times faster than comparing ordinary strings, where α is the number of characters packed into a single word. Consequently, the optimal time complexity for pattern searching using packed strings is $\mathcal{O}(m/\alpha)$. Several studies have investigated accelerating pattern searching in tries using packed strings [22,23]; however, no previous work has applied this technique to DFAs with a theoretical evaluation. This research gap arises primarily due to the structural incompatibility between packing techniques and DAG structures.

In this paper, we introduce a *packed ADFA* (PADFA), the first approach to apply the packing technique to ADFA. Given an ADFA $\mathcal{A}$ accepting $\mathcal{S}$, our proposed method extracts some specific paths, called *heavy paths*, from $\mathcal{A}$ by *symmetric centroid path decomposition* (SymCPD). Then, the method compiles them into a single packed string and maintains the remaining edges as *biased search tree* (BST). Using the obtained packed string and BST, the method performs the pattern searching in $\mathcal{O}(m/\alpha + \log k)$ time. We theoretically show that a PADFA for any ADFA achieves the time-optimal pattern searching, i.e., $\mathcal{O}(m/\alpha)$, if m is sufficiently long compared to k. Additionally, we demonstrate that a PADFA for any minADFA consumes fewer bits of memory than a trie if it has a sufficiently large number of states compared to k. PADFAs are useful not only for pattern searching but also for pattern matching, a task finding a pattern string from a text string. A *directed acyclic word graph* (DAWG) [8], an ADFA representing all suffixes of the text, is an indexing structure for pattern matching, and we can obtain a packed DAWG in the same manner as a general PADFA. The packed DAWG provides the time-optimal pattern matching when m is sufficiently long. We also conducted experiments with real-world datasets. The empirical results indicate that PADFA improves both space and time efficiencies of pattern searching.

1.1 Contribution

We here organize our theoretical contributions. Let $\mathcal{S}$ be a set of k distinct strings with the alphabet size σ, and A be the packed ADFA, our proposed indexing structure, obtained from an ADFA representing $\mathcal{S}$ with n states. We assume the word RAM model with machine words of length ω and define $\alpha \triangleq \omega/\lceil\log_2 \sigma\rceil$. Then, the pattern-searching problem on A for a given pattern string P of length m is solved in the following time and space complexity.

Theorem 1. *The time complexity of pattern searching on a PADFA for any ADFA is* $\mathcal{O}(m/\alpha + \log k)$.

Theorem 2. *The space consumption of a PADFA for any minADFA is* $n(1 + \lceil\log_2 \sigma\rceil) + \mathcal{O}(k(\log n + \log \sigma)) + o(n)$ *bits.*

Thus, A generally has an additive overhead of $\log k$ for the time-optimal pattern searching $\mathcal{O}(m/\alpha)$. However, by introducing an assumption that m is sufficiently long, the time complexity can be improved as follows.

Corollary 1. *Given PADFA* A *for any ADFA and any query pattern of length* $m \in \Omega(\alpha \log k)$, *the pattern searching on* A *takes* $\mathcal{O}(m/\alpha)$ *time, which is optimal for pattern searching with packed strings.*

Additionally, we can improve the space complexity by assuming an appropriate condition that k is relatively smaller than n as follows.

Corollary 2. *Given PADFA* A *of any minADFA satisfying* $\max\{k \log n, k \log \sigma\} \in o(n)$, A *consumes* $n(1 + \lceil \log_2 \sigma \rceil) + o(n)$ *bits of space.*

In other words, A consumes fewer bits of memory than the trie because a trie of n vertices requires almost $n(2+\log_2 \sigma)$ bits, which is at most $n(1+\lceil \log_2 \sigma \rceil)$ bits. We believe that the above two conditions are sufficiently realistic. This is because situations where more efficient pattern searching is desired typically involve handling large dictionaries and long patterns. In such cases, we expect that the conditions will generally be met. Thus, the above theoretical results highlight the advantage of PADFAs over tries.

PADFAs are useful not only for pattern searching but also for substring pattern matching, determining whether a pattern string P of length m occurs in a text string T of length n. It is known that a DAWG of T, which is an ADFA representing all $k = n + 1$ suffixes of T (including the empty suffix), has only $\mathcal{O}(n)$ vertices and edges. We can construct packed DAWG in the same manner as general PADFAs and derive the following corollary.

Corollary 3. *For any string T of length n and any pattern string P of length m, the packed DAWG for T consumes $\mathcal{O}(n(\log n + \log \sigma))$ bits of space and provides substring pattern matching in $\mathcal{O}(m/\alpha + \log n)$ time.*

Moreover, using Corollary 1, we can derive that the packed DAWG achieves the optimal time complexity $\mathcal{O}(m/\alpha)$ when m is sufficiently long.

1.2 Related Work

In the context of text indexing using graph structures, treating specific paths as strings stored in contiguous memory is a crucial technique for achieving efficient pattern searching. *Compact tries* (also known as *patricia tries*) [18] and *minimal prefix tries* [3] handle unary paths in a trie as single edges labeled by strings These techniques reduce memory consumption and enhance search efficiency, and they can also be applied to ADFA [14].

The *heavy path decomposition* (also known as *centroid path decomposition*) [21] is a powerful tool for performing efficient queries on trees by extracting specific paths from the tree. It handles more edges within packed strings than other methods that optimize only unary paths, while it offers theoretical guarantees on query time. Some studies have adapted this technique to achieve efficient tries [11,16].

Several variants of heavy path decomposition for general DAGs have recently been proposed [7, 15]. This technique has been applied to various areas, including efficient random access for grammars [7], and accelerating query processing for *compacted directed acyclic word graphs* (CDAWGs) [6]. However, no research has yet applied heavy path decomposition to ADFAs or utilized it to accelerate pattern searching.

2 Preliminaries

We start by defining the pattern-searching problem that this paper addresses. After that, we describe the definition of ADFAs and present several techniques that serve as crucial components of PADFAs proposed in Sect. 3.

2.1 The Pattern-Searching Problem in the Word RAM Model

An alphabet Σ is an ordered set of σ distinct characters. A *string* is a finite sequence of characters drawn from Σ. For any string S, $|S|$ represents its length, and $S[i]$ denotes its ith character, where $1 \leq i \leq |S|$. The empty string has zero length and is denoted by ε, i.e., $|\varepsilon| = 0$. Additionally, let \$ and $\#$ be special characters:\$ can appear only at the end of non-empty strings, indicating the end of a string, while $\#$ never appears in the input strings and is used solely for our algorithm. For string $S = xyz$ $(x, y, z \in \Sigma^*)$, the strings x, y, and z are referred to as a *prefix*, *substring*, and *suffix* of S, respectively. For $1 \leq i \leq j \leq |S|$, $S[i..j] \triangleq S[i] \cdots S[j]$ denotes the substring of S that starts at position i and ends at position j. For convenience, we define $S[i..j] \triangleq \varepsilon$ if $j < i$. The length of the *longest common prefix* (LCP) of two strings S and T is denoted by LCP$(S, T) \triangleq \max(\{0\} \cup \{i \mid S[1..i] = T[1..i]\})$.

Let $\mathcal{S} \triangleq \{S_1, \ldots, S_k\}$ be a set of k distinct strings, called a *dictionary*. The *pattern-searching problem* involves evaluating whether a given pattern string P is an element of the dictionary $\mathcal{S}$, i.e., $P \in \mathcal{S}$. We consider the indexing version of the above problem. In other words, $\mathcal{S}$ is given in advance, and the goal is to preprocess $\mathcal{S}$ into an appropriate indexing structure for efficiently executing the pattern-searching queries for various pattern strings provided online.

This paper focuses on the pattern-searching problem in the *word RAM model* [13] with a machine word length of ω. The model provides several constant-time instructions, including random access, word-wise logical and arithmetic operations, and word-wise comparison which returns the position of the first miss-matched bit. Let $\ell \triangleq \max_{S \in \mathcal{S}} |S|$ be the length of the longest string in the dictionary $\mathcal{S}$. We assume that $\sigma, k, \ell < 2^\omega$; namely, each character and its position in each string can be represented by a single word. Define $\alpha \triangleq \omega / \lceil \log_2 \sigma \rceil$. Each character is then represented by $\lceil \log_2 \sigma \rceil$ bits, and a string of length n is represented using $n \lceil \log_2 \sigma \rceil$ bits, stored in $\lceil n/\alpha \rceil$ contiguous words. Strings stored in contiguous memory are called *packed strings*. Using the constant-time word-wise comparison of the word RAM model, we can obtain LCP(X, Y) for any packed strings X and Y in $\mathcal{O}(|\text{LCP}(X, Y)|/\alpha)$ time.

Algorithm 1 Determining whether $P \in \mathcal{S}$ in $\mathcal{A}$ accepting $\mathcal{S}$

```
1: v_0 ← r
2: for i in 1, ..., m do
3:     v_i ← δ(v_{i−1}, P[i])          ▷ move along one edge
4:     return false if v_i = ⊥
5: return v_m ∈ F
```

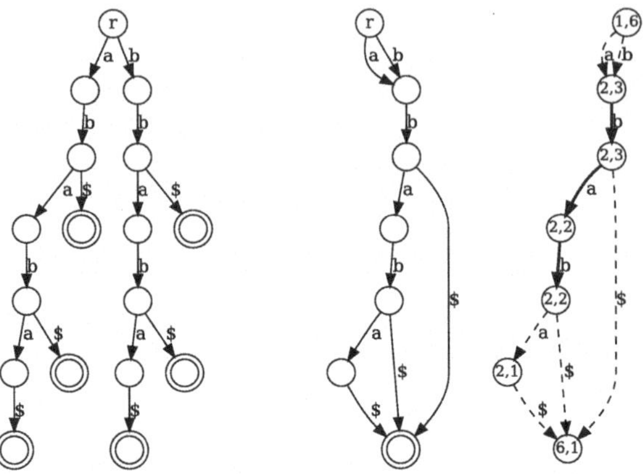

Fig. 1. The trie (left), the minADFA (center), and its SymCDP (right), for a dictionary $\mathcal{S} = \{\text{ab\$}, \text{abab\$}, \text{ababa\$}, \text{bb\$}, \text{bbab\$}, \text{bbaba\$}\}$. For the trie and minADFA, a vertex labeled r represents a start state, and double circles represent accepting states. For SymCPD, bold and dashed edges represent heavy and light edges, respectively. Each vertex is labeled by the two integers indicating $\pi(r, v)$ and $\pi(v, W)$.

2.2 Acyclic Deterministic Finite Automatons (ADFAs)

A finite automaton (FA) is a tuple $\mathcal{A} \triangleq \langle V, E, F, r \rangle$, where V is a set of vertices (states), $E \subseteq V \times V \times \Sigma$ is a set of directed labeled edges (transitions), $F \subseteq V$ is the set of accepting states, and $r \in V$ is the initial state. An FA is *deterministic* if the out-edges of the same vertex are labeled with distinct characters. A deterministic FA (DFA) is *complete* if each vertex has the out-edge labeled by each character from Σ, and is *partial* otherwise [8]. A complete DFA may contain a vertex with no directed path to any accepting state, whereas an equivalent partial DFA can be obtained by removing such redundant states. For any partial DFA $\mathcal{A}$, let $\delta : V \times \Sigma \to V \cup \{\bot\}$ be its transition function such that $\delta(u, c) \triangleq v$ if $(u, v, c) \in E$ and $\delta(u, c) \triangleq \bot$ otherwise; namely, $\mathcal{A}$ immediately halts when it reaches $\bot$. Throughout this paper, we assume that any given DFA is partial and contains no redundant state. For any $v \in V$, let $d_{\text{in}}(v)$ and $d_{\text{out}}(u)$ be the in and out degree of v, respectively. v is *source* if $d_{\text{in}}(v) = 0$, and u is *sink* if $d_{\text{out}}(u) = 0$. A DFA is *acyclic* if it has no cycles. For any $U \subseteq V$, define $d_{\text{in}}(U) = \sum_{u \in U} d_{\text{in}}(u)$ and $d_{\text{out}}(U) = \sum_{u \in U} d_{\text{out}}(u)$. Without loss of generality, we assume that an acyclic DFA (ADFA) has a unique source, which is the initial state r.

An ADFA $\mathcal{A}$ *accepts* a string P of length m iff there exists a sequence of states $(v_0, \ldots, v_m)$ such that $v_0 = r$, $v_m \in F$, and $\delta(v_{i-1}, P[i]) = v_i$ for all $1 \leq i \leq m$. Let

$\mathcal{S}$ be the set of all strings accepted by $\mathcal{A}$, and then, $\mathcal{S}$ is finite since $\mathcal{A}$ is acyclic and $|V|$ is finite. Algorithm 1 shows the pattern-searching process in $\mathcal{A}$ and takes $O(m \log \sigma)$ time, independently of $|V|$. A *trie* for $\mathcal{S}$ is a tree-formed ADFA accepting $\mathcal{S}$, where its accepting states correspond to its leaves when every string in $\mathcal{S}$ ends with $\$$. Let $\mathcal{A}$ be the ADFA obtained by merging isomorphic subtrees of the trie. Then, $\mathcal{A}$ is minimal and has exactly one accepting state, which is its unique sink. Figure 1 illustrates an example of a trie and its corresponding minimal ADFA (minADFA).

2.3 Building Blocks of PADFAs

We here introduce three techniques, a *symmetric centroid path decomposition* (SymCPD) [15], a *biased search tree* (BST) [5], and a *fully indexable dictionary* (FID) [20], that are employed to implement our PADFA.

SymCPD is a technique for decomposing a DAG into disjoint paths, serving as a generalization of the well-known *heavy path decomposition* [21] for a tree. Consider a DAG with a vertex set V and a labeled edge set E, assuming a unique source $r \in V$ and a set of sinks $W \subseteq V$. For any $u, v \in V$, let $\pi(v, u)$ denote the number of directed paths from v to u, where $\pi(v, v) \triangleq 1$. For any $U \subseteq V$, define $\pi(v, U) \triangleq \sum_{u \in U} \pi(v, u)$. Additionally, let $\lambda(v) \triangleq (\lfloor \log_2 \pi(r, v) \rfloor, \lfloor \log_2 \pi(v, W) \rfloor)$. SymCPD then divides E into two disjoint sets H and L, where $H \triangleq \{(u, v, c) \in E \mid \lambda(u) = \lambda(v)\}$ is the set of *heavy edges*, and $L \triangleq E \setminus H$ is the set of *light edges*. Using these sets, H and L, we can derive the following useful properties [15, Lemma 2.1].

Property 1. The edge-induced subgraph $\langle V, H \rangle$ forms a set of disjoint paths.

Property 2. Any path on the DAG contains at most $2 \log_2 \lfloor \pi(r, W) \rfloor$ light edges.

The right figure of Fig. 1 gives an example of SymCPD.

A BST is a data structure storing a subset of an ordered set and various operations, including access, insert, delete, and more. Consider an ordered universe $\Sigma = \{c_1, \ldots, c_\sigma\}$ of σ items with the total order $c_1 < \cdots < c_\sigma$. Let $w : \Sigma \to \mathbb{N}_+$ be a weight function over Σ, and define $w(C) = \sum_{c \in C} w(c)$ for any $C \subseteq \Sigma$. We denote by $\mathcal{B}(C, w)$ a BST storing C. The access operation in $\mathcal{B}(C, w)$ with a query item c, denoted by access$(\mathcal{B}(C, w), c)$, returns the position index of c if $c \in C$, and NULL otherwise. $\mathcal{B}(C, w)$ is implemented as a biased binary search tree, with the following space complexity and time complexity for an access operation.

Property 3. A succinct representation of $\mathcal{B}(C, w)$ consumes $(2 + \lfloor \log \sigma \rfloor)|C| + o(|C|)$ bits [19].

Property 4. An access operation access$(\mathcal{B}(C, w), c)$ takes $\mathcal{O}(1 + \log(w(C)/w(c)))$ time if $c \in C$, and $\mathcal{O}(\log w(C))$ time otherwise [5].

The factor of 1 that appears in the above time complexity is introduced to avoid the case where $w(C) = w(c)$ results in $\log(w(C)/w(c)) = 0$, which occurs when $|C| = 1$. In general, an access operation in a balanced binary search tree requires $\mathcal{O}(\log |C|)$ time. Consequently, by introducing an appropriate weight function w, a BST can reduce the average access time for query items provided online.

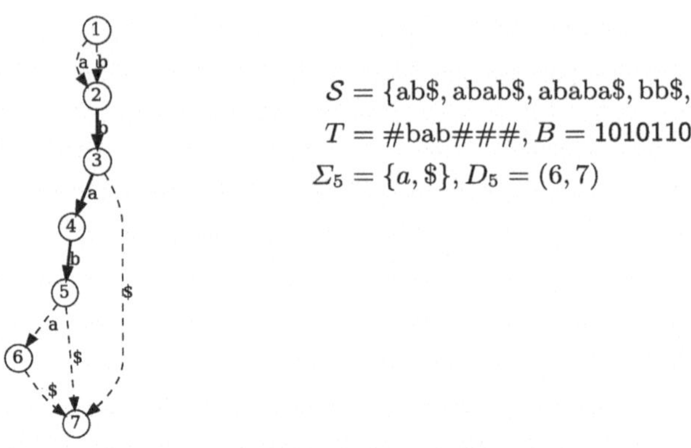

Fig. 2. The minADFA $\mathcal{A}$ accepting $\mathcal{S}$ (left), and $\mathcal{S}, T, B, \Sigma_5$ and D_5 of the PADFA A representing $\mathcal{A}$ (top-right). A has four biased search trees corresponding to the vertices $1, 3, 5$, and 6. The vertex 5 has two light edges $(5, 6, \text{a})$ and $(5, 7, \$)$.

A FID is a succinct data structure representing a bit string, a sequence of 0 s and 1 s, providing the rank and select operations. Let $\mathcal{D}$ be a FID representing a bit string B of length n. Given any $1 \leq i \leq n$, a rank operation computes the number of 1 s in the prefix $B[1, i]$, and a select operation returns the position of the ith 1 in B. The space and time complexity of a FID is as follows.

Property 5. $\mathcal{D}$ consumes $n + o(n)$ bits and performs constant-time rank and select operations.

3 Packed ADFAs (PADFAs)

A PADFA is an implementation of an ADFA that accelerates pattern searching using packed strings. We begin by defining a PADFA and explaining its pattern-searching process. Then, we theoretically analyze the time and space complexities of PADFAs.

3.1 The Definition of PADFAs

Consider a dictionary $\mathcal{S}$ of k distinct strings, where each string in $\mathcal{S}$ is assumed to end with a special symbol \$. Given an ADFA $\mathcal{A}$ accepting $\mathcal{S}$ with n states, let H and L be the set of heavy and light edges obtained by applying SymCPD to $\mathcal{A}$. Thus, $\mathcal{A}$ has at most k sinks, denoted by $W \subseteq V$. Here, we represent each vertex in V as an integer in $\{1, \ldots, n\}$ such that $r = 1$ and $v = u + 1$ for any $(u, v, c) \in H$. Such a vertex ordering always exists since H forms disjoint paths. For any $v \in V$, let L_v denote the light edges out-going from v. Given L_v, define $D_v \triangleq \{u \mid (v, u, c) \in L_v\}$ as its destination list, and $\Sigma_v \triangleq \{c \mid (v, u, c) \in L_v\}$ as its label list. Additionally, let B be the binary string of length n such that $B[v] = 1$ iff $L_v \neq \emptyset$, and let $n_L \triangleq \sum_{v \in V} B[v]$. In other words, L is factorized into n_L disjoint edgesets $\{L_v \mid B[v] = 1\}$.

A PADFA represents H as a single packed string and L as an array of BSTs with some ancillary data structures. Let T be a packed string such that $T[v] = c$ if $(v, v +$

Algorithm 2 Determining whether $P \in \mathcal{S}$ in A accepting $\mathcal{S}$

```
1:  v ← 1, j ← 1
2:  while j ≤ m do
3:      l ← LCP(T[v..|T|], P[j..m])
4:      v ← v + l, j ← j + l                ▷ move along j heavy edges
5:      break if j > m
6:      return false if B[v] = 0
7:      p ← access(𝓑(v), P[j])
8:      return false if p = NULL.
9:      v ← D_v[p], j ← j + 1                ▷ move along one light edge
10: return true
```

$1, c) \in H$ and # otherwise; in other words, all heavy edges are packed together in T. Define $w_v : \Sigma \to \mathbb{N}_+$ such that $w_v(c) = \pi(u, W)$ if $(v, u, c) \in L_v$ and $w_v(c) = \infty$ otherwise. Additionally, let $\mathcal{B}(v) \triangleq \mathcal{B}(\Sigma_v, w_v)$ be a BST with universe Σ. Then, L_v is represented by a pair of $\mathcal{B}(v)$ and D_v because the destination of the light edge labeled by c in L_v is stored in $D_v[i]$, where $i = \text{access}(\mathcal{B}(v), c)$. Define $\mathcal{B} \triangleq \{\mathcal{B}(v) \mid B[v] = 1\}$ and $D \triangleq \{D_v \mid B[v] = 1\}$. Also, let $\mathcal{D}$ be a FID of B. In summary, the PADFA A of the given $\mathcal{A}$ is the quadruplet of the packed string T, the BSTs $\mathcal{B}$, the destination lists D, and the FID $\mathcal{D}$. Figure 2 gives an example of minADFA and some elements of PADFA.

Algorithm 2 presents a pattern-searching process on A with a query string P. In each iteration, $j - 1$ indicates the number of characters already matched. First, the algorithm computes $l = \text{LCP}(T[v..|T|], P[j..m])$ using the packed technique, where l represents the number of characters that can be traversed along the heavy edges. After traversing along the heavy edges as long as possible, it checks whether v has a light edge $(v, u, P[j])$ using B and $p = \text{access}(\mathcal{B}(v), P[j])$. If such a light edge exists, it retrieves u from $D_v[p]$. The process repeats until a termination condition is met.

3.2 The Time and Space Complexities of PADFAs

Firstly, we discuss the time complexity of Algorithm 2. Let $\mathcal{A}$ denote any ADFA accepting $\mathcal{S}$, and let A denote the PADFA of $\mathcal{A}$. Note that $\mathcal{A}$ does NOT need to be minimal. Define I be the number of iterations of the algorithm. Let v_i, j_i, and l_i be the value of v, j, and l after 4, and p_i be the value of p after 7 of the ith iteration for any $1 \leq i \leq I$. Additionaly, define $c_i = P[j_i]$ and $u_i = D_{v_i}(p_i)$. Then, the upper bound of I is given as follows.

Lemma 1. *The number of iterations I in Algorithm 2 is at most* $1 + 2\lfloor \log_2 k \rfloor$.

Proof. The algorithm alternates between traversing several consecutive heavy edges and a single light edge, so the number of iterations is, at most, the number of light edges in one path plus one. Since $\pi(r, W) = k$ stands, any path in $\mathcal{A}$ contains at most $2\lfloor \log_2 k \rfloor$ light edges by Property 2. Consequently, $I \leq 1 + 2\lfloor \log_2 k \rfloor$. □

Next, we introduce two lemmas for traversing heavy edges and light edges.

Lemma 2. *The total time complexity of traversing heavy edges in Algorithm 2 is at most* $\mathcal{O}(m/\alpha + \log k)$.

Proof. Since the algorithm traverses at most m heavy edges, it follows that $\sum_{i=1}^{I} l_i \leq m$ stands. Given that the time complexity of LCP(S_1, S_2) is $\mathcal{O}(\lceil \text{LCP}(S_1, S_2)/\alpha \rceil)$, the total complexity obtaining all l_i is $\mathcal{O}(I + m/\alpha)$, as derived below.

$$\sum_{i=1}^{I} \left\lceil \frac{l_i}{\alpha} \right\rceil \leq I + \sum_{i=1}^{I} \frac{l_i}{\alpha} = I + \frac{m}{\alpha}$$

Since traversing l_i heavy edges can be done in constant time through a simple addition, as shown in 4, the total time complexity for traversing heavy edges is $\mathcal{O}(m/\alpha + \log k)$. □

Lemma 3. *The total time complexity of traversing light edges in Algorithm 2 is at most* $\mathcal{O}(\log k)$.

Proof. Because traversing a light edge requires an access operation access$(\mathcal{B}(v_i), c_i)$, the total cost for traversing light edges reaches its maximum when I access operations are executed, with the final access operation returning NULL. Since v_{i+1} must be a descendants of u_i, $\pi(v_{i+1}, W) \leq \pi(u_i, W)$ holds for all i. First, a look-up time of $\mathcal{B}(v_i)$ from $\mathcal{B}$ is constant because the FID $\mathcal{D}$ of B provides a constant-time select operation. By substituting the definition of w_{v_i} to Property 4, the time complexity of access$(\mathcal{B}(v_i), c_i)$ becomes $\mathcal{O}(1 + \log(\pi(v_i, W)/\pi(u_i, W)))$ if p_i is not NULL, and $\mathcal{O}(\log \pi(v_i, W))$ otherwise. Then, the total cost of I access operations is computed as follows.

$$\sum_{i=1}^{I-1} \left(1 + \log \frac{\pi(v_{i-1}, W)}{\pi(u_{i-1}, W)}\right) + \log \pi(v_m, W) = I - 1 + \log \frac{\prod_{i=1}^{I} \pi(v_i, W)}{\prod_{i=1}^{I-1} \pi(u_i, W)}$$
$$\leq I + \log \frac{\prod_{i=1}^{I} \pi(v_i, W)}{\prod_{i=2}^{I} \pi(v_i, W)} = I + \log \pi(v_1, W)$$

By combining facts that $\pi(v_1, W) \leq k$ and $I \leq 1 + 2\lfloor \log_2 k \rfloor$, the total access time is at most $\mathcal{O}(\log k)$. Lastly, the traversing time of a light edge is constant because its destination u_i is obtained by accessing $D_v[p_i]$. Consequently, the total time complexity of traversing light edges is at most $\mathcal{O}(\log k)$. □

From the two lemmas above, we immediately obtain the following theorem.

Theorem 1. *The time complexity of pattern searching on a PADFA for any ADFA is* $\mathcal{O}(m/\alpha + \log k)$.

The above theorem shows A demonstrates pattern searching in near-optimal time, with an overhead of $\mathcal{O}(\log k)$. Assuming $m \in \Omega(\alpha \log k)$, we obtain the following corollary.

Corollary 1. *Given PADFA* A *for any ADFA and any query pattern of length* $m \in \Omega(\alpha \log k)$, *the pattern searching on* A *takes* $\mathcal{O}(m/\alpha)$ *time, which is optimal for pattern searching with packed strings.*

Second, we discuss the space complexity of the PADFA for a minADFA. Let $\mathcal{A}$ denote the minADFA accepting $\mathcal{S}$, and let A denote the PADFA of $\mathcal{A}$. First, we analyze the number of light edges L in A. Define $V^{\text{out}}_{=1} \triangleq \{v \in V \mid d_{\text{out}}(v) = 1\}$ and $V^{\text{out}}_{\geq 2} \triangleq \{v \in V \mid d_{\text{out}}(v) \geq 2\}$. In the same manner, we introduce $V^{\text{in}}_{=1}$ and $V^{\text{in}}_{\geq 2}$.

Lemma 4. *For any minADFA, $d_{\text{out}}(V^{\text{out}}_{\geq 2})$ and $d_{\text{in}}(V^{\text{in}}_{\geq 2})$ are upper-bounded by $2k$.*

Proof. Assume a given ADFA $\mathcal{A}$ forms a trie with vertices V, edges E, and sinks W. For simplicity, define $n = |V|$, $m = |E|$, $n_1 = |V^{\text{out}}_{=1}|$, $n_2 = |V^{\text{out}}_{\geq 2}|$, and $d = d_{\text{out}}(V^{\text{out}}_{\geq 2})$. Then, by definition, $n = m+1$, $|W| = k$, $n = k+n_1+n_2$, and $m = n_1+d$ hold. By combining these facts, we derive $d - n_2 = k - 1$. Using the fact that $d \geq 2n_2$, $n_2 \leq k - 1$ holds. Now, assume $d \geq 2k$. Then, $n_2 \geq k + 1$ would hold, which contradicts the earlier fact that $n_2 \leq k - 1$. Therefore, $d < 2k$ stands for trie-formed ADFAs. Since a minADFA is obtained by merging some isomorphic subtrees of the trie-formed ADFAs, we conclude that $d_{\text{out}}(V^{\text{out}}_{\geq 2}) < 2k$ also holds for any minADFA $\mathcal{A}$. Similarly, $d_{\text{in}}(V^{\text{in}}_{\geq 2}) < 2k$ can also be proven in the same manner. □

From the above lemma, we derive the following upper bound on $|L|$.

Lemma 5. *For any minADFA, the number of light edges $|L|$ is $\mathcal{O}(k)$.*

Proof. By the definition of SymCPD, any edge that does not share its starting or ending points with any other edge must be categorized as a heavy edge. Thus, $|L|$ is bounded by the number of edges that share at least its starting or ending vertex with others. Consequently, $|L| \leq d_{\text{out}}(V^{\text{out}}_{\geq 2}) + d_{\text{in}}(V^{\text{in}}_{\geq 2}) \leq 4k$ holds by Lemma 4. □

Finally, we obtain the following theorem.

Theorem 2. *The space consumption of a PADFA for any minADFA is $n(1+\lceil\log_2 \sigma\rceil)+\mathcal{O}(k(\log n + \log \sigma)) + o(n)$ bits.*

Proof. The PADFA A consists of four data structures: the packed string T, the BSTs $\mathcal{B}$, the destination lists D, and the FID $\mathcal{D}$. The packed string T consumes $n\lceil\log_2 \sigma\rceil$ bits. The BSTs $\mathcal{B}$ consumes $\mathcal{O}(k \log \sigma)$ bits because a single BST $\mathcal{B}(v)$ consumes $\mathcal{O}(|\Sigma_v| \log \sigma)$ bits by Property 3, and $\sum_v |\Sigma_v| = |L|$, which is upper bounded by $\mathcal{O}(k)$. The destination lists D stores all destinations of L and requires $\mathcal{O}(k \log n)$ bits. According to Property 5, the FID $\mathcal{D}$ consumes $n + o(n)$ bits. By summing up all space complexities, the theorem follows. □

If k is relatively small compared to n, the PADFA A can be represented in a more compact space, as shown below.

Corollary 2. *Given PADFA A of any minADFA satisfying $\max\{k \log n, k \log \sigma\} \in o(n)$, A consumes $n(1 + \lceil\log_2 \sigma\rceil) + o(n)$ bits of space.*

This corollary demonstrates the advantage of PADFA over trie. The information-theoretic lower bound of a trie of n' vertices is $n'(2 + \log_2 \sigma)$ bits, which is greater that $n'(1 + \lceil\log_2 \sigma\rceil)$ bits. Since n is typically smaller than n', A consumes less memory than the trie, even though A forms a DAG. Furthermore, FID $\mathcal{D}$ representing B

Table 1. The characteristics for each dataset and the size of tries and ADFA.

	dictionary				Trie $\mathcal{A}_{\text{trie}}$		PADFA $\mathbf{A}_{\text{min}}$	
	σ	k	total len.	ave. len.	$\|V\|$	$\|E\|$	$\|V\|$	$\|E\|$
url	93	862,665	72,540,387	84.089	10,146,553	10,146,552	1,612,336	2,040,555
city	78	177,030	1,970,082	11.183	846,550	846,549	198,195	333,800
prot	25	157,237	46,687,247	295.046	35,028,185	35,028,184	32,905,500	33,030,196

can be compressed using the zeroth-order empirical entropy of each string while still supporting constant-time operations [20]. Since the number of 1 s in B is $\mathcal{O}(k)$, this representation achieves significant compression when k is sufficiently small with respect to n. Then, the space complexity of A can be regarded as $n\lceil\log_2 \sigma\rceil + o(n)$ bits.

Lastly, we show an application of our PADFA for substring pattern matching, determining whether a pattern string P of length m occurs in a text string T of length n. A DAWG of T is an ADFA representing all $n+1$ suffixes of T and consumes $\mathcal{O}(n)$ vertices and edges [8]. We can obtain a packed DAWG in the same manner as a general PADFA and the following corollary by directly applying Theorem 1 and 2 to DAWGs.

Corollary 3. *For any string T of length n and any pattern string P of length m, the packed DAWG for T consumes $\mathcal{O}(n(\log n + \log \sigma))$ bits of space and provides substring pattern matching in $\mathcal{O}(m/\alpha + \log n)$ time.*

In addition, according to Corollary 1, the packed DAWG achieves the optimal time complexity $\mathcal{O}(m/\alpha)$ when m is sufficiently long.

4 Experiments

We implemented various ADFAs and applied them to multiple real-world datasets to demonstrate that PADFAs achieve better space and time efficiency compared to ADFAs that do not utilize packed strings. We begin by describing our experimental settings, followed by a presentation of the experimental results.

4.1 Experimental Settings

We used three real-world datasets, url, city, and prot as input dictionaries. url consists of URLs from a crawl of the `.eu` domain conducted in 2005 [2,9]. city is a list of cities with a population of 500 or more dumped by GeoNames [1]. prot contains the first 50 MiB of protein sequences downloaded from the Pizza&Chilli Corpus [12]. In all dictionaries, each character was represented using one byte (8 bits). Table 1 summarizes characteristics of these dictionaries.

We implemented five types of ADFAs: $\mathcal{A}_{\text{trie}}$, $\mathbf{A}_{\text{pref}}$, $\mathbf{A}_{\text{path}}$, $\mathcal{A}_{\text{min}}$ and $\mathbf{A}_{\text{min}}$. $\mathcal{A}_{\text{trie}}$ is a simple trie. $\mathbf{A}_{\text{pref}}$ is a *minimal prefix trie* [3] that stores only the minimal prefixes needed to identify each string and represents the remaining suffixes as packed strings. $\mathbf{A}_{\text{path}}$ is a *path-decomposed trie* [11] that stores heavy paths of $\mathcal{A}_{\text{trie}}$ as packed strings, which is essentially equivalent to the PADFA for a simple trie. $\mathcal{A}_{\text{min}}$ is the minADFA

Table 2. The memory consumption and the computing times of ADFAs.

	Memory [MiB]					Time [ms]				
	$\mathcal{A}_{\mathrm{trie}}$	$\mathbf{A}_{\mathrm{pref}}$	$\mathbf{A}_{\mathrm{path}}$	$\mathcal{A}_{\mathrm{min}}$	$\mathbf{A}_{\mathrm{min}}$	$\mathcal{A}_{\mathrm{trie}}$	$\mathbf{A}_{\mathrm{pref}}$	$\mathbf{A}_{\mathrm{path}}$	$\mathcal{A}_{\mathrm{min}}$	$\mathbf{A}_{\mathrm{min}}$
url	59.269	20.751	13.625	11.676	4.773	5911.507	5056.601	1046.047	5691.689	1219.044
city	4.945	2.045	1.618	1.910	1.270	221.616	179.772	133.276	233.288	161.726
prot	204.609	38.729	34.130	189.000	32.757	2614.497	363.963	175.183	2780.915	313.974

obtained from $\mathcal{A}_{\mathrm{trie}}$. $\mathbf{A}_{\mathrm{min}}$ is our PADFA for $\mathcal{A}_{\mathrm{min}}$. Consequently, $\mathcal{A}_{\mathrm{trie}}$ and $\mathcal{A}_{\mathrm{min}}$ do not use packed strings, while $\mathbf{A}_{\mathrm{pref}}$, $\mathbf{A}_{\mathrm{path}}$, and $\mathbf{A}_{\mathrm{min}}$ utilize packed strings.

In the experiments, some data structures described in Sect. 3 were replaced with more practical alternatives to improve the practical performance of ADFAs. First, we use *two-stage heavy path decomposition* (two-stage HPD) [7] instead of SymCPD. The two-stage HPD also decomposes an edge set into heavy and light edges. Technically, the sets H and L obtained by the two-stage HPD are slightly different from those obtained by the SymCPD, and the set L in the two-stage HPD is smaller compared to that in the SymPCD. However, all theoretical results presented in our papers remain valid when employing the two-stage HPD. Thus, we used the two-stage HPD in the experiments. Secondly, we implemented all branches of ADFAs as a simple edge list and accessed them by a simple binary search instead of BSTs. This change was made because a memory-efficient implementation of BSTs is empirically slow and its time-efficient implementation consumes more space than simple binary search. We selected this implementation because of time-space tradeoffs.

We first constructed the five types of ADFAs for three input dictionaries and measured their memory consumption. Next, for each ADFA, we performed pattern searching using all strings in the dictionary as queries and recorded the total computation times. All programs were implemented in C++ and compiled with GCC 12.2.0 using the -O3 option[1]. All experiments were performed on a machine running Debian 12, equipped with an Intel(R) Xeon(R) CPU 2.20GHz processor, 32GiB of memory, and a register (word) size $\omega = 64$ bits, meaning that a word can store $\alpha = 8$ characters.

4.2 Experimental Results

Table 2 shows the memory consumption and computing times of each combination of ADFAs and dictionaries. For the prot, which is composed of long strings, we can see that the $\mathcal{A}_{\mathrm{trie}}$ and $\mathcal{A}_{\mathrm{min}}$, which do not use packed strings, are dramatically larger in size compared to the $\mathbf{A}_{\mathrm{pref}}$, $\mathbf{A}_{\mathrm{path}}$, and $\mathbf{A}_{\mathrm{min}}$, which utilize packed strings. This demonstrates that using packed strings is particularly beneficial for dictionaries composed of long strings.

The table also shows that $\mathbf{A}_{\mathrm{min}}$ achieved the best memory efficiency and the second-best time efficiency, while $\mathbf{A}_{\mathrm{path}}$ achieved the second-best memory efficiency and the best time efficiency across all dictionaries. $\mathbf{A}_{\mathrm{min}}$ and $\mathbf{A}_{\mathrm{path}}$ can be regarded as our

[1] The source code is available at https://github.com/shibh308/Packed_ADFA.

PADFAs for the minADFA $\mathcal{A}_{\min}$ and trie $\mathcal{A}_{\text{trie}}$, respectively. When comparing the pairs $(\mathcal{A}_{\text{trie}}, \mathcal{A}_{\min})$ and $(\mathbf{A}_{\text{path}}, \mathbf{A}_{\min})$, we observe that applying our packing technique to tries and minADFAs consistently improves both space and time efficiency. Furthermore, these results suggest that one can manage the trade-off between time and space efficiency by selecting either a minADFA or trie as input for the PADFA.

5 Conclusion

We proposed PADFA, a general framework for packing any ADFA, which empirically reduces both the time and space complexity of pattern searching. Theoretically, we demonstrated that pattern searching in a PADFA can be performed in $\mathcal{O}(m/\alpha + \log k)$ time, achieving time-optimal searching for sufficiently long patterns. We also proved that a PADFA constructed from a minADFA consumes less space than a trie when the dictionary size is relatively smaller than the size of the minADFA. Furthermore, we empirically show that PADFAs for both the trie and minADFA achieved the best space and time efficiency for real-world datasets. The results also suggest that the generality of PADFA allows for controlling the trade-off between speed and memory in pattern searching.

References

1. Geonames dump. https://download.geonames.org/export/dump/
2. Laboratory for web algorithmics. https://law.di.unimi.it/datasets.php
3. Aoe, J.: An efficient digital search algorithm by using a double-array structure. IEEE Trans. Software Eng. **15**(9), 1066–1077 (1989). https://doi.org/10.1109/32.31365
4. Appel, A.W., Jacobson, G.J.: The world's fastest scrabble program. Commun. ACM **31**(5), 572–578 (1988). https://doi.org/10.1145/42411.42420
5. Bent, S.W., Sleator, D.D., Tarjan, R.E.: Biased search trees. SIAM J. Comput. **14**(3), 545–568 (1985). https://doi.org/10.1137/0214041
6. Bille, P., Gørtz, I.L., Skjoldjensen, F.R.: Deterministic indexing for packed strings. In: Kärkkäinen, J., Radoszewski, J., Rytter, W. (eds.) 28th Annual Symposium on Combinatorial Pattern Matching, CPM 2017, 4–6 July 2017, Warsaw, Poland. LIPIcs, vol. 78, pp. 6:1–6:11. Schloss Dagstuhl - Leibniz-Zentrum für Informatik (2017). https://doi.org/10.4230/LIPIcs.CPM.2017.6
7. Bille, P., Landau, G.M., Raman, R., Sadakane, K., Satti, S.R., Weimann, O.: Random access to grammar-compressed strings and trees. SIAM J. Comput. **44**(3), 513–539 (2015). https://doi.org/10.1137/130936889
8. Blumer, A., Blumer, J., Haussler, D., Ehrenfeucht, A., Chen, M.T., Seiferas, J.I.: The smallest automaton recognizing the subwords of a text. Theor. Comput. Sci. **40**, 31–55 (1985). https://doi.org/10.1016/0304-3975(85)90157-4
9. Boldi, P., Codenotti, B., Santini, M., Vigna, S.: Ubicrawler: a scalable fully distributed web crawler. Softw. Pract. Exp. **34**(8), 711–726 (2004)
10. Daciuk, J., Mihov, S., Watson, B.W., Watson, R.E.: Incremental construction of minimal acyclic finite state automata. Comput. Linguist. **26**(1), 3–16 (2000). https://doi.org/10.1162/089120100561601

11. Ferragina, P., Grossi, R., Gupta, A., Shah, R., Vitter, J.S.: On searching compressed string collections cache-obliviously. In: Lenzerini, M., Lembo, D. (eds.) Proceedings of the Twenty-Seventh ACM SIGMOD-SIGACT-SIGART Symposium on Principles of Database Systems, PODS 2008, 9–11 June 2008, Vancouver, BC, Canada, pp. 181–190. ACM (2008). https://doi.org/10.1145/1376916.1376943
12. Ferragina, P., Navarro, G.: Pizza&chili corpus. https://pizzachili.dcc.uchile.cl/
13. Fredman, M.L., Willard, D.E.: Surpassing the information theoretic bound with fusion trees. J. Comput. Syst. Sci. **47**(3), 424–436 (1993). https://doi.org/10.1016/0022-0000(93)90040-4
14. Fujita, Y., Ichihashi, Y., Kanda, S., Morita, K., Fuketa, M.: Full-text search using double-array CDAWG. Int. J. Future Comput. Commun. **5**(6), 237–240 (2016). https://doi.org/10.18178/ijfcc.2016.5.6.478
15. Ganardi, M., Jez, A., Lohrey, M.: Balancing straight-line programs. J. ACM **68**(4), 27:1–27:40 (2021). https://doi.org/10.1145/3457389
16. Kanda, S., Köppl, D., Tabei, Y., Morita, K., Fuketa, M.: Dynamic path-decomposed tries. ACM J. Exp. Algorithmics **25**, 1–28 (2020). https://doi.org/10.1145/3418033
17. Knuth, D.E.: The Art of Computer Programming, Volume III: Sorting and Searching. Addison-Wesley (1973)
18. Morrison, D.R.: PATRICIA - practical algorithm to retrieve information coded in alphanumeric. J. ACM **15**(4), 514–534 (1968). https://doi.org/10.1145/321479.321481
19. Navarro, G., Sadakane, K.: Fully functional static and dynamic succinct trees. ACM Trans. Algorithms **10**(3), 16:1–16:39 (2014). https://doi.org/10.1145/2601073
20. Raman, R., Raman, V., Satti, S.R.: Succinct indexable dictionaries with applications to encoding k-ARY trees, prefix sums and multisets. ACM Trans. Algorithms **3**(4), 43 (2007). https://doi.org/10.1145/1290672.1290680
21. Sleator, D.D., Tarjan, R.E.: A data structure for dynamic trees. J. Comput. Syst. Sci. **26**(3), 362–391 (1983). https://doi.org/10.1016/0022-0000(83)90006-5
22. Takagi, T., Inenaga, S., Sadakane, K., Arimura, H.: Packed compact tries: a fast and efficient data structure for online string processing. IEICE Trans. Fundam. Electron. Commun. Comput. Sci. **100-A**(9), 1785–1793 (2017). https://doi.org/10.1587/transfun.E100.A.1785
23. Tsuruta, K., et al.: c-trie++: a dynamic trie tailored for fast prefix searches. Inf. Comput. **285**(Part), 104794 (2022). https://doi.org/10.1016/j.ic.2021.104794

Holey Graphs: Very Large Betti Numbers are Testable

Dániel Szabó(✉) and Simon Apers

Université Paris Cité, CNRS, IRIF, Paris, France
{szabo,apers}@irif.fr

Abstract. We show that the graph property of having a (very) large k-th Betti number β_k (over $\mathbb{Z}_2$) for constant k is testable with a constant number of queries in the dense graph model. More specifically, we consider a clique complex defined by an underlying graph and prove that for any $\varepsilon > 0$, there exists $\delta(\varepsilon, k) > 0$ such that testing whether $\beta_k \geq (1-\delta)d_k$ for $\delta \leq \delta(\varepsilon, k)$ reduces to tolerantly testing $(k+2)$-clique-freeness, which is known to be testable. This complements a result by Elek (2010) showing that Betti numbers are testable in the bounded-degree model. For our result we consider simplicial complexes as combinatorial objects, and combine matroid theory and the graph removal lemma.

Keywords: Betti number · Property testing · Simplicial matroid

1 Introduction

There has been an increasing interest in the notion of "topological data analysis" (TDA), where data is interpreted as a topological object called a "simplicial complex", and the homology of the complex is used to robustly classify the data. An important example of such a complex is the Vietoris-Rips or flag complex – this complex is obtained by associating a graph to the input data (e.g., with edges indicating similarity between data points), and the complex consisting of all subsets that induce cliques in the graph. An important feature of this simplicial complex is the so-called Betti number, or intuitively the number of high dimensional "holes". In particular, the so-called persistent Betti numbers have been useful for applications because they capture a scale-independent global property of the data set [6,20,23]. Unfortunately, calculating Betti numbers exactly is computationally hard (more precisely QMA1-hard [11]), and so the task of estimating them has been in the focus of interest [4,5,7,19,21]. In this work we use the lense of "(graph) property testing" to further our understanding of Betti number estimation.

In graph property testing we wish to decide whether a graph has a certain property, or whether it is "far" from having that property [16]. In the *dense* graph model, an n-vertex graph is ε-far from having a property if we have to add or remove more than εn^2 edges for the graph to have the property. A *tester*

R. Královič and V. Kůrková (Eds.): SOFSEM 2025, LNCS 15539, pp. 298–310, 2025.
https://doi.org/10.1007/978-3-031-82697-9_22

for a given property is a randomized algorithm that, given query access to the adjacency matrix of a graph G, can distinguish with constant success probability whether G has that property or is ε-far from having it. A graph property is said to be *testable* if there exists a tester that makes a number of queries that is a function only of ε, and so independent of the graph size. Examples of testable properties are bipartiteness, triangle-freeness and, more generally, monotone (closed under removing edges) and hereditary (closed under removing vertices) graph properties [2,3].

In this work we prove the following theorem (for a more formal statement see Theorem 2).

Theorem 1 (Informal). *The property of having a (very) large k-th Betti number is testable for constant k.*

Strictly speaking, we consider the k-th Betti number of the *clique complex* associated to G. This is the simplicial complex defined by the family of all vertex subsets that induce a clique in G. On an intuitive level, the k-th Betti number β_k of this complex counts the number of independent k-dimensional "holes" in the complex, which is bounded by the number of k-cliques d_k in G. More formally, β_k equals the rank of the k-th *homology group*. We prove Theorem 1 by showing that, for any constant k and $\varepsilon > 0$, there exists $\delta(\varepsilon, k) > 0$ such that testing whether $\beta_k \geq (1-\delta)d_k$ for any $\delta \leq \delta(\varepsilon, k)$ reduces to tolerantly testing $(k+2)$-clique-freeness. We prove the result for

$$\delta(\varepsilon, 0) = \sqrt{2\varepsilon}\ , \qquad \delta(\varepsilon, 1) = \varepsilon/3\ , \qquad \delta(\varepsilon, k) = 1/\text{tower}(k^4 \log(1/\varepsilon))\ \ (k > 1).$$

Here the tower(ℓ)-function denotes a height-ℓ tower of powers of 2's – this explains the extra quantifier in "*(very)* large Betti numbers". Nonetheless, this property is neither trivial for constant k and ε nor monotone or hereditary. To see this, consider the $(k+1)$-partite graph which has $d_k = (\frac{n}{k+1})^{k+1}$ and $\beta_k = (\frac{n}{k+1} - 1)^{k+1}$ [5, Proposition 1], and so $\beta_k/d_k = 1 - O(k^2/n)$. This shows that for any k and $\delta > 0$ there exist graphs with the property $\beta_k \geq (1-\delta)d_k$. (Compare this to our Proposition 4 which states that any large clique complex that has few k-faces is close to having a large k-th Betti number.) Moreover, the quantity β_k/d_k increases as a function of n, so the property cannot be monotone or hereditary.

Betti Numbers in the Bounded-Degree Model. This work was partly motivated by the result of Elek [12] in the *bounded-degree model*. In this model, the graph is assumed to have a constant bound d on the vertex degrees, and a query reveals the (at most d) neighbours of a vertex. Elek showed that, for any $\varepsilon > 0$ and with a number of queries only dependent on ε, it is possible to return an estimate $\hat{\beta}_k$ satisfying $\hat{\beta}_k = \beta_k \pm \varepsilon n$ for $k < d$ (for $k \geq d$ necessarily $d_k = \beta_k = 0$). The proof is based on (sparse) graph limits, and is very different from our approach.

Unfortunately, such a result in the dense graph model is not possible: returning an estimate $\hat{\beta}_k = \beta_k \pm \varepsilon n$ (or even $\hat{\beta}_k = \beta_k \pm \varepsilon d_k$) requires $\Omega(n)$ many queries in the dense graph model. To see this, consider the case $k = 0$ for which

$d_0 = n$ and β_0 equals the number of connected components of the graph. The cycle graph has $\beta_0 = 1$ while any graph with $\leq n/2$ edges has $\beta_0 \geq n/2$. However, it takes $\Omega(n)$ queries to distinguish these graphs in the dense model. This motivates the weaker formulation of large Betti number testability that we use in our work.[1]

Quantum Algorithms for Large Betti Numbers. Another motivation came from a recent stream of works on quantum algorithms for estimating Betti numbers (see e.g. [11,18,19,21]). Under certain conditions (e.g., a form of well-conditionedness), these works suggest an exponential quantum speedup over classical algorithms, returning an estimate $\hat{\beta}_k = \beta_k \pm \varepsilon n^k$ in time poly$(n, k, 1/\varepsilon)$. Such an estimate is only relevant for very large Betti numbers (β_k scaling with n^k), and a stringent question is whether "typical" graphs can have such large Betti numbers. In a recent work [4] a classical benchmark algorithm was proposed based on path-integral Monte Carlo. The algorithm has a polynomial runtime in certain regimes, narrowing down the conditions for a potential exponential quantum speedup.

The current work investigates this question from a new (property testing) perspective, and it yields a tool to investigate whether typical graphs can have a (very) large Betti number.

Open Questions. Our work raises a number of open questions. The most obvious one is whether our results can be pushed further. E.g., it might be possible to test more moderately sized Betti numbers, or Betti numbers for non-constant k (the case of interest for quantum algorithms).

Another open direction is to generalize the framework of graph property testing to simplicial complexes more generally. By limiting ourselves to clique complexes, we could phrase our results in the graph property testing language, but this might not be the most natural approach.

Outline of Proof and Paper. In Sect. 2 we formally introduce property testing, simplicial complexes, and the necessary matroid theory.

The first contribution appears in Sect. 3. We use the matroid notion of independence to relate the Betti number β_k to the number of "independent" K_{k+2} cliques in the graph, and we bound the total number of cliques as a function of the number of independent cliques.

Finally, in Sect. 4, we build on these tools to reduce the problem of testing large Betti numbers to that of (tolerantly) testing clique-freeness, which is known to be testable. In particular, we show that having a large Betti number implies that the graph is close to being K_{k+2}-free, while being far from having a large Betti number implies that the graph is far from being K_{k+2}-free.

[1] Note also that the contrapositive, having a small Betti number, is trivial to test. E.g., a graph cannot be far from having small Betti number β_0 since we can always add a cycle, thereby setting $\beta_0 = 1$.

2 Preliminaries

In this section we introduce necessary but well-known preliminaries on graph property testing, simplicial complexes and matroid theory.

2.1 Property Testing and Subgraph Freeness

In graph property testing we want to decide if an n-vertex input graph G has a property P or it is ε-far from any graph satisfying P. In the dense graph model we have query access to elements of the adjacency matrix of G, and we want to minimise the number of queries we make. This model was introduced in [17].

Definition 1 (ε-far). *The distance of two graphs G and G' is defined as*

$$D(G, G') = \frac{\min_\pi \{G \triangle \pi(G')\}}{n^2},$$

where π is any permutation of the vertices and $\triangle$ denotes the symmetric difference of the two edge sets. We say that G is ε-far *from property P if $D(G, G') > \varepsilon$ for all G' satisfying P.*

Definition 2 (Property tester). *In the dense graph model, a randomised algorithm A is a property testing algorithm for property P if given query access to the adjacency matrix of the input graph G, it satisfies the following.*

- *If G satisfies P then A returns "YES" with probability $\geq 2/3$.*
- *If G is ε-far from P then A returns "NO" with probability $\geq 2/3$.*

A tolerant *property testing algorithm satisfies a stronger constraint: for some $\varepsilon_1 < \varepsilon_2$, it distinguishes between (i) G being ε_1-close to P, and (ii) G being ε_2-far from P.*

We call a property P *(tolerantly) testable* if there is a (tolerant) property testing algorithm such that the number of queries it makes is independent of the input length.

The following well-known lemma (that can be proved using the Szemerédi regularity lemma [24]) has been central to proving many testability results, and we will also use it.

Lemma 1 (Graph removal lemma, [15]). *For any graph H and any $\varepsilon > 0$ there exists a $\delta > 0$ such that the following holds: any n-vertex graph G ($|V(H)| < n$) that contains at most $\delta n^{|V(H)|}$ copies of H as subgraphs can be made H-free by removing at most εn^2 edges (i.e. G is ε-close to being H-free).*

It follows almost directly from this result that, for any constant-sized graph H, the property of being H-free is testable [1]. Combined with the fact that every testable property in the dense graph model is also *tolerantly* testable [13][2], we get the following lemma which we are going to use later.

[2] More precisely, they prove that for every testable property there is a distance approximation algorithm. This implies tolerant testability.

Lemma 2. *For any graph H the property of H-freeness is tolerantly testable in the dense graph model. The number of queries depends only on the distance parameters $\varepsilon_1, \varepsilon_2$ and on $|H|$.*

2.2 Simplicial Complexes

A simplicial complex is a downward closed set family over a set V of vertices. As such, it can be thought of as a higher-dimensional generalisation of graphs (albeit more restrictive than hypergraphs).

Definition 3 (Simplicial complex). *An (abstract) simplicial complex Δ is a set of subsets of the vertex set V, such that if $S \in \Delta$ and $S' \subset S$ then also $S' \in \Delta$. The sets in Δ with cardinality $k+1$ are called the k-faces of Δ.*

We are going to denote the set of k-faces of a complex by $F_k(\Delta) = \{S \in \Delta, |S| = k+1\}$ and its size by $d_k(\Delta) = |F_k(\Delta)|$. When it is clear from the context which simplicial complex is being considered, we will write only F_k and d_k. If the largest subset of V that is in the complex is of cardinality $D+1$ then we say that the complex is D-dimensional.

A *clique complex* is a special case of a simplicial complex, and is defined by some underlying graph G. The sets in the clique complex associated to G are exactly the cliques of G. This implies for instance that a size-$(k+1)$ subset $S \subseteq V$ is in the complex if (and only if) all the size-k subsets of S are in the complex.

The k-chain group C_k of a simplicial complex Δ over an Abelian group $\mathbf{G}$ is defined as $C_k = \{\sum_{i=1}^{d_k} \alpha_i S_i\}$ where $S_i \in F_k(\Delta)$ and $\alpha_i \in \mathbf{G}$. Many sources consider integer coefficients ($\mathbf{G} = \mathbb{Z}$), but for our purpose it suffices to pick binary coefficients $\mathbf{G} = \mathbb{Z}_2$. This lies closest to our combinatorial interpretation of homology, which is based on *unoriented* faces and for which the chain group $C_k = 2^{F_k}$ is then simply the set of all subsets of F_k. As a consequence, we will refer to the elements of C_k either as a sum of k-faces or as a set of k-faces – the two are equivalent.

For each $k > 0$, the k-th boundary operator δ_k is a homomorphism that maps a k-face to the sum of the $(k-1)$-faces that "surround" the k-face.

Definition 4 (Boundary operator). *Let $k \geq 1$. The k-th boundary operator is a homomorphism $\delta_k : C_k \to C_{k-1}$. For $S \in F_k(\Delta)$ and $S = \{v_1, v_2, \ldots, v_{k+1}\}$ it is defined by $\delta_k(S) = \sum_{i=1}^{k+1} S \setminus \{v_i\}$.*

It follows from this definition that the boundary of a boundary is always zero, i.e., $\delta_k(\delta_{k+1}(.)) = 0$. Sometimes we are going to use the same notation for *boundary vectors*: for $S \in F_k$, $\delta_k(S) \in \{0,1\}^{d_{k-1}}$, where a coordinate is 1 iff the corresponding $(k-1)$-face appears in the boundary of S.

Now we can define the Betti numbers, that are at the center of interest in this article.

Definition 5 (Betti number). *The k-th Betti number β_k of a simplicial complex Δ is the rank of the k-th homology group:*

$$\beta_k(\Delta) = \mathrm{rk}(\ker(\delta_k)/\mathrm{im}(\delta_{k+1})).$$

More intuitively, define a k-dimensional hole as a subset $H \subseteq F_k(\Delta)$ that *has* no boundary and *is* no boundary. Equivalently, $\delta_k(H) = 0$ (and so $H \in \ker(\delta_k)$) and there exists no $H' \subseteq F_{k+1}(\Delta)$ such that $\delta_{k+1}(H') = H$ (and so $H \notin \mathrm{im}(\delta_{k+1})$). Then β_k counts the number of "independent" k-dimensional holes in the complex – where we formalize the notion of independence in the next section.

2.3 Matroids

The appropriate notion of independence of faces and of holes comes from matroid theory. A matroid is a downward closed set family with an additional property[3] called the exchange property.

Definition 6 (Matroid). *A matroid M over ground set E is a family of subsets $I \subseteq 2^E$ called the independent subsets of E, and which satisfies the following properties.*

1. $\emptyset \in I$.
2. *If $A \in I$ and $B \subseteq A$ then $B \in I$.*
3. *If $A, B \in I$ and $|B| < |A|$ then $\exists v \in A \setminus B$ such that $B \cup \{v\} \in I$.*

The easiest example of a matroid is a graph. In this case, the ground set E in the matroid is the edge set of the graph, and we call a subset of edges independent if it is cycle-free. Matroids that can be defined this way by a graph are called *graphic matroids* or cycle matroids.

Another important example is linear independence of vectors. The elements of E are vectors from a vector space, and a subset of them is called independent if the vectors are linearly independent (over a field F). Matroids that can be defined in this way are called *linear matroids* (or representable over F).

The *simplicial matroid* (or simplicial geometry) $M_k(\Delta)$ associated to a simplicial complex Δ is a linear matroid defined as follows. It appears in e.g. [9,10].

Definition 7 (Simplicial matroid). *The k-simplicial matroid $M_k(\Delta)$ associated to a simplicial complex Δ is the linear matroid whose ground set is the set of boundary vectors $\delta_k(S) \in \{0,1\}^{d_{k-1}}$ for $S \in F_k(\Delta)$.*

Motivated by this, we call a subset of k-faces independent if the corresponding boundary vectors are linearly independent (over the field $\{0,1\}$).

A maximal independent set of a matroid M is called a *basis*. It is well known that all the bases of a matroid have the same size, equal to the *rank* $\mathrm{rk}(M)$ of the matroid. The full k-simplicial matroid $M_k(\Delta_k^{\mathrm{full}})$ is the k-simplicial matroid associated to the full complex $\Delta_k^{\mathrm{full}} = \{S \subseteq V, |S| \leq k+1\}$ that contains all the $k+1$-subsets as k-faces, but it does not have any higher dimensional face.

[3] In this sense matroids are a specialisation of simplicial complexes, although we are going to use them in a different way.

Proposition 1 (e.g. [9], Proposition 6.1.5). $\mathrm{rk}(M_k(\Delta_k^{\mathrm{full}})) = \binom{n-1}{k}$.

For a construction, fix a vertex u of Δ_k^{full} and take the set of k-faces that contain u. It is easy to see that this set of size $\binom{n-1}{k}$ is a basis of the matroid.

3 Betti Numbers via Independent Faces

In this section we connect the number of independent k-faces with the total number of k-faces, and connect the Betti number β_k to the number of independent k- and $(k+1)$-faces in the complex.

The notion of independence of faces in Definition 7 leads to the following useful observation. It is a direct consequence of the fact that a set of k-faces has zero boundary if and only if the sum of the corresponding boundary vectors is the zero vector.

Proposition 2. *In a k-dimensional simplicial complex (i.e., $|F_{k+1}| = 0$), a set of k-faces is independent iff no subset of them forms a k-dimensional hole.*

The independence of holes is defined similarly. A k-dimensional hole is a set of k-faces (an element of C_k), and associated to it is a characteristic vector over $\{0,1\}^{d_k}$. We associated the same kind of (boundary) vectors to $(k+1)$-faces: in this sense a k-dimensional hole can be seen as the boundary of a virtual $(k+1)$-dimensional object. This way, a set of holes is independent if the corresponding vectors are linearly independent (over field $\{0,1\}$). An analogue of Proposition 2 tells us that a set of k-dimensional holes is independent iff no subset of them (as virtual $(k+1)$-faces) forms a $(k+1)$-dimensional hole.

Let us denote the rank $\mathrm{rk}(M_k(\Delta))$ of the k-simplicial matroid (i.e. the size of a maximal independent set of k-faces in Δ) by $r_k(\Delta)$. We are going to need a lower bound on this value in terms of the total number of k-faces $d_k(\Delta)$. For the sake of completeness, we also include an upper bound in the statement.

Lemma 3. *For any $0 \le k < n$ and any simplicial complex Δ*

$$\frac{k+1}{n} d_k(\Delta) \le r_k(\Delta) \le \min\left\{d_k(\Delta), \binom{n-1}{k}\right\}.$$

Proof. Trivially, $r_k(\Delta) \le d_k(\Delta)$. Moreover, the set of k-faces $F_k(\Delta)$ of any complex Δ can be obtained from that of the full complex $F_k(\Delta_k^{\mathrm{full}})$ by removing faces, and this can only decrease the rank. Combined with Proposition 1 we hence get $r_k(\Delta) \le \binom{n-1}{k}$.

Now let us prove the main part of the claim, which is the lower bound. We use a similar argument to the one below Proposition 1. Let u be a vertex in Δ that is included in a maximum number of k-faces (i.e., the vertex with the highest "k-face-degree"). These k-faces that contain u are independent because each contains a $(k-1)$-face that the others do not (the one without u), and this is a non-zero element in their boundary vector. As there are d_k many k-faces in Δ, each incident to $k+1$ vertices, the average "k-face-degree" of a vertex is $(k+1)d_k/n$. Thus the independent set of k-faces defined by u has at least this many k-faces, and so $r_k \ge (k+1)d_k/n$.

The next lemma shows a nice connection between the rank, the number of faces and the Betti number. For $k = 0$ the formula gives the well-known graph formula $c = n - t$, where t is the number of edges in a spanning forest, n is the number of vertices, and c is the number of connected components. When $k = 1$ and the underlying graph is connected and planar, it gives the Euler formula $n + f = e + 2$ (with n the number of vertices, f the number of faces surrounded by edges and e the number of edges) because $\beta_1 = f - 1$, $d_1 = e$, $r_1 = n - 1$ and $r_2 = 0$.

Lemma 4 (**[22] Proposition 3.13.**). *For any simplicial complex,*

$$\beta_k = d_k - r_k - r_{k+1}.$$

Proof. By applying the rank–nullity theorem to δ_k, we get $d_k = \dim\ker\delta_k + \dim \mathrm{im}\delta_k$. Now notice that $\dim \mathrm{im}\delta_k = r_k$ because the independence of k-faces is defined through their boundary vectors (Definition 7), thus we have $\dim\ker\delta_k = d_k - r_k$. From Definition 5 we can see that $\beta_k = \dim\ker\delta_k - \dim \mathrm{im}\delta_{k+1}$. Substituting what we got before we obtain $\beta_k = (d_k - r_k) - r_{k+1}$.

For the interested reader we also we also give an alternative, combinatorial proof of this lemma in Appendix A, which ties closer to the spirit of this work.

In the special case where $k = 0$ and the graph defined by the vertices and edges of Δ is connected, we have $\beta_0 = 1$, $d_0 = n$ and $r_{k+1} = n - 1$ (a spanning tree of the graph is a maximal independent edge set). Thus, r_0 has to be defined as 0, which makes sense if we think about r_k as $r_k = \dim(\mathrm{im}(\delta_k))$.

Remark 1. Let $T_\ell = \{S \subseteq V, |S| = \ell+1\}$ and $A \subseteq T_k$. In early works [8,10] only complexes of the form $\Delta^{\mathrm{full}}_{k-1} \cup A$ are analysed in detail. This family of complexes is not enough to express the k-th Betti number of an arbitrary simplicial complex. For example, for this restricted class of complexes Cordovil [8, Proposition 1.2] showed that $r_k = d_k - \beta_k$, which is only a special case of Lemma 4 (with $r_{k+1} = 0$).

An easy consequence of Lemma 4 is the following statement.

Proposition 3. *For any simplicial complex Δ and $k \geq 1$, $\beta_k(\Delta) \leq \binom{n-1}{k+1}$.*

Proof. In Δ^{full}_k we have $\beta_k = d_k - r_k - 0 = \binom{n}{k+1} - \binom{n-1}{k} = \binom{n-1}{k+1}$, and removing k-faces or adding $(k+1)$-faces cannot increase this value.

4 Testing Large Betti Numbers

In this section we turn to our main result, proving that we can test whether a Betti number is large. We now state our main theorem again.

Theorem 2 (**formal version of Theorem** 1). *Consider a clique complex Δ. For any constant k and $\varepsilon > 0$, there exists $\delta(\varepsilon, k) > 0$ such that the property of having k-th Betti number $\beta_k(\Delta) \geq (1 - \delta)d_k$ (over $\mathbb{Z}_2$) is testable for any $\delta \leq \delta(\varepsilon, k)$ (with distance parameter ε).*

Even though our results in Sect. 3 hold for general simplicial complexes, the main theorem is restricted to clique complexes. The reason for this is that we wish to phrase our results in the well-established setting of graph property testing. By constraining ourselves to clique complexes, having a large Betti number becomes a graph property (of the underlying graph) rather than a property of an abstract simplicial complex. Also, this way we can use some previous results from graph property testing, like the tolerant testability of subgraph freeness (Lemma 2).

4.1 Warm-Up: Testing Many Components

The 0-th Betti number β_0 of a clique complex Δ equals the number of connected components of the underlying graph G. As an informal warm-up and a blueprint for the general case, we show how to test whether $\beta_0 \geq (1-\delta)n$. The argument involves two reductions.

First, we argue that having a large 0-th Betti number is equivalent to having few independent edges. From Lemma 4 we get that $\beta_0 = n - r_1 - r_0$ where $r_0 = 0$. Thus, $\beta_0 \geq (1-\delta)n$ is equivalent to $r_1 \leq \delta n$, and so testing large β_0 reduces to testing whether G has a small number of independent edges.

Now comes the second reduction, in which we argue that testing whether G has few independent edges can be reduced to *tolerantly* testing edge-freeness. For this, note that if G has $r_1 \leq \delta n$ independent edges then the total number of edges $|E| \leq \binom{\delta n+1}{2} < \delta^2 n^2/2 + O(n)$,[4] and so G must be $1.1\delta^2/2$-close to being edge-free. On the other hand, if G is ε-far from having $r_1 \leq \delta n$, then G must also be ε-far from having $r_1 = 0$, i.e. from being edge-free. So we reduced the problem of testing $\beta_0 \geq (1-\delta)n$ to that of tolerantly testing edge-freeness (with parameters $\varepsilon_1 = 1.1\delta^2/2$ and $\varepsilon_2 = \varepsilon$). It remains to note that edge-freeness is tolerantly testable by Lemma 2.

4.2 General Case

We now turn to proving our general result (Theorem 2), that having a Betti number $\beta_k \geq (1-\delta)d_k$ is testable for constant k. Following the blueprint from the previous section, we first reduce the problem to testing whether there are few independent $(k+1)$-faces, and then reduce testing few independent $(k+1)$-faces to tolerantly testing $(k+2)$-clique freeness.

We consider a clique complex Δ with underlying graph G. From Lemma 4 we get that

$$d_k - r_{k+1} \geq \beta_k = d_k - r_{k+1} - r_k, \tag{1}$$

from which we can prove the following lemma.

[4] From Proposition 3 we would get $|E| \leq r_1 n/2 \leq \delta n^2/2$. We get the better bound by noticing that if there are δn independent edges, then we have a maximum number of edges if all the independent edges are in the same connected component and this component is a $K_{\delta n+1}$.

Lemma 5 (Large Betti number $\preceq$ few independent cliques). *In a clique complex Δ with underlying graph G, if $\beta_k \geq (1-\delta)d_k$ then $r_{k+1} \leq \delta d_k$. If G is ε-far from having $\beta_k \geq (1-\delta)d_k$ in Δ then it is $\varepsilon/2$-far from having $r_{k+1} = 0$, i.e., G is $\varepsilon/2$-far from K_{k+2}-freeness.*

Proof. The first part of the claim is clear from Eq. (1). For the second part (being ε-far) we will use the definition of ε-far (Definition 1). I.e. we want to prove that if every graph that is ε-close to G has $\beta_k < (1-\delta)d_k$ then every graph that is $\varepsilon/2$-close to G satisfies $r_{k+1} > 0$.

For contradiction assume that there is a particular H that is $\varepsilon/2$-close to G but has $r_{k+1} = 0$. Since H is ε-close to G, it has $\beta_k < (1-\delta)d_k$, or equivalently $r_k + r_{k+1} > \delta d_k$ (using Lemma 4). Because of this, we have $d_k < r_k/\delta \leq \binom{n-1}{k}/\delta$ (by Proposition 3). With the construction of Proposition 4 below, we can modify H to get an H' that is $\alpha = \varepsilon/2$-close to H (thus still ε-close to G) and that has $\beta_k \geq (1-\delta)d_k$. This contradicts the assumption that every graph that is ε-close to G satisfies $\beta_k < (1-\delta)d_k$.

Proposition 4. *Consider a graph $H = (V, E)$ with $|V| = n$ sufficiently large, and assume that H has at most $\binom{n-1}{k}/\delta$ k-faces. Then for any constant proximity parameter α, there is another graph H' that is α-close to H and has $\beta_k \geq (1-\delta)d_k$.*

Proof. We give a construction that modifies H to get H'. Let us choose any vertex set $S \subseteq V$ of size $|S| = \alpha n$. We delete all the edges that go between S and $V \setminus S$, and we modify the edges within S to construct a complete $(k+1)$-partite subgraph. This modifies at most αn^2 edges, and so yields a graph H' that is α-close to H.

The subgraph of H' induced by S is a complete $(k+1)$-partite graph with $\left(\frac{\alpha n}{k+1}\right)^{k+1}$ many k-faces. Thus, in H' we have at most this amount plus the number of original k-faces of H, i.e. $d_k(H') \leq \left(\frac{\alpha n}{k+1}\right)^{k+1} + \binom{n-1}{k}/\delta$. The number of independent k-holes in the subgraph of H' induced by S is $\left(\frac{\alpha n}{k+1} - 1\right)^{k+1}$ (see for instance [5, Proposition 1]), so in H' it is at least this much: $\beta_k(H') \geq \left(\frac{\alpha n}{k+1} - 1\right)^{k+1}$. Clearly $\beta_k(H')/d_k(H') = 1 - O(1/n)$. For any $\delta > 0$ this is at least $1-\delta$ for n sufficiently large.

Remark 2. In Lemma 5 the second proximity parameter is not necessarily half of ε, it can be arbitrarily close to it. E.g. it could be 0.99ε, but then we have to use the construction of Proposition 4 with $\alpha = 0.01\varepsilon$ instead of $\varepsilon/2$.

For our second reduction, we use Proposition 3, which tells us that if $r_{k+1} \leq \delta d_k$ then

$$d_{k+1} \leq \frac{\delta}{k+2} n d_k \leq \frac{\delta}{k+2} n \binom{n}{k+1} \leq \frac{\delta}{(k+2)!} n^{k+2}.$$

Combined with Lemma 5 we get that $\beta_k \geq (1-\delta)d_k$ implies $d_{k+1} \leq \frac{\delta}{(k+2)!} n^{k+2}$. We see that a large Betti number implies a small number of K_{k+2}'s in the graph, while being far from having a large Betti number implies being far from K_{k+2}-freeness (by Lemma 5).

In fact, by the graph removal lemma (Lemma 1), a small number of K_{k+2}'s implies that the graph is close to being K_{k+2}-free. More specifically, for any $\varepsilon' > 0$ there exists $\delta = \delta(k, \varepsilon') > 0$ such that if G has at most $\frac{\delta}{(k+2)!} n^{k+2}$ many K_{k+2}'s then G is ε'-close to being K_{k+2}-free. By picking (say) $\varepsilon' = \varepsilon/2$, it follows that we can test whether $\beta_k \geq (1-\delta) d_k$ by tolerantly testing whether G is $\varepsilon/2$-close or ε-far from K_{k+2}-freeness. By Lemma 2 we know that K_{k+2}-freeness is indeed tolerantly testable, and this proves our main Theorem 2.

To finish, we comment on the scaling of $\delta(k, \varepsilon)$ (the complexity of the algorithm is dominated by $1/\delta$). The current best upper bound in the graph removal lemma requires $\delta(k, \varepsilon) \leq 1/\text{tower}(5(k+2)^4 \log(1/\varepsilon))$ [14], where tower(i) is a tower of twos of height i (e.g., tower(3) $= 2^{2^2}$). For the case of $k = 0$ we could avoid this: recall from Sect. 4.1 that $r_1 \leq \delta n$ implies that G is $\delta^2/2$-close to being edge-free. Similarly, for $k = 1$ we can get a better bound: $r_2 \leq \delta n^2$ implies that G is 3δ-close to being triangle-free. Indeed, if we remove all the edges of a maximal independent triangle set (at most $3\delta n^2$ edges), then any remaining triangle in the graph would contradict the maximality of the chosen set. We leave the extension of similar arguments to higher k for future work. In conclusion, we get a tester that distinguishes $\beta_k \geq (1-\delta) d_k$ from being ε-far under the constraints

$$\delta < \sqrt{2\varepsilon} \;\; (k=0), \quad \delta < \varepsilon/3 \;\; (k=1), \quad \delta < 1/\text{tower}(5(k+2)^4 \log(1/\varepsilon)) \;\; (k>1).$$

Disclosure of Interests. The authors have no competing interests to declare that are relevant to the content of this article.

A An Alternative, Combinatorial Proof of Lemma 4

Proof. The proof goes by induction. Let Δ denote the simplicial complex being considered and let us take a basis of the k-simplicial matroid over Δ. For the base case, we consider the subcomplex where this is the set of all k-faces and all the higher dimensional faces are removed, in which case $r_k = d_k$ and $\beta_k = r_{k+1} = 0$ and so the formula holds. In the inductive step we will put back all of the removed faces. We start by adding the rest of the k-faces one by one, and we argue that each added face creates exactly one new independent hole.

First, note that adding a dependent k-face S to the complex creates at least one hole (otherwise we could have added it to the basis by Proposition 2). Moreover, the hole is independent of the previous ones because it contains the face S, which no other hole contains so far.

Then, we prove that adding a k-face creates at most one hole. By contradiction, assume that there is a k-face S such that when added to the set, more than one new independent holes are created. We consider two of them, $\{S, R_1, \ldots, R_p\}$ and $\{S, T_1, \ldots, T_q\}$, which we call the "R-hole" and the "T-hole". Necessarily they have zero boundary (we denote the boundary vectors the same way as the

k-faces):

$$S + R_1 + \cdots + R_p = 0$$
$$S + T_1 + \cdots + T_q = 0.$$

Adding the equations shows that $\{R_1, \ldots, R_p, T_1, \ldots, T_q\}$ must also be a hole, call it the "RT-hole". It does not contain S, and so must have been present before adding S. However, by construction, the R-, T- and RT-holes are not independent, and so we get a contradiction.

Let Δ_k denote the complex we have now: it contains exactly the faces of Δ up to dimension k, and no faces of higher dimension. So far we proved that $r_k = d_k - \beta_k(\Delta_k)$. Let us consider the set of "potential k-holes" in Δ_k, i.e. sets H of cardinality $k+2$ where all the $(k+1)$-subsets of H are in Δ_k. These are those holes of Δ_k that may be filled by $(k+1)$-faces in Δ.

Now we continue the induction by adding to Δ_k the $(k+1)$-faces of Δ one by one (and in the end the higher dimensional faces as well) to get back Δ. Each $(k+1)$-face fills a potential hole, and it is independent of the previously added ones if and only if the hole being filled is independent of the previously filled ones (as they are the same subset). Thus, every time Δ gains an independent $(k+1)$-face it loses an independent k-hole. This finishes the proof, as adding faces of dimension larger than $k+1$ does not change any parameter in the claim.

References

1. Alon, N., Duke, R.A., Lefmann, H., Rodl, V., Yuster, R.: The algorithmic aspects of the regularity lemma. J. Algorithms **16**(1), 80–109 (1994). https://doi.org/10.1006/jagm.1994.1005
2. Alon, N., Shapira, A.: Every monotone graph property is testable. In: Proceedings of the Thirty-Seventh Annual ACM Symposium on Theory of Computing, pp. 128–137 (2005). https://doi.org/10.1145/1060590.1060611
3. Alon, N., Shapira, A.: A characterization of the (natural) graph properties testable with one-sided error. SIAM J. Comput. **37**(6), 1703–1727 (2008). https://doi.org/10.1137/06064888X
4. Apers, S., Gribling, S., Sen, S., Szabó, D.: A (simple) classical algorithm for estimating Betti numbers. Quantum **7**, 1202 (2023). https://doi.org/10.22331/q-2023-12-06-1202
5. Berry, D.W., et al.: Analyzing prospects for quantum advantage in topological data analysis. PRX Quantum **5**, 010319 (2024). https://doi.org/10.1103/PRXQuantum.5.010319
6. Bukkuri, A., Andor, N., Darcy, I.K.: Applications of topological data analysis in oncology. Front. Artif. Intell. **4** (2021). https://doi.org/10.3389/frai.2021.659037
7. Chazal, F., Fasy, B., Lecci, F., Michel, B., Rinaldo, A., Wasserman, L.: Subsampling methods for persistent homology. In: Bach, F., Blei, D. (eds.) Proceedings of the 32nd International Conference on Machine Learning. Proceedings of Machine Learning Research, vol. 37, pp. 2143–2151. PMLR, Lille, France (2015). https://proceedings.mlr.press/v37/chazal15.html

8. Cordovil, R.: Sur les géometries simpliciales. C. R. Hebd. Seances Acad. Sci. **286**(25), 1219–1222 (1978)
9. Cordovil, R., Lindström, B.: Simplicial matroids. In: Combinatorial Geometries. Cambridge University Press (1987). https://doi.org/10.1017/CBO9781107325715
10. Crapo, H.H., Rota, G.C.: On the foundations of combinatorial theory II. Combinatorial geometries. Stud. Appl. Math. **49**(2), 109–133 (1970). https://doi.org/10.1002/sapm1970492109
11. Crichigno, M., Kohler, T.: Clique homology is QMA1-hard. Nat. Commun. **15** (2024). https://doi.org/10.1038/s41467-024-54118-z
12. Elek, G.: Betti Numbers are Testable, pp. 139–149. Springer Berlin Heidelberg Berlin, Heidelberg (2010). https://doi.org/10.1007/978-3-642-13580-4_6
13. Fischer, E., Newman, I.: Testing versus estimation of graph properties. SIAM J. Comput. **37**(2), 482–501 (2007). https://doi.org/10.1137/060652324
14. Fox, J.: A new proof of the graph removal lemma. Ann. Math. **174**, 561–579 (2011). https://doi.org/10.4007/annals.2011.174.1.17
15. Füredi, Z.: Extremal hypergraphs and combinatorial geometry. In: Proceedings of the International Congress of Mathematicians, pp. 1343–1352. Birkhäuser Basel (1995). https://doi.org/10.1007/978-3-0348-9078-6_129
16. Goldreich, O.: Introduction to testing graph properties. In: Goldreich, O. (eds.) Property Testing: Current Research and Surveys, pp. 105–141 (2010). https://doi.org/10.1007/978-3-642-16367-8
17. Goldreich, O., Goldwasser, S., Ron, D.: Property testing and its connection to learning and approximation. J. ACM **45**(4), 653–750 (1998). https://doi.org/10.1145/285055.285060
18. Gyurik, C., Cade, C., Dunjko, V.: Towards quantum advantage via topological data analysis. Quantum **6**, 855 (2022). https://doi.org/10.22331/q-2022-11-10-855
19. Hayakawa, R.: Quantum algorithm for persistent Betti numbers and topological data analysis. Quantum **6**, 873 (2022). https://doi.org/10.22331/q-2022-12-07-873
20. Krishnapriyan, A.S., Montoya, J., Haranczyk, M., Hummelshøj, J., Morozov, D.: Machine learning with persistent homology and chemical word embeddings improves prediction accuracy and interpretability in metal-organic frameworks. Sci. Rep. **11**(1), 8888 (2021). https://doi.org/10.1038/s41598-021-88027-8
21. Lloyd, S., Garnerone, S., Zanardi, P.: Quantum algorithms for topological and geometric analysis of data. Nat. Commun. **7**(1), 1–7 (2016). https://doi.org/10.1038/ncomms10138
22. Nanda, V.: Computational algebraic topology lecture notes. https://people.maths.ox.ac.uk/nanda/cat/TDANotes.pdf
23. Pranav, P., et al.: The topology of the cosmic web in terms of persistent Betti numbers. Mon. Not. R. Astron. Soc. **465**(4), 4281–4310 (2016). https://doi.org/10.1093/mnras/stw2862
24. Szemerédi, E.: Regular partitions of graphs. In: Problèmes combinatoires et théorie des graphes (Colloq. Internat. CNRS, Univ. Orsay, Orsay, 1976), Colloq. Internat. CNRS, vol. 260, pp. 399–401. CNRS, Paris (1978)

Warm-Started QAOA with Aligned Mixers Converges Slowly Near the Poles of the Bloch Sphere

Reuben Tate(✉) and Stephan Eidenbenz

CCS-3: Information Sciences, Los Alamos National Laboratory, Los Alamos, NM, USA
{rtate,eidenben}@lanl.gov

Abstract. In order to boost the performance of the Quantum Approximate Optimization Algorithm (QAOA) to solve problems in combinatorial optimization, researchers have leveraged the solutions returned from classical algorithms in order to create a warm-started quantum initial state for QAOA that is biased towards "good" solutions. Cain et al. showed that if the classically-obtained solutions are mapped to the poles of the Bloch sphere, then vanilla QAOA with the standard mixer "gets stuck". If the classically-obtained solution is instead mapped to within some angle θ from the poles of the Bloch sphere, creating an initial product state, then QAOA with optimal variational parameters is known to converge to the optimal solution with increased circuit depth if the mixer is modified to be "aligned" with the warm-start initial state. Leveraging recent work of Benchasattabuse et al., we provide theoretical lower bounds on the circuit depth necessary for this form of warm-started QAOA to achieve a desired change $\Delta\lambda$ in approximation ratio; in particular, we show that for small θ, the lower bound on the circuit depth roughly scales proportionally with $\Delta\lambda/\theta$.

Keywords: Quantum Computing · Combinatorial Optimization · Quantum Approximate Optimization Algorithm

1 Introduction

The Quantum Approximate Optimization Algorithm (QAOA) [6] and its greatly generalized sibling Quantum Alternating Operator Ansatz (also QAOA) [9] is a leading quantum algorithm or perhaps a quantum heuristic to find approximate solutions to combinatorial optimization problems, such as Traveling Salesperson, Satisfiability, or – and this is the most commonly studied case – Maximum Cut. While a diversity of QAOA variations have emerged over the past decade [3], the concept of using a quantum state representation of a high-quality classically obtained approximate solution as starting state for QAOA is our focus for analysis. More precisely, we study a case where the classically-obtained initial solution is mapped to within some angle θ from the poles of the Bloch sphere, creating

R. Královič and V. Kůrková (Eds.): SOFSEM 2025, LNCS 15539, pp. 311–323, 2025.
https://doi.org/10.1007/978-3-031-82697-9_23

an initial product state; this QAOA variant with optimal variational parameters is known to converge to the optimal solution with increased circuit depth if the mixer is modified to be "aligned" with the warm-start initial state.

Provable approximation ratio results for QAOA are still very rare. Our work focuses on the effect that the parameter θ has on the behavior of the QAOA circuit, thus giving new insights into the theory of QAOAs. More specifically, we provide θ-dependent lower bounds on the required circuit depth needed to achieve a desired change in approximation ratio. The key takeaway is the following: *In the context of warm-started QAOA (with "aligned" mixers), one should avoid initializing the warm-start very close to poles of the Bloch sphere (i.e. θ near 0).* Some preliminary numerical simulations [5,12] had already suggested that initializing near the poles was not ideal; this work solidifies such a result with a theoretical backing. While our results do not show a worst-case advantage of warm-start QAOA over either the traditional QAOA approach (e.g. approximation ratio of 0.6924 for 3-regular MAX-CUT [6]), the practical advantage of warm-started QAOA in NISQ and numerical experiments should be explored further.

This paper is organized as follows: in Sect. 2, we set up the necessary notation for this work and briefly review the standard QAOA algorithm, in Sect. 3, we give a more detailed description of the initial warm-start used and the corresponding "aligned" mixer, in Sect. 4, we consider an illustrative simple toy problem that provides a geometric intuition for why one should expect initializations near the poles to perform poorly, and in Sect. 5, we provide a theoretical analysis that shows that such an intuition holds more generally as well by proving lower bounds on the circuit depth. We provide a discussion and conclude in Sect. 6.

2 Notation and Background

For the purposes of this work, we assume throughout that we are working with some classical objective function $c : \{0,1\}^n \to \mathbb{Z}^{\geq 0}$ to be maximized, which maps bitstrings of a particular length to a non-negative integer. We also assume that c is not identically zero everywhere; together with non-negativity, this ensures that $c_{\max} := \max_{x \in \{0,1\}^n} c(x) > 0$. Let $\mathcal{A}$ be some (possibly randomized) algorithm and let $\mathcal{A}(c)$ be the (expected) objective value obtained by running $\mathcal{A}$ with an instance corresponding to objective function c, we then define the corresponding (instance-specific) approximation ratio[1] as, $\lambda_{\mathcal{A}}(c) = \frac{\mathcal{A}(c)}{c_{\max}}$; when the context is clear, we will often write $\lambda_{\mathcal{A}}(c)$ as $\lambda_{\mathcal{A}}$ or even just λ.

We will often use the MAX-CUT problem as an example which is as follows: given a graph $G = (V, E)$, partition the vertices V into two groups, S and $V \setminus S$, so that the number of edges between the groups is maximized. If $|V| = n$, the

[1] In the classical optimization literature, the term *approximation ratio* is often only used to refer to *worst-case* ratio (of the expected algorithm objective value to the optimal objective value) amongst some class of objective functions. In recent years, the term has also been used to refer to the ratio for individual instances, which is how the term will be used in this work.

partitioning of the vertices can be expressed as a bitstring x whose ith bit is 0 if the ith vertex is in S and 1 otherwise. For any fixed choice of graph G, the corresponding objective function for this problem is: $c(x) = \sum_{(i,j)\in E} \frac{1}{2}(1 - z_i z_j)$ where $z_i = 2x_i - 1 \in \{-1, +1\}$ for all i.

2.1 Standard QAOA

We next review the QAOA algorithm and set up the needed notation that will be used throughout this work. We use X, Y, Z to denote the standard Pauli matrices. For a multi-qubit system, we use X_j, Y_j, Z_j to denote the operation of applying X, Y, or Z to the jth qubit respectively. We use I and $\mathbf{0}$ to denote the identity and all-zeros matrix respectively; the dimensions of such matrices will be clear from context. For any square matrix M and scalar t, we define the corresponding matrix $U(M, c) := e^{-itM}$. For a combinatorial optimization problem determined by a cost function $c : \{0,1\}^n \to \mathbb{Z}^{\geq 0}$ on n-length bitstrings, we define the corresponding cost Hamiltonian as the matrix C such that $C|b\rangle = c(b)|b\rangle$.

Given a classical objective function c on n-length bitstrings (with corresponding cost Hamiltonian C), a Hermitian matrix B (called the mixing Hamiltonian) of appropriate size, and initial quantum state $|\psi_i\rangle$ in a 2^n-dimensional Hilbert space, a circuit depth p, and variational parameters $\gamma = (\gamma_1, \ldots, \gamma_p), \beta = (\beta_1, \ldots, \beta_p)$, we define the following variational waveform of depth-p QAOA as follows:

$$|\psi_p(\gamma, \beta)\rangle := U(B, \beta_p)U(C, \gamma_p) \cdots U(B, \beta_1)U(C, \gamma_1) |\psi_i\rangle . \tag{1}$$

For unconstrained optimization problems, the mixing Hamiltonian B is usually taken to be the transverse field mixer, which, for an n-qubit system, is defined as

$$B_{\text{TF}} = \sum_{j=1}^{n} X_j. \tag{2}$$

Additionally, the starting state is usually taken to be an equal superposition of all 2^n bitstrings of length n:

$$|\psi_i\rangle = |+\rangle^{\otimes n} = \frac{1}{\sqrt{2^n}} \sum_{b\in\{0,1\}^n} |b\rangle . \tag{3}$$

In the case where $|\psi_i\rangle$ is the most-excited state of H_B, there exists a choice of angles γ and β for which the QAOA circuit can be viewed as a Trotterization of the Quantum Adiabatic Algorithm which is known to, under mild assumptions (see [2,6]), converge to the optimal solution given enough time; in other words, for a cost function c that we wish to maximize, we have that:

$$\lim_{p\to\infty} \max_{\gamma,\beta} \left[\langle\psi_p(\gamma, \beta)| C |\psi_p(\gamma, \beta)\rangle \right] = \max_{x\in\{0,1\}^n} c(x),$$

where $\langle\psi_p(\gamma, \beta)| C |\psi_p(\gamma, \beta)\rangle$ is the expected cost value obtained from measuring the output state of QAOA.

3 Initial Product States and Aligned Mixers

For the standard QAOA algorithm [6] and many of its variants, the equal superposition $|\psi_i\rangle = |+\rangle^{\otimes n}$ is used as the initial state. In the context of MAX-CUT for example, quantum measurement of $|+\rangle^{\otimes n}$ produces a uniform distribution of all 2^n cuts in the graph; put another way, each vertex, independent of the other vertices, has probability 1/2 of being on one side of the cut or the other.

However, one can consider modifying the QAOA algorithm by using a different initial state $|\psi_i\rangle$. Often, such initial states are constructed as a function of *classically* obtained solutions. This method is parametrized by a parameter θ which we refer to as the *initialization angle*; to this end, we first introduce some helpful notation.

3.1 Construction of Warm-Started States

Let $\boldsymbol{n} = (x, y, z)$ be a unit vector written in Cartesian coordinates. We let $|\boldsymbol{n}\rangle$ denote the single-qubit quantum state whose qubit position on the Bloch sphere is $\boldsymbol{n}$. For $\boldsymbol{n} = (x, y, z)$, we define the following single-qubit operation:

$$B_{\boldsymbol{n}} = xX + yY + zZ, \tag{4}$$

and let $B_{\boldsymbol{n},j}$ denote the operation of applying the operation $B_{\boldsymbol{n}}$ on the jth qubit. The unitary $U(B_{\boldsymbol{n},j}, \beta)$ can be geometrically interpreted as a single-qubit rotation by angle 2β about the axis that points in the $\boldsymbol{n}$ direction [8].

For an n-qubit product state $|s\rangle = \bigotimes_{j=1}^{n} |\boldsymbol{n}_j\rangle$, we define an n-qubit operation $B_{|s\rangle}$ in terms of the single-qubit operations above:

$$B_{|s\rangle} = \sum_{j=1}^{n} B_{\boldsymbol{n}_j,j}. \tag{5}$$

When $|s\rangle$ is a product state, one can show that $|s\rangle$ can be prepared and that $B_{|s\rangle}$ can be implemented in most quantum devices with a constant-depth circuit using standard single-qubit rotation gates about the x, y, and z axes. Moreover, they [12] remark that $|s\rangle$ is a ground state of $B_{|s\rangle}$ for any product state $|s\rangle$. Thus, as remarked in [12], as long as none of the qubits of the initial product state $|\psi_i\rangle$ are initialized at the poles, running QAOA with the standard phase separator C and mixer $B_{|\psi_i\rangle}$ will yield the optimal solution as the circuit depth goes to infinity (assuming that γ and β are chosen optimally) [12]. In general, whenever the initial state of QAOA is the ground state of the mixer, we say that the mixer is *aligned* with the initial state, and refer to this category of QAOA variants as *QAOA with aligned mixers*.

Prior approaches for warm-started QAOA considered initial states of the form $|\psi_i\rangle = \bigotimes_{j=1}^{n} |\boldsymbol{n}_j\rangle$ where $\boldsymbol{n}_1, \ldots, \boldsymbol{n}_n$ are obtained by some classical procedure. In the work by Egger et al. [5], for problems whose corresponding QUBO (Quadratic Unconstrained Binary Optimization) formulation satisfies certain properties, they solve a relaxation of the QUBO and map the solutions to states in,

$$\mathbf{Arc} = \{\cos(\theta/2)\,|0\rangle + \sin(\theta/2)\,|1\rangle : \theta \in (0, \pi)\}, \tag{6}$$

i.e., points on the Bloch sphere that intersect with the xz-plane with non-negative x coordinate; the blue arc in Fig. 1 corresponds to the possible qubit positions. For the MAX-CUT problem, Tate et al. [11,12] consider higher-dimensional relaxations of the Max-Cut problem (i.e. the Burer-Monteiro relaxation and the relaxation used in the Goemans-Williamson algorithm) which, after a possible projection, yield points all over the surface of the Bloch sphere; however, any advantages this method has over others cannot be attributed to simply utilizing more of the Bloch sphere's surface as seen in the following remark.

Remark 1. Tate et al. [12] show that for every initial product state $|\psi_i\rangle = \bigotimes_{j=1}^{n} |\boldsymbol{n}_j\rangle$, there exists a different initial product state $|\psi_i'\rangle = \bigotimes_{j=1}^{n} |\boldsymbol{n}_j'\rangle$ with $\boldsymbol{n}_j' \in \mathbf{Arc}$ for all j, such that, up to a global phase, QAOA with aligned mixers and initial state $|\psi_i\rangle$ returns the same state as QAOA with aligned mixers and initial state $|\psi_i'\rangle$. This property holds for any optimization problem and corresponding cost Hamiltonian C. More specifically, if $|\boldsymbol{n}_j\rangle = \cos(\theta_j/2)\,|0\rangle + e^{i\phi_j}\sin(\theta_j/2)\,|1\rangle$ is an arbitrary qubit on the Bloch sphere with polar angle $0 \le \theta_j \le \pi$ and azimuthal angle $0 \le \phi_j < 2\pi$, then one can choose $|\boldsymbol{n}_j'\rangle = \cos(\theta_j/2)\,|0\rangle + \sin(\theta_j/2)\,|1\rangle$.

We now explicitly define the warm-starts used in this work. First, for any product state $|s\rangle = \bigotimes_{j=1}^{n} |s_j\rangle$, we define θ_j and ϕ_j to be the polar and azimuthal angle of $|s_j\rangle$ on the Bloch sphere, i.e., $|s_j\rangle = \cos(\theta/2)\,|0\rangle + e^{i\phi}\sin(\theta/2)\,|1\rangle$. For each $j \in [n]$, we use $\hat{\theta}_j$ to measure the angle that the jth qubit is to the nearest pole, i.e., $\hat{\theta}_j = \min(\theta_j, \pi - \theta_j)$. This work focuses on the restriction that each qubit is at most angle $\theta \in [0, \pi/2]$ away from one of the poles, i.e., $\hat{\theta}_j \le \theta$ for all $j \in [n]$; we refer to such warm-starts as *within-θ-warm-starts*. In light of Remark 1 above, in the context of warm-started QAOA with within-θ-warm-starts with aligned mixers, it suffices to only consider warm-starts of the form $|\psi_i\rangle = \bigotimes_{j=1}^{n} |\theta_j\rangle$ where $|\theta_j\rangle := \cos(\theta/2)\,|0\rangle + \sin(\theta/2)\,|1\rangle \in \mathbf{Arc}$ with $\hat{\theta}_j \le \theta$ for all $j \in [n]$.

In addition, we also consider a subclass of within-θ-warm-starts where each qubit is *exactly* some fixed angle from the poles of the Bloch sphere, i.e., $\theta_j = \theta$ for all $j \in [n]$; we refer to such warm-starts as *at-θ-warm-starts*. As seen in Fig. 1, this forces each qubit to be in one of two positions on the Bloch sphere; this naturally induces a *corresponding bitstring* $b \in \{0,1\}^n$ with the property that at $\theta = 0$, measuring the warm-start yields exactly the state $|b\rangle$.

For the MAX-CUT problem, previous works have considered QAOA with at-θ-warm-starts. Tate et al. [10] provided some theoretical results for at-θ-warm-starts in the case of single-round Max-Cut QAOA on 3-regular graphs with aligned mixers. Egger et al. [5] consider a QAOA-variant (different than their QUBO-relaxation variant) with at-θ-warm-starts with $\theta = \pi/3$, but with an

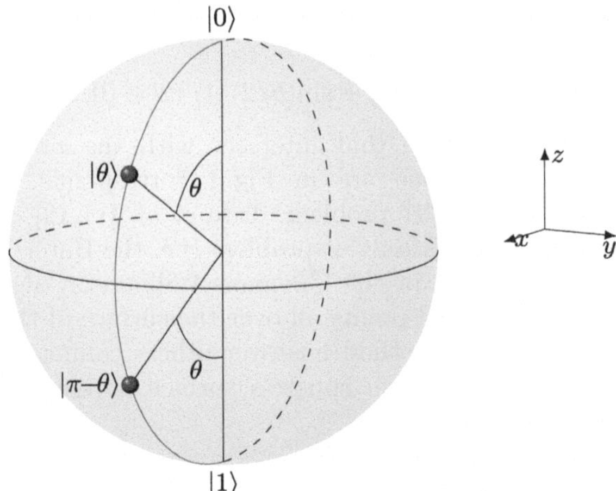

Fig. 1. A geometric depiction of the states $|\theta\rangle$ and $|\pi - \theta\rangle$ on the Bloch sphere; these two positions correspond to qubit positions found in at-θ-warm-starts. The blue half-circle, **Arc**, in the xz-plane denotes all the possible positions for $|\theta\rangle$ as θ varies from 0 to π.(Color figure online)

unaligned mixer that has the property that there exists parameters γ and β such that depth-1 QAOA yields the bitstring b corresponding to the warm-start (i.e. it returns exactly $|b\rangle$). Cain et al. [4] considered Max-Cut QAOA with at-θ-warm-starts with $\theta = 0$ with the transverse field mixer; it was found that when the cut associated with the corresponding bitstring b was a "good" cut, then this variant of QAOA yields little to no improvement regardless of the circuit depth used.

4 A Toy Problem

Before giving the formal result, we first consider a toy problem to give the reader intuition for why one might expect warm-started QAOA to perform poorly if the warm-start is initialized near the poles. This toy problem consists of a single qubit and the classical objective function is given by $c(x) = x$ for $x \in \{0, 1\}$. This corresponding cost Hamiltonian is technically $C = \frac{1}{2}(I + Z)$ but we can instead[2] use the simpler Hamiltonian $C = Z$.

For general problems, if the corresponding bitstring b of a at-θ-warm-start is non-optimal, then the QAOA circuit effectively needs to "correct" certain qubits so that they are on the opposite hemisphere of the Bloch sphere from where they started. For our toy problem, since $\max_x c(x) = 1$, if we initialize with the warmstart $|\psi_i\rangle = |\theta\rangle$ with θ small, then the QAOA circuit needs to move the qubit from (near the) north pole of the Bloch sphere to the south pole.

[2] A similar replacement technique, obtained by shifting and scaling the cost objective, is discussed in Sect. 5.

For our toy problem, both QAOA unitaries, $U(B_{|\psi_i\rangle}, \beta)$ and $U(C, \gamma) = U(Z, \gamma)$ correspond to single-qubit rotations on the Bloch sphere, with the former rotating about the original qubit position and the latter rotating about the z-axis of the Bloch sphere. To simplify matters, we assume that the variational parameters γ and β are non-zero and are chosen so that the resulting (single-qubit) state stays in the xz-plane of the Bloch sphere after each QAOA unitary (i.e. rotation); this assumption puts an upper bound on the required circuit depth needed to achieve some change in approximation ratio since there may perhaps be better choices for γ and β. When θ is small, it takes several layers of the QAOA circuit to significantly move the qubit away from its original position since the axes of rotation for both QAOA unitaries are so close to another as seen in Fig. 2. In particular, the QAOA circuit evolves the state as follows:

$$|\theta\rangle \to |-\theta\rangle \to |3\theta\rangle \to |-3\theta\rangle \to |5\theta\rangle \to \cdots . \tag{7}$$

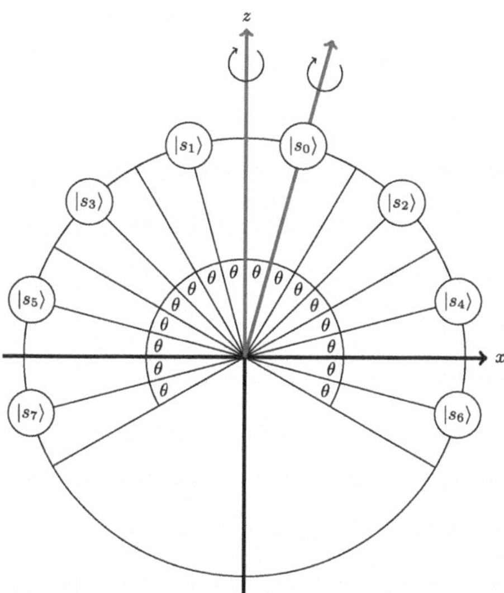

Fig. 2. An illustration of the evolution of quantum state as each unitary of the QAOA circuit is applied. The large circle represents the intersection of the Bloch sphere with the xz-plane. The two alternating unitaries of the QAOA circuit correspond to a single-qubit rotation, represented by the two red arrows. Starting with the warm-start initial state $|s_0\rangle$, the state $|s_j\rangle$ ($j = 0, 1, \ldots, 7$) represents the quantum state after j unitary operations of the QAOA; when j is even, then $|s_j\rangle$ represents the result of a depth-p QAOA circuit with $p = j/2$. (Color figure online)

In general, after p layers of the QAOA circuit, the resulting quantum state will be $|(2p+1)\theta\rangle$. Since $\max_x c(x) = 1$, then the resulting initial and final approximation ratios are $\lambda_i = \sin^2(\theta/2)$ and $\lambda_f = \sin^2((2p+1)\theta/2)$ respectively.

For very small values of θ, $\lambda_i \approx 0$, and thus the change in approximation is approximately,

$$\Delta\lambda \approx \lambda_f = \sin^2((2p+1)\theta/2). \tag{8}$$

If $\Delta\lambda$ is held constant, then the right-hand side of Eq. 8 is also constant, which can only occur if $2p+1$ and θ are inversely proportional to one another, i.e., the required circuit depth scales as $p = O(1/\theta)$ for small θ.

While the toy example gives a geometric intuition for why small initialization angles are bad, this intuition is difficult to directly generalize for multi-qubit systems since the Bloch sphere is not suitable for modeling entangled systems. Instead, we consider a more sophisticated approach in the next section, allowing us to obtain *lower* bounds on the required circuit depth needed to achieve certain changes in approximation ratio.

5 Bounds on Required Depth for Small Initialization Angle

Recent work by Benchasattabuse et al. [1] provide lower bounds on the number of rounds of QAOA that are required to obtain guaranteed approximation ratios; moreover, such bounds are applicable across a wide variety of optimization problems, phase separators, and mixing Hamiltonians. We now analyze such bounds in the context of warm-started QAOA and eventually show (Theorem 4) that for a desired change of approximation ratio $\Delta\lambda$, the required number of rounds scales roughly as $\Omega(\Delta\lambda/\theta)$ for small θ approaching zero.

Benchasattabuse et al. [1] obtained the QAOA circuit-depth bounds by utilizing one of the lower bounds on quantum annealing times found by García-Pintos et al. [7] which we summarize below in Theorem 1.

Theorem 1 (García-Pintos et al. [7]). *Let $H(t) = (1 - g(t))H_0 + g(t)H_1$ be a quantum annealing annealing schedule where $g(0) = 0$ and $g(t_f) = 1$, t_f is the total annealing time, and H_0 and H_1 are Hamiltonians with zero ground state energies. For $t \in [0, t_f]$, let $|\psi_t\rangle$ be the quantum state at time t of the annealing process. If $|\psi_0\rangle$ is a ground state of H_0, then the following inequality holds:*

$$t_f \geq \frac{\langle H_0\rangle_{t_f} + \langle H_1\rangle_0 - \langle H_1\rangle_{t_f}}{\|[H_1, H_0]\|}, \tag{9}$$

where for any appropriately-sized Hamiltonian H and $t \in [0, t_f]$, $\langle H\rangle_t := \langle\psi_t| H |\psi_t\rangle$.

Benchasattabuse et al. [1] are able to translate these bounds on annealing times into bounds on QAOA circuit depth by using the fact that QAOA can be interpreted as quantum annealing with a "bang-bang" schedule, i.e., H_1 and H_0 correspond to the phase separator and mixing Hamiltonian respectively, the annealing schedule is such that $g(t) \in \{0, 1\}$ for all $t \in [0, t_f]$, and the annealing time is $\sum_{j=1}^{p}(|\beta_j| + |\gamma_j|)$.

Remark 2. The above paragraph should **not** be misinterpreted to mean that the results of this paper only work for QAOA with specific parameter choices (e.g. choosing γ and β in a way corresponding to a linear annealing schedule). The results in this paper hold for *any* choice of γ and β parameters.

The QAOA circuit-depth bounds obtained by Benchasattabuse et al. (Theorem 2 of [1]) assume that the initial state is the equal superposition state $|+\rangle^{\otimes n}$. It is not too difficult to adjust the bounds (and the corresponding proof) to account for arbitrary initial states. We state the adjusted bounds in Theorem 2 and prove that such an adjustment is correct.

Theorem 2. *Given a classical objective function $c(x)$ for a maximization task, represented by the Hamiltonian C, encoded into a phase separator Hamiltonian $H_1 = c_{max}I - C$ and a mixing Hamiltonian H_0, where all Hamiltonians are 2π periodic with zero ground state energies. Let c_{max} denote the global maximum of $c(x)$. Let $|\psi_i\rangle$ be a ground state of H_0 such that measurement of $|\psi_i\rangle$ yields an approximation ratio of λ_i. If a QAOA protocol with p rounds driven by H_0 and H_1 that starts from $|\psi_i\rangle$ and reaches a state $|\psi_f\rangle$ with approximation ratio λ_f, then,*

$$p \geq \frac{\langle\psi_f| H_0 |\psi_f\rangle + \Delta\lambda \cdot c_{max}}{4\pi||[C, H_0]||}, \tag{10}$$

where $\Delta\lambda = \lambda_f - \lambda_i$.

Proof. In the proof by [1], they show that the above bound holds if $\Delta\lambda \cdot c_{\max}$ is instead replaced with $\langle\psi_i| H_1 |\psi_i\rangle - \langle\psi_f| H_1 |\psi_f\rangle$; this is true even if $|\psi_i\rangle \neq |+\rangle^{\otimes n}$. However, these two expressions are equal:

$$\begin{aligned}\langle\psi_i| H_1 |\psi_i\rangle - \langle\psi_f| H_1 |\psi_f\rangle &= \langle\psi_f| C |\psi_f\rangle - \langle\psi_i| C |\psi_i\rangle &(11)\\ &= c_{\max}\lambda_f - c_{\max}\lambda_i &(12)\\ &= \Delta\lambda \cdot c_{\max}, &(13)\end{aligned}$$

hence showing that this adjusted version of the bound holds.

The original bound,

$$p \geq \frac{\langle\psi_f| H_0 |\psi_f\rangle + c_{\max}\lambda_f - c_{\text{avg}}}{4\pi||[C, H_0]||}. \tag{14}$$

in the statement of Theorem 2 of [1], which assumes that $|\psi_i\rangle = |+\rangle^{\otimes n}$, can be obtained from our adjusted bounds by observing that,

$$\lambda_i c_{\max} = \frac{\langle+|^{\otimes n} C |+\rangle^{\otimes n}}{c_{\max}} c_{\max} = c_{\text{avg}}. \tag{15}$$

For convenience, we consider a looser-version of the bound in Corollary 1 and use $p_{\min}$ to denote the value of this looser bound.

Corollary 1. *The bound in the statement of Theorem 2 can be replaced with*

$$p \geq p_{min} := \frac{\Delta\lambda \cdot c_{max}}{4\pi||[C, H_0]||}. \tag{16}$$

Proof. This follows from Theorem 2 and that $\langle\psi_f| H_0 |\psi_f\rangle \geq 0$ (as H_0 is assumed to have a zero ground state energy).

Before continuing, we point out certain important conditions in Theorem 2. First, all of the mixers need to be 2π periodic. In particular, this will hold for the cost Hamiltonian if the corresponding classical objective function $c : \{0,1\}^n \to \mathbb{R}$ has integer objective values, i.e., $c(x) \in \mathbb{Z}$ for all $x \in \{0,1\}^n$. Now consider the transverse field mixer B_{TF} and a warm-start mixer $B_{|\psi_i\rangle}$ that is aligned with the warm-start $|\psi_i\rangle$. It can be shown that both of these are 2π periodic; however, Theorem 2 also requires that the mixers also have zero energy which is not the case for B_{TF} and $B_{|\psi_i\rangle}$. In particular, for *any* product state $|s\rangle$, the mixer $B_{|s\rangle}$ has energies ranging from $-n$ to n (see [12] for a proof); however, we can shift the mixers to have energies ranging from 0 to n as so: $\hat{B}_{|s\rangle} := \frac{1}{2}(nI - B_{|s\rangle})$. Choosing $|s\rangle$ appropriately, we have $\hat{B}_{\text{TF}} := \frac{1}{2}(nI - B_{\text{TF}})$ and $\hat{B}_{|\psi_i\rangle} := \frac{1}{2}(nI - B_{|\psi_i\rangle})$, where $\hat{B}_{\text{TF}}$ and $\hat{B}_{|\psi_i\rangle}$ have energies ranging from 0 to n. Moreover, now $|+\rangle^{\otimes n}$ and $|\psi_i\rangle$ are *ground* states of $\hat{B}_{\text{TF}}$ and $B_{|\psi_i\rangle}$ respectively as required by Theorem 2. With some appropriate rescaling of the variational parameters γ and β, one can show that replacing B_{TF} with $\hat{B}_{\text{TF}}$ (or $B_{|\psi_i\rangle}$ with $\hat{B}_{|\psi_i\rangle}$) in the QAOA circuit will result in the same final state. Nonetheless, it is important to consider the versions of these mixers with zero-ground state energy as the commutator term $[C, H_0]$ in Theorem 2 and Corollary 1 assume the mixer is written in such a form. The following lemma will be helpful in regards to analyzing such commutators.

Lemma 1. *Let* $|\psi_i\rangle = \bigotimes_{j=1}^n |\theta_j\rangle = \bigotimes_{j=1}^n |\boldsymbol{n}_j\rangle$ *be a within-θ-warm-start. Then,* $[B_{\boldsymbol{n}_j,j}, C] = \sin(\hat{\theta}_j)[X_j, C]$, *where* $\hat{\theta}_j$ *is the angle that the* j*th qubit is away from one of the poles of the Bloch sphere (defined in Sect. 3.1).*

Proof. Recall that $B_{|\psi_i\rangle} = \sum_{j=1}^n B_{\boldsymbol{n}_j,j}$, where $B_{\boldsymbol{n}_j,j} = x_j X_j + y_j Y_j + z_j Z_j$, where $\boldsymbol{n}_j = (x_j, y_j, z_j)$ is the position of the jth qubit in Cartesian coordinates. Given $|\psi_i\rangle = \bigotimes_{j=1}^n |\theta_j\rangle$, we can calculate the jth qubit position as:

$$\boldsymbol{n}_j = (\sin(\theta_j), 0, \cos(\theta_j)) \tag{17}$$

and thus, we have that, $B_{\boldsymbol{n}_j} = \sin(\theta)X_j + \cos(\theta)Z_j$. Since $[Z_j, C] = 0$ for all $j \in [n]$, we have that $[B_{\boldsymbol{n}_j,j}, C] = \sin(\theta_j)[X_j, C]$.

To complete the proof, it suffices to show that $\sin(\theta_j) = \sin(\hat{\theta}_j)$; this can be seen by recalling the identity $\sin(x) = \sin(\pi - x)$ and the fact that $\hat{\theta}_j = \min(\theta_j, \pi - \theta_j)$.

Using the commutator relation in Lemma 1, we now present our main result which places a lower bound on the required number of angles for warm-started QAOA with within-θ-warm-starts.

Theorem 3. *Fix a phase separator* $H_1 = c_{max}I - C$ *(where* C *is a Hamiltonian that corresponds to some classical objective function* $c(x)$*), choice of* $\Delta\lambda$, *and a within-θ-warm-start $|\psi_i\rangle$. Let $p_{min}(\hat{B}_{|\psi_i\rangle})$ correspond to the bound p_{min} in Corollary 1 where H_0 is replaced with $H_0 = \hat{B}_{|\psi_i\rangle}$. Then,*

$$p_{min}(\hat{B}_{|\psi_i\rangle}) \geq \frac{\Delta\lambda}{\sin(\theta)}\mathcal{F}(c), \tag{18}$$

where $\mathcal{F}(c)$ is a function that depends on the classical objective c and is independent of $\Delta\lambda, |\psi_i\rangle$, and θ.

Proof. Recall from Remark 1, that we may assume (without loss of generality) that $|\psi_i\rangle$ has the form $|\psi_i\rangle = \bigotimes_{j=1}^{n} |\theta_j\rangle$ where $\hat{\theta}_j$ (the angle that the jth qubit of $|\psi_i\rangle$ is from the nearest pole on the Bloch sphere) lies in the interval $[0, \theta]$.

Now, recall that $B_{|\psi_i\rangle} = \sum_{j=1}^{n} B_{\boldsymbol{n}_j,j}$, where $\boldsymbol{n}_j = (x_j, y_j, z_j)$ is the position of the jth qubit in Cartesian coordinates. Also recall from Lemma 1 that $[B_{\boldsymbol{n}_j,j}, C] = \sin(\hat{\theta}_j)[X_j, C]$. Thus, it follows that,

$$[B_{|\psi_i\rangle}, C] = \sum_{j=1}^{n}[B_{\boldsymbol{n}_j,j}, C] = \sum_{j=1}^{n}\sin(\hat{\theta}_j)[X_j, C]. \tag{19}$$

Taking norms on both sides yields,

$$\|[B_{|\psi_i\rangle}, C] = \left\|\sum_{j=1}^{n}\sin(\hat{\theta}_j)[X_j, C]\right\| \leq \sum_{j=1}^{n}\sin(\theta)\,\|[X_j, C]\| = \frac{\sin(\theta)}{\mathcal{F}(c)}\cdot\frac{c_{\max}}{2\pi},$$

where,

$$\mathcal{F}(c) = \frac{c_{\max}}{2\pi}\cdot\left(\sum_{j=1}^{n}\|[X_j, C]\|\right)^{-1}. \tag{20}$$

Using basic commutator properties, it is easy to show that $[\hat{B}_{|\psi_i\rangle}, C] = -\frac{1}{2}[B_{|\psi_i\rangle}, C]$. Finally, from the definition of $p_{\min}$ in Corollary 1, we have

$$p_{\min}(\hat{B}_{|\psi_i\rangle}) = \frac{\Delta\lambda\cdot c_{\max}}{4\pi\|[C, \hat{B}_{|\psi_i\rangle}]\|} = \frac{\Delta\lambda\cdot c_{\max}}{2\pi\|[C, B_{|\psi_i\rangle}]\|} \geq \frac{\Delta\lambda}{\sin(\theta)}\mathcal{F}(c).$$

Note that for small θ, $\sin(\theta) \approx \theta$, so in such a case, the required number of rounds of warm-started QAOA, as a function of initialization angle θ and desired change in approximation ratio $\Delta\lambda$, scales as $p = \Omega(\Delta\lambda/\theta)$. In particular, the number of required rounds approaches infinity as θ approaches 0, which suggests that the method of warm-started QAOA discussed in this work is not recommended for very small choices of θ if one wishes to obtain an appreciable value of $\Delta\lambda$.

In the special case of at-θ-warm-starts, where each qubit in the warm-start is *exactly* angle θ away from one of the poles, we have a slightly stronger result that directly relates the lower bounds on the required number of rounds required for warm-started QAOA vs standard QAOA.

Theorem 4. *Fix a phase separator $H_1 = c_{max}I - C$ (where C is a Hamiltonian that corresponds to some classical objective function $c(x)$), choice of $\Delta\lambda$, and let $|\psi_i\rangle$ be an at-θ-warm-start for some $0 \leq \theta \leq \pi/2$. Let $p_{min}(\hat{B}_{TF})$ and $p_{min}(\hat{B}_{|\psi_i\rangle})$ correspond to the bound p_{min} in Corollary 1 where H_0 is replaced with $H_0 = \hat{B}_{TF}$ and $H_0 = \hat{B}_{|\psi_i\rangle}$ respectively. Then,*

$$p_{min}(\hat{B}_{|\psi_i\rangle}) = \frac{1}{\sin\theta} \cdot p_{min}(\hat{B}_{TF}). \tag{21}$$

Proof. Recall from Remark 1, that we may assume (without loss of generality) that $|\psi_i\rangle$ has the form $|\psi_i\rangle = \bigotimes_{j=1}^{n} |\theta_j\rangle$. Since $|\psi_i\rangle$ was an at-θ-warm-start, then $\hat{\theta}_j$ (the angle that the jth qubit of $|\psi_i\rangle$ is from the nearest pole on the Bloch sphere) is equal to θ for all $j \in [n]$.

As a result of Lemma 1 we have $[B_{n_j,j}, C] = \sin(\hat{\theta}_j)[X_j, C] = \sin(\theta)[X_j, C]$. Thus, it follows that,

$$[B_{|\psi_i\rangle}, C] = \sum_{j=1}^{n} [B_{n_j,j}, C] = \sin(\theta) \sum_{j=1}^{n} [X_j, C] = \sin(\theta) \cdot [B_{\mathrm{TF}}, C]. \tag{22}$$

The result then follows from the commutator relation above and the definition of $p_{\min}$ in Corollary 1.

One should be careful in interpreting Theorem 4; since $p_{\min}(\hat{B}_{|\psi_i\rangle}) \geq p_{\min}(\hat{B}_{\mathrm{TF}})$, at first glance, it appears that it is always preferable to run standard QAOA. However, this is not necessarily the case. First, the inequality is a result of comparing lower bounds which are not necessarily tight, the relationship between the actual number of required rounds (between standard QAOA and warm-started QAOA) may actually be much different. Second, even if the bounds were tight, this relationship between $p_{\min}(\hat{B}_{|\psi_i\rangle})$ and $p_{\min}(\hat{B}_{\mathrm{TF}})$ assumes that $\Delta\lambda$ fixed; if instead the target approximation ratio λ_f is fixed and λ_i is allowed to vary, then this relationship does not necessarily hold. In particular, numerical evidence such as those found in [5,12], suggest that for many instances and for a suitable choice of θ, the number of required rounds to achieve a particular target approximation ratio λ_f for warm-started QAOA is much less compared to standard QAOA.

6 Discussion and Conclusion

This work is consistent with preliminary numerical simulations [5,12] which suggests that there is very little change in the approximation ratio when the initialization angle θ is near zero; there were previous geometrical intuitions for why this was the case, but this work shows definitively that small choices of initialization angles are not suitable for warm-started QAOA with aligned mixers.

It should be noted our results are fundamentally a consequence of QAOA with *aligned* mixers being able to be interpreted as a quantum-annealing protocol with a bang-bang schedule; for QAOA with non-aligned mixers, it is likely the case that an entirely different type of analysis is necessary to obtain similar lower bounds on the circuit depth.

Acknowledgments. This work was supported by the U.S. Department of Energy through the Los Alamos National Laboratory. Los Alamos National Laboratory is operated by Triad National Security, LLC, for the National Nuclear Security Administration of U.S. Department of Energy (Contract No. 89233218CNA000001). LANL report LA-UR-24-29913.

References

1. Benchasattabuse, N., Bärtschi, A., García-Pintos, L.P., Golden, J., Lemons, N., Eidenbenz, S.: Lower bounds on number of QAOA rounds required for guaranteed approximation ratios. arXiv preprint arXiv:2308.15442 (2023)
2. Binkowski, L., Koßmann, G., Ziegler, T., Schwonnek, R.: Elementary proof of QAOA convergence. arXiv preprint arXiv:2302.04968 (2023)
3. Blekos, K., et al.: A review on quantum approximate optimization algorithm and its variants. Phys. Rep. **1068**, 1–66 (2024)
4. Cain, M., Farhi, E., Gutmann, S., Ranard, D., Tang, E.: The QAOA gets stuck starting from a good classical string. arXiv preprint arXiv:2207.05089 (2022)
5. Egger, D.J., Mareček, J., Woerner, S.: Warm-starting quantum optimization. Quantum **5**, 479 (2021)
6. Farhi, E., Goldstone, J., Gutmann, S.: A quantum approximate optimization algorithm. arXiv preprint arXiv:1411.4028 (2014)
7. García-Pintos, L.P., Brady, L.T., Bringewatt, J., Liu, Y.K.: Lower bounds on quantum annealing times. Phys. Rev. Lett. **130**(14), 140601 (2023)
8. Glendinning, I.: Rotations on the bloch sphere (2010). https://doi.org/10.13140/RG.2.2.27566.25922
9. Hadfield, S., Wang, Z., O'Gorman, B., Rieffel, E.G., Venturelli, D., Biswas, R.: From the quantum approximate optimization algorithm to a quantum alternating operator ansatz. Algorithms **12**(2) (2019). https://doi.org/10.3390/a12020034, https://www.mdpi.com/1999-4893/12/2/34
10. Tate, R., Eidenbenz, S.: Guarantees on warm-started QAOA: single-round approximation ratios for 3-regular maxcut and higher-round scaling limits. arXiv preprint arXiv:2402.12631 (2024)
11. Tate, R., Farhadi, M., Herold, C., Mohler, G., Gupta, S.: Bridging classical and quantum with SDP initialized warm-starts for QAOA. ACM Trans. Quant. Comput. **4**(2), 1–39 (2023)
12. Tate, R., Moondra, J., Gard, B., Mohler, G., Gupta, S.: Warm-started QAOA with custom mixers provably converges and computationally beats Goemans-Williamson's max-cut at low circuit depths. Quantum **7**, 1121 (2023). https://doi.org/10.22331/q-2023-09-26-1121

Online Busy Time Scheduling with Untrusted Prediction

Rick van de Bovenkamp and Alison Hsiang-Hsuan Liu(✉)

Utrecht University, Utrecht, The Netherlands
h.h.liu@uu.nl

Abstract. In this work, we study the online busy time scheduling problem with infinite processors, where each job j has a release time r_j, a processing time p_j, and a deadline d_j. Busy time scheduling aims to use multiple processors to schedule jobs concurrently to minimize the time a machine has to process jobs. We consider the case proposed by Koehler and Khuller [20], where a single machine has unlimited processors. Moreover, we consider the case where the online algorithm can access prediction, which might be imperfect, on when the machine should be active. We present an algorithm, Multiplier, that dynamically adjusts its strategy for reserving time windows for potential future jobs according to the prediction it receives and how much it trusts the prediction. We show that Multiplier is $(1+\frac{4}{1+\lambda})$-consistent and $(1+\frac{4}{1-\lambda})$-robust with a trust parameter $\lambda \in [0,1)$.

Keywords: online busy time scheduling · untrusted prediction · consistency · robustness · adaptive strategy

1 Introduction

In 2010, Flammini et al. introduced the busy time scheduling (BTS) problem [16], which concerns the energy consumption in data centers [1,10,19], cloud computing systems [25,29], optical network design [14,15], and more [18]. In the BTS problem, we are given a set of jobs, where each job j has a processing time and has to be executed within its feasible time interval on multiple machines, each with one or more processors that can execute at most one job at a time. The objective is to finish all jobs feasibly so that the total time the machines are actively processing jobs is minimized.

Formally, we have a set of jobs J and a set of machines M. Each job $j \in J$ has a *release time* $r_j \geq 0$, a required *processing time* $p_j > 0$, and a *deadline* $d_j \geq r_j + p_j$. Each machine in M has $g \geq 1$ *processors*, which can be interpreted as the capacity of the maximum number of jobs that can be executed simultaneously on the same machine. The *busy time* of a machine is the total length of the time when it is executing at least one job. Any feasible algorithm must decide what the starting time $\mathbf{s}_j$ is such that $\mathbf{s}_j + p_j \leq d_j$. The objective is to minimize the total busy time of all machines while feasibly executing all the jobs in J.

R. Královič and V. Kůrková (Eds.): SOFSEM 2025, LNCS 15539, pp. 324–336, 2025.
https://doi.org/10.1007/978-3-031-82697-9_24

The offline version of the BTS problem is NP-hard even when $g = 2$, which can be implied by the work of Wintekler et al. about wavelength assignment minimization in fiber-optical networks [31]. On the other hand, the online (time) setting considers the situation where the jobs arrive at their release time, and the online algorithm has to assign the job to a machine (and a processor) upon the arrival of the job without any knowledge of future jobs.

It is common to use *competitive analysis* to measure the performance of an online algorithm ALG. Let ALG(J) denote the cost of ALG on jobs J and OPT(J) denote the optimal solution cost on jobs J. The *competitive ratio* of ALG is at most c if $\text{ALG}(J) \leq c \cdot \text{OPT}(J) + a$ for any input instance J, where a is a non-negative constant. Moreover, such an algorithm is called c-competitive.

Previous Results on Online BTS. The online BTS problem was first studied in the *rigid* setting, where each job j has $d_j = r_j + p_j$. Namely, the job cannot be "shifted" within its feasible interval $[r_j, d_j)$, and the challenge is how to partition the jobs to the machines (each with a capacity of g) wisely. Shalom et al. [28] showed that no deterministic online algorithm could be better than g-competitive and provided a $5 \log p_{\max}$-competitive algorithm, where $p_{\max}$ is the largest processing time of the jobs. The authors also studied different special instances. When the feasible intervals $[r_j, d_j)$ of jobs j in the input instance form a clique, there exists a $2(1 + \phi)$-competitive algorithm, where ϕ is the golden ratio. The competitive ratio can be further reduced to ϕ if the feasible intervals form a one-sided clique. In 2020, Ren et al. introduced a setting that closely resembles the problems faced in cloud computing [27]. They use a model in which machines may have different numbers of processors (that is, g is a variable on different machines) and varying costs associated with them. Their algorithm is also non-clairvoyant, meaning the processing times are unknown until a job has finished running. The competitive ratio of their algorithm is $\Theta(\log \frac{p_{\max}}{p_{\min}})$.

The *flexible* setting, where jobs may have wider feasible intervals concerning their processing times, was initialized in 2017 [17,20,27]. In this setting, $d_j \geq r_j + p_j$ for every job j in the input instance. The current best results are from Koehler and Khuller [20]. The authors proposed a $O(\log \frac{p_{\max}}{p_{\min}})$-competitive algorithm for the case where each machine has a bounded number g of processors. The authors then focused on the challenge of when to start a job when there are infinite possibilities and turned to the setting where machines have infinite processors (that is, $g \to \infty$). They proposed a 5-competitive algorithm, DOUBLER, for this setting[1] and showed that no deterministic online algorithm could be better than ϕ-competitive. Interestingly, all three papers [17,20,27] present online algorithms that are very similar. The algorithms wait until some unscheduled job reaches its latest possible starting time, at which point it is scheduled. Then, this job is marked as a "primary" job or a "flag" job. The algorithm by Koehler and Khuller [20] then reserves an open window relative to the processing time of the scheduled job, under which all jobs that fit in it are also scheduled. The other

[1] Fong et al. claimed a 4-competitive result in [17] but the analysis is proved to be incorrect according to Koehler and Khuller [20] and van de Bovenkamp [11].

two algorithms define a minimum overlap ratio with the primary/flag job, for which all jobs that satisfy this condition are scheduled [17,26].

For more results for the BTS problem, we direct the readers to a survey by Chau and Li [13].

Online Algorithms with Untrusted Prediction. The pure online model where competitive analysis is pessimistic in that the uncertainty controlled by the adversarial input is over-powerful; the worst case does not happen often but dominates the algorithm's performance. A new model has arisen in which the online algorithm can access a prediction about future events computed by machine-learning techniques. However, the prediction is imperfect and can even be malicious. Thus, blindly trusting the prediction may cause disastrous results.

Since the predictions that are made may contain large errors, we consider it to be untrusted prediction. Ideally, we want to design an online algorithm that is both *consistent* (that is, the algorithm's performance improves as the prediction quality improves, and the algorithm performs close to the optimal solution when the prediction is perfect) and *robust* (that is, the algorithm should not be much worse than the best-known online algorithm if the prediction is very bad or even adversarial).

For scheduling problems with untrusted prediction (or machine-learned prediction), people have worked on makespan minimization [21,22,24], flow-time minimization [5,6], scheduling with speed scaling [4,8], contract scheduling [3], peak demand minimization [23], and more [7,9,12].

There are multiple different formal definitions of consistency and robustness of an online algorithm. Let $\textsc{Alg}(J, \mathcal{S})$ be the cost incurred by the $\textsc{Alg}$ algorithm on input J with prediction $\mathcal{S}$ and $\mathrm{OPT}(J)$ the optimal cost on J. In this work, we follow the definition in [2]:

Definition 1. (Consistency). *The* consistency *of algorithm* $\textsc{Alg}$ *with untrusted prediction* $\mathcal{S}$ *is the worst-case competitive ratio given that the prediction is as good as possible. Formally,* $\textsc{Alg}$ *is* r*-consistent where* $r = \sup_J \inf_{\mathcal{S}} \frac{\textsc{Alg}(J,\mathcal{S})}{OPT(J)}$.

Definition 2. (Robustness). *The* robustness *of algorithm* $\textsc{Alg}$ *with untrusted prediction* $\mathcal{S}$ *is the worst-case competitive ratio, irrespective of the prediction. Formally,* $\textsc{Alg}$ *is* w*-robust where* $w = \sup_J \sup_{\mathcal{S}} \frac{\textsc{Alg}(J,\mathcal{S})}{OPT(J)}$.

Our Contribution. This paper studies the online BTS problem on a single machine with unlimited processors. Moreover, we consider the model where untrusted prediction about when the machine should be active is available. With this prediction, we propose an online algorithm, $\textsc{Multiplier}$, that is $(1 + \frac{4}{1+\lambda})$-consistent and $(1+\frac{4}{1-\lambda})$-robust, where $\lambda \in [0, 1)$ is the trust parameter indicating how much the algorithm trust the prediction.

The $\textsc{Multiplier}$ algorithm mimics the behavior of the $\textsc{Doubler}$ algorithm by Koehler and Khuller [20]. The difference is that unlike $\textsc{Doubler}$, which always reserves a time window with a size twice the job triggering the procedure (see Sect. 3 for a recap of DOUBLER), MULTIPLIER may have different sizes of the reserved time window that dependents on the trust parameter λ and the prediction.

Intuitively, we want the MULTIPLIER algorithm to degrade to the DOUBLER algorithm when it does not trust the prediction at all and follows the prediction as much as possible when it completely trusts the prediction. Moreover, the length of the reserved time window should be a factor of the (primary) job's processing time, without any additive term, to be embedded into the proof framework. To this end, we carefully design two factors, α and β, which are both functions of λ, to limit the range of size of the reserved time window according to the job that triggers the scheduling procedure and the prediction received by the algorithm. More specifically, the β factor indicates the boundary behavior of MULTIPLIER when it is not possible to incorporate the prediction into the current schedule, and the α factor indicates how suspicious MULTIPLIER is even when the prediction can be incorporated into the current schedule.

The MULTIPLIER algorithm dynamically decides the size of the reserved time window considering the prediction, using α and β. The variable factor of the reserved time window size increases the complexity of the analysis. Our design of how α and β are functions of the trust parameter λ allows MULTIPLIER to smoothly morph from DOUBLER to an algorithm that completely trusts the prediction. Moreover, we show that MULTIPLIER guarantees the worst case robustness even for malicious predictions that are designed against it.

Paper Organization. We will first introduce the necessary preliminaries in Sect. 2. For completeness, we will recap the competitive analysis of the DOUBLER algorithm by Koehler and Khuller [20] in Sect. 3. In Sect. 4, based on DOUBLER, we propose our MULTIPLIER algorithm and analyze its robustness and consistency. Finally, we will conclude and reflect upon our research in Sect. 5.

2 Preliminaries

We consider the online BTS problem with untrusted prediction of when the machine should be on. The following are notations and definitions needed throughout the paper.

Problem Definition. We are given a machine with unlimited processors and a set of jobs J. Each job $j \in J$ with *processing time* p_j is revealed to the online algorithm at its *release time* r_j and needs to be finished by its *deadline* d_j. For job j, the time interval $[r_j, d_j)$ is its *feasible interval.*

An online scheduling algorithm ALG has to decide the *starting time* $\mathsf{s}_j^{\text{ALG}}$ for each job $j \in J$. Denoted $\mathsf{e}_j^{\text{ALG}} = \mathsf{s}_j^{\text{ALG}} + p_j$ as the *end time* of job j in the schedule by ALG, the time interval $[\mathsf{s}_j^{\text{ALG}}, \mathsf{e}_j^{\text{ALG}})$ is called the *execution interval* of j. A schedule ALG is *feasible* if, for each job $j \in J$, its execution interval is contained within its feasible interval. Equivalently, ALG is feasible if $\mathsf{s}_j^{\text{ALG}} \in$

$[r_j, d_j - p_j)$ for all jobs $j \in J$. A machine is *busy* at time t according to a schedule ALG if there exists at least one job j such that $t \in [\mathtt{s}_j, \mathtt{e}_j)$. We define the *busy time interval(s)* concerning a schedule ALG on an input J as the union of time t when there is at least one job $j \in J$ executed at t and denote it by $\mathcal{T}^{\text{ALG}}(J)$. Formally, $\mathcal{T}^{\text{ALG}}(J) = \bigcup_{j \in J} [\mathtt{s}_j^{\text{ALG}}, \mathtt{e}_j^{\text{ALG}})$. Note that $\mathcal{T}^{\text{ALG}}$ consists of one or multiple contiguous time intervals. We denote the *length* of a set of time intervals $\mathcal{T}$ by $\mu(\mathcal{T})$. Thus, the length of the busy time interval(s) produced by a schedule ALG on input J is denoted by $\mu(\mathcal{T}^{\text{ALG}}(J))$. Similarly, we denote $\mathtt{s}_j^*$ and $\mathcal{T}^*(J)$ as the starting time of job j and busy time interval(s) on input J of an optimal offline solution, which knows the complete input in advance and has unlimited computation power, respectively. When the context is clear, we drop the parameters ALG and/or J.

Definitions for Algorithms. We denote ℓ_j as the *latest starting time* of a job j, which is defined by $d_j - p_j$. In both DOUBLER and MULTIPLIER algorithms, the jobs assigned at their latest starting time are called *primary* jobs. Given an algorithm ALG, we let $\mathtt{A}_t^{\text{ALG}}$ denote all *available* jobs at time t, namely, the jobs j with $r_j \leq t$ that are not yet scheduled by ALG by time t. When the value of t is clear in the given context, we will refer to $\mathtt{A}_t^{\text{ALG}}$ as $\mathtt{A}$. For convenience, we also denote T as the set of *times in the problem instance*, which start at 0 and ends at $\max_{j \in J}\{d_j + p_j\}$.

Definitions for Analysis. Given two jobs, i and j, we note that their executions *overlap* in a schedule A if and only if $\mathtt{s}_i \leq \mathtt{s}_j + p_j$ and $\mathtt{s}_j \leq \mathtt{s}_i + p_i$. The *interval graph* of the schedule ALG is then defined as the intersection graph of the execution intervals $[\mathtt{s}_j^{\text{ALG}}, \mathtt{e}_j^{\text{ALG}})$ for all $j \in J$. More specifically, in the interval graph of ALG on input J, $G_{\text{ALG}} = (V, E_{\text{ALG}})$, each job $j \in J$ has a corresponding vertex $v_j \in V$, and two vertices v_i and v_j are adjacent if and only if the corresponding jobs i and j overlap in ALG.

A *connected component* in the interval graph of a schedule ALG is a maximal set $C \subseteq V$ of vertices where each pair of vertices in C are connected via vertices in C. We abuse the term connected component to describe the set of jobs corresponding to the vertices in a connected component C. Since a schedule ALG may contain multiple connected components, we denote the set of connected components as $\mathcal{C}^{\text{ALG}}$, where C_i^{ALG} is the i-th connected component in $\mathcal{C}^{\text{ALG}}$.

Let $J(C_i)$ denote the set of jobs that are inside the connected component C_i. We also denote $\mathtt{s}(C_i)$ as the *start time of the connected component* C_i, which is defined by $\min\{\mathtt{s}_j : j \in J(C_i)\}$. Similarly, we denote $\mathtt{e}(C_i)$ as the *end time of the connected component* C_i, which is defined by $\max\{\mathtt{e}_j : j \in J(C_i)\}$.

3 Recap the DOUBLER Algorithm [20]

Our work heavily relies on the DOUBLER algorithm proposed by Koehler and Khuller [20]. For completeness, we recap the DOUBLER algorithm and its analysis in this section.

DOUBLER algorithm. The DOUBLER algorithm is a "lazy" algorithm in the sense that it is triggered only at the latest start time of some available job. More specifically, DOUBLER keeps a set of time intervals $\mathcal{T}^D$, which is open for scheduling jobs. Initially, $\mathcal{T}^D$ is empty. At the time t where there exists a job $k \in \mathtt{A}_t^{\text{DOUBLER}}$ with $\ell_k = t$, DOUBLER executes k and updates $\mathcal{T}^D$ by merging it with a *reserved time window* $[t, t+2 \cdot p_k)$, which is twice k's processing time p_k[2]. Then, the time moves on to $t' \geq t$ (since we are in the online time model), and DOUBLER executes any available job that can be fit within $\mathcal{T}^D$. Formally, the job $j \in \mathtt{A}_{t'}^{\text{DOUBLER}}$ with $p_j + t' \in \mathcal{T}^D$ will be executed at time t'. Note that $\mathtt{A}_{t'}^{\text{DOUBLER}}$ may contain new jobs released at or after t, and that j becomes a primary job as well if $\ell_j = t'$.

Analysis Recapped. We recap the 5-competitiveness analysis of DOUBLER by Koehler and Khuller [20]. Firstly, note that $\mathcal{T}^{\text{DOUBLER}}(J) \subseteq \mathcal{T}^D$ at the end, and the cost of DOUBLER is upper-bounded by the size of the final $\mathcal{T}^D$. Namely, $\mu(\mathcal{T}^{\text{DOUBLER}}(J)) \leq \mu(\mathcal{T}^D)$. Recall that $\mathcal{T}^*$ denotes the busy time intervals in the optimal solution. The general idea is to partition $\mathcal{T}^D$ into $\mathcal{T}^D \bigcap \mathcal{T}^*$ and $\mathcal{T}^D \setminus \mathcal{T}^*$ and analyze these two parts separately.

Consider the primary jobs in $\mathtt{P}$. We partition $\mathtt{P}$ into $\mathtt{P}_1$ and $\mathtt{P}_2$, where $\mathtt{P}_1 = \{j \in \mathtt{P} : [\mathtt{s}_j^{\text{DOUBLER}}, \mathtt{s}_j^{\text{DOUBLER}} + 2 \cdot p_j) \subseteq \mathcal{T}^*\}$ and $\mathtt{P}_2 = \mathtt{P} \setminus \mathtt{P}_1$. We abuse the notation and let $\mathtt{P}_1$ denote the intervals $\bigcup_{j \in \mathtt{P}_1} [\mathtt{s}_j, \mathtt{s}_j + 2 \cdot p_j)$ and $\mathtt{P}_2$ denote the intervals $\bigcup_{j \in \mathtt{P}_2} [\mathtt{s}_j, \mathtt{s}_j + 2 \cdot p_j)$. Since $\mathtt{P}_1 \subseteq \mathcal{T}^*$, $\mu(\mathtt{P}_1) \leq \mu(\mathcal{T}^*)$. It follows that $\mu(\mathcal{T}^D) \leq \mu(\mathcal{T}^*) + \mu(\mathtt{P}_2)$.

The rest of the analysis is to bound $\mu(\mathtt{P}_2)$. To this end, the authors in [20] partition the jobs in $\mathtt{P}_2$ by the connected components of the optimal solution they belong to. Formally, let $\mathtt{P}_2^h$ be the jobs in $\mathtt{P}_2 \bigcap C_h^*$, where C_h^* is the h-th connected component in the optimal schedule. Then, $\mu(\mathtt{P}_2) \leq \Sigma_h \Sigma_{j \in \mathtt{P}_2^h} 2p_j$. Order the jobs $j \in \mathtt{P}_2^h$ by non-decreasing $\mathtt{s}_j^{\text{DOUBLER}}$ and denote $\sigma(i)$ be the i-th job in this ordering. The authors then show that $\mu(\mathtt{P}_2) \leq 4 \cdot \mu(\mathcal{T}^*)$ by arguing that

1. for any $i \geq 1$, the job $\sigma(i)$ is available at the time when the first job $\sigma(1)$ is executed by DOUBLER (formally, $r_{\sigma(i)} \leq \mathtt{s}_{\sigma(1)}^{\text{DOUBLER}}$), and
2. the jobs $\sigma(1), \sigma(2), \cdots$ have doubling processing times (formally, $p_{\sigma(i+1)} \geq 2 \cdot p_{\sigma(i)}$ for all i).

Thus, $\mu(\mathcal{T}^{\text{DOUBLER}}(J)) \leq \mu(\mathcal{T}^D) \leq \mu(\mathtt{P}_1) + \mu(\mathtt{P}_2) \leq \mu(\mathcal{T}^*) + 4 \cdot \mu(\mathcal{T}^*)$, and the desired 5-competitiveness is proven.

4 The MULTIPLIER Algorithm

In this section, we consider the setting where prediction on when the machine should be on is available to the online algorithm. Formally, the prediction $\mathcal{S}$

[2] The algorithm's name DOUBLER comes from the fact that it always reserves a time window with a size twice of the primary job.

is a set of time intervals at which the machine should be active. However, the prediction can be imperfect or even adversarial. The challenge is how to trust the prediction such that the online algorithm's performance is improved when the prediction is correct while the online algorithm does not fail dramatically when the prediction is malicious.

Based on the DOUBLER algorithm by Koehler and Khuller [20], we propose an algorithm, MULTIPLIER, which incorporates the prediction $\mathcal{S}$ with the DOUBLER algorithm. Recall that in the DOUBLER algorithm, each primary job j triggers reserving a time window of size twice p_j. Let $\lambda \in [0, 1)$ be a *trust parameter* of MULTIPLIER, where $\lambda \to 1$ means the algorithm fully trusts the prediction and $\lambda = 0$ means the algorithm does not trust the prediction at all. Instead of always reserving a window of size of twice p_j, the MULTIPLIER algorithm uses λ to decide the length of the time window that should be reserved when there is a primary job. More specifically, for every primary job j, MULTIPLIER reserves a window with a length of w_j, which is a function of the primary job's processing time p_j, the prediction $\mathcal{S}$, and the trust parameter λ. We carefully design the function of w_j to make sure that it is between $\frac{2}{1+\lambda}$ and $\frac{2}{1-\lambda}$.

Algorithm Description. First, we let $\alpha = \frac{1-\lambda}{2}$ and $\beta = \frac{1+\lambda}{2}$. The MULTIPLIER algorithm is triggered at time t if there is a job $k \in \mathtt{A}_t^{\text{MULTIPLIER}}$ with $\ell_k = t$. It keeps a set of time intervals $\mathcal{T}^M$, which is reserved for scheduling jobs. Initially, $\mathcal{T}^M$ is empty. At the time t where there exists a job $k \in \mathtt{A}_t^{\text{MULTIPLIER}}$ with $\ell_k = t$, MULTIPLIER executes k.

Now, we want to decide the length of the reserved time window when an available job j reaches its latest start time. First, we let $\hat{w}_j$ be the length of the longest interval starting from the current time t to some $\hat{t}$ such that $[t, \hat{t}) \subseteq I \in \mathcal{S}$. If no such interval in $\mathcal{S}$ that contains t, we set $\hat{t} = t$ and $\hat{w}_j = 0$. The MULTIPLIER algorithm then reserves a time window of length $w_j \cdot p_j$, where $w_j := \max\{\frac{1}{\beta}, \min\{\frac{1}{\alpha}, \frac{\hat{w}_j}{p_j}\}\}$, starting at time t, and updates $\mathcal{T}^M$ by merging it with the reserved time window $[t, t + w_j \cdot p_j)$.

Then, as the time moves on to $t' \geq t$, MULTIPLIER executes any available job that can be fit within $\mathcal{T}^M$. Formally, the job $j \in \mathtt{A}_{t'}^{\text{MULTIPLIER}}$ with $p_j + t' \in \mathcal{T}^M$ will be executed at time t'. (See Algorithm 1 for a pseudo code.)

By how the function w_j is designed, we observe that:

Observation 3. (Clamped window size). *The window size factor w_j incurred by job j according to* MULTIPLIER *is always between $\frac{1}{\beta}$ and $\frac{1}{\alpha}$. More specifically,* MULTIPLIER *exactly follows the prediction only when the prediction tells it to open a reserved time window with a factor between $\frac{1}{\beta}$ and $\frac{1}{\alpha}$.*

Note that the smaller λ is, the smaller the difference between $\frac{1}{\alpha}$ and $\frac{1}{\beta}$ is. More specifically, when $\lambda = 0$ (that is, MULTIPLIER does not trust the prediction at all), $\frac{1}{\beta} = \frac{1}{\alpha} = 2$, and $w_j = 2$. It implies that MULTIPLIER behaves the same with DOUBLER when $\lambda = 0$. On the other hand, when $\lambda \to 1$ (that is, MULTIPLIER fully trusts the prediction), $\frac{1}{\beta} \to 1$ and $\frac{1}{\alpha} \to \infty$. It implies that MULTIPLIER

Algorithm 1. MULTIPLIER$_\lambda$

Input: a set of jobs J and prediction $\mathcal{S}$, which is a set of time intervals
The set of primary jobs $\mathtt{P}$ and the set of open time intervals $\mathcal{T}^M$ are initially empty
Let $\alpha = \frac{1-\lambda}{2}$ and $\beta = \frac{1+\lambda}{2}$
while at time $t \in T$ **do**
  while there exists some job available at t with $\ell_j = t$ **do**
    Execute j
    $\mathtt{P} \leftarrow \mathtt{P} \bigcup \{j\}$
    Let $\hat{w}_j = \max_{\hat{t} \geq t : [t,\hat{t}) \subseteq I \in \mathcal{S}} \{\hat{t} - t\}$
    Let $w_j = \max\{\frac{1}{\beta}, \min\{\frac{1}{\alpha}, \frac{\hat{w}_j}{p_j}\}\}$
    $\mathcal{T}^M \leftarrow \mathcal{T}^M \bigcup \{[t, t + w_j \cdot p_j)\}$
  end while
  Schedule every job j available at t and with $[t, t+p_j) \subseteq \mathcal{T}^M$
end while

reserves a time window as long as the interval indicated by $\mathcal{S}$. Even more, when there is no interval in the prediction $\mathcal{S}$ that fully contains the execution interval of the primary job j, $[t, t+p_j)$, the trusting MULTIPLIER only executes the triggering primary job j at time t but not open any reserved window.

4.1 Analysis of MULTIPLIER

The analysis of MULTIPLIER largely follows the analysis of DOUBLER by Koehler and Khuller [20], which we recap in Sect. 3. The challenge of our analysis is that, instead of reserving time windows with a constant factor of the processing time of the corresponding primary jobs, the reserved time windows in MULTIPLIER can have different lengths with respect to the corresponding primary jobs.

Notations. First, we (re)define the notations for convenience. We redefine $\mathtt{P}$ as the primary jobs produced by MULTIPLIER on input J, trust parameter λ, and prediction $\mathcal{S}$. For simplicity, we denote $\mathtt{s}_j$ as the start time of job j by MULTIPLIER and $\mathtt{s}^*_j$ the start time of j in the optimal schedule.

Recall that $\mathcal{T}^*$ is the busy time interval of the optimal schedule on input J. Similarly to the analysis of DOUBLER, $\mathtt{P}$ is partitioned into $\mathtt{P}_1 = \{j \in \mathtt{P} : [\mathtt{s}_j, \mathtt{s}_j + w_j \cdot p_j) \subseteq \mathcal{T}^*\}$ and $\mathtt{P}_2 = \mathtt{P} \setminus \mathtt{P}_1$. We abuse the notation so that $\mathtt{P}_1$ and $\mathtt{P}_2$ also denote $\bigcup_{j \in \mathtt{P}_1} [\mathtt{s}_j, \mathtt{s}_j + w_j \cdot p_j)$ and $\bigcup_{j \in \mathtt{P}_2} [\mathtt{s}_j, \mathtt{s}_j + w_j \cdot p_j)$, respectively.

Let $\mathcal{C}^* = \{C^*_1, C^*_2, \cdots\}$ be the set of connected components in the optimal schedule. We denote $\mathtt{P}^h_2$ as the set of jobs $\mathtt{P}_2 \bigcap J(C^*_h)$. Consequently, $\mu(\mathtt{P}_2) \leq \sum_h \sum_{j \in \mathtt{P}^h_2} w_j \cdot p_j$.

The analysis then focuses on bounding $\mu(\mathtt{P}^h_2)$ for all h. We denote L as the number of jobs in $\mathtt{P}^h_2$. We order the jobs in $\mathtt{P}^h_2$ in a non-decreasing order of their starting time by MULTIPLIER. Formally, $\sigma(i)$ is the job in $\mathtt{P}^h_2$ with the i-th earliest starting time. That is, $\mathtt{s}_{\sigma(1)} \leq \mathtt{s}_{\sigma(2)} \leq \cdots \leq \mathtt{s}_{\sigma(L)}$.

Analysis Framework. Recall that $\mathcal{T}^M = \bigcup_{j \in \mathsf{P}} [\mathsf{s}_j, \mathsf{s}_j + w_j \cdot p_j)$. Since $\mathcal{T} \subseteq \mathcal{T}^M$, we bound $\mu(\mathcal{T})$ by bounding $\mu(\mathcal{T}^M)$. By the definitions of $\mathcal{T}^M$, P_1, and P_2, $\mu(\mathcal{T}^M) \leq \mu(\mathsf{P}_1) + \mu(\mathsf{P}_2) \leq \mu(\mathcal{T}^*) + \mu(\mathsf{P}_2)$. Finally, since $\mu(\mathsf{P}_2) \leq \sum_h \mu(\mathsf{P}_2^h)$, we will bound $\mu(\mathsf{P}_2^h)$ by $\mu(C_h^*)$ for each h separately.

We first show that similar to DOUBLER, the algorithm MULTIPLIER on J with prediction $\mathcal{S}$ preserves the desired properties of jobs in P_2^h.

Lemma 4. (Primary job availability). *All (primary) jobs in P_2^h are available at the starting time of job $\sigma(1)$. Formally, $r_{\sigma(i)} \leq \mathsf{s}_{\sigma(1)}$ for all $i \geq 1$.*

Proof. First, we show that $r_{\sigma(i)} + p_{\sigma(i)} \leq \mathsf{e}(C_h^*)$. According to the definitions, job $\sigma(i) \in \mathsf{P}_2^h = \mathsf{P}_2 \bigcap C_h^*$. Since $\sigma(i) \in C_h^*$, job $\sigma(i)$ is scheduled inside C_h^* in the optimal schedule, and its end time in the optimal schedule is $\mathsf{s}^*_{\sigma(i)} + p_{\sigma(i)} \leq \mathsf{e}(C_h^*)$. As $\mathsf{s}^*_{\sigma(i)} \geq r_{\sigma(i)}$, we have $r_{\sigma(i)} + p_{\sigma(i)} \leq \mathsf{e}(C_h^*)$.

Next, we show that $\mathsf{s}_{\sigma(1)} + w_{\sigma(1)} \cdot p_{\sigma(1)} \geq \mathsf{e}(C_h^*)$. Since job $\sigma(1) \in \mathsf{P}_2 \bigcap C_h^*$, job $\sigma(1)$ is a P_2 job, and $[\mathsf{s}_{\sigma(1)}, \mathsf{s}_{\sigma(1)} + w_{\sigma(1)} \cdot p_{\sigma(1)}) \not\subseteq [\mathsf{s}(C_h^*), \mathsf{e}(C_h^*))$. This implies that at least one of $\mathsf{s}_{\sigma(1)} < \mathsf{s}(C_h^*)$ and $\mathsf{s}_{\sigma(1)} + w_{\sigma(1)} \cdot p_{\sigma(1)} > \mathsf{e}(C_h^*)$ must be true. We show in the following that $\mathsf{s}_{\sigma(1)} \geq \mathsf{s}(C_h^*)$ and thus we get $\mathsf{s}_{\sigma(1)} + w_{\sigma(1)} \cdot p_{\sigma(1)} > \mathsf{e}(C_h^*)$. Suppose on contrary that $\mathsf{s}_{\sigma(1)} < \mathsf{s}(C_h^*)$. The latest start time $\ell_{\sigma(1)} = \mathsf{s}_{\sigma(1)} < \mathsf{s}(C_h^*) \leq \mathsf{s}^*_{\sigma(1)}$, where the first equality is due to that $\sigma(1)$ is a primary job. Since $\ell_{\sigma(1)} < \mathsf{s}^*_{\sigma(1)}$, job $\sigma(1)$ in OPT starts later than its latest start time, which is a contradiction. Thus $\mathsf{s}_{\sigma(1)} \geq \mathsf{s}(C_h^*)$.

Finally, we combine the two arguments in the previous paragraphs and conclude $r_{\sigma(i)} \leq \mathsf{s}_{\sigma(1)}$. Suppose $r_{\sigma(i)} > \mathsf{s}_{\sigma(1)}$. We can assign job $\sigma(i)$ to its release time, i.e., $\mathsf{s}_{\sigma(i)} = r_{\sigma(i)}$. In this way, the end time of $\sigma(i)$ is $r_{\sigma(i)} + p_{\sigma(i)} \leq \mathsf{e}(C_h^*) \leq \mathsf{s}_{\sigma(1)} + w_{\sigma(1)} \cdot p_{\sigma(1)}$, where the inequalities come from the previous derivations. Now we have $[r_{\sigma(i)}, r_{\sigma(i)} + p_{\sigma(i)}) \subseteq [\mathsf{s}_{\sigma(1)}, \mathsf{s}_{\sigma(1)} + w_{\sigma(1)} \cdot p_{\sigma(1)})$. This means $\sigma(i)$ is completely inside job $\sigma(1)$'s window and $\sigma(i)$ is not a primary job, which is a contradiction. Thus $r_{\sigma(i)} \leq \mathsf{s}_{\sigma(1)}$. □

Lemma 5. (Primary jobs size growth). *The processing time of each primary job increases by at least a factor of the length of their corresponding reserved time windows. Formally, $p_{\sigma(i+1)} \geq w_{\sigma(i)} \cdot p_{\sigma(i)}$ for all $i \in \{1, 2, \cdots, L-1\}$.*

Proof. By Lemma 4, job $r_{\sigma(i+1)} \leq \mathsf{s}_{\sigma(1)} \leq \mathsf{s}_{\sigma(i)}$, and by definition, $\mathsf{s}_{\sigma(i+1)} \geq \mathsf{s}_{\sigma(i)} \geq r_{\sigma(i)}$. Assume, for the contrary, that $p_{\sigma(i+1)} \leq w_{\sigma(i)} \cdot p_{\sigma(i)}$. Then the job $\sigma(i+1)$ must have been scheduled at $\mathsf{s}_{\sigma(i)}$ under the window $w_{\sigma(i)}$, which contradicts the fact that $\sigma(i+1) \in \mathsf{P}_2^h \subseteq \mathsf{P}$ is a primary job. □

By the definition of connected components, Observation 3, and Lemma 5, we note that any job $\sigma(i) \in \mathsf{P}_2^h$ has the following properties:

[P1] $p_{\sigma(i)} \leq \mu(C_h^*)$,
[P2] $\frac{1}{\beta} \leq w_{\sigma(i)} \leq \frac{1}{\alpha}$, and
[P3] $p_{\sigma(i)} > w_{\sigma(i-1)} \cdot p_{\sigma(i-1)}$ if $i \geq 2$.

Next, we develop an arithmetic tool for the later analysis of robustness and consistency:

Lemma 6. (A bound on P_2^h).

$$\sum_{j\in P_2^h} w_j \cdot p_j \leq w_{\sigma(L)} \cdot \mu(C_h^*) + \mu(C_h^*) \cdot \sum_{j=1}^{L-1} \frac{1}{\prod_{i=j+1}^{L-1} w_{\sigma(i)}}$$

Proof. Since $\sigma(L)$ is in C_h^* in the optimal schedule, $p_{\sigma(L)} \leq \mu(C_h^*)$. By **[P1]**, $p_{\sigma(L)} \leq \mu(C_h^*)$. Thus, $\sum_{j\in \mathtt{P}_2^h} w_j \cdot p_j \leq w_{\sigma(L)} \cdot \mu(C_h^*) + \sum_{i=1}^{L-1} w_{\sigma(i)} \cdot p_{\sigma(i)}$.

By **[P3]**, we can claim that $w_{\sigma(i)} \cdot p_{\sigma(i)} < \mu(C_h^*)$ for all $1 \leq i \leq L-1$, no matter what the value of $w_{\sigma(i)}$ actually is. Then, by applying **[P2]**, $w_{\sigma(i)} \cdot w_{\sigma(i-1)} \cdot p_{\sigma(i-1)} < w_{\sigma(i)} \cdot p_{\sigma(i)} < \mu(C_h^*)$ for all $2 \leq i \leq L$. That is, $w_{\sigma(i-1)} \cdot p_{\sigma(i-1)} \leq \frac{\mu(C_h^*)}{w_{\sigma(i)}}$. Recursively, we generalize the claim to $w_{\sigma(i)} \cdot p_{\sigma(i)} \leq \mu(C_h^*) \cdot \frac{1}{\prod_{i=j+1}^{L-1} w_{\sigma(i)}}$. The lemma is then proven by summing up all the terms. □

Robustness. For robustness, we bound the performance of MULTIPLIER against the optimal schedule OPT, given any input instance J and advice $\mathcal{S}$.

Theorem 7. MULTIPLIER *is* $(1 + \frac{4}{1-\lambda})$*-robust for all* $\lambda \in [0, 1)$.

Proof. By Lemma 6, $\mu(\mathtt{P}_2^h) \leq w_{\sigma(L)} \cdot \mu(C_h^*) + \mu(C_h^*) \cdot \sum_{j=1}^{L-1} \frac{1}{\prod_{i=j+1}^{L-1} w_{\sigma(i)}}$. By Observation 3, $w_{\sigma(L)} \leq \frac{1}{\alpha}$. Thus, $w_{\sigma(L)} \cdot \mu(C_h^*) \leq \frac{1}{\alpha} \cdot \mu(C_h^*)$. Also by Observation 3, $w_{\sigma(i)} \geq \frac{1}{\beta}$. Therefore, $\mu(C_h^*) \cdot \sum_{j=1}^{L-1} \frac{1}{\prod_{i=j+1}^{L-1} w_{\sigma(i)}} \leq \mu(C_h^*) \cdot \sum_{j=1}^{L-1} \beta^{j-1} \leq \mu(C_h^*) \cdot \sum_{j=0}^{\infty} \beta^j$. By the definition where $\alpha = \frac{1-\lambda}{2}$ and $\beta = \frac{1+\lambda}{2}$, $\mu(\mathtt{P}_2^h) \leq \frac{1}{\alpha} \cdot \mu(C_h^*) + \mu(C_h^*) \cdot \sum_{j=0}^{\infty} \beta^j = \frac{4}{1-\lambda} \cdot \mu(C_h^*)$. The theorem is then proven with $\mu(\mathtt{P}_1) \leq \mu(\mathcal{T}^*)$.

Consistency when $\lambda < 1$. To prove consistency, we need to determine what the "best possible advice" is for an algorithm, which is difficult. Instead, we start out by providing an advisor ADV that produces some advice $\mathcal{S}_{\mathtt{ADV}}$ according to the behavior of MULTIPLIER. By bounding the worst case ratio of the cost of MULTIPLIER on input J and advice $\mathcal{S}_{\mathtt{ADV}}$ to the optimal solution's cost on J, we can provide a bound on the consistency of MULTIPLIER.

Design Advice. Given the job instances J, let ADV(J, t, MULTIPLIER) denote the (partial) advice provided to MULTIPLIER at time t. Denote $C^*(j)$ as the time interval of the connected component in the optimal schedule that contains job j. We also abuse the notation P and denote it as the union of all (current) primary jobs' execution intervals. We set ADV(J, t, MULTIPLIER) $= [t, t+\mu(C^*(j))$ if $t = \ell_j$ for some job $j \in \mathtt{A}_t^{\text{MULTIPLIER}}$ and $[t, t+p_j)$ is not "covered" by the union of execution intervals of the current primary jobs. For any other t, we set ADV(J, t, MULTIPLIER) $= \emptyset$. Note that the available jobs $\mathtt{A}_t^{\text{MULTIPLIER}}$ and the set of (current) primary jobs P are well defined in this case since MULTIPLIER is deterministic.

Denoting ADV as the union of ADV(J, t, MULTIPLIER) for all t, we can use this advice to bound the consistency of MULTIPLIER from above. First, we bound the cost incurred by the last job in $\mathtt{P}_2^h$.

Lemma 8. (Cost incurred by $\sigma(L)$). *Consider running* MULTIPLIER *on input* J *with the advice* ADV*. The cost incurred by the reserved window of the last job* $\sigma(L) \in P_2^h$ *is* $w_{\sigma(L)} \cdot p_{\sigma(L)} \leq \frac{1}{\beta} \cdot \mu(C_h^*)$.

Proof. First, recall that according to the construction of ADV, $\hat{w}_{\sigma(i)} = \mu(C_h^*)$. We show this lemma by a case distinction based on the job size of $p_{\sigma(L)}$.

- If $\beta \cdot \mu(C_h^*) \leq p_{\sigma(L)} \leq \mu(C_h^*)$, then $w_{\sigma(L)} = \max\{\frac{1}{\beta}, \min\{\frac{1}{\alpha}, \frac{\hat{w}_{\sigma(L)}}{p_{\sigma(L)}}\}\} \leq \max\{\frac{1}{\beta}, \min\{\frac{1}{\alpha}, \frac{\hat{w}_{\sigma(L)}}{\beta \cdot \hat{w}_{\sigma(L)}}\}\} = \frac{1}{\beta}$. In this case, $w_{\sigma(L)} \cdot p_{\sigma(L)} \leq \frac{1}{\beta} \cdot \mu(C_h^*)$.
- If $\alpha \cdot \mu(C_h^*) < p_{\sigma(L)} < \beta \cdot \mu(C_h^*)$, then $\frac{1}{\beta} < \frac{\hat{w}_{\sigma(L)}}{p_{\sigma(L)}} < \frac{1}{\alpha}$. Therefore, $w_{\sigma(L)} = \frac{\hat{w}_{\sigma(L)}}{p_{\sigma(L)}}$. In this case, $w_{\sigma(L)} \cdot p_{\sigma(L)} \leq \mu(C_h^*)$.
- If $p_{\sigma(L)} < \alpha \cdot \mu(C_h^*)$, then $\frac{\hat{w}_{\sigma(L)}}{p_{\sigma(L)}} \geq \frac{1}{\alpha}$. Therefore, $w_{\sigma(L)} = \frac{1}{\alpha}$. In this case, $w_{\sigma(L)} \cdot p_{\sigma(L)} \leq \mu(C_h^*)$.

The lemma is proven since $\frac{1}{\beta} = \frac{2}{1+\lambda} \in [1, 2]$. □

Lemma 9. *Consider running* MULTIPLIER *on input* J *with the advice* ADV*. Any other job* $j \in P_2^h \backslash \{\sigma(L)\}$ *has a reserved time window with a length of* $\frac{1}{\alpha}$*. Formally,* $w_{\sigma(i)} = \frac{1}{\alpha}$ *for all* $i < L$.

Proof. By Observation 3, $w_{\sigma(i)} \in [\frac{1}{\beta}, \frac{1}{\alpha}]$. Therefore, we assume for the contrary that $w_{\sigma(i)} = \frac{1}{\alpha} - x$ where $x \in (0, \frac{1}{\alpha} - \frac{1}{\beta}]$. According to MULTIPLIER and ADV, it implies that $\frac{1}{\alpha} - x = \frac{\hat{w}_{\sigma(i)}}{p_{\sigma(i)}} = \frac{\mu(C_h^*)}{p_{\sigma(i)}}$, which further implies that $\mu(C_h^*) = p_{\sigma(i)} \cdot (\frac{1}{\alpha} - x)$. By Lemma 5, $w_{\sigma(L-1)} \cdot p_{\sigma(L-1)} < p_{\sigma(L)}$. Thus, $w_{\sigma(L-1)} \cdot p_{\sigma(L-1)} < \mu(C_h^*)$. However, as mentioned, $\mu(C_h^*) = p_{\sigma(i)} \cdot (\frac{1}{\alpha} - x) = w_{\sigma(i)} \cdot p_{\sigma(i)} = w_{\sigma(L-1)} \cdot p_{\sigma(L-1)}$ when $i = L - 1$, which leads to a contradiction for $i = L - 1$.

On the other hand, for all $i < L - 1$, we know that $p_{\sigma(i)} < p_{\sigma(L-1)}$, which means that $\frac{\mu(C_h^*)}{p_{\sigma(i)}} \geq \frac{1}{\alpha}$, since the same holds for $p_{\sigma(L-1)}$. It proves the lemma. □

Theorem 10. MULTIPLIER *is* $(1 + \frac{4}{1+\lambda})$*-consistent when* $\lambda \in [0, 1)$.

Proof. We focus on $\mu(P_2^h) = \sum_{i=1}^{L} w_{\sigma(i)} \cdot p_{\sigma(i)}$. By Lemmas 8 and 9, we can rewrite the bound on the cost of P_2^h to $\sum_{j \in P_2^h} w_j \cdot p_j \leq \frac{1}{\beta} \cdot \mu(C_h^*) + \mu(C_h^*) \cdot \sum_{j=1}^{L-1} \frac{1}{\prod_{i=j+1}^{L-1} w_{\sigma(i)}} \leq \frac{1}{\beta} \cdot \mu(C_h^*) + \mu(C_h^*) \cdot \sum_{j=0}^{\infty} (\frac{1}{\alpha})^j$. By how we set α and β, $\mu(P_2^h) \leq \frac{4}{1+\alpha} \cdot \mu(C_h^*)$. It proves the theorem since $\mu(P) \leq \mu(P_1) + \sum_h \mu(P_2^h) \leq \mu(T^*) + \frac{4}{1+\alpha} \cdot \sum_h \mu(C_h^*) = \mu(T^*) + \frac{4}{1+\alpha} \cdot \mu(T^*)$.[3] □

Consistency when $\lambda = 1$. If $\lambda = 1$, by the definitions, $\alpha = \infty$ and $\beta = 1$. Similarly, we use ADV as our advisor. Once $t = \ell_j$ for some job j, $\hat{w}_j = \mu(C^*(j))$ at the beginning of the optimal connected component to which j belongs. Then, since $\alpha = \infty$, all jobs that fit inside this connected component are scheduled because $w_j \cdot p_j = \mu(C^*(j))$. Therefore, the cost inside the connected components is the same as the optimal cost. That is, the MULTIPLIER algorithm is 1-consistent when $\lambda = 1$.

[3] Recall that C_h^* are connected components in the optimal schedule.

5 Locally Updateable Prediction and Concluding Remark

Consistency and robustness are quantifiable measures of the quality of an online algorithm with machine-learned prediction. However, for scheduling problems and other problems with potentially infinitely large instances, we can see that depending on the required format of the prediction, the prediction may also be infinitely large. Such prediction may be infeasible to compute.

Moreover, if we use the online-time model for an online algorithm, we may see that it is easier to predict things about the near future rather than the far future. Anecdotally, we can look at weather prediction to see that this may be true, and literature confirms this. For reference, see the paper about uncertainty in weather and climate prediction by Slingo et al. [30].

Therefore, we want to discuss the notion of *locally updateable online algorithms*. An online algorithm is locally updateable if it accommodates the machine-learned prediction about the future, which can be revised as time progresses. Allowing the prediction to be updated in an online algorithm allows the problem instance to be potentially infinitely large, and it may help the advisor provide more accurate prediction.

Recall that, in the MULTIPLIER algorithm, we consider the prediction a series of time intervals. We have formulated the algorithm so that prediction has to be given at the beginning of the problem instance. However, we can reformulate the algorithm to query the advisor each time an available job reaches its last starting time. We require the prediction to be some number that denotes the advised reserved window. This allows the advisor to give prediction that contains a finite number of bits and to use all the local information about the problem instance. Moreover, this does not affect the consistency and robustness of MULTIPLIER. Therefore, the MULTIPLIER algorithm can be used as a locally updateable online algorithm.

References

1. Amur, H., Cipar, J., Gupta, V., Ganger, G.R., Kozuch, M.A., Schwan, K.: Robust and flexible power-proportional storage. In: SoCC, pp. 217–228 (2010)
2. Angelopoulos, S., Dürr, C., Jin, S., Kamali, S., Renault, M.P.: Online computation with untrusted advice. In: ITCS (2020)
3. Angelopoulos, S., Kamali, S.: Contract scheduling with predictions. J. Artif. Intell. Res. **77**, 395–426 (2023)
4. Antoniadis, A., Ganje, P.J., Shahkarami, G.: A novel prediction setup for online speed-scaling. In: SWAT (2022)
5. Azar, Y., Leonardi, S., Touitou, N.: Flow time scheduling with uncertain processing time. In: STOC (2021)
6. Azar, Y., Leonardi, S., Touitou, N.: Distortion-oblivious algorithms for minimizing flow time. In: SODA (2022)
7. Balkanski, E., Périvier, N., Stein, C., Wei, H.: Energy-efficient scheduling with predictions. In: NeurIPS (2023)
8. Bamas, É., Maggiori, A., Rohwedder, L., Svensson, O.: Learning augmented energy minimization via speed scaling. In: NeurIPS (2020)

9. Bampis, E., Dogeas, K., Kononov, A.V., Lucarelli, G., Pascual, F.: Scheduling with untrusted predictions. In: IJCAI (2022)
10. Baptiste, P.: Scheduling unit tasks to minimize the number of idle periods: a polynomial time algorithm for offline dynamic power management. In: SODA, pp. 364–367 (2006)
11. van de Bovenkamp, R.: Online busy time scheduling (2024)
12. Boyar, J., Favrholdt, L.M., Kamali, S., Larsen, K.S.: Online interval scheduling with predictions. In: WADS (2023)
13. Chau, V., Li, M.: Active and busy time scheduling problem: a survey. In: Complexity and Approximation (2020)
14. Chen, S., Ljubic, I., Raghavan, S.: The regenerator location problem. Networks **55**(3), 205–220 (2010)
15. Flammini, M., Marchetti-Spaccamela, A., Monaco, G., Moscardelli, L., Zaks, S.: On the complexity of the regenerator placement problem in optical networks. IEEE/ACM Trans. Netw. **19**(2), 498–511 (2011)
16. Flammini, M., et al.: Minimizing total busy time in parallel scheduling with application to optical networks. Theor. Comput. Sci. **411**(40–42), 3553–3562 (2010)
17. Fong, K.C.K., Li, M., Li, Y., Poon, S., Wu, W., Zhao, Y.: Scheduling tasks to minimize active time on a processor with unlimited capacity. In: TAMC (2017)
18. Gerards, M.E.T., Hurink, J.L., Hölzenspies, P.K.F.: A survey of offline algorithms for energy minimization under deadline constraints. J. Sched. **19**(1), 3–19 (2016)
19. Irani, S., Pruhs, K.: Algorithmic problems in power management. SIGACT News **36**(2), 63–76 (2005)
20. Koehler, F., Khuller, S.: Busy time scheduling on a bounded number of machines (extended abstract). In: WADS (2017)
21. Lattanzi, S., Lavastida, T., Moseley, B., Vassilvitskii, S.: Online scheduling via learned weights. In: SODA (2020)
22. Lavastida, T., Moseley, B., Ravi, R., Xu, C.: Learnable and instance-robust predictions for online matching, flows and load balancing. In: ESA (2021)
23. Lee, R., Maghakian, J., Hajiesmaili, M.H., Li, J., Sitaraman, R.K., Liu, Z.: Online peak-aware energy scheduling with untrusted advice. In: e-Energy (2021)
24. Li, S., Xian, J.: Online unrelated machine load balancing with predictions revisited. In: ICML (2021)
25. Oprescu, A., Kielmann, T.: Bag-of-tasks scheduling under budget constraints. In: CloudCom, pp. 351–359 (2010)
26. Ren, R., Tang, X.: Online flexible job scheduling for minimum span. In: SPAA (2017)
27. Ren, R., Tang, X.: Busy-time scheduling on heterogeneous machines. In: IPDPS (2020)
28. Shalom, M., Voloshin, A., Wong, P.W.H., Yung, F.C.C., Zaks, S.: Online optimization of busy time on parallel machines. Theor. Comput. Sci. **560**, 190–206 (2014)
29. Shi, W., Hong, B.: Resource allocation with a budget constraint for computing independent tasks in the cloud. In: CloudCom, pp. 327–334 (2010)
30. Slingo, J., Palmer, T.: Uncertainty in weather and climate prediction. In: Phil. Trans. R. Soc. A.3694751–4767 (2011)
31. Winkler, P., Zhang, L.: Wavelength assignment and generalized interval graph coloring. In: SODA (2003)

Maximal α-Gapped Repeats in a Fibonacci String

Kazuma Yamane[1], Yuto Nakashima[2](✉), Kazuhisa Seto[3], and Takashi Horiyama[3]

[1] Graduate School of Information Science and Technology, Hokkaido University, Sapporo, Japan
elms-yamane1227@eis.hokudai.ac.jp
[2] Department of Informatics, Kyushu University, Fukuoka, Japan
nakashima.yuto.003@m.kyushu-u.ac.jp
[3] Faculty of Information Science and Technology, Hokkaido University, Sapporo, Japan
{seto,horiyama}@ist.hokudai.ac.jp

Abstract. A gapped repeat is a substring of the form uvu where u is any nonempty string called arm, and v is any string called gap. A gapped repeat is maximal if the characters on the left to both arms differ and those on the right to both arms differ. For any real number $\alpha \geq 1$, an α-gapped repeat is a gapped repeat such that $|u|+|v| \leq \alpha|u|$. One of the fundamental problems for repetitive structures on strings is analyzing the number of substrings that have such structures in a string. Kolpakov et al. [CPM 2014] showed that $O(\alpha^2 n)$ upper bound and $\Omega(\alpha n)$ lower bound on the maximum number of maximal α-gapped repeats in a string of length n, and the current best upper bound is $3(\pi^2/6+5/2)\alpha n$ by I and Köppl [TCS 2019]. Another interesting problem in this line of research is revealing the structures in well-known repetitive strings. In this paper, we investigate the maximal α-gapped repeats in a Fibonacci string. We show an interesting characterization of the form of an arm. More precisely, there are two possible cases in the form of an arm: Fibonacci arms and palindromic arms. By using these structures, we can obtain an upper bound of the maximum number of maximal α-gapped repeats in the k-th Fibonacci string. We can also show that the Fibonacci strings are a new example that contains $\Omega(\alpha n)$ maximal α-gapped repeats.

Keywords: repetitive structures · maximal gapped repeats · Fibonacci strings

1 Introduction

A gapped repeat is an extension of squares (or tandem repeats) and has also been well-studied. For any nonempty string u and any string v, the gapped repeat is defined as a repetition of the form uvu, where u and v are called arm and gap, respectively. For any real number $\alpha \geq 1$, a gapped repeat is called

R. Královič and V. Kůrková (Eds.): SOFSEM 2025, LNCS 15539, pp. 337–350, 2025.
https://doi.org/10.1007/978-3-031-82697-9_25

α-gapped repeat if $|u| + |v| \leq \alpha|u|$, where $|u|$ and $|v|$ are the length of u and v, respectively. Furthermore, a gapped repeat is maximal if the characters on the left and the right of its arms differ. This structure, which is also called a maximal pair, is motivated by biological sequence analysis, such as repetitive DNA and RNA structures and interspersed repeats (c.f. Sect. 7.11.2 of [8] and references in [9]). In many applications in computational biology, identical substrings (two arms) need to be located close to each other. That is a natural reason why we consider a bounded gap by α for maximal repeats.

The α-gapped repeat was first introduced by Kolpakov et al. [13] (but there are several earlier works for maximal gapped repeats in the context of maximal pairs [3,8,16]). For the problem of finding all maximal α-gapped repeats in a given string, there are optimal $O(\alpha n)$ time solutions for constant alphabets [4,19] and integer alphabets [6]. In fact, the running time of the algorithm by Tanimura et al. [19] is $O(\alpha n + M)$ where M is the number of outputs (i.e., the number of maximal α-gapped repeats in the input string). The upper bound $O(\alpha n)$ of M, which was shown in the later [4], implies that Tanimura et al.'s algorithm also runs in $O(\alpha n)$ time. As this fact shows, combinatorial properties on the structures contribute to developing efficient algorithms, and there are several studies on the maximum number of maximal α-gapped repeats. Kolpakov et al. [13,14] first showed $O(\alpha^2 n)$ upper bound and $\Omega(\alpha n)$ lower bound. Crochemore et al. [4] improved this upper bound to $O(\alpha n)$. As a result, the asymptotic upper and lower bounds were equivalent. Gawrychowski et al. [6] investigated the exact values of the upper bound on maximal α-gapped repeats and showed that it is $18\alpha n$. I and Köppl [9] improved it to $3(\pi^2/6 + 5/2)\alpha n$ ($\approx 12.43\alpha n$).

Another interesting problem in this line of research is revealing the structures in well-known repetitive strings. For example, if we focus on Fibonacci strings, combinatorial structures and properties on Fibonacci strings are well-studied, and Fibonacci strings have many interesting properties (e.g., [5,10,12,15,17]). In particular, properties of repetitive structures on Fibonacci strings give new insights and deeper understandings, such as lower bounds of the complexity of algorithms (e.g., [7,11,18]). In this paper, to further understand the maximal α-gapped repeats, we investigate the maximal α-gapped repeats in a (finite) Fibonacci string. A Fibonacci string is the string defined by $f_1 = b, f_2 = a; f_i = f_{i-1}f_{i-2}$ ($i \geq 3$). We show that there are only two possible cases in the form of an arm. In more detail, we obtain the following characterization of the maximal α-gapped repeats in a Fibonacci string; (1) the arm of every maximal α-gapped repeat in a Fibonacci string is either a Fibonacci string or the characteristic palindromic prefix of a Fibonacci string (i.e., the string that can be obtained by removing the last two characters from a Fibonacci string is a palindrome), (2) if the arm is a Fibonacci string, then the maximal α-gapped repeat is a suffix of the whole Fibonacci string. Moreover, we can see that several α-gapped repeats in a Fibonacci string cannot deeply overlap each other since their arms are Fibonacci strings or Fibonacci-like palindromes. We believe that our new knowledge of repetitive structures in Fibonacci strings is of independent interest and can be used for analysis. By using these structures, we can give a non-trivial upper bound on the number of maximal α-gapped repeats in the k-th Fibonacci string of length n (Theorem 3). We can also show that the Fibonacci strings are a new example that contains $\Omega(\alpha n)$ maximal α-gapped repeats.

Due to the space limitation, we omit some proofs (marked $\star$). These materials will be in the full version of this paper.

2 Preliminaries

Strings and Gapped Repeats. Let $\Sigma = \{a, b\}$ be a binary alphabet. An element of Σ is called a *character*. Σ^* denotes the set of the all finite *strings* over Σ. The length of a string $w \in \Sigma^*$ is denoted by $|w|$. The empty string is the string of length 0. For $w = xyz$, then x, y and z are called a *prefix*, *substring*, and *suffix* of w, respectively. They are called a *proper prefix* and a *proper suffix* of w if $x \neq w$ and $z \neq w$, respectively. Consider any positions i, j in a string w with $1 \leq i \leq j \leq |w|$. We denote by $w[i]$ the i-th character of w, and by $w[i..j]$ the substring of w starting at position i and ending at position j. We denote by w^n the string with w repeated n times. We say that a string v is a *border* of w if v is both a (possibly empty) proper prefix and suffix of w. The string obtained by flipping characters in a string w is denoted by $\overline{w}$ (e.g., $\overline{aab} = bba$). For any suffix x of a string w, the string obtained by removing the suffix x from w is denoted by wx^{-1} (e.g., $aabab(ab)^{-1} = aab$). For any string w, w^R denotes the reversal of w, i.e., $w^R = w[|w|]w[|w| - 1] \cdots w[1]$. A string w is a palindrome if $w = w^R$. A position i in a string w is called an occurrence of a substring u of w if i is a beginning position of u. For any two substrings $w[i..j]$ and $w[i'..j']$ of w, we say that they overlap if there exists an integer k such that $i \leq k \leq j$ and $i' \leq k \leq j'$.

For any non-empty string u and any string v, a string of the form uvu is called a *gapped repeat*. If $|u| + |v| \leq \alpha|u|$ for any real $\alpha \geq 1$, uvu is called α-*gapped repeat*. For any gapped repeat uvu, the left u, denoted by u_ℓ, is called the *left arm*, and the right u, denoted by u_r, is called the *right arm*. Let w' be a gapped repeat in w and $u_\ell = w[i..j]$ and $u_r = w[i'..j']$. w' is a *maximal* gapped repeat if u_ℓ and u_r satisfy the following two conditions: (1) $w[i-1] \neq w[i'-1]$ if $1 < i$, (2) $w[j+1] \neq w[j'+1]$ if $j' < n$. For example, we consider the string $w = baabaabbbbabaaa$. Let $u_\ell = u_r = aba$, and $v = abbbb$. The substring $w' = u_\ell v u_r$ is a gapped repeat, but both the right characters to u_ℓ and u_r are a. This violates the maximality; thus, $w' = abaabbbaba$ is not a maximal gapped repeat. Let $u_\ell = u_r = abaa$, and $v = bbbb$. The string $w' = abaabbbbabaa$ satisfies the maximality.

Fibonacci Strings. A *Fibonacci string* is the binary string over an alphabet $\{a, b\}$ defined as follows: $f_1 = b, f_2 = a; f_i = f_{i-1}f_{i-2}$ $(i \geq 3)$. For example, $f_3 = ab, f_4 = aba, f_5 = abaab, f_6 = abaababa, f_7 = abaababaabaab$. There is a well-known integer sequence called the *Fibonacci sequence* that is defined as follows: $F_0 = 0, F_1 = 1; F_i = F_{i-1} + F_{i-2}$ $(i \geq 2)$. The length of f_i is equal to the i-th term of the Fibonacci sequence. In general,

$$F_i = |f_i| = \frac{1}{\sqrt{5}} \left\{ \left((1+\sqrt{5})/2\right)^i - \left((1-\sqrt{5})/2\right)^i \right\} \ (i \geq 1).$$

We can also explain Fibonacci strings by using a string morphism ϕ such that $\phi(a) = ab$, $\phi(b) = a$. It is known that $f_i = \phi^{i-1}(b)$ $(i \geq 1)$ (cf. [10]).

3 Properties on Fibonacci Strings

In this section, we introduce properties on Fibonacci strings which we will use later. Firstly, we explain several basic properties. Lemmas 2 and 3 can be obtained by the definition of the Fibonacci string.

Lemma 1 ([2]). *For every i, f_i satisfies the following properties.*

(a) aa and bb cannot appear at the beginning and end of f_i [2, Lemma 7].
(b) aaa and bb cannot appear in f_i [2, Lemma 7].
(c) Any string of the form u^k $(k \geq 4)$ cannot appear in f_i [2, Lemma 9].
(d) Any string of the form $(ab)^k, (ba)^k$ $(k \geq 3)$ cannot appear in f_i [2, Lemma 8].

Lemma 2. *For every i, f_i satisfies the following properties.*

(a) f_i has ab as a prefix $(i \geq 3)$.
(b) f_i has ba as a suffix if i is even, ab otherwise $(i \geq 3)$.
(c) f_i has $abaa$ as a prefix $(i \geq 5)$.
(d) f_i has $aababa$ as a suffix if i is even, aab as a suffix otherwise $(i \geq 5)$.

Based on property (b) of Lemma 2, we define the function δ as $\delta(i) = ba$ if i is even and $\delta(i) = ab$ otherwise. Let $f_i = P_i\delta(i)$ for every $i \geq 4$. It is also known that P_i is a palindrome.

Lemma 3 (Lemma 2.8 of [10]). *For every $i \geq 3$, $f_i = f_{i-2}f_{i-3} \cdots f_1\delta(i)$.*

Corollary 1. *For every $i \geq 6$, $f_i = f_{i-2}f_{i-2}f_{i-5}f_{i-4}$.*

Corollary 2. *For every $i \geq 5$, $f_i = f_{i-2}f_{i-3}f_{i-3}f_{i-4}$.*

It is also known that the longest border of f_i is f_{i-2} for every $i \geq 4$ [10, Lemma 2.5]. This implies that the following corollary.

Corollary 3. *The set of all borders of f_i is $\{f_{i-2\ell} \mid 1 \leq \ell \leq \lfloor \frac{i}{2} \rfloor - 1\}$.*

The next three lemmas explain characterizations for occurrences in Fibonacci strings of characteristic substrings related to Fibonacci strings. The first one can be easily obtained by the definition.

Lemma 4. *For every $i \geq 6$ and j satisfying $4 \leq j \leq i-2$, f_j^2 is a prefix of f_i.*

Lemma 5 (Lemma 37 of [18]). *For every k and i satisfying $4 \leq i \leq k$, every substring w of f_k of length $|f_i|$ that has f_{i-1} as a prefix is either $w = f_{i-1}f_{i-2}$ or $w = f_{i-2}f_{i-1}$.*

Lemma 6 ($\star$). *For every $i \geq 7$ and j satisfying $3 \leq j \leq i-3$, $f_{j+1}^2 f_{j-2} f_{j-1}$ is a suffix of f_i if i's parity and j's parity are distinct, $f_{j+1}^2 f_j$ as a suffix otherwise.*

Roughly speaking, we will show that each arm of maximal gapped repeats is either a Fibonacci string or a palindrome P_i.

Lemma 7. *For every $k \geq 4$, $P_{k+1} = \phi(P_k)a$.*

Proof. By the definition of P_k, $P_{k+1} = f_{k+1}[1..F_{k+1} - 2] = f_{k+1}[1..F_{k+1} - 3]a$. Since $|\phi(\delta(k))| = 3$ for any k, $f_{k+1} = \phi(f_k) = \phi(P_k) \cdot \phi(\delta_k)$ holds. This implies that $f_{k+1}[1..F_{k+1} - 3] = \phi(P_k)$. Therefore, $P_{k+1} = \phi(P_k)a$ holds.

Lemma 8 ($\star$). *For every $k \geq 5$, all borders of P_k are $P_4, \ldots, P_{k-1}$.*

We finally introduce an important property about the Fibonacci sequence to estimate the number of maximal α-gapped repeats in Fibonacci strings. Let ψ be the reciprocal sum of the Fibonacci sequence.

Lemma 9 ([1]). *ψ is the irrational number that is approximately*

$$\psi = \sum_{i=1}^{\infty} \frac{1}{F_i} = \frac{1}{1} + \frac{1}{1} + \frac{1}{2} + \frac{1}{3} + \frac{1}{5} + \frac{1}{8} + \frac{1}{13} + \frac{1}{21} + \cdots = 3.3598856662431\cdots$$

4 Maximal α-Gapped Repeats in a Fibonacci String

In this section, we first characterize the maximal α-gapped repeats in Fibonacci strings. Based on the characterization, we also prove an upper bound of the maximum number of the maximal α-gapped repeats in a Fibonacci string.

The key ideas of the characterizations are as follows. We consider two types of maximal gapped repeats in a Fibonacci string. The first type is a maximal gapped repeat that appears as a suffix of the whole Fibonacci string, and the second type is a maximal gapped repeat that is not a suffix of the whole Fibonacci string. We can show that the arms of maximal gapped repeats of the first type are a Fibonacci string. In the second case, the arms are a string obtained by removing the last two characters (δ) of a Fibonacci string. We will explain these characterizations in Sect. 4.1. In addition, we give an upper bound for each case (Sect. 4.2).

4.1 Classification of Arms

We consider maximal gapped repeats in the m-th Fibonacci string f_m. Let $\mathcal{SR}$ denote the set of maximal α-gapped repeats in f_m that is a suffix of f_m and $\mathcal{NSR}$ denote the set of maximal α-gapped repeats in f_m that is not a suffix of f_m ($\mathcal{SR}$ and $\mathcal{NSR}$ indicate *suffix repeats* and *non-suffix repeats*). Clearly, the set $\mathcal{A}$ of maximal α-gapped repeats in f_m satisfies $\mathcal{A} \subseteq \mathcal{SR} \cup \mathcal{NSR}$.

Suffix Maximal Gapped Repeats. We first explain the case that a maximal α-gapped repeat is a suffix of the whole Fibonacci string (Theorem 1). We use two auxiliary lemmas to prove it. The first lemma shows a characterization of suffix maximal gapped repeats.

Lemma 10. *For every $i \geq 5$, every suffix s of f_i that has ab as a prefix satisfies either (i) $s = f_j$ for some $j \leq i$ or (ii) s has $f_k f_{k+1}$ as a prefix for some k satisfying $3 \leq k \leq i-3$.*

Proof. We prove this lemma by induction on i. If $i = 5$ (i.e., $f_5 = abaab$), every suffix of f_5 that has ab as a prefix is a Fibonacci string. If $i = 6$ (i.e., $f_6 = abaababa$), proper suffixes of f_6 that have ab as a proper prefix are aba and $ababa$. Since $aba = f_4$ and $ababa = f_3 f_4$, the statement holds for the two cases. Suppose that the statement holds for $i = c, c+1$ for some $c \geq 5$. We show that the statement also holds for $i = c+2$. Since $f_c[1] = a$, $f_{c+2}[F_{c+1}..F_{c+1}+1] \neq ab$. If $|s| \leq F_c$, s holds the conditions by the induction hypothesis. Otherwise (i.e., $|s| \geq F_c + 2$), by the induction hypothesis for $i = c+1$, s can be written as either $s = f_j$ for some $j \leq c+1$, or $s = f_k f_{k+1} u f_c$ for some $u \in \Sigma^*$ and integer k satisfying $2 \leq k \leq c-2$. In the latter case, s holds the second condition. We consider the former case. If $j = c+1$, $s = f_{c+2}$ implies that s holds the first condition. If $j < c+1$, the parity of j and the parity of $c+1$ should be the same. Hence, s can be written as $s = f_j f_{j+1} v$ for some $v \in \Sigma^*$ and holds the second condition. Thus, the statement holds.

The second lemma explains that the preceding character of a suffix corresponding to (ii) of Lemma 10 is uniquely determined based on the parity of k. This lemma can be used to discuss the maximality of gapped repeats.

Lemma 11. *For every $k \geq 3$, a substring $f_k f_{k+1}$ of f_m is preceded by a if k is odd, b if k is even.*

To prove Lemma 11, we show the following lemma (we will actually use Corollary 4) about the occurrences of substrings $f_k f_{k+1}$.

Lemma 12. *For every $k \geq 3$, if $f_k f_{k+1}$ occurs at position i in f_m, then $f_{k-1} f_k$ occurs at position j in f_{m-1} such that $|\phi(f_{m-1}[1..j-1])| + 1 = i$.*

Proof. Suppose that k is even. Then, $f_k f_{k+1}$ has ab as both a prefix and a suffix by properties (a) and (b) of Lemma 1. By the definition of the morphism ϕ, every substring ab in f_m has to be replaced from b in f_{m-1}. This implies that $f_k f_{k+1}$ at position i in f_m is produced from $f_{k-1} f_k$ at position j in f_{m-1} such that $|\phi(f_{m-1}[1..j-1])| + 1 = i$.

Suppose that k is odd. Then, $f_k f_{k+1}$ has $ababa$ as a suffix since $f_3 f_4 = ababa$ (when $k = 3$) and f_{k+1} has $ababa$ as a suffix for every odd $k \geq 5$. If $f_k f_{k+1}$ is not a suffix of f_m, the succeeding character of $f_k f_{k+1}$ should be a by property (d) of Lemma 1. This implies that $f_k f_{k+1}$ at position i in f_m is produced from $f_{k-1} f_k$ at position j in f_{m-1} such that $|\phi(f_{m-1}[1..j-1])| + 1 = i$.

Therefore, the statement holds for every $k \geq 3$.

Corollary 4. *For any integer $k \geq 3$, if $f_k f_{k+1}$ occurs at position i in f_m, then $f_2 f_3 = aab$ occurs at position j in f_{m-k+2} that satisfies $|\phi^{k-2}(f_{m-k+2}[1..j-1])| + 1 = i$.*

Proof (of Lemma 11). Let $f_m = x \cdot f_k f_{k+1} \cdot y$ for some $x, y \in \Sigma^*$. Due to Corollary 4, there exist x' and y' such that $\phi^{k-2}(x') = x$, $\phi^{k-2}(y') = y$, and $f_{m-k+2} = x' \cdot f_2 f_3 \cdot y'$. Since $f_2 f_3 = aab$ is not a prefix of f_{m-k+2}, x' is not the empty string. Moreover, the last character of x' is b by property (b) of Lemma 1. This fact implies that x has $\phi^{k-2}(b) = f_{k-1}$ as a suffix. Therefore, the lemma holds.

Now, we are ready to prove the following theorem which characterizes the arms of maximal gapped repeats belonging to $\mathcal{SR}$.

Theorem 1. *Let σ be a maximal gapped repeat in f_m that is a suffix of f_m. Then, the arm of σ is a Fibonacci string.*

Proof. Let $\sigma = uvu$ be a maximal gapped repeat that is a suffix of f_m. If $\sigma = f_m$, the arms are a Fibonacci string by Cororally 3. Since both a and b are Fibonacci strings, we assume that $\sigma \neq f_m$ and $|u| \geq 2$ (actually, there is no maximal gapped repeat with arm b by property (b) of Lemma 1). We first show that u has ab as a prefix. If b is a prefix of u, by property (b) of Lemma 1, the preceeding characters of u are a. This contradicts the maximality of σ. If aa is a prefix of u, by Lemma 1, the preceeding characters of u are b. This also contradicts the maximality of σ. Thus, the prefix of length 2 of u must be ab. Assume on the contrary that u is not a Fibonacci string. Then, by Lemma 10, there exists $f_k f_{k+1}$ for some k satisfying $3 \leq k \leq m-3$ that begins with the left arm u. By Lemma 11, the preceeding character of $f_k f_{k+1}$ in f_m is uniquely determined based on the parity of k. Thus, both the preceding characters of arms u are the same. This contradicts the maximality of σ. Hence, u must be a Fibonacci string.

Non-suffix Maximal Gapped Repeats. We next explain the case that a maximal gapped repeat is not a suffix of the whole Fibonacci string (Theorem 2). Firstly, we consider a restriction of arms in this case. In the proof of Theorem 1, we saw that the arm u of length more than 2 has neither b nor aa as a prefix. In the case of non-suffix maximal gapped repeats, the arm u of length more than 2 also has neither b nor aa as a suffix. This restriction can be shown in a symmetric way. This is a powerful restriction which is satisfied in Fibonacci strings. We also use the property to prove the following lemma which characterizes the arms of maximal gapped repeats belonging to $\mathcal{NSR}$.

Theorem 2. *Let σ be a maximal gapped repeat in f_m that is not a suffix of f_m. Then, the arm of σ is in $\mathcal{P}_m = \{P_4, \ldots, P_m\}$ where $P_m = f_m \delta(m)^{-1}$ $(m \geq 4)$.*

Proof. We will prove the following facts: (1) By the inverse morphism of ϕ, any maximal gapped repeat in f_i that is not a suffix of f_m can be reduced to a maximal gapped repeat of arm a in f_k for some $k < i$. (2) By the morphism ϕ, the arm of any maximal gapped repeat in f_i that is not a suffix of f_m produced from a maximal gapped repeat of arm a in f_k for some $k < i$ is in $\{P_4, \ldots, P_i\}$. These facts imply that any maximal gapped repeat whose right arm is not a suffix in f_i has an arm in $\{P_4, \ldots, P_i\}$ (since a is the only possible arm of $f_3 = aba$).

We first prove (1). Let σ be an arbitrary maximal α-gapped repeat in f_i and u be the arm of σ whose length is greater than 1. As stated in the proof of Theorem 1, the prefix of u must be ab. By a similar argument, the suffix of u also must be ba. Then, we can write $u = au'a$ for some string u' that has b as a border ($u' = b$ is also possible). There exist two cases such that σ does not contradict maximality: the characters on both sides of each arm are (i) identical or (ii) different. Let u_ℓ be the left arm and u_r be the right arm. Note that $u_\ell = u_r$.

(i) Without loss of generality, we assume that a is the character on both sides of u_ℓ and b is the character on both sides of u_r. We consider the strings $au_\ell a = a \cdot au'a \cdot a$ and $bu_r b = b \cdot au'a \cdot b$. By the definition of morphism ϕ, $aau'aa$ and $bau'ab$ should be produced by substrings $bau''ba$ and $aau''a$ in f_{i-1} where $\phi(au'') = au'$ (u'' is possiblly the empty string), respectively. More precisely, $u_\ell = au'a = \phi(au''b)$ and $u_r = au'a$ is a prefix of $\phi(au''a) = au'a \cdot b$. Thus, there exists a maximal gapped repeat in f_{i-1} such that the arms are au''. (ii) Without loss of generality, we assume that a is the character on the left to u_ℓ and the right to u_r, and b is the character on the right to u_ℓ and the left to u_r. We consider the strings $au_\ell b = a \cdot au'a \cdot b$ and $bu_r a = b \cdot au'a \cdot a$. By the definition of morphism ϕ, $aau'ab$ and $bau'aa$ should be produced by substrings $bau''a$ and $aau''ba$ in f_{i-1} where $\phi(au'') = au'$ (u'' is possiblly the empty string), respectively. More precisely, $u_\ell = au'a$ is a prefix of $\phi(au''a) = au'a \cdot b$ and $u_r = au'a = \phi(au''b)$. Thus, there exists a maximal gapped repeat in f_{i-1} such that the arms are au''.

By the above arguments, we can see that any maximal gapped repeat $au'a \cdot v \cdot au'a$ in f_i corresponds to a maximal gapped repeat $\phi^{-1}(au') \cdot \phi^{-1}(av) \cdot \phi^{-1}(au')$ in f_{i-1}. By repeating this operation, any maximal gapped repeat is reduced to a maximal gapped repeat of arm a.

We next prove (2). Let u_l and u_r be the left and right arms. Since b cannot be any arm by Lemma 1, the shortest arm is $P_4 = a$. We assume that $u = P_k$ for any $k \geq 4$.

(i) Without loss of generality, we assume that a is the character on both sides of u_ℓ and b is the character on both sides of u_r. We consider the strings $au_\ell a = a \cdot P_k \cdot a$ and $bu_r ba = b \cdot P_k \cdot ba$ in f_i since $bu_r b$ is succeeded by a. By the definition of morphism ϕ and Lemma 7, $aP_k a$ and $bP_k ba$ produce substrings $ab\phi(P_k)ab = abP_{k+1}b$ and $a\phi(P_k)aab = aP_{k+1}ab$ in f_{i+1}, respectively. Thus, there exists a maximal gapped repeat in f_{i+1} such that the arms are P_{k+1}. (ii) Without loss of generality, we assume that a is the character on the left to u_ℓ and the right to u_r, and b is the character on the right to u_ℓ and the left to u_r. We consider the strings $au_\ell ba = a \cdot P_k \cdot ba$ and $bu_\ell a = b \cdot P_k \cdot a$ in f_i since $au_r b$ is succeeded by a.

By the definition of morphism ϕ and Lemma 7, $aP_k ba$ and $bP_k a$ produce substrings $ab\phi(P_k)aab = abP_{k+1}ab$ and $b\phi(P_k)a = aP_{k+1}b$ in f_{i+1}, respectively.

Thus, there exists a maximal gapped repeat in f_{i+1} such that the arms are P_{k+1}. By the above argument, we conclude the proof of (2). Notice that the case (i) cannot appear since $a \cdot P_4 \cdot a = a^3$ for the initial case does not occur in f_i.

These reductions for both directions use corresponding functions. Therefore, the lemma holds.

4.2 Upper Bound on the Number of Maximal α-Gapped Repeats

By using the characterizations, we give an upper bound on the number of maximal α-gapped repeats in a Fibonacci string.

Theorem 3. *An upper bound on the number of maximal α-gapped repeats in the k-th Fibonacci string of length n is $2.1\alpha n + 0.73\alpha \log n + 4.2\alpha$.*

We give an upper bound for each of $|\mathcal{SR}|$ and $|\mathcal{NSR}|$ (resp. Lemmas 14 and 18), separately.

Upper bound of suffix maximal α-gapped repeats. We show a relation between distinct suffix maximal gapped repeats such that they share the right arms. More precisely, we can prove that the left arms of such repeats cannot overlap each other because the arms are Fibonacci strings.

Lemma 13. *Let σ and σ' be suffix maximal gapped repeats in f_m such that the arms are f_k ($k \geq 3$). The left arms cannot overlap each other.*

Proof. Let $f_m[i..j]$ and $f_m[i'..j']$ be left arms of σ and σ' ($i < i' \leq j < j'$), respectively, and $f_m[i..j] = f_m[i'..j'] = f_k$ for some $k \leq m-2$. Let us denote the overlap $f_m[i'..j]$ by w. Since w is a border of f_k, we can write $w = f_{k-2\ell}$ for some ℓ by Corollary 3. By the definition of Fibonacci strings, f_k has $f_{k-2\ell-1}f_{k-2\ell}$ as a suffix. Hence, the left arm of σ' is preceded by $f_{k-2\ell-1}$. We can also see that f_m has $f_{k-1}f_k$ as a suffix by the definition of Fibonacci strings. These facts imply that both the preceding characters of arms of σ' are the same. This contradicts the maximality of σ'.

Thanks to the non-overlapping structures of the arms, we can then obtain a non-trivial upper bound of $|\mathcal{SR}|$ as follows.

Lemma 14. $|\mathcal{SR}| \leq 0.73\alpha \log n - 0.02\alpha$.

Proof. Let $f_k v f_k$ be a suffix α-gapped repeat in f_m for some $v \in \Sigma^*$. By Lemma 13, such repeats can exist at most $\lfloor \alpha \rfloor$ for each k. Moreover, the length of the arm of any α-gapped repeat has to be shorter than $|f_m|$, and the parity of a suffix Fibonacci string of f_m has to be the same as the parity of f_m. These facts imply that $|\mathcal{SR}| \leq (\lfloor \frac{m}{2} \rfloor - 1)\alpha$. By giving an upper bound of m by $n = |f_m|$, we can approximate an upper bound by the text length n. Since $((1-\sqrt{5})/2)^m$ is maximum when $m = 2$,

$$\frac{1}{\sqrt{5}}\left(\left((1+\sqrt{5})/2\right)^m - \left((1+\sqrt{5})/2\right)^2\right) \leq n$$

holds. This implies that $m < 1.441 \log n + 1.966$. Due to the upper bound, we can obtain the lemma.

Upper bound of non-suffix maximal α-gapped repeats. Somewhat similar to the previous case, we first show a relation regarding overlaps between distinct maximal gapped repeats such that they are sharing the left arms P_i. To do so, we observe that the structure of certain overlaps of two P_i in Lemma 15, and then we will discuss the relation in Lemma 16.

Lemma 15 ($\star$). *For any $i \geq 6$, $P_{i+1} = f_{i-1} \cdot P_i = P_i \cdot f_{i-1}^R = f_{i-1} \cdot P_{i-2} \cdot f_{i-1}^R$.*

Lemma 16. *Let σ and σ' be maximal gapped repeats in f_m such that the arms are P_k $(k \geq 6)$ and these two repeats begin at the same position. If the right arms $f_m[i..j]$ of σ and $f_m[i'..j']$ of σ' overlap (i.e., $i < i' \leq j < j'$), then, $f_m[i..j'] = P_{k+1}$ holds.*

Proof. Let $w = f_m[i..j'] = w_1 f_m[i'..j] w_2$ for some strings w_1 and w_2. Since $f_m[i'..j]$ is a border of P_k, $f_m[i'..j] \in \{P_4, \ldots, P_{k-1}\}$ holds by Lemma 8. Hence, we can write $f_m[i'..j]$ as $f_m[i'..j] = P_\ell$ for some integer ℓ satisfying $4 \leq \ell < k$. Assume on the contrary that $f_m[i..j'] \neq P_{k+1}$ holds. This assumption implies that $\ell \neq k-2$ by Lemma 15. We consider the following two cases on ℓ: (i) $4 \leq \ell \leq k-3$, (ii) $\ell = k-1$. (i) Since $f_\ell = f_{\ell-2} f_{\ell-1}[1..|f_{\ell-1}|-2]\overline{\delta(\ell)}$ holds, f_k can be written as $f_k = x \cdot f_{\ell+1}^2 P_\ell \cdot \delta(k)$ for some string x by Lemma 6. This implies that $P_k = x f_{\ell+1}^2 P_\ell$. Moreover, by Lemma 4, P_k has a prefix $f_{\ell+1}^2$. Thus, there exists $f_{\ell+1}^4$ in f_m, a contradiction. (ii) By the definitions, we can see that $w = f_{k-2} P_{k-1} w_2 = f_{k-2} f_{k-1} w_2'$ for some suffix $w_2' = w_2[3..|w_2|]$. This implies that the right-arm of σ is succeeded by $\overline{\delta(k-1)[1]}$. From the maximality of σ, the succeeding character of the left-arm of σ is $\delta(k-1)[1]$. Let us consider the substring $f_{k-1} w_2' w_3$ of length $|f_k|$ (i.e., $|w_3| = 2$) that has the right-arm P_k of σ' as a prefix. By Lemma 5, this substring is equal to either $f_{k-1} f_{k-2} (= f_k)$ or $f_{k-2} f_{k-1}$. If the former case, the right-arm of σ' is secceeded by $\delta(k)[1] = \overline{\delta(k-1)[1]}$. This contradicts the maximality of σ'. Assume that the substring is $f_{i-2} f_{i-1}$. This implies that there exists a substring $w \cdot w_3 = f_{k-2} f_{k-2} f_{k-1}$ in f_m. As discussed in the proof of Lemma 11, $f_{k-2} f_{k-1}$ is preceded by f_{k-3}. However, this contradicts the fact that f_{k-3} is a suffix of f_{k-2} (which is a prefix of $w \cdot w_3$). Therefore, the lemma holds.

The final lemma for our result explains that there should be a short distance between two left arms P_k if the left arms do not overlap each other.

Lemma 17 ($\star$). *Let $\sigma = f_m[i..j]$ and $\sigma' = f_m[i'..j']$ $(i < i' \leq j < j')$ be maximal gapped repeats in f_m such that the arms are P_k $(k \geq 4)$ and $|\sigma| = |\sigma'|$. If $i' - i \geq |P_k|$ (i.e., the left arms do not overlap), then $i' - i \geq |P_k| + 2$ holds.*

Now, we can obtain the following upper bound for the number of non-suffix maximal α-gapped repeats.

Lemma 18. *$|\mathcal{NSR}| \leq 2.1\alpha n + 4.2\alpha$.*

Proof. For any integer $k \geq 4$, let $\mathcal{NSR}_k$ be the set of maximal α-gapped repeats in f_m such that the arms are P_k and the repeats are not a suffix of f_m. To give an upper bound $|\mathcal{NSR}_k|$, we consider the exact β-gapped repeats defined as follows:

a gapped repeat uvu is called an exact β-gapped repeat if $|u|+|v| = \beta|u|$ (for any rational number β). Let us denote the set of maximal exact β-gapped repeats in f_m such that the arms are P_k and the repeats are not a suffix of f_m by $\mathcal{EGR}_k(\beta)$. Since the arms of non-suffix repeats which we want to count here are P_k, we can analyze $|\mathcal{NSR}_k|$ by $|\mathcal{NSR}_k| = \sum_{\beta \in B} |\mathcal{EGR}_k(\beta)|$, where $B = \{1, \frac{|P_k|+1}{|P_k|}, \ldots, \lfloor\alpha\rfloor\}$ (e.g., the set of rational numbers less that α with denominator $|P_k|$). We give an upper bound $|\mathcal{NSR}_k|$ for every k by considering the following three cases: (i) $k = 4$, (ii) $k = 5$, (iii) $k \geq 6$.

(i) Let $\sigma, \sigma' \in \mathcal{EGR}_4(\beta)$ for some β. It is easy to see that β has to be an integer since the length of the arm is 1, and the length of the repeats is $\beta + 1$. Moreover, the left arms σ and σ' cannot overlap. By combining with Lemma 17, we can obtain the following upper bound:

$$|\mathcal{NSR}_4| = \sum_{\beta=1}^{\lfloor\alpha\rfloor} |\mathcal{EGR}_4(\beta)| \leq \sum_{\beta=1}^{\lfloor\alpha\rfloor} \left\lceil \frac{n-(\beta+1)}{3} \right\rceil < \frac{n+2}{3}\alpha$$

where $n = |f_m|$.

(ii) Let $\sigma, \sigma' \in \mathcal{NSR}_5$ such that the repeats begin at the same position. Firstly, we observe whether the right arms can overlap or not (because Lemma 16 stands for any $k \geq 6$). The only possibility is the overlap with a border a. Since the succeding character of the right-arm of σ is b, the succeding character of the left-arm of σ is a. This implies that the succeeding character of the right-arm of σ' is also b. Then, there is an occurrence of $(ab)^3$, a contradiction. Hence, the right arms cannot overlap. Suppose that σ is a maximal exact $\frac{3\ell}{3}$-gapped repeat for some integer $\ell \geq 1$. Due to the above discussion, there is no maximal exact $\frac{3\ell+c}{3}$-gapped repeat for any $c \in \{1, 2\}$ such that the repeats begin at the same position. Thus, it is enough to give an upper bound of $|\mathcal{EGR}_5(\beta)|$ for every integer β for obtaining an upper bound of $|\mathcal{NSR}_5|$. By Lemma 17, the maximum number of the beginning positions for disjoint left arms of maximal exact β-gapped repeats is at most $\lceil \frac{n-3(\beta+1)}{5} \rceil$. For each such left arm, there is a possibility of another left-arm that begins in the left-arm. Hence, $|\mathcal{NSR}_5| = \sum_{\beta=1}^{\lfloor\alpha\rfloor} |\mathcal{EGR}_5(\beta)| \leq \sum_{\beta=1}^{\lfloor\alpha\rfloor} (2\lceil(n-3(\beta+1))/5\rceil) < 2 \cdot (n+2)/5 \cdot \alpha$.

(iii) Let $\sigma, \sigma' \in \mathcal{NSR}_k$ such that the repeats begin at the same position. Because of Lemma 16, there is only one candidate of the overlap (P_{k-2}) of the right arms. Due to a similar discussion for the previous case, there exists only the possibility of $c \in \{1, \ldots, |P_k| - 1\}$ such that the rigth-arm of the maximal exact $\frac{|P_k|\ell+c}{|P_k|}$-gapped repeat overlap with the right-arm of σ. Thus, it is enough to give an upper bound of $2|\mathcal{EGR}_k(\beta)|$ for every integer β for obtaining an upper bound of $|\mathcal{NSR}_k|$. By Lemma 17, the maximum number of the beginning positions for disjoint left arms of maximal exact β-gapped repeats is at most $\lceil \frac{n-|P_k|(\beta+1)}{|P_k|+2} \rceil$. For each such left arm, there are two possibilities (borders P_{k-1} and P_{k-2}) of the other left arms that begin in the left arm of σ (we can see the candidates from the discussion for case (i) in the proof of Lemma 16). However, we now show that the two possibilities cannot occur at the same time. We consider a substring $P_{k+1} = P_k \cdot f_{k-1}^R = f_{k-1}P_k = f_{k-1}P_{k-2}f_{k-1}^R$ (by Lemma 15). Since f_k

is a prefix of P_{k+1}, the prefix P_k of P_{k+1} is succeded by $\delta(k)$. Assume that there exists an occurrence of P_k in P_{k+1} such that $P_{k+1} = x \cdot P_k$ where $P_k = x \cdot P_{k-1}$. By the assumption, this P_{k-1} is succeded by $\delta(k-1)$ (since f_{k-1} is a prefix of P_k). Though these $\delta(k)$ and $\delta(k-1)$ are substrings at the same position, this contradicts $\delta(k) \neq \delta(k-1)$. Thus, for each fixed left arm, there is a possibility of another left arm that begins in the left arm. Hence,

$$|\mathcal{NSR}_k| = \sum_{\beta=1}^{\lfloor\alpha\rfloor} |\mathcal{EGR}_k(\beta)| \leq \sum_{\beta=1}^{\lfloor\alpha\rfloor} \left(2 \cdot 2 \left\lceil \frac{n - |P_k|(\beta+1)}{|P_k|+2} \right\rceil\right) < 4 \cdot \frac{n+2}{F_k}\alpha.$$

Finally, we summarize the above upper bounds by using Lemma 9.

$$\begin{aligned}
|\mathcal{NSR}| &\leq |\mathcal{NSR}_4| + |\mathcal{NSR}_5| + |\mathcal{NSR}_6| + \cdots \\
&= \frac{n+2}{3}\alpha + 2 \cdot \frac{n+2}{5}\alpha + 4(n+2)\alpha \cdot \left(\frac{1}{8} + \frac{1}{13} + \frac{1}{21} + \cdots\right) \\
&\leq \frac{11n+22}{15}\alpha + 4(n+2)\alpha \cdot (3.36 - 3.03) \leq 2.1\alpha n + 4.2\alpha.
\end{aligned}$$

We can obtain the upper bound.

Overall, we now obtain Theorem 3.

4.3 Lower Bound for a Fibonacci String

We can also show that the Fibonacci strings are a new example that contains $\Omega(\alpha n)$ maximal α-gapped repeats. In order to prove this fact, we count the following maximal gapped repeats. We choose two occurrences, the i-th and j-th occurrences ($i \leq j$), of substring aa. It is clear that aa is preceded and succeeded by b. Thus, the substring that begins with the first (resp., second) a of the i-th aa and ends with the second (resp., first) a of the j-th aa is a maximal gapped repeat with single-character arms. By counting such maximal gapped repeats, we can obtain the following bound.

Lemma 19 ($\star$). *For every $m \geq 5$, f_m contains at least $0.04\alpha n - 1.04n - 2.64$ maximal α-gapped repeats.*

An idea for counting such repeats is given as follows. We consider the string $f'_m = ab \cdot (aabab)^k \cdot x \cdot aababa$ of length $n = F_m$ for some possibly empty string x of length $n - 5k - 8$. This string can be obtained by the following simple structures on a Fibonacci string: the length of a substring between two consecutive occurrences of aa is at most three. Then, the string contains the above repeats with the single-character arms, but the uniform occurrences of aa in f'_m are sparser than that in f_m. Hence, the number of such repeats in f'_m gives a lower bound.

A further work for the lower bound for a Fibonacci string is counting repeats based on the characterization which we proposed in this paper.

Acknowledgments. This work was supported by JSPS KAKENHI Grant Numbers JP21K17705 (YN), JP23H04386 (YN), JP22H00513 (KS), JP20H05964 (TH), and JP22H03549 (TH).

References

1. André-Jeannin, R.: Irrationalité de la somme des inverses de certaines suites récurrentes. CR Acad. Sci. Paris Sér. I Math. **308**(19), 539–541 (1989)
2. Apostolico, A., Brimkov, V.E.: Fibonacci arrays and their two-dimensional repetitions. Theoret. Comput. Sci. **237**(1–2), 263–273 (2000)
3. Brodal, G.S., Lyngsø, R.B., Pedersen, C.N.S., Stoye, J.: Finding maximal pairs with bounded gap. In: Crochemore, M., Paterson, M. (eds.) Combinatorial Pattern Matching, 10th Annual Symposium, CPM 99, Warwick University, UK, July 22-24, 1999, Proceedings. Lecture Notes in Computer Science, vol. 1645, pp. 134–149. Springer (1999). https://doi.org/10.1007/3-540-48452-3_11
4. Crochemore, M., Kolpakov, R., Kucherov, G.: Optimal bounds for computing α-gapped repeats. In: Language and Automata Theory and Applications, pp. 245–255. Springer (2016)
5. Fraenkel, A.S., Simpson, J.: The exact number of squares in Fibonacci words. Theor. Comput. Sci. **218**(1), 95–106 (1999). https://doi.org/10.1016/S0304-3975(98)00252-7
6. Gawrychowski, P., I, T., Inenaga, S., Köppl, D., Manea, F.: Tighter bounds and optimal algorithms for all maximal α-gapped repeats and palindromes: finding all maximal α-gapped repeats and palindromes in optimal worst case time on integer alphabets. Theory Comput. Syst. **62**, 162–191 (2018)
7. Giuliani, S., Inenaga, S., Lipták, Z., Romana, G., Sciortino, M., Urbina, C.: Bit catastrophes for the Burrows-Wheeler transform. In: Drewes, F., Volkov, M. (eds.) Developments in Language Theory - 27th International Conference, DLT 2023, Umeå, Sweden, June 12-16, 2023, Proceedings. Lecture Notes in Computer Science, vol. 13911, pp. 86–99. Springer (2023). https://doi.org/10.1007/978-3-031-33264-7_8
8. Gusfield, D.: Algorithms on Strings, Trees, and Sequences: Computer Science and Computational Biology. Cambridge University Press (1997)
9. I, T., Köppl, D.: Improved upper bounds on all maximal α-gapped repeats and palindromes. Theor. Comput. Sci. **753**, 1–15 (2019). https://doi.org/10.1016/J.TCS.2018.06.033
10. Iliopoulos, C.S., Moore, D., Smyth, W.F.: A characterization of the squares in a Fibonacci string. Theoret. Comput. Sci. **172**(1–2), 281–291 (1997)
11. Inoue, H., Matsuoka, Y., Nakashima, Y., Inenaga, S., Bannai, H., Takeda, M.: Factorizing strings into repetitions. Theory Comput. Syst. **66**(2), 484–501 (2022). https://doi.org/10.1007/S00224-022-10070-3
12. Kishi, K., Nakashima, Y., Inenaga, S.: Largest repetition factorization of Fibonacci words. In: Nardini, F.M., Pisanti, N., Venturini, R. (eds.) String Processing and Information Retrieval - 30th International Symposium, SPIRE 2023, Pisa, Italy, September 26-28, 2023, Proceedings. Lecture Notes in Computer Science, vol. 14240, pp. 284–296. Springer (2023). https://doi.org/10.1007/978-3-031-43980-3_23

13. Kolpakov, R., Podolskiy, M., Posypkin, M., Khrapov, N.: Searching of gapped repeats and subrepetitions in a word. In: Symposium on Combinatorial Pattern Matching, pp. 212–221. Springer (2014)
14. Kolpakov, R., Podolskiy, M., Posypkin, M., Khrapov, N.: Searching of gapped repeats and subrepetitions in a word. J. Discrete Algorithms **46-47**, 1–15 (2017). https://doi.org/10.1016/J.JDA.2017.10.004
15. Kolpakov, R.M., Kucherov, G.: On maximal repetitions in words. In: Ciobanu, G., Paun, G. (eds.) Fundamentals of Computation Theory, 12th International Symposium, FCT '99, Iasi, Romania, August 30 - September 3, 1999, Proceedings. Lecture Notes in Computer Science, vol. 1684, pp. 374–385. Springer (1999). https://doi.org/10.1007/3-540-48321-7_31
16. Kolpakov, R.M., Kucherov, G.: Finding repeats with fixed gap. In: de la Fuente, P. (ed.) Seventh International Symposium on String Processing and Information Retrieval, SPIRE 2000, A Coruña, Spain, September 27-29, 2000, pp. 162–168. IEEE Computer Society (2000). https://doi.org/10.1109/SPIRE.2000.878192
17. Melançon, G.: Lyndon factorization of sturmian words. Discret. Math. **210**(1), 137–149 (2000). https://doi.org/10.1016/S0012-365X(99)00123-5
18. Navarro, G., Ochoa, C., Prezza, N.: On the approximation ratio of ordered parsings. IEEE Trans. Inf. Theory **67**(2), 1008–1026 (2021). https://doi.org/10.1109/TIT.2020.3042746
19. Tanimura, Y., Fujishige, Y., I, T., Inenaga, S., Bannai, H., Takeda, M.: A faster algorithm for computing maximal α-gapped repeats in a string. In: Iliopoulos, C.S., Puglisi, S.J., Yilmaz, E. (eds.) String Processing and Information Retrieval - 22nd International Symposium, SPIRE 2015, London, UK, September 1-4, 2015, Proceedings. Lecture Notes in Computer Science, vol. 9309, pp. 124–136. Springer (2015). https://doi.org/10.1007/978-3-319-23826-5_13

Author Index

R. Královič and V. Kůrková (Eds.): SOFSEM 2025, LNCS 15539, pp. 351–353, 2025.
https://doi.org/10.1007/978-3-031-82697-9

The manufacturer's authorised representative in the EU is Springer Nature Customer Service Centre GmbH, Europaplatz 3, 69115 Heidelberg, Germany. If you have any concerns regarding our products, please contact ProductSafety@springernature.com

Printed and bound by CPI Group (UK) Ltd, Croydon, CR0 4YY
15/07/2026
02167637-0003